ELECTROMAGNETIC SHIELDING HANDBOOK FOR WIRED AND WIRELESS EMC APPLICATIONS

THE KLUWER INTERNATIONAL SERIES IN ENGINEERING AND COMPUTER SCIENCE

ELECTROMAGNETIC SHIELDING HANDBOOK FOR WIRED AND WIRELESS EMC APPLICATIONS

by

Anatoly Tsaliovich

Kluwer Academic Publishers
Boston / London / Dordrecht

Distributors for North, Central and South America:
Kluwer Academic Publishers
101 Philip Drive
Assinippi Park
Norwell, Massachusetts 02061 USA
Tel: 781-871-6600
Fax: 781-871-6528
E-mail: kluwer@wkap.com

Distributors for all other countries:
Kluwer Academic Publishers Group
Distribution Centre
Post Office Box 322
3300 AH Dordrecht, THE NETHERLANDS
Tel: 31 78 6392 392
Fax: 31 78 6546 474
E-mail: orderdept@wkap.nl

Electronic Services: http://www.wkap.nl

Library of Congress Cataloging-in-Publication Data

Tsaliovich, A. B. (Anatoliĭ Borisovich)
   Electromagnetic shielding handbook for wired and wireless EMC
applications / by Anatoly Tsaliovich.
      p.   cm -- (The Kluwer international series in engineering and
   computer science ; SECS 462)
   Includes bibliographical references and index.
   ISBN: 0-412-14691-6
   1. Shielding (Electricity)   I. Title.   II. Series.
TK7867.8.T7823   1999   98-37792
621.382'24--dc21   CIP

Copyright © 1999 AT&T. All rights reserved.
Published by Kluwer Academic Publishers

All rights reserved. No part of this publication may be reproduced, stored in a retrieval system or transmitted in any form or by any means, mechanical, photo-copying, recording, or otherwise, without the prior written permission of the publisher, Kluwer Academic Publishers, 101 Philip Drive, Assinippi Park, Norwell, Massachusetts 02061

Printed on acid-free paper.
Printed in the United States of America

Contents

Foreword xiii

Preface xvii

1. **INTRODUCTION TO SYSTEM EMI AND EMC**

1.1 Electronic System In An EMI Environment 1
  1.1.1 System EMI: A Necessary Evil 1
  1.1.2 EMC—The Science Of Electronic System Coexistence 6
  1.1.3 Electronic System EMI Synthesis and Analysis 14
1.2 System EMI Controlling Factors 17
  1.2.1 Basic EMI Study Approaches and Coupling Path 18
  1.2.2 Meet EMI Emitters / Sensors 18
  1.2.3 What You Always Wanted to Know (But were Afraid to Ask) about Differential and Common Mode Transmission 26
  1.2.4 Meet EMI Generators / Receptors 29
1.3 Defining And Modeling EMC Performance Parameters 44
  1.3.1 Basic Emissions and Immunity Units 44
  1.3.2 Typical EMC Performance Models 45
  1.3.3 Elementary Source = Radiator + Generator 53
  1.3.4 Electronic System as an Active Antenna Array 55
1.4 EMI Suppression: the important objective of EMC 59
  1.4.1 EMI Mitigation Techniques 59
  1.4.2 Comparing Different EMI Suppression Techniques: Cable Crosstalk Example 64

2. UNDERSTANDING ELECTROMAGNETIC SHIELDING

2.1 Shielding at the EMC "Front line" 71
2.2 A Shield Is a Shield Is a Shield 73
    2.2.1 Shielding Enclosures for Electronic Products 74
    2.2.2 Cable Shielding 80
    2.2.3 Architectural Shielding 81
    2.2.4 "Mobile" Shielding: How Good Shield is a Taxicab 84
    2.2.5 System Shielding: Limitations of System EMI Analysis 84
    2.2.6 Shielding, Grounding, and All That 'Jazz' 85
    2.2.7 Connectors: Don't Miss the Link 87
    2.2.8 Importance of Shielding System Integrity:
        Shielding System for Electrostatic Discharge Protection 89
    2.2.9 Ground Plane in Printed Circuit Board: Parallel Shielding Effect 91
2.3 This Confusing World Of Electromagnetic Shielding 92
    2.3.1 What Is Electromagnetic Shielding, Anyhow? 92
    2.3.2 Do You Speak "Shieldinese" ? 94
    2.3.3 Transmission Theory of Shielding: User Alert 96
    2.3.4 What We Have Learned and What's Missing 103
2.4 Shielding—This Is Not Very Simple 104
    2.4.1 A Fresh Look at a Familiar Problem 105
    2.4.2 Transmission Line vs. Plane Wave Shielding Models ...110
    2.4.3 Plane Shield / Plane Wave Model Constraints 113
2.5 A System View On Electromagnetic Shielding 116
    2.5.1 The Roads We Take 116
    2.5.2 "Parallel Bypass" Model of Energy transfer through the Shield 119
    2.5.3 "Series" Model of System Electromagnetic Shielding ..121
    2.5.4 Reciprocity Principle in Electromagnetic Shielding 127
2.6 So, What Is Electromagnetic Shielding, After All? 129
    2.6.1 System Definitions of Electromagnetic Shielding 129
    2.6.2 Shielding Model Generality Vs Relevancy 130
    2.6.3 What's Next? 132

3. TRANSFER PARAMETERS OF ELECTROMAGNETIC SHIELDS AND ENCLOSURES

3.1 Figure of Merit of electromagnetic Shield Performance 133
    3.1.1 Transfer Impedance - Shield Transfer Parameter of Choice 133
    3.1.2 Universal Set of Shield Transfer Parameters: from Cables to
        Small and Large Products to Architectural Constructions 137
    3.1.3 Finding EMI Voltage: A Union of Field and Circuit Theories 143
3.2 Thin Homogeneous Solid Shield Performance 149
    3.2.1 Current and Field Distribution in Homogeneous Metallic Tube -
        Coaxial Cable Shield / Outer Conductor 149

          3.2.2   Thin Homogeneous Solid Current-Carrying Plane or "Other" Shield
                  Shapes Have Identical to Cylindrical Shield Field Distribution 155
          3.2.3   Transfer Parameters of Thin Solid Homogeneous Shields 156
          3.2.4   Thin Homogeneous Solid Shield Performance Analysis 163
    3.3 A "Second Look" At Transfer Parameters 172
          3.3.1   Circuit Theory/Reactive Near Field Nature of Transfer
                  Parameters.. 172
          3.3.2   Shielding "from Currents" vs Shielding "from Fields" 173
          3.3.3   The Role of Coupling in Shield Transfer Function 174
          3.3.4   High Frequency Current Return Path in the Shield Mainly Follows
                  the Direct Path 180
    3.4 Practical Non-Uniform Shields 182
          3.4.1   The "Troubling World" of Shield Non-Uniformities 182
          3.4.2   Spiral Shields 185
          3.4.3   Shields with a "Longitudinal" Seam and/ or
                  Electromagnetic Gasket 191
          3.4.4   Foil Shields: Effects of Seam and Overlap 195
          3.4.5   Mesh Shields 205
    3.5 Braided Shield 209
          3.5.1   Geometry and Design Parameters of Braided Shield 209
          3.5.2   Physical Processes in Braided Shield: Noah's Ark of Shielding
                  Problems
                 10
          3.5.3   Engineering Model of Braided Shield 214
          3.5.4   Braided Shield Performance 219
    3.6 Multilayer Shields 220
          3.6.1   Multilayer Homogeneous Shields 221
          3.6.2   Multilayer Nonhomogeneous Shield Performance 228
          3.6.3   Calculating "Derivative" Shields 240

4. **EMI ENVIRONMENT AND ELECTROMAGNETIC COUPLING IN SHIELDING**

    4.1 Defining Shielding Coupling 245
          4.1.1   Shielding To Decouple And Coupling To Shield 245
          4.1.2   Electromagnetic Environment and Coupling in Shielding Model..246
          4.1.3   Definitions: What Is "Shielding" Coupling? 249
    4.2 A Roadmap To Shielding Coupling Mechanisms: EMI Signal Propagation
          Regimes...252
          4.2.1   A Roadmap to Roadmap: Maxwell's Equations and
                  Coupling Mechanisms.. 252
          4.2.2   Signal Propagation Media Parameters 253
          4.2.3   Role of Time Variations: Solid Homogeneous Shield
                  in a Wide Frequency Band 254
          4.2.4   Coupling Regimes in Conductors and Dielectrics 256

    4.2.5 "Electrical Geometry" in Near-Zone / Far-Zone Radiating Fields .260
    4.2.6 Shielding Coupling / Propagation Regimes: Summary 263
 4.3 Induction Field Coupling 264
    4.3.1 Static and Stationary Field EMI Environment and Coupling 264
    4.3.2 Quasi-Stationary Field Coupling 266
 4.4 Crosstalk EMI Environment and Coupling 269
    4.4.1 Crosstalk Interactions 269
    4.4.2 Transverse Reactive Field Environment and Coupling.................270
    4.4.3 Common Impedance (Galvanic) Coupling 278
    4.4.4 Coupling via Third Circuit 279
    4.4.5 Coupling Between Shielded or Coaxial Lines 280
    4.4.6 Crosstalk Coupling Between Balanced vs Unbalanced Lines:
           Common and Differential Mode Regimes 282
    4.4.7 Crosstalk Coupling Reduction Techniques 284
 4.5 EMI Environment And Shielding Coupling In Radiating Fields 295
    4.5.1 Problems and Strategies 295
    4.5.2 Radiating EMI Environment: Time and Space Domain
           Complications..298
    4.5.3 How Do Radiating Fields Couple? 300
    4.5.4 Antenna Currents and Radiation Resistance of Lines and
           Shields 305
    4.5.5 Aperture Coupling in Radiation Field 311
 4.6 Transient Response of Shielding 316

5. SHIELDING EFFECTIVENESS FOR EMI PROTECTION

 5.1 The Technical Bottom Line of Shielding Performance 319
 5.2 Shielding effectiveness in Static and Stationary Fields 324
    5.2.1 Problems of Statics 324
    5.2.2 Electrostatic Shielding 326
    5.2.3 Magnetostatic Shielding 331
    5.2.4 Non-Conductive Statics Shielding 336
    5.2.5 Complexities and Specifics of Static Shield Design 338
 5.3 Shielding For Crosstalk Protection: From Mils To Miles 340
    5.3.1 "Compensation" Shielding from Low-Frequency Crosstalk 340
    5.3.2 Crosstalk between Shielded Lines in a Wide Frequency Band 344
    5.3.3 Shielding for Crosstalk Protection in Printed Circuit Boards
           and Integrated Circuits 352
    5.3.4 Crosstalk in Coaxial and Shielded Transmission Lines 360
 5.4 Shielding In Radiating Fields 365
    5.4.1 Shielding from Radiation vs Shielding from Crosstalk 365
    5.4.2 Deterministic vs Statistical Nature of EMI Environment:
           Indoor Propagation of PCS Signal 367

Contents ix

5.5 From Transfer Impedance to Shielding Effectiveness and Crosstalk Attenuation 369
5.6 EMC And Shielding Performance Of Typical Electronic Cables 374
5.7 Ground Planes And Radiation: From EMC Test Site To PCB 378
5.8 Shielding Enclosures For Radiating Field Protection 386
    5.8.1 Critical Parameters of Shielding Enclosures 387
    5.8.2 Radiated EMI Environment Issues 387
    5.8.3 Radio Wave Penetration Through Apertures, Seams, and Slots: Navigating Through Literature 394
    5.8.4 "Mobile" Shielding Enclosures 404
5.9 Wireless Product Shielding 410

6. SHIELDING MEASUREMENT TECHNIQUES AND APPARATUS: THE TOOLS OF THE TRADE

6.1 The proof of the shielding 417
    6.1.1 Measurement Objectives 417
    6.1.2 Test Space Environment 419
    6.1.3 Shielding Measurement Rationale 421
    6.1.4 Roadmap to Shielding Measurements 426
6.2 Global System Shielding Effectiveness Measurements 426
    6.2.1 System Shielding Effectiveness Measurements in RF Radiating Fields 426
    6.2.2 System Measurements in Magnetostatic Fields 433
    6.2.3 Near Field RF and Microwave Measurements 434
6.3 Shielding Assembly Measurements 434
    6.3.1 Specifics of Shielding Assembly Measurements 434
    6.3.2 Shielding Assembly Measurements in Radiating and Magnetostatic Fields 435
    6.3.3 Antenna Current Measurements 438
    6.3.4 Shielding Effectiveness via Crosstalk Measurements 443
6.4 Transfer Impedance and Capacitive Coupling Impedance Measurements ...445
    6.4.1 Coaxial Structures: Is There a Sextaxial in the Cards? 445
    6.4.2 Terminated Triaxial 455
    6.4.3 Line Injection Shielding Effectiveness Measurements 461
6.5 Testing shielding system elements 464
    6.5.1 Shielding Enclosures and Building Structures 464
    6.5.2 How to Test Shielded Cables and Connectors? 470
    6.5.3 Testing Shielding Materials 471
    6.5.4 Electromagnetic Gasket Test Specifics 478
6.6 Testing in Time Domain 479
6.7 Test Result Correlation And Interpretation 482
    6.7.1 "Apples and Oranges" 482
    6.7.2 Validating Test Procedures 487

7. SHIELDING ENGINEERING

- 7.1 System Approach to Shielding Engineering 491
  - 7.1.1 Shielding Engineering Problems 491
  - 7.1.2 To Shield or Not to Shield: the First and the Last Questions 495
  - 7.1.3 "Black Box" Model of Product and System Shielding 497
- 7.2 Methods and Techniques for Shielding Design 505
  - 7.2.1 Basic Shielding Design Principles 505
  - 7.2.2 Design by Constraints 506
  - 7.2.3 Optimal Design 508
- 7.3 Cable Shielding Design for EMC Performance 509
  - 7.3.1 Local Problems of Boundary Design 509
  - 7.3.2 Local Optimization 515
  - 7.3.3 Global Braid Optimization by Cost Criterion 519
- 7.4 Shielding Enclosure Design for emc performance 525
  - 7.4.1 Shielding Enclosure Design Issues 525
  - 7.4.2 Aperture vs Cavity Resonances 527
  - 7.4.3 Shielded Cabinets, Frames, and Shelves 530
  - 7.4.4 Shielding Enclosure "Hardware" 537
- 7.5 Transmission Effects In Shielding Circuit 543
- 7.6 Shielding System Grounding, Termination, And Partition 546
  - 7.6.1 What Is Grounding Really About: Myth and Reality 546
  - 7.6.2 To Ground or Not To Ground ? 550
  - 7.6.3 Challenges of Shielded Cable Assembly 556
  - 7.6.4 Shielding and Ground Loops 565
  - 7.6.5 Shielding and Grounding System Topology and Partition 571
  - 7.6.6 Designing Shielding System 574
- 7.7 Shielding Performance Stability and Reliability 575
  - 7.7.1 Electromagnetic Shield in Physical Environment 575
  - 7.7.2 Environmental Stability and Aging of Electromagnetic Shielding .577
  - 7.7.3 Effect of Manufacturing Tolerances on the Shield Performance Variability 582
  - 7.7.4 Sneaky Problem of a "Rusty Bolt": Intermodulation 587

8. THIS BRAVE NEW OLD WORLD OF ELECTROMAGNETIC SHIELDING

- 8.1 Shielding Unlimited 589
- 8.2 Leaky Shielding 590
- 8.3 "Plastic" Shielding 594
- 8.4 Ferromagnetic Absorptive Shielding 600
- 8.5 Superconductive Shields 604
- 8.6 Chiral Shields 607
- 8.7 "Moving Up" The Frequency Spectrum 614

8.8 Shielding From High Voltage Discharge 616
8.9 Shielding "On-demand" 618

Epilogue 620

9 APPENDIX: Selected Topics In Electromagnetics And Circuit Theory

A.1 "He Who Would Search For Pearls, Must Dive Below" 621
A.2 Maxwell's Equations survival kit 622
    A.2.1 Differential Form of Maxwell's Equations 622
    A.2.2 Integral Form of Maxwell's Equations 631
A.3 Poynting Vector and Poynting Theorem 633
A.4 Circuit Theory survival kit 634
    A.4.1 Lumped-Element Networks 634
    A.4.2 Reciprocity Theorem 636
    A.4.3 Differential Equations of a Transmission Line 638
A.5 Analogy Between Wave Propagation In Transmission Lines And Free Space 639
A.6 Numerical Techniques: Are We There Yet? 645

Bibliography 649

Index 677

Foreword

Dear Reader:

Probably, one of the most important things an author can do to promote his book, is to explain its rationale: what is this book about, why was it written, who needs it, how to use it, how different it is from other books in the field? As a rule, this is done in the Foreword, the Preface, and the Introduction to the book, which are usually compiled *after* the main body of the book is written. The following rationale takes full advantage of such retrospect review and analysis of this book material. Hopefully, it will be useful in answering these questions, as well as "instilling" in readers the electromagnetic shielding-related curiosity and desire to apply the described old and new principles of this practically important and intellectually stimulating fascinating discipline.

Scope

This handbook spans a wide range of practical electromagnetic shielding problems generated by modern wired and wireless technologies: low voltage electronics and power transmission, remote control and telecommunications, civil and military, consumer and medical products, geophysics and space explorations, - you name it! These technologies cover time and frequency domain, ranging from picoseconds to "infinity" and from fraction of Hz to many GHz.

Contents

Page-by-page, the book will lead you from fundamental ideas of electromagnetic compatibility to shielding application objectives to EMI environment, to electromagnetic coupling mechanisms and shield transfer parameters, to methods and techniques for identification, formulation, evaluation, and measurement of shielding characteristics, to advanced design techniques in important areas: shielded cables and interconnection lines, electronic product enclosures, printed wiring board and IC circuit EMI hardening techniques, shield terminations and grounding for EMI protection.

Methodology

The system model of electromagnetic shielding is defined with regard to electromagnetic compatibility (EMC) objectives. The model is used to identify, formulate, and analyze the figures of shielding merit, applicable to the respective levels of shielding system. The study of shielding phenomena heavy relies on the use of the electromagnetic compatibility approaches, antenna-based models, crosstalk theory, and system analysis. Basic concepts of signal propagation mode / electromagnetic coupling analysis, "parallel" review and "combined use" of the field and circuit theory interpretations are combined with original treatment of common and differential transmission mode EMI effects, and use of optimization theory for shield and shielding system design.

Book History (and Pre-History)

This handbook can be viewed both as a "brand-new" book and as a second, expanded edition of the previous book by the same author, "Cable Shielding for Electromagnetic Compatibility". Compared with this previous book, the new handbook addresses the broader needs of those interested in product and system shielding, as well as in a more general view at electromagnetic shielding discipline, along with the cable shielding topics.

Here's how it happened.

The 'Cable Shielding' book was published just three years ago . Yet, even for the short time elapsed since its publication, many significant changes and innovations occurred in the age-old electromagnetic shielding discipline, which had to be reflected in the book's contents. This fact, combined with continuously growing applications, with the interest in the subject among professionals and students, and with the apparent success of the previous book on cable shielding, prompted both the publisher and the author to look into the book's second edition.

As usual, the book's update was used to introduce the latest research and publications, to clarify some of the issues by incorporating the reader's comments, to expand certain topics, to optimize the text structure, and to eliminate the identified typographic errors. The update also gave an opportunity to answer the questions posed by the readers "at large" and at the engineering educational courses where the book was used as a textbook. Along with expanding on "old" topics, often the new material was introduced on the same subject. In particular, such subjects were affected as shielding applications, the effect of the shield on the circuit parameters, 'leaky coax', etc.

However in the process, the text that was supposed to become the second edition of the "cable shielding" book was transformed into a new handbook, covering much wider grounds. There were several important reasons behind such transformation. To start with, numerous responses came in from the readers, which indicated that the subject of electromagnetic shielding, as it is treated in the book, is of interest to a much wider audience than implied by the book's title. While originally the book was conceived mainly as a monograph dedicated to *cable* shielding, during the writing it has overgrown such scope. In fact, the complexity of the cable shielding subject required the application of the most general concepts of electromagnetics and circuit theory. Coincidentally, those appear to be the same concepts which could be used also in a large variety of other shielding applications: shielding rooms, office cabinets, desktop conducting enclosures ("boxes"), screens, PCB ground planes, gaskets, partial shields, etc. This didn't go unnoticed by the readers.

Next, the same factors that prompted the writing of the book on cable shielding, were at work demanding a text on these, other, shielding applications to the benefit of electromagnetic compatibility. The proliferation of electronics in business and residence, emergence of new technologies - from MRI to cellular, miniaturization accompanied by smaller weight, size, lower signal levels, higher frequencies, the realization of problems caused by the electromagnetic pollution and biological safety considerations, adoption of strict electromagnetic compatibility regulations - all resulted in fast growing applications of electromagnetic shielding techniques. Correspondingly, the demand grew for quality information about technical and economical aspects of such techniques. It was felt that many professionals would benefit from the book "expansion" into these diverse applications.

And last but not least, the book actually was already being used by many professionals not only in cable shielding domain, but also to address specific "non-cable" shielding problems. The "de-facto" broadening of the book's scope was possible because it heretofore contained the background and the fundamental principles, as well as practical approaches necessary for such applications. Missing, however, was a broader view at the book's philosophy and principles: the concepts, which were previously used to investigate the *cable* shielding problems, had had to be generalized onto a much *wider basis*, including mathematical formulation and analysis to cover these new, "non-cable", shielding applications.

Under these pressures, the priorities in preparing the book shifted. Along with updating and upgrading the already existing cable-related material, a number of absolutely new topics were introduced, reflecting important "non-cable" applications. In turn, their analysis often required a respective expansion of the underlying principles, to provide a much wider "philosophical" perspective for both "new" and "old" topics.

Indeed, one of important specifics of cable shielding stems from the problem geometry: as a rule, cables possess large length, compared with often electrically infinitesimal other dimensions. Therefore, cable shielding often could be treated as a *one-dimensional* electromagnetics problem, at least, in a first approximation and under certain conditions. It also facilitated the use of much simpler models based on the circuit theory. Not so in general case of shielding "at large": a finite size product shielding enclosure may require *two-* and *three-dimensional* models and account for the "edge" effect distortion and diffraction at the enclosure boundaries. Also, in multi-dimensional problems the shielded conductors do not necessarily run *parallel* to the shielding enclosure surfaces (confines), as it could often be assumed in shielded cable applications.

But now, with the cable shielding topic losing its "exclusivity" in the book's architecture, with the inclusion of many new shielding topics, and with the wider level of interpretation of the background information, — the very "spirit" of the book has changed away from cable shielding monograph. Then, the original book title would not any more reflect its "thrust" and contents. Also, because of such transformation, even if on the first sight certain analyses in the new book look just like before, their sense and interpretation might be essentially different!

So, What Do You Get?

Such rationale culminated in a handbook on shielding, which would cover a wide range of practical shielding problems, while utilizing and expanding on the innovative principles and approaches that marked and distinguished the previous book and facilitated its usefullness, as well as professional success. To be sure that nothing would be lost, many useful features of the previous book were selectively and discriminately transfered into the new book and often expanded in the process. The new book further develops such fundamental ideas as the guiding role of protected system purpose and configuration and EMI environment in formulation and evaluation of basic shielding objectives and parameters: definitions, figures of merit, problem analysis and formulation, measurement and evaluation techniques. Basic concepts of signal propagation mode / electromagnetic coupling analysis, "parallel" review and "combined use" of the field and circuit theory interpretations, original treatment of common and differential transmission mode EMI effects, use of optimization theory for shield design - and many other, - were expanded and further clarified to cover new areas of electromagnetic shielding theory. It appeared both possible and expedient to apply to the new book the same format, including the "author's notes" which proved itself as a highly useful feature.

I hope, the introduced changes have helped to improve on and expand the handbook's utility and readability. Anyway, the best way to form an opinion is to read the new book. And since it is never late and never enough to improve further, the reader's comments, responses, and suggestions will be greatly appreciated.

ANATOLY TSALIOVICH

Preface

"The mathematical theory of wave propagation along a conductor with an external coaxial return is very old, going back to the work of Rayleigh, Heaviside and J.J. Thomson."

These words were written by S.A. Schelkunoff back in 1934. Indeed, those early works were performed at an extremely high sientific and technical level, often hardly exceeded in the next hundred years (much to the surprise, and sometimes even embarassment, of many of our proud and knowledgeable contemporaries).

Of significance to us: along with the signal propagation in the line, these original works dealt with electromagnetic shielding of the environment inside and/or outside the metallic enclosures. Maxwell himself developed pioneering studies of single-layer shielding shells, while a paper with such "modern" title as "On the Magnetic Shielding of Concentric Spherical shells" was presented by A. W. Rucker as early as 1893 (by the way, quite informative and entertaining, though not easy reading)![*]
Such "state of the art" shielding theory created in the XIX century is even more amazing, if you think that almost at the same time (namely, in 1860s) a manuscript of Jules Verne's book "Paris in the XX Century" was rejected by publisher because it predicted such "outrageously incredible" electro- technology like, for example, FAX service by wires and electrical chair (with the last invention, I suspect many readers would rather have Jules Verne be wrong!).

However, though the beginning of *electromagnetic shielding* theory dates back more than a century, this dynamic field keeps constantly growing driven by practical applications. From the earliest tasks of crosstalk reduction in telephone cables to complex noise, system reliability, and even health hazard problems in modern electronic systems, electromagnetic shielding always was and still remains one of the most radical ways to deal with the electromagnetic interference (EMI). Thus, it evolved into an important engineering discipline.

Today as in the past, many system designers and product development engineers are confronted with necessity to select or design shielded facilities, enclosures, and interconnections. The latest surge of interest in shielding is generated by the needs to assure electromagnetic compatibility (EMC) of computer, telecommunications, and cable television systems in a host of consumer, military, and industrial applications. Ironically, even state-of-the-art wireless (cordless telephone, cellular, PCN/PCS), satellite communications, and fiber optic electronics are among the users of shielded products. However, meeting the industry needs is far from trivial. In par-

---

[*] *Phil. Mag.*, S.5, vol. 37, No. 224, Jan. 1894, pp. 95-130

ticular, the discipline of electronic electromagnetic shielding presents three specific challenges. First, electromagnetic shielding theory and concepts are based on serious physical and mathematical background. Specialized knowledge is required to apply these principles toward practical implementation. Second, modern advances in shielding technology and applications may often call for a revision of previously adopted theories and assumptions. And the last but not least, economical considerations must be an integral part of the system and system component optimization process, because the use of shielded product versions may lead up to 30 - 40% increase in cost. However, the usual "shielding economics" rarely goes beyond a "good enough / too expensive" elementary considerations, instead of concentrating on a system approach to shielding technical effectiveness and economical efficiency. So, the need for efficient shielding is even more imperative.

For these reasons, a wide-spread and pressing demand exists for information on electromagnetic shield design, utilization, and evaluation techniques, as pertains to electromagnetic compatibility. This demand fuels the expansion of the corporate and government R&D programs, development of new national and international standards, steady flow of original publications in trade and professional magazines and conference proceedings. As a result, the problem is not so much in the lack of information, but in the difficulties of sorting through a large number of isolated sources with sometimes conflicting contents and providing their interpretation for specific applications. The point is that notwithstanding the achieved progress, the shielded room, product enclosure, shielded cable, connector, and cable assembly shield evaluation still remains the subject of debates over very basic questions: what to evaluate, how to test, and how to interpret the measurement data. There is no consensus even about the very definitions of the electromagnetic shielding effectiveness. The available shield evaluation parameters and test techniques are so numerous and different in approaches that selection of and correlation between them present serious difficulties.

The diversity in shielding performance parameters and test techniques reflects fundamental differences in system EMI and EMC requirements, shield design and working environments, frequency range, grounding conditions, measured values (i.e., dynamic range), and other specifics. So far, no shielding parameter, and for that matter no shield evaluation method, were created to address simultaneously all the complex electromagnetic shielding problems. As a result, no shielding test standard is comprehensive enough to become universal, in spite of all the efforts and much to the frustration of many professional organizations. An author of a paper-review summarized many years of activities in his international standard development organization as "... the years that the group has been meeting without apparently coming to a definitive conclusion."

For analysis, it is convenient to group the available information sources on shielding in four main categories. First, there are *fundamental works* which were written sometimes many years ago and have now become classical. At the roots of the contemporary electromagnetic shielding theory are the names of S.A. Schelkunoff and H. Kaden. In particular, Schelkunoff is credited with the electromagnetic theory of homogeneous cylindrical shields and the introduction of the surface transfer impedance concept into the theory of shielding. Kaden developed original approaches to the non-homogeneous shield analysis: spiral shields, shields with apertures, multilayer shields. Without their works (e.g., see ref[P.1, P.2]), any library on electromagnetic shielding will be incomplete. These works give an excellent and rigorous treatment of electromagnetic shielding theory in general as well as specific applications to shielded systems and their components in particular, which never lost its actuality. Of course, for obvious reasons these sources do not address many electromagnetic compatibility problems of modern electronic systems, as well as the latest technological advances. On the other hand, many new professionals which just enter the field of electromagnetic shielding, are often not familiar with these sources since they went out of print many years ago.

The second category consists of *books* published or updated within the *last two decades* or so. There are not too many such sources providing a comprehensive treatment of the electromagnetic shielding. Several book titles (by no means a complete list) are ref [P.3-P.7]. Usually, these books are dedicated mainly to specific electromagnetic shielding objectives, or based on a certain limited set of assumptions, or use empirical approaches which are not conducive to theoretical analysis.

Within the third category, the numerous papers in relatively recent conference *proceedings* and *articles* in trade magazines together provide a most up to date coverage of the diverse electromagnetic shielding topics. As an example, the IEEE Transactions on Electromagnetic Compatibility prepared two special issues on electromagnetic shielding: in 1968 (March 1968, volume EMC-10, Number 1) and in 1988 (August 1988, volume 30, Number 3). Lately, multiple publications appeared exploiting diversified numerical techniques for computer analysis of important practical shielding problems. All these publications must be certainly watched by professionals familiar with the discipline and wishing to keep abreast of new developments. However, they cannot serve as a reference source and substitute for a systematic analysis and study for those who just enter the field as well as for experienced engineers.

Shielding standards constitute the fourth category, and they can be a valuable source of information. International and national standard organizations, civil and military authorities concerned with the uniformity of practices and techniques of shield evaluation, work on the recommended test procedures and guides. As a rule, these documents are available to the public and are often worth to read and study.

It follows that no single source or even a group of sources address all the complex shielding problems in a comprehensive way. Such work is still to be done (although recognize the fact that "nothing can be everything for everybody" in any field!). Neither it is the goal of this book. Rather the book is conceived as a "Why- and How-To" reference and/or textbook - a serious monograph which addresses electromagnetic shielding as applied to solving EMC problems in electronic systems. The book emphasizes both theory and the "nuts and bolts" of electromagnetic shield design, evaluation, and application in the system. Step by step, it leads the reader consistently from the most general ideas of electromagnetic compatibility to electromagnetic energy coupling and transfer mechanisms in shielded circuits to design and selection methods of optimal shields which are sufficiently effective, yet still not prohibitively expensive and possible to terminate to state of the art shield evaluation techniques and EMC performance experimental data to new horizons in this important field. Basic information is complemented with innovative approaches and abundant experimental data from real life practice, much of each has never been published before.

The material is structured in eight chapters and reference Appendix at the end of the book.

Chapter I, "INTRODUCTION TO SYSTEM EMI AND EMC" provides an overview of the basics and definitions of electromagnetic interference (EMI) and electromagnetic compatibility (EMC), addresses electronic system problems in the EMI environment, the role of system elements and interconnections, and identifies the major EMI controlling factors. While common and established EMC principles are only briefly reviewed for reference purposes, the discussion concentrates on ideas to be used in further analysis of electromagnetic shielding. First, the concepts of the electronic system EMI synthesis and analysis are suggested and then applied to develop the principles of electronic system and system element EMC performance, to generate models of electromagnetic emissions and immunity. The system representation as an active antenna array permits the analysis of the system EMC performance in terms of and using the mathematical apparatus of the antenna theory. Throughout the book, this material serves as a foundation to clarify the role, objectives, and principles of electromagnetic shielding, to review numerous electromagnetic shielding philosophies and develop novel approaches based on common positions, and to identify the shield design, evaluation, and application criteria and limitations.

Chapter II, "UNDERSTANDING ELECTROMAGNETIC SHIELDING" introduces shielding as just one, although important EMI mitigation technique. It addresses the effects and problems of electromagnetic shielding as they relate to EMC applications, as well as the physical principles and mathematical fundamentals of shielding phenomena. These are illustrated using experimental data and different, sometimes controversial and conflicting theories. The specifics of and differences between various electromagnetic shielding effectiveness definitions, theories, and parameters are discussed. Both older and new concepts and results are reviewed. Finally, a comprehensive approach to electromagnetic shielding is suggested and shielding models are defined with regard to the selected system level and adopted requirements to the model generality and relevancy.

Chapter III, "TRANSFER PARAMETERS OF ELECTROMAGNETIC SHIELDS AND ENCLOSURES" introduces *differential* (that is applicable to an electrically small size and/or length) parameters which characterize energy transfer across the shield. The electrical and mathematical models are developed, addressing the shield performance both as a shield and as a high frequency conductor. This is one of the most important topics in system EMC. The provided analysis extends through all popular shield designs: homogeneous and non-homogeneous, single- and multi-layer; foil and mesh, apertures, serve and spiral tapes, braid; magnetic and non-magnetic, linear and non-linear materials.

It considers specifics of energy transfer in such important practical applications as cables, ground and power planes in printed circuit boards, shelves and office cabinets, shielding rooms, electromagnetic gaskets. To cover other important specifics, like apertures, edge effects in finite (electrical) size enclosures respective practical evaluation concepts and techniques are introduced. In particular, the presented analysis is widely based on and draws upon the concept of transfer impedance.

Chapter IV, "EMI ENVIRONMENT AND COUPLING IN ELECTROMAGNETIC SHIELDING" addresses the shield-to-external and shield-to-internal field coupling, the interaction of the coupling and transfer functions, as well as the impact of the shield on coupling between shielded circuits and lines. In *statics*, as well as relatively *close* to the source in *induction* and (partially) *near zone radiation* regimes the fields have distinctly electric and/or magnetic character, which is determined by the impedances of the respective EMI sources, receivers, and signal (EMI) propagation channels. For *fast varying* fields and at relatively *large distance* from the source, typical is the *far* zone radiation field, where the energy is evenly distributed between the electric and magnetic components. Respective assumptions are useful to simplify the coupling, as well as the transfer function problem solutions. However as you will see, the concepts of *slow* and *fast*, *far* and *near*, are largely dependent on the geometry and physical properties of the particular interferers and propagation media, even for the same rate of signal variations (frequency). This is especially important to remember when *different* media are involved simultaneously, where the same signal will appear, say, slow in one medium and fast - in the other. A typical example of such complications is shielding with its inherent interface between dielectric and metal.

Chapter V, "SHIELDING EFFECTIVENESS FOR EMI PROTECTION" makes use of the developed comprehensive shielding system model to assemble the shield transfer and coupling functions into respective figures of merit to evaluate shielding effectiveness in important practical applications. It addresses such subjects as shielding effectiveness of product enclosures, architectural structures, and mobile objects, the crosstalk attenuation between shielded lines, printed circuit board traces, and cables, the rationale, specifics, and problems of shielding system grounding and termination. The presented factual data, test results, and examples enhance the understanding of complex shielding problems and provide useful reference material.

Chapter VI, "MEASUREMENT TECHNIQUES AND APPARATUS: THE TOOLS OF THE TRADE" starts with a discussion of electromagnetic shielding measurement philosophies, which parallel the approaches to shielding effectiveness definitions and evaluation. The particular measurement techniques and their practical implementations are derived from these basic principles. The importance and techniques for shield measuring equipment calibration are emphasized. Although as a rule the detailed design specifications of the described apparatus are omitted, the supplied information is usually adequate to meet practical needs.

Chapter VII, "SHIELDING ENGINEERING" addresses the practical aspects important in shielding design, manufacturing, and utilization. First, "low" and "high" level shielding models are formulated and reviewed. At the low level, the "bypass" shield model clarifies the roles of shield imperfections, connectors, gaskets, and ground loops in the shield EMC performance. At the high level, the concept of the shielding system topology is considered, which consists of separate, but interacting conducting enclosures and cable shields, which protect the system elements and subsystems. The guidance on shield performance modeling and optimization is illustrated with practical examples. Other topics include mechanical parameters and environmental stability, aging, shielding performance specification problems, electromagnetic shielding economics.

Chapter VIII, "THIS BRAVE NEW WORLD OF ELECTROMAGNETIC SHIELDING " serves as a "happy end" of the electromagnetic shielding saga. It summarizes some potentially promising developments in the electromagnetic shielding industry. Some of these developments are based on old enough principles driven by new applications, whereas other are quite innovative. From expansion of the frequency band of "traditional" electronic electromagnetic shielding to hundreds of GHz to conductive plastics to "leaky" and "smart" shields to chiral shields - the progress and innovation burgeon in all electromagnetic shielding strategic areas: applications, designs, materials. The future sure looks bright for electronic electromagnetic shielding. The presented experimental data is useful both as reference material for practical applications, and as an illustration of fundamental theoretical principles.

Appendix, "SELECTED TOPICS IN ELECTROMAGNETICS AND CIRCUIT THEORY" provides a reference background in electromagnetics and circuit theory. The most widely used mathematical representations of the Maxwell's Equations, as well as the Ampere's and Faraday's laws are treated at a relatively simple level. The "math" is often complemented with respective physical considerations and examples from the shielding applications. Only the subjects and results are discussed, which are used in the main body of the book. This is the "bare necessity minimum" which is absolutely necessary to understand electromagnetic shielding. The chapter also adresses, albeit briefly, the use of numerical methods and techniques in solving shielding problems. The readers are strongly encouraged not to limit themselves to this material. Some suggested references are quoted in the Bibliography section, but much-much more is available.

The book capitalizes on the author's more than 30 years of research, industrial, and teaching experience in the fields of electromagnetic compatibility and shielding, cable industry, wire and wireless telecommunications, and system optimization. The majority of the set forth principles were first-hand tested and validated in numerous practical applications. Also, the choice of addressed topics, as well as the format and structure of the presented material were recurrently verified on and improved through interaction with many electronics hardware and system designers, EMC specialists, wire and cable engineers, and wired and wireless communications professionals in the course of the author's scientific and engineering career and during educational courses and lectures given by the author. It is hoped that the book will be of interest to an engineer concerned with electronic system EMC design and EMI troubleshooting, to the electronic cable designer and user, and as a textbook to a student just learning the exciting world of electromagnetic shielding.

A Note about Presentation

To address the diverse needs of engineers and technicians who need a hands-on approach to practical shielding problems while providing a platform for a rigorous discussion to those interested in a deeper insight, an original "theory at the background" way of presentation is adopted. The "elementary" physics of the considered phenomena, the results of quantitative analysis, and the factual and experimental data, are presented in the main body of each chapter and can stand on their own. The theoretical details and mathematical derivations have been relegated to an "author's notes" format distinguished by a different, smaller typeface and enclosed in pairs of "dingbat" symbols (✦). Thus, the book can serve alike as an introduction to electromagnetic shielding for novices and as a useful reference for experienced engineers. Nevertheless, it is in no way an "easy reading", especially for those without proper experience in the shielding area and/or necessary background. It requires attentive and tedious study, brushing up on theoretical skills, referring to the original literature sources, and above all, trying to keep an open, independent, and flexible mind.

ACKNOWLEDGEMENTS

There's not much left, as to repeat the respective section of the "Cable Shielding" book. Here it is (with corrected typos):

I would like to express my deep appreciation to numerous people who contributed to this book in many ways. From my Ph.D. dissertation mentor professor N. D. Kurbatov (Bonch-Bruyevich Telecommunications Institute, Leningrad / Sanct Petersburg, Russia) who first involved me into the electromagnetic shielding field, to the people who read and reviewed this book and whose comments were invaluable. Special thanks to Dr. B. Szentkuti (Swiss PTT). My appreciation also goes to the companies where some of the described research was conducted by the author: AT&T Bell Laboratories, Belden, Thomas & Betts, and others. Thanks also to Kluwer Academic Publishers, the publisher of this book, to Ms. Patricia Lincoln for tedious work looking through the manuscript, to my family who so patiently suffered from my busy schedules and inconsistencies during the writing of this book - to all of them my gratitude for help and apologies for the inflicted inconveniences.

Adding to this honor list, my warmest thanks go to Mr. P. Inglis (FCC) and Dr. Koon Hoo Teo (Nortel) for their encouragement and advice during the writing of the book, and for valuable comments about its contents and title.

1

Introduction to System EMI and EMC

1.1 ELECTRONIC SYSTEM IN AN EMI ENVIRONMENT

1.1.1 System EMI: A Necessary Evil

In modern times, there hardly is a force that can stop or even slow down the ever-accelerating progress of radio and electronic technology, except for...radio and electronics. Indeed, many of the useful and user-friendly devices that have penetrated all facets of our life and so dramatically changed it (hopefully, for the better) are also known to be hostile to each other and, in some cases, to human health. The list of such devices, systems, and services is almost endless, and every day brings to light new sources of concern. Just to name "a dozen" of these, we have:

- computers and, for that matter, all types of digital devices
- radio and TV broadcast services
- hi-fi audio and video recording, reproduction, and storage (including multi-media computing)
- radar, maritime and aeronautical communication and radio-navigation, standard frequencies
- wired and wireless telecommunications networks and devices
- cellular telephone, mobile radio, and personal telecommunications services
- educational and scientific electronics equipment

- military electronics
- medical and biological apparatus
- information technology equipment (ITE)
- household appliances, home-automation
- electrical power equipment, industrial electronics, lights and fluorescent lamps

The problem is that each of these devices, systems, and services generate electromagnetic fields which may affect any other radio and electronic systems, as well as living cells. The harmful effects may range from inter- and intra-system interference when the signals occupy the same or overlapping electromagnetic spectra to the distortion of the transmitted information and operator annoyance to equipment malfunction and long ranging biological consequences. Thus, from nuisances like "hum" at the radio and "snow" at the TV image to cash register errors and hearing aid distortions to heart pacemaker malfunction, cancers, and air traffic accidents - electromagnetic interference is almost always among the "usual suspects".

For some of these devices and systems, the generation of external fields and currents is the main function. These are called *intentional radiators:* (e.g., radio communications devices). For other devices, the external fields are merely unwanted byproducts of their function. These devices are referred to as *unintentional radiators.* One important and typical unintentional radiator is the digital computer.

As an example, Fig. 1.1 shows a desktop computer system electromagnetic emission spectrum measured @ 3 m distance in the frequency band 30–220 MHz. It is compared with the ambient environment at the open-area test site where the computer emissions were measured.

The computer emissions are wide band - emission harmonics occupy the entire investigated band. On the other hand, the ambients can be both *digital* and *analog, wideband* and *narrowband.* The narrowband ambients (from several kilohertz to several hundred kilohertz wide) are typical of carrier systems, wireless communications, or broadcast radio and television stations. Because in this case the goal is to achieve as strong as possible signal and the transmission energy is concentrated within a narrow band of frequencies, these signals are prevalent in the ambient EMI signature. Thus, in the open-area test site, the observed ambient amplitudes were significantly higher than the digital computer system emissions. The important fact is that the computer emissions and open-area ambients occupy the same frequency band.

How wide can the system emission spectrum be? Expanding on the previous example, the computer emissions may occupy the spectrum from direct current (dc) to ionizing radiation (10^{16} Hz—see Fig. 1.2).

As another example, Fig. 1.3 shows the frequency bands occupied by commercial aircraft communication and navigation signals, ranging from kHz to GHz range. Add to the effects of man-made systems some natural electromagnetic phenomena—atmospheric (e.g., lightning), solar (e.g., high sunspot cycles), and galactic—and you get the picture!

Figure 1.4 features important concepts of an electronic system and the electromagnetic interference (EMI) environment. In a most general sense, an electronic system can be defined as a functionally related group of interacting electical elements (e.g., subsystems or components) assembled together to achieve a particular common

Introduction to System EMI and EMC 3

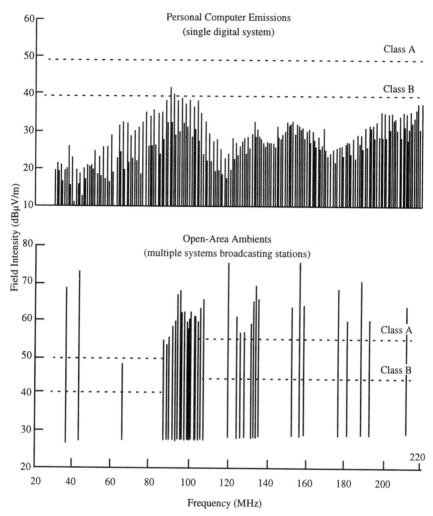

FIGURE 1.1 Electronic system spectra

objective. So, for example, a telecommunications local area network (LAN) may be assumed to consist of interconnected switches, terminals, power supplies, and so forth, with each of these devices housed in different physical blocks. Each block is assembled of printed wiring boards (PWBs) and so on, down to a single IC level or even to an IC's component parts. A wide variety of interconnections between the parts of a LAN can be used, including various kinds of PCB traces, electronic and fiber optic cables as well as wireless media (e.g., microwave, infrared, and so on).

The combined activity of all man-made electronic and radio systems and natural sources of radiation results in an electromagnetic environment that depends on specific technical features of the working systems and subsystems. It may vary with geographical, meteorological, sociological, military, and even political (e.g., jamming) conditions. The electromagnetic environment manifests itself as a background

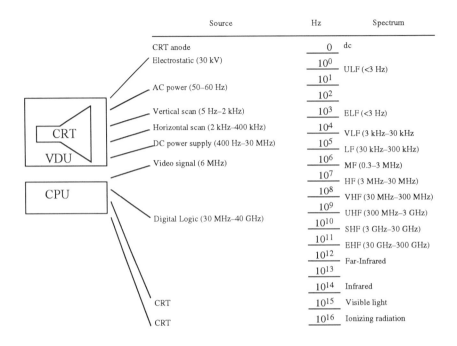

FIGURE 1.2 Computer emission spectra

noise that interferes with the useful signals utilized by any system. Thus, it should be viewed as *electromagnetic interference (EMI) environment*.

From the point of view of all other systems, the contribution to the EMI environment from any specific system or subsystem can be characterized as either electromagnetic environmental pollution or system-specific EMI. If such contamination of the electromagnetic environment by undesirable emissions is considered without re

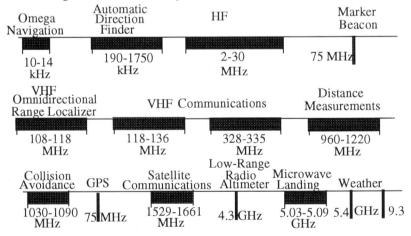

FIGURE 1.3 Aircraft electromagnetic spectrum

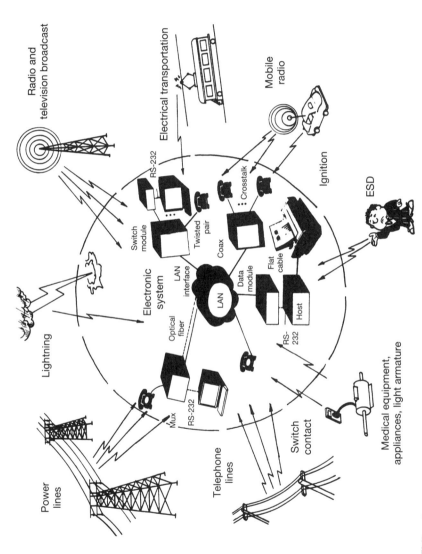

FIGURE 1.4 EMI environment

gard to specific recipients of interference, then it is qualified as electromagnetic pollution. On the other hand, the system-specific EMI takes place between the different systems (inter-system EMI) as well as between subsystems of the same system (intra-system EMI).

✦ Unfortunately, the electromagnetic emission "vice" is inherent to functionality: both intentional and unintentional radiators need to generate electric currents, voltages, and electromagnetic fields—the same phenomena that may cause interference to other devices and systems and affect biological organisms. Whenever an electrical signal is transmitted in a wired or wireless media, an electromagnetic field is generated in the surrounding space. This field induces currents and electromotive forces in other circuits. Also, electronic systems are usually galvanically connected to other systems via input/output (I/O) and power supply lines or cables. Thus, these negative effects are imminent; that is, they present a "necessary evil"—as long as there are electrical signals in a system, the EMI is the "flip side of the coin." However, even if these negative effects cannot be eliminated in principle, they can be controlled. For instance, various man-made systems can be designed in such a way as to occupy different parts of electromagnetic spectrum or to observe some kind of spectrum access and utilization etiquette. Really, there seems to be "enough room for everybody," given that frequencies from fractions of a hertz to hundreds of gigahertz are presently available to the radio and electronics. This being the case, it should be noted that there are still too many electronic systems that are "hungry" and fighting for this available spectrum. In other words, the electromagnetic spectrum is a limited resource. That's why EMI reduction remains one of the most important means of providing for different system coexistence and increasing spectrum utilization efficiency.✦

1.1.2 EMC—The Science Of Electronic System Coexistence

In real life, an electronic system acts simultaneously as an EMI source and receptor, and both of these properties affect the coexistence between systems. System electromagnetic compatibility (EMC) can be defined (e.g., see the IEEE Standard Dictionary of Electrical and Electronics Terms, IEEE Std. 100-1992) as the

ability of a system to function satisfactorily in its electromagnetic environment without introducing intolerable disturbance to that environment.

In essence, this definition emphasizes three identifiable main manifestations of EMI interactions: system *emissions*, system *immunity*, and *crosstalk* between systems (see Fig. 1.5). Instead of the term immunity, its antithesis, susceptibility, is often used as a measure of system vulnerability to EMI.

Figure 1.6 uses a computer system to illustrate four physical phenomena associated with EMC interactions: (1) radiated electromagnetic field emissions generated in the space around the source, (2) conducted current and voltage emissions produced by the system in its power cords and distributed through the power network to other systems (also, a more general conducted emissions principle is often used, which accounts, likewise, for the EMI generating conducted currents and voltages in the input/output cabling, not just in the power cord), (3) system radiated immunity - performance (reliability) under the impact of the interfering electromagnetic fields, and (4) system conducted immunity - reliability under the impact of the interfering conducted currents and voltages injected into the system.

Introduction to System EMI and EMC 7

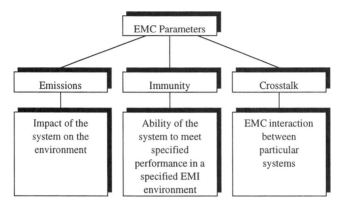

FIGURE 1.5 EMC parameters in an electronic system

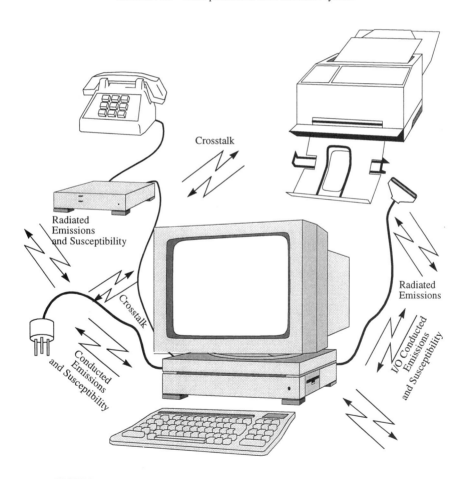

FIGURE 1.6 EMC phenomena: conducted and radiated emissions and susceptibility

Correspondingly, in the EMC discipline it is customary to characterize the EMI phenomena by four basic EMC parameters:
1. System radiated emissions are measured as the electric, E_e, or magnetic, H_e, field intensity generated by the system. Usually, the radiated emissions from the system are specified in the far-field zone. Although there is no fixed boundary between the radiated field far and near zones, we will consider here the far radiated field zone starting at a distance, D, larger than the signal wavelength, λ : $D > \lambda$. The units for radiated emissions evaluation are the same as for the field intensity: μV/m or dB μV/m at a specified distance from the source.
2. System conducted emissions are measured as voltage at the corresponding system terminals. The units for conducted emissions evaluation are μV or dBμV.
3. System radiated immunity represents the measure of the system ability to meet required performance when exposed to a specified electromagnetic field. Usually, the system radiated immunity is measured as the maximum field intensity, E_i, (in μV/m or dBμV/m) that the system can withstand without loosing its specified performance.
4. System conducted immunity represents the measure of the system's ability to meet performance requirements when a specified current or voltage is injected into certain system terminals. The system radiated immunity is measured as the maximum injected current or voltage that the system can withstand without failing to achieve its specified performance.

✦ This is not to say that the stated EMC parameter set is universal, or the only one possible. For example, some regulations may limit the scope of conducted emissions concept only to power cords, or not regulate the immunity for certain products (like the FCC Regulations in the USA), while other may introduce additional requirements (e.g., see Table 1.1 for a sampling of several international standard titles).
Moreover, while the regulations often concern only with the *inter*-system EMI, the *intra*-system EMI may seriously affect the system ability to perform and quality, for instance, the crosstalk between the susbsystems and elements in Fig. 1.6. Even if the intra-system EMI will often (but not always- think of examples!) not affect the system compliance with EMC regulations, it may make the system disfunctional or reduce its quality. For these reasons we will dedicate enough time to shielding for protection from both, inter- and intra-system EMI.✦

✦ Figure 1.6 relates both to the *intra*-system EMI and crosstalk between the subsystems and elements of the same system, as well as to the *inter*-system EMC between different systems, and this book does address both subjects .✦

Emission parameters represent the measure of the system impact on the environment (electromagnetic pollution) without regard to the potential receptor of this EMI. On the other hand, immunity parameters represent the measure of the system's ability to meet required performance in a specified environment, without regard to the actual source of this EMI. If both the EMI source and receptor are identified, and the cause and effect of interference are measured in the same units, it is often convenient to evaluate EMI directly as crosstalk between the specific systems.

Table 1.1: INTERNATIONAL AND EUROPEAN EMC STANDARDS

International	European	Test Conditions	Residential/ Commercial/Light Industry	Industrial Environment
EMISSIONS			EN 50081-1	EN 50081-2
CISPR 22	EN 55022 Class B	Rad: 30-1000 MHz Cond: 0.15-30 MHz	30-37 dB(QP) 66-56 dB(QP) 56-46 dB(AV)	
CISPR 11	EN 55011 Class B	Rad: 30-1000 MHz Cond: 0.15-30 MHz		30-37 dB(QP) 79-73 dB(QP) 66-60 dB(AV)
CISPR 14	EN 55014	0-2 kHz	discontinuous interference	class A,B,C,D
IEC 1000-3-2	EN 60555-2			
IEC 1000-3-3	EN 60555-3			
IMMUNITY			EN 50082-1	EN 50082-2
IEC 1000-4-2	EN 61000-4-2	ESD Contact Air	4 kV 8 kV	4 kV 8 kV
IEC 1000-4-3	ENV 50140	80-1000 MHz, 80% AM	3 V/m	10 V/m
	ENV 50204	900 MHz, 50%/200 Hz		
IEC 1000-4-4	EN 61000-4-4	EFT/Burst 5/50 ns	1 kV AC power line 0.5 kV control line	2 kV AC power line 1 kV control line
IEC 1000-4-5	EN 61000-4-5	Voltage Surge 1.2/50 μs(8/20μs)	2 kV common mode 1 kV differential mode	4 kV common mode 2 kV differential mode
IEC 1000-4-8	EN 61000-4-8	50 Hz magnetic field	3 A/m	30 A/m
IEC 1000-4-11	EN 61000-4-11	Voltage dips, interruptions	30% / 10 ms 60% / 100 ms >95% / 5000 ms	

For example, crosstalk attenuation, A_c, between two cables can be measured as a ratio of the signal voltage, V_s, in the source cable to the induced EMI voltage, V_r, in the receptor cable:

$$A_c = 20\log_{10}\left|\frac{V_s}{V_r}\right|, \text{dB} \tag{1.1a}$$

Other units of crosstalk are also used, e.g., induced EMI voltage as a percentage (%) of the source voltage:

$$C = \frac{V_r}{V_s} \times 100, \ \% \tag{1.1b}$$

As a rule (there are also important exceptions), V_s is measured at the end of the circuit that is connected to the signal generator. V_r can be measured at the generator (near) end of the line or at the other (far) end of the line. In the first case, the result of Eqs. (1.1a and 1.1b) is called near-end crosstalk, or fast crosstalk. In the latter case, the result is far-end crosstalk. When testing systems for EMI compliance, the measured values of the EMC parameters are compared with the limits spelled out in applicable regulations (e.g., see Refs. [1.1–1.7]), and a pass / fail judgement is made.

Different regulating authorities have specified electronic and radio system emissions and immunity in different frequency bands and at various levels. For instance, in Fig. 1.7 the radiated emission requirements are shown for four popular regulations.

One important note: These limits are not applicable to separate devices or system components but to the whole systems that are configured according to explicit rules.

In principle, an electronic product (system) must usually meet a proper *set* of EMC requirements. As an example, Fig. 1.8 shows a typical set of EMC requirements for a cellular station: immunity I to the ambient environment, emissions E - radiated field (power) limitations in- and out-of-band, crosstalk X (say, between the signal areas covered by different station antennas at the tower, or co-channel interference between different cells), biological safety B - in terms of radiated power density and / or SAR (specific absorption rate).

✦ It is presumed that the reader is familiar with power, voltage, and field intensity units. Just as a reminder, the watt (W), milliwatt (mW), and microwatt (μW); volt (V), millivolt (mV) and microvolt (μV); and volt per meter (V/m) and microvolt per meter (μV/m) are units of power, voltage, and field intensity, respectively. These units are widely accepted in the EMI community. The power, P_l (in milliwatts) is a linear unit which is equal to 10^{-3} of the corresponding unit in watts. Voltage, V_l, in microvolts, and field intensity, E_l (in microvolts per meter) are linear units, which are equal to 10^{-6} of the corresponding units in volts and volts per meter. The logarithmic units, P_d in dBm, E_d in dBμV/m, and V_d in dBμV, are calculated from the linear units as $10 \log(P_l/1 \text{ mW})$, $20 \log(E_l/1 \text{ μV/m})$ and $20 \log(V_l/1 \text{ μV})$, respectively. Of course, P_l, V_l, and E_l must be also expressed in mW, μV, and μ V/m, respectively. These are only *basic* units, and many more are actually in use. For instance, a set of logarithmic units deals with relative quantities, e.g., dBc is defined by the ratio of a signal of interest to the level of the carrier, dBi is defined by the gain of a given antenna in comparison to the "gain" of an isotropic antenna, etc.✦

Introduction to System EMI and EMC 11

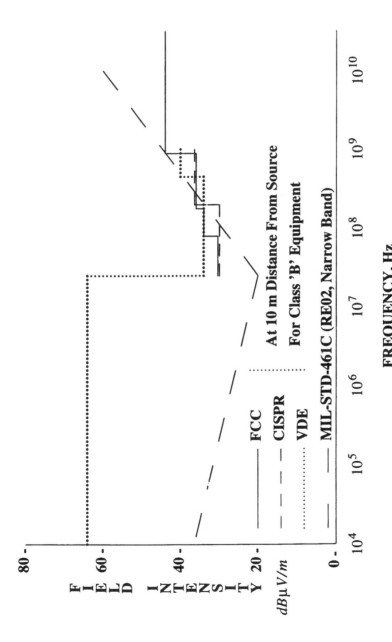

FIGURE 1.7 Radiated emission limits

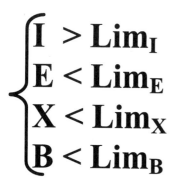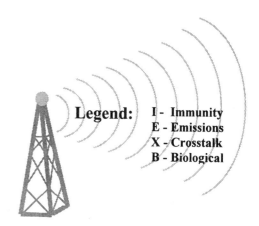

FIGURE 1.8 EMC equation set

♦ In principle, the voltages V_s and V_r in Eqs. (1.1a and 1.1b) can be related to the source system emission fields and receptor system immunity. In this way, crosstalk may be viewed as a secondary EMC parameter that can be derived from the basic emissions and immunity parameters. Such a parameter presents a very convenient characteristic which almost ideally fits the EMC evaluation problem per the definition of EMC. However, the relationship between the basic EMC parameters and crosstalk depends on numerous factors: system configurations, electrical characteristics, mutual arrangement and geometry, and environment properties. Therefore, the correlation between the emissions/immunity set of parameters and crosstalk may be not trivial. In the general case, when the source (in the immunity case) and the receiver (in the emissions case) of EMI are not known (or can change or act in a complex environment), it may be cumbersome, impractical, or just not possible to relate basic EMC parameters (i.e., emissions and immunity) to the crosstalk.

However, in the case of EMI in a cable, the correlation difficulties may be not as severe, and significant advantages often can be realized by relating the emissions/immunity and crosstalk. Historically, the concepts of crosstalk were developed mainly for induction field applications. But it is in just such fields that the evaluation of system emissions and immunity often present serious difficulties! This is because, in induction and near-zone radiating fields, the field configurations and effects

Introduction to System EMI and EMC 13

depend not merely on the interference source or receptor properties but on the properties of both. Thus, using crosstalk EMI parameter for system (and especially cabling) interaction in the induction and near fields may present not just an alternative but the only "right choice." Therefore, in this book we will address the problems of cable coupling in the induction and near radiating fields and provide useful relationships on field/crosstalk correlation.✦

✦ The concept of compliance with specified limits looks quite logical and relatively simple to implement. However, the development of the limits themselves presents serious technical challenges and may have far-reaching economic, social, and even political consequences. Indeed, one of the most important electronic system qualitative characteristics is signal-to-noise (S/N) ratio or its logarithmic measure (in dB) at a certain point in the system receiver. While the signal level (S) at this receiver point is determined by the system transmission power (perceived as EMI emissions by other systems), the interference at the same point is determined by the EMI emissions from other systems. Given the *required* transmission quality (i.e., required S/N), the higher the system immunity level, the higher EMI levels it can tolerate, and vice versa. If the EMI immunity level of the system is specified, then its required S/N ratio is generally determined by the EMI environment background noise. Hence, the limits to the background noise in general, and/or to the contribution to it (electromagnetic pollution) of particular systems, can be derived, and emission limits thereby obtained. On the other hand, if the EMI environment is specified, then the required S/N ratio is determined by a system's immunity level, the limits of which can be determined in this case.

However, if neither immunity nor background noise levels are fixed (which is the general case for limit development), the same required S/N ratio can be obtained in many ways, by simultaneously increasing (decreasing) the system immunity level (χ) and increasing (decreasing) EMI level ($\wp$), as long as the ratio between these characteristics exceeds a certain limit $\Re$:

$$\frac{\chi}{\wp} > \Re \qquad (1.2)$$

With regard to Eq. (1.2), the concept of crosstalk is especially useful, because it facilitates direct calculation of S/N ratio.

In the 1980s, the general regulations were concentrating mainly on emission limits [1.1–1.4]. Which means that the system designers had certain flexibility in selecting immunity levels. Only several industry-specific standards were regulating the product and services immunity (e.g., military [1.5], TV receivers, automotive, and some others). In the 1990s, the center gravity has shifted towards specifying system immunity levels, especially within European Community [1.6-1.7]. Thus, the present trend is toward balancing the emission and immunity requirements.✦

Because the primary objective of this book is to review the electromagnetic shielding phenomena, at this time it is at least worth mentioning where shielding fits into this EMC parameter set. In principle, electromagnetic shielding primarily affects radiated emissions from and immunity of electronic systems. However, it may also have indirect effects on conducted emissions and immunity. There are two physical mechanisms involved:

1. The EMI induced in the interconnections (power cord, I/O cables, shieldedcircuits and lines) can be conducted to the line load, thus creating a secondary source of conducted EMI.

2. Shielding may change the transmission properties of the circuits and cable conductors, thus changing conditions of conducted EMI propagation.

Of course, we have the whole book ahead to clarify this subject in a great deal of detail.

1.1.3 Electronic System EMI Synthesis and Analysis

Modern electronic systems incorporate a large number of fast digital and analog devices assembled in functionally interactive circuits and packaged in separate units interconnected by cables. These include printed wiring boards (PWB), shelves, boxes, racks, and so on. In the general case, each of the system ingredients (elements) has a local functional objective of its own, which purpose is to support the achievement of a higher level purpose of the whole system—the global functional objective. For example, the global objective of a LAN is to provide the specified communications capabilities with required quality. Such system-oriented definition provides for flexibility in selecting the system limits and scope, and at the same time facilitates quantitative characterization of the system and its elements.

Simultaneously, either as a direct function (in intentional radiators) or as a by-product of its useful activities (in unintentional radiators), each of the system elements generates its own local EMI signature which is the source of interference with other elements within and outside of the given system. The total of all these local emissions is perceived by the outside world as the system global radiated emissions (e.g., see Fig. 1.9).

Such interpretation of system EMI results in two practical EMC-related study objectives: electronic system EMI synthesis and electronic system EMI analysis. The goal of the system EMI synthesis is to predict the whole system emissions (or immunity), using known EMC performance parameters of the system elements: subsystems, subassemblies, components. This problem is tied into the system EMC performance evaluation. The solution permits us to evaluate the performance of system elements instead of the whole system, which often can be impossible, undesirable, or expensive.

The goal of the EMI analysis is to determine the limits of element contributions into the whole system emissions (or immunity), using the known or desired EMC performance of the system. This problem is tied into the system EMC design. The solution of this problem permits us to specify the EMC performance of system elements, which will guarantee the whole system performance.

For instance, by knowing the EMC limits of the whole system, the EMC parameters of the system elements can be selected in an optimal way with regard to certain adopted criteria. The system element EMC parameters (but as a rule, not their values!) may be the same as for the whole system (e.g., emissions and immunity), or they may be component-specific (e.g., cable shielding effectiveness in conjunction with the specified EMI currents in the cable). Both EMI synthesis and EMI analysis problems can be based on an important concept of system element EMC perfor-

Introduction to System EMI and EMC 15

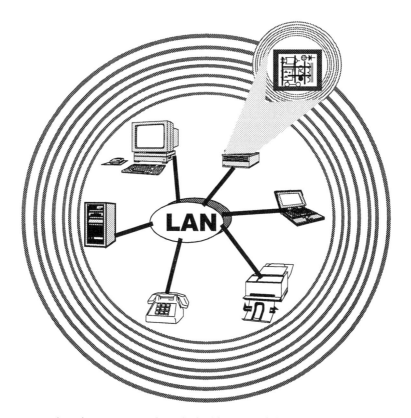

An active antenna array is synthesized from a set of elementary sources:
- I/O and power cabling
- Printed wire board and backplane lines
- Flat metallic surfaces
- 3-D metallic elements

FIGURE 1.9 Electromagnetic system as a source of complex radiating electromagnetic field

mance, as a measure of the element contribution to the undesirable emissions and susceptibility of the whole system.

Along with the system EMC design and EMC prediction, the system element EMC performance parameters can be used for component and subassembly design and evaluation, and to create a unified approach to EMI models' generation. So far, research has been done mainly on separate component-level EMI: IC devices, PWB interconnections, cables.

However, relatively little is known about the "synergy" of these component emissions within a system, which can be a crucial factor in defining the EMI profile of a complex system as a whole.

Consider an electronic system consisting of individual elements. The emissions from each system element can be modeled by an elementary EMI source (ES), containing its own signal generator and signal emitter of EMI signal. Similarly, the im-

munity of each system element can be modeled by an elementary receiver (ER) containing its own signal receptor and signal sensor of EMI signal. Both the ES and ER can be of radiated or conducted nature; the radiated signal emitters are called radiators. In general, here we will use the following terminology:
- emitters — the physical structures propagating EMI from the EMI source into the ambient environment or other systems ("victims") ;
  radiators — field emitters
- sensors — the physical structures accepting EM from the ambient environment or other systems ("culprits")
- generators — the functional system elements feeding the EMI source emitters the signal,
- receptors — the functional system elements receiving the EMI signal.

In our applications, we will understand emitters / sensors as both radiated and conducted "producers" of EMI.

Although these terms are by no means widely used, we do need some terminology to refer to! Some typical radiated emission generators are clock oscillators, line and bus drivers, buffers, analog (narrowband) signal sources, and power supply switching elements. Typical radiators are PWB traces and wiring, IC packages, cables and connectors, and improperly grounded shielding enclosures, frames, and racks.

The system radiated EMI profile is determined by superposition of emissions from individual components and subassemblies as well as by their interaction. Thus, the system can be modeled by a set of ES components. To account for ES interaction, it is assumed that in the system each generator can feed one or several radiators, and that the radiators can be mutually coupled. In this manner, an electronic system can be defined as an active antenna array consisting of an interacting set of generators/receptors and radiators/sensors. Such an approach yields the problem to analysis and gives access to a well developed antenna theory and test methodology. In particular, the system radiated field pattern can be found as a combination of the array element patterns. Elsewhere [1.10], similar models have been utilized to synthesize and analyze real electronic system radiating fields, explain and interpret EMC test results, and arrive at important practical recommendations.

In the case of conducted emissions (immunity), the elementary sources can be modeled as consisting of proper signal generators (receptors) within the system under test and a transmission line—cable that connects a generator (receptor) to the power network (to the source of injected EMI).

Because of the reciprocity of electromagnetic processes, the properties of radiators/sensors and generators/receptors are often assumed to be reciprocal. However, care should be exercised in formulating the "system" for such reciprocity to hold. Whereas, in general, the reciprocity principle is one of the most basic laws of nature, it is applicable only under certain conditions: the systems must be passive and linear. Moreover, application of the reciprocity principle to EMC should be done in the context of the system EMI. For example, compare the cases of system immunity and emissions. It is true that the behaviors of any given (nonmagnetic) wire as a radiator and as a sensor are reciprocal. However, this does not mean that the strongest radiator (dominant radiator) in the system is also simultaneously the most sensitive sensor (dominant sensor). Indeed, one possible reason for the radiator's dominance may be a high signal level within it. (This, of course, is determined by the properties of respective generators. Another reason for the "dominancy" consists of the radiation properties of the radiator itself.) But as a direct result of the high signal level, the signal-to-noise ratio in this particular part of the system may be better (higher) in relation to particular EMI conditions. Thus, the system will be more immune to the EMI! We will discuss the application of the reciprocity principle to the shielding problems in chapter 2 and the Appendix.

1.2 SYSTEM EMI CONTROLLING FACTORS

Following the described above models of elementary EMI sources, their interactions can be conceptually viewed as an *EMI channel* (see Fig. 1.10).

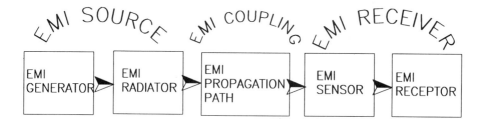

FIGURE 1.10   EMI channel

Based on EMI channel representation, we will consider three groups of system EMI controlling factors:

1. Factors characterizing system elements as emitters / sensors
2. Factors characterizing system elements as generators / receptors
3. Factors characterizing the transmission path (that is, "*coupling*" path) from generators to receptors, or, on a more limited scale, from emitters to sensors.

18 Electromagnetic Shielding Handbook

1.2.1 Basic EMI Study Approaches and Coupling Path

✦ In section 1.1.3, we gave definitions of the emitters / sensors and generators /receptors. It appears that depending on specifics of the EMI environment, *the same* emitters / sensors and generators /receptors can act (even simultaneously!) in the path of both radiated and conducted EMI. Thus, the coupling mode between the given electronic system and the environment (emissions and immunity case), or between the interacting electronic systems (crosstalk) may determine how we "look at" and study the emitters / sensors and generators /receptors. For this reason, we start the discussion of the EMI controlling factors with the last item.

Table 1.2 provides a classification of the basic EMI study approaches and respective coupling paths. It is based on the definitions of the EMC parameters and on at least preliminary understanding of coupling mechanisms

Because we haven't built yet the necessary background knowledge, we present the Table 1.2 without explanations, and at this time it may look to some readers somewhat "cryptic". But we do need some kind of EMI process classification to keep in mind, while discussing the specific emitters / sensors and generators /receptors. In chapter 5 this subject will be covered to a great extent.✦

Table 1.2 Basic EMI Study Approaches and Coupling Mechanisms

EMI Mode	EMC Effects	Coupling Mechanism	EMI "Carrier"	Study Methodology
Radiation	Emissions/ Immunity	Radiated Far / Near Field Coupling	Electromagnetic Field	Radio Wave Propagation, Electromagnetic Field
Induction	Crosstalk	Reactive Near Field Coupling	Electric / Magnetic Reactive Fields	Circuit Theory, Electric / Magnetic Field
Conduction	Conducted Emissions/ Crosstalk	Galvanic (Conductive) Coupling	Voltage, Current	Circuit Theory

1.2.2 Meet EMI Emitters / Sensors

Clearly, the emitter / sensor efficiency depends on its physical, geometrical, and electrical properties, which combine to create a certain kind of *working regime*. Thus, the working regime of a cable EMI conductor or radiator is defined by (1) circuit type; (2) circuit coupling to other circuits, metallic masses, and ground; and (3) circuit transmission mode (common and/or differential).

Interconnections as EMI Sources in Electronic System
We will scope the term electronic interconnections in a broad sense:
   not only designed into the system interconnections, like cables, PCB traces, etc., should be accounted for, but also any metallic element which carries currents, can radiate, couple, or conduct EMI.

Introduction to System EMI and EMC 19

Such definition includes both intentional and unintentional emitters / sensors, the last being often unexpected and unaccounted for. Electronic interconnections and especially electronic cables present a typical case of EMI emitting /sensing that very often results in failure to meet EMC regulations. In the system, the interconnection EMI potential depends on its function. Indeed, the line function in the system (whether designed in or unexpected) is determined by the generators feeding the line and the respective cable or interconnection features and mode of operation. These factors affect the emitter / radiator efficiency. Depending on the type of transmitted signal, the cables and interconnections can be divided into two large groups: (1) information-carrying, whether designed or not, and (2) power supply and control. The main objective of the designed information-carrying interconnection is to preserve signal integrity. As a rule, these lines are designed to transmit relatively low power, produce negligible radiation, and operate in an extremely broad frequency band. Of course, as always, the rules are not without exceptions, e.g., high-energy antenna feeders, "leaky" coaxial cables, and low-frequency or dc control lines. In contrast, the main objective of the power line is to provide the necessary energy to the load. The functional currents and/or voltages in the power cables may have relatively large amplitudes and low frequencies: dc (direct current) and ac (sinusoidal, 50 or 60 Hz). However, higher frequencies—up to several hundred kilohertz, sinusoidal or pulse—are also being used, as in switching power supplies.

> ✦ So, a power supply cable will conduct and radiate only low-frequency, strong currents and fields, and the interconnection cables will conduct and radiate higher frequency but "weaker" currents and fields, right?
> Wrong! The problem is that, because of the coupling between different circuits as well as between the circuits and external fields, the low-frequency signals (e.g., switching power supply noise) may couple into and appear on interconnection cables. As a result, the interconnections may propagate and radiate low-frequency signals induced from the power supply circuits. By the same token, the high-frequency signal may appear and propagate on the power supply and control cables.
>
> This fact has very serious consequences. For example, the power supply cords are not designed to transmit information signals and, unless special measures are taken, their construction may facilitate efficient radiation at higher frequencies. Therefore, with regard to EMI emissions and immunity, these cables are often among the biggest "offenders."
> By the way, this is one of the main reasons why shielded power cords are used: shielding can reduce the induction on and radiated emissions from the power cord, as well as crosstalk between the power cord and other power and information carrying circuits, both at "power supply" and "interconnection" frequencies. Of course, to achieve that goal, shielded power supply cables must be properly designed, connectorized, and utilized. Otherwise, the results may be just opposite of what is expected. But that is why you are taking time to read this book! We will address the effects of power cord shielding in more detail later in the book.✦

Within an electronic system at large, a whole hierarchy of interconnections can be established (see Fig. 1.11). This classification, although not mandatory, it is widely accepted in several variations) embraces interconnections from the "chip" level and up to global networks. Electronic cabling (or wiring) can be used at any of these interconnection levels, thereby contributing to the system EMI. This cabling also can be shielded.

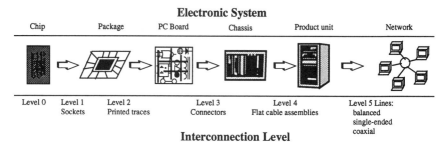

FIGURE 1.11   Interconnection levels in electronic equipment

To illustrate the effect of the cables on system radiated emissions, a computer system (Fig. 1.12) was tested for compliance with FCC regulations. Three tests were performed under different I/O cabling conditions:

1. I/O cables connected
2. I/O cables disconnected
3. I/O cables connected but shielded

The experimental data is presented in Fig. 1.13. At least three important conclusions follow from the presented data. First, the effect of cabling disconnection and shielding was dramatic, actually making the difference between the system meeting and failing the FCC regulations (CFR 47, Part 15). Second, shielding the cables produced effects seemingly identical to cable disconnection: graphs (b) and (c) in Fig. 1.13 are absolutely identical! Third, almost all emission reductions occurred below approximately 100 MHz. Of course, such dramatic effects are observed only if the cables are dominant radiators in the system. As a matter of fact, easy to see that in the setup in Fig. 1.12, the cables were dominant radiators only in the low-frequency range.

How does emissions/immunity depend on the number of cables in a system? Considering the cables to be elements of an active antenna array that models the system, the combined action of a number of cables can be evaluated theoretically using proper antenna element and array factors as suggested by antenna theory. Because these factors depend on system specifics and environment, these conditions must be identified before such calculations can be made. A certain theoretical insight into this problem will be given in the following sections. Some experimental data are presented in Fig. 1.14. This data is generic, based on statistical processing of measurement results from approximately 80 systems which were tested for compliance with the FCC regulations. Only systems with dominant cable radiation are represented in Fig. 1.14. As shown, the system emissions are almost proportional to the number of cables when the number is small (approximately below 5). With further increase in the number of cables, the increments in emissions become smaller.

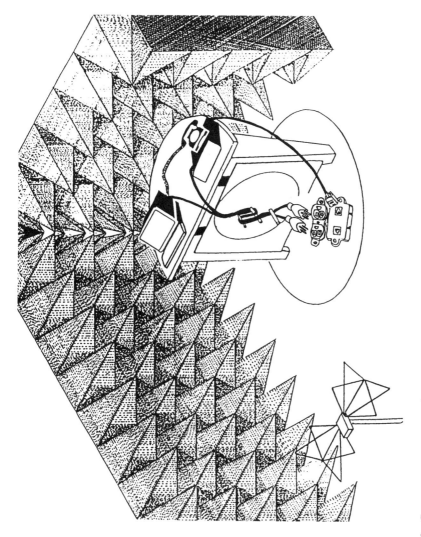

FIGURE 1.12 Computer system test setup in an absorber lined chamber

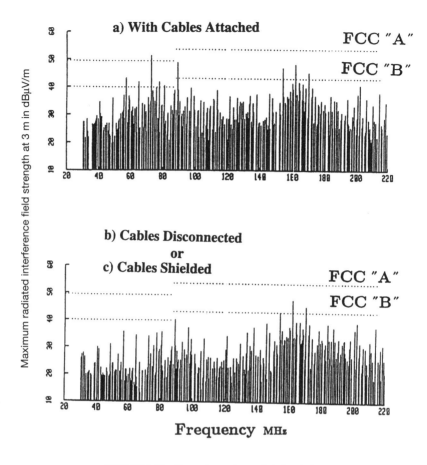

FIGURE 1.13 Computer system radiated emissions

✦ Two topics discussed in the previous paragraph warrant a detailed explanation.
1. A dominant radiator obscures the emissions from other, secondary radiators. It is only after the emissions from a dominant radiator are suppressed that the next strongest radiator can be identified, which then becomes the dominant radiator, and so on. Because of this, suppressing a dominant radiator will reduce overall EMI only to the level of the next strongest radiator. After suppressing several radiators, it may be necessary to return to the first radiator.

    The concept of the dominant radiator can be used to explain the coinciding results in the tests (b) and (c) in Fig. 1.13. These results do not mean that disconnecting and shielding the cables is always equivalent. They mean that, just in this particular case, both the cable disconnection and shielding have suppressed the cable-related emissions below the level of the next strongest radiators, which could be other system elements present in the setup, e.g., the computer, printer, and modem themselves, or their power cords. If these radiators were suppressed, the residual emissions from the shielded cables could have been identified. With regard to Fig. 1.14, unless some additional tests were made (which is not trivial, since the data is statistical), the smaller slope of the curve in Fig. 1.13, after the number of cable exceeded 5, can also be at least partially explained by the increasing weight of secondary emissions.

2. The box in Fig. 1.14, which contains actual measurement statistics, gives some indication that with the increase of the number of cables, the lower frequencies of emissions are emphasized.

Introduction to System EMI and EMC 23

The "lowering" of emission frequencies with the growing number of cables can be explained based on antenna theory. In particular, in Ref. [1.8], a general electronic system radiating pattern simplification principle was formulated. According to this principle, "a complex system consisting of multiple elementary sources may act as an active antenna array consisting of a smaller number (or even a single) larger size system radiators." Naturally, the larger size of the radiators leads to emphasizing the lower radiated frequencies and simpler radiation patterns. A more detailed discussion of this subject is given in Ref. [1.8].✦

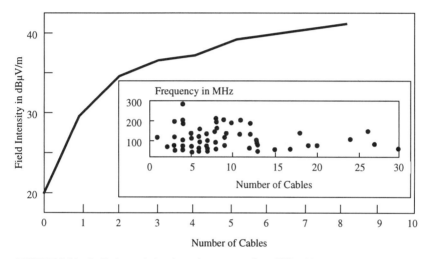

FIGURE 1.14 Radiating emission dependence on number of I/O cables in an electronic system

Interconnection Circuit Types and Physical Designs

The circuit type is determined by its physical shape and electrical symmetry with regard to the circuit conductors carrying the direct and return circuit currents. Several examples of typical interconnection circuits are shown in Fig. 1.15. In the balanced circuit, both conductors create an electrically symmetrical system with identical direct and return current paths. Such a configuration results in direct and return currents that are equal in value and opposite in direction (i.e., the phase shift between the direct and return currents is 180°). In the unbalanced circuit, this electrical symmetry is violated, which leads to inequality of the values of the currents in the conductors, and other than 180° phase difference. This does not mean, however, that direct and return currents are different. They are different only in these two conductors, which means that the current also will find some other paths and create new radiators in the system—quite an undesirable event! The "ultimate" in the circuit asymmetry is a single-ended circuit, which has "ground" (or any large metallic conductor, for that matter) as a return path.

The coaxial circuit is also unbalanced but, because of the special shape of the external conductor, it is "self-shielded" and under certain conditions can "protect itself" from EMI. This will be shown later.

To enhance the symmetry, the balanced circuits are often twisted. In this way, the existing nonuniformities of the balanced circuit are compensated by the cancelling

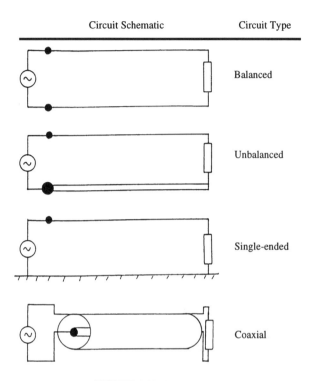

FIGURE 1.15 Interconnection circuit types

effects of the neighboring twisted sections. A smaller twist pitch (period) produces better theoretical EMI compensation, but it also increases mechanical strain in the twisted conductors, which generates a nonuniformity between separate sections. Therefore, twisting cable circuits involves certain limitations. The optimal twist pitch (period) depends on the cable conductor diameter and mechanical properties (namely, tensile strength, elongation at mechanical load, and resilience at twisting) and is usually selected as one period per 1 to 10 cm of cable length.

In practical terms, the circuits are realized as insulated conductors assembled in a cable core (see Fig. 1.16), traces on printed circuit boards (usually abbreviated as PCB, - also the term printed wiring board - PWB, may be used). This is not to count the *unintentional* circuits, which could be just about any conductor or conductive element combination in the system - PCB ground and power planes, shielding enclosures, rack structures, etc. More details on unintentional "third" circuit formation will follow in chapter 4.

In a cable core, the conductors or pairs can be situated in one plane (flat cables) or have predominantly circular cross section (round cables). A cable can contain as few as one conductor (in this case, it is called a wire) or as many as 5000 pairs, as in some telephone cables.

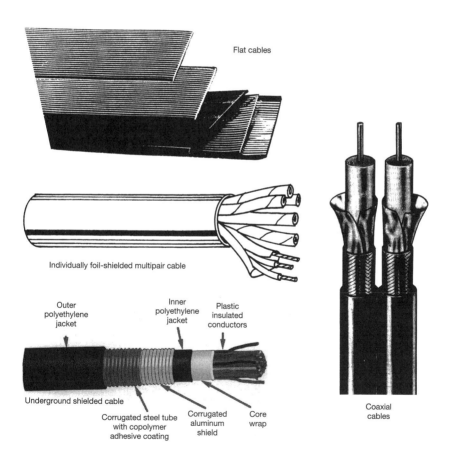

FIGURE 1.16 Examples of electronic cable designs

Each conductor (and/or pair, and/or any group of conductors or pairs, and/or the whole cable) may or may not have one or more conductive or magnetic shielding layers, a plastic jacket, and some other elements such as mechanical protection layers, messenger for aerial suspension, anti-corrosive coating and so on. The presence of the other conductors, and especially the shields, may drastically change the circuit transmission and EMI characteristics. Specific problems arise, such as crosstalk between the circuits in the same cable.

Similar considerations are true for the other type interconnections, e.g., traces on PCB and backplanes. There is a large diversity of EMI emitters / sensors at the higher interconnection levels, as identified by the "icons" in Fig. 1.11. These elements are related to different types of enclosures, frames / racks, ground planes, and others. These topics will be a subject of detailed discussion in following chapters.

1.2.3 What You Always Wanted to Know (But were Afraid to Ask) about Differential and Common Mode Transmission

There are various ways to introduce the common and differential mode concepts. Consider first the balanced circuit shown in Fig. 1.17.

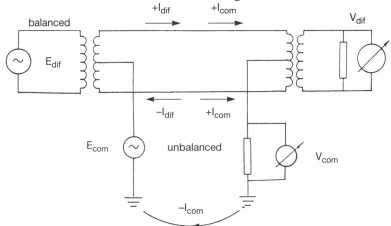

FIGURE 1.17 Differential- and common-mode transmission

At its ends, the circuit is fed by and loaded on transformers. With regard to the source E_{dif}, the circuit works in a balanced mode so that the currents I_{dif} in the direct and return circuit conductors are equal in value and opposite in direction. This regime is called differential mode transmission. In principle, there is no need to have transformers for the circuit to be balanced. But transformers may be helpful to retain the balanced mode transmission if a balanced circuit is connected to an unbalanced source or load, or carries otherwise unbalanced current (in this case, the transformer is called 'balun').. With regard to the E_{com} source, the currents I_{com} in both circuit direct and return conductors are equal in value but have the same direction. This regime is called common mode transmission. Again, in principle, there is no need for a transformer to provide for a common mode transmission.

As seen from the diagram, the differential and common mode currents do not affect each other's loads. However, the combined effect of the differential and common mode transmission appears as a difference in the currents in the direct and return current circuit conductors. In general, if the currents in the direct and return conductors of the circuit are different, they can be represented by two components: differential mode currents and common mode currents.

Now, consider *crostalk* between two lines running *in parallel*. At the surface, it seems that the common mode transmission should result in larger crosstalk than differential mode transmission. This is because in the first the effects of coupling from each of the source line conductors *add,* while in the last - *deduct*. And indeed, such considerations are correct in many practical applications. However in general case, and especially with *radiating* circuits, the situation is much more complicated than superficial intuition may predict. We will have an extended discussion on the subject in chapters 4 and 5.

✦ We selected this particular approach for two reasons: to give some historic perspective and to illustrate possible "useful" aspects of common mode transmission. Back at the dawn of the electronics era, the utilization of schemes similar to that of Fig. 1.17 was the techology's answer to the need for multiplexing. Indeed, the shown circuit permits two virtual lines, using two physical conductors. By using another common/differential line combination instead of the ground connection, the common mode circuit was also made balanced—what was called a *phantom circuit*. Now, four conductors were utilized to organize three balanced circuits. This process could have been repeated to organize additionally a super-phantom circuit.

Later, when the frequency carrier systems were introduced, these techniques lost their economic importance and technical appeal. In particular, the phantom circuits, especially of the nonbalanced kind, exhibited extremely poor crosstalk characteristics. It is mainly because of crosstalk problems that the utilization of the balanced cables for long-distance telecommunications was limited just to several hundred kilohertz. Theories of crosstalk via the third circuits were developed and used to explain and mitigate the respective interference effects in balanced and coaxial cables. Though also prone to third-circuit crosstalk (see chapter 4), the coaxial cable lines are inherently shielded. Thus, the crosstalk problems in coaxial lines are not as drastic and, for this reason, these cables were used up to very high frequencies. ✦

In modern times, using digital modulation techniques, advanced cable shielding, and fiber optics, crosstalk problems in general, and those related to common mode propagation in particular, have partially lost their urgency for long-distance telephony (and so have the third-circuit theories - see author's note above). Also, great advances were made to reduce the crosstalk, as well as system susceptibility to crosstalk, generated in the high frequency electronic cables and between the printed circuit and backplane traces.

But, even as engineers try to forget this "nasty" problem, it has "re-incarnated" as a radiation problem associated with electromagnetic compatibility issues. Now this "old enemy" has come under the "disguise" of common mode emissions and susceptibility.

However, there are inherent distinctions between the common mode effects on crosstalk between the closely spaced lines versus the radiated emissions of common mode signal-carrying currents. For example, in the process of crosstalk between closely spaced lines, the common mode return path can always be identified (e.g., in the shields or the ground planes, if not in other conductors of the line). One can imagine, that when moving "far enough" from such common mode source, it will look to the observer more like a differential mode source, just due to this identifiable return path.

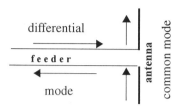

FIGURE 1.18 Radiating antenna is a perfect differential-to- common mode transducer

But where is the "return path" of the radiating current? As Fig. 1.18 illustrates, an antenna can be thought of as an "absolute" transformer of the differential mode currents into common mode currents. Again, using your imagination, you can move as close *to* or as far as you want *from* the antenna in Fig. 1.18 (i.e., perpendicular to the antenna in its plane, and neglecting the antenna feeder radiation), but the radiating common mode transmission will never "transform" into differential mode. The antenna common mode currents generate radiating fields, which then "live independently ever after" in space " far" from the antenna, whereas the near field-related common mode has a "limited domain" only with regard to the reactive fields "tied to" two conductors under consideration. As we will see in the following chapters, the majority of such specifics are the result of fundamental differences in the nature, properties, and propagation of the near-zone reactive fields and far-zone radiating fields. We will also see that currents in allegedly "pure" differential mode real life structures, e.g., a balanced line, in reality exhibit not exactly 180° phase difference and therefore also generate radiating fields!

Effects of Coupling and Grounding on Transmission Mode
At first glance, differential mode transmission is "a natural" for balanced circuits, and common mode transmission is inherent to single-ended circuits, whereas, in general, a nonbalanced circuit will have a combination of both modes. However, the transmission regime depends not just on the circuit physical construction but also on the grounding conditions and interactions with other circuits. As an illustration, in Fig. 1.19 two cases are presented when a perfectly balanced circuit develops common mode currents.

In cases "a" and "b", in a perfectly balanced circuit generator V_d produces differential currents that are equal in value and opposite in direction in direct (I_{d1}) and return (I_{d2}) current conductors. In Fig. 1.19a, a single-ended circuit with current I_e induces same-direction electromotive force in the balanced circuit conductors. As shown, the balanced circuit is coupled to ground. Therefore, the induced electromotive force will result in the same direction currents, I_{d1} and I_{c2}, in the balanced circuit conductors—that is, common mode currents. The total currents in the balanced circuit conductors will be $I_1 = I_{c1} + I_{d1}$ and $I_2 = I_{c2} - I_{d2}$.

If the single-ended circuit were symmetrically situated with regard to the balance circuit, induced currents I_{c1} and I_{d2} would be equal. Then, the resulting differential mode current that the balanced circuit load will "see" is $I_d = I_1 - I_2 = I_{d1} - I_{d2}$. However, if the single-ended circuit were asymmetrically situated with regard to the balance circuit (e.g., closer to the direct path conductor than to the return path conductor currents $I_{c1} > I_{c2}$), the resulting differential mode current that the balanced circuit load will see is $I_d = I_1 - I_2 = I_{d1} - I_{d2} + (I_{c1} - I_{c2})$. This means that the differential (i.e., "useful") current in the balanced circuit has also changed due to EMI.

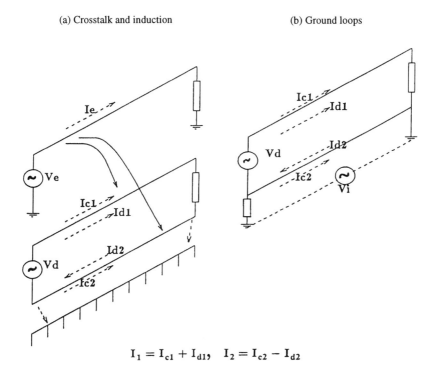

FIGURE 1.19 Common-mode sources

In Fig. 1.19b, another typical source of common mode currents is shown: ground loops. Again, a perfectly balanced circuit is galvanically connected to the ground, which has potential difference V_i between the two points. It is easy to see from the diagram how, in this case, common mode currents I_{c1} and I_{c2} are generated. The consequences of this are similar to the previously discussed case.

In Fig. 1.20, the common impedance coupling via the grounding connection of an RS 232C chip, acts as a source of common mode potential for the chip circuitry and any interconnection leading to it, including (and especially) the I/O cable, transforming them into an antenna (a far field radiation pattern of an electrically long cable radiator is shown as an illustration).

1.2.4 Meet EMI Generators / Receptors

While the circuit balance and transmission mode determine the emitter / sensor properties of EMI sources, the signal parameters are basic EMI generator / receptor characteristics. For example, the Fig. 1.21 illustrates the effect of the clock wave shapes (presented in the oscillograms in black boxes) on the frequency characteris-

30 Electromagnetic Shielding Handbook

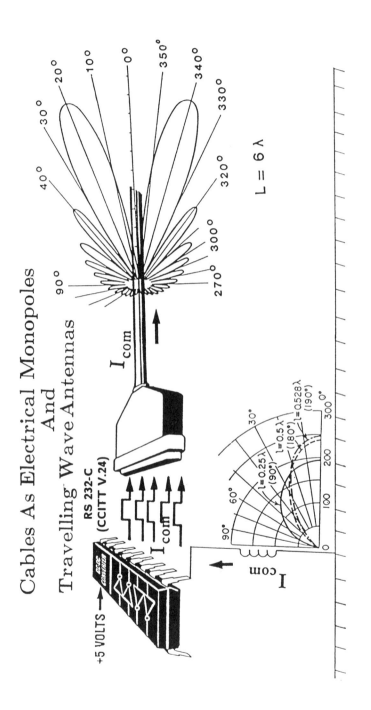

FIGURE 1.20 Voltage drop in grounding system as a source of common mode emissions

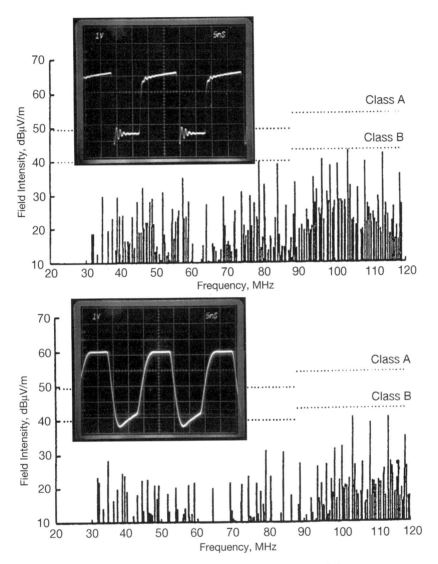

FIGURE 1.21 Clock pulse shape effect on system emissions

tics of the system emissions. As shown, the "smoother" clock pulses result in significantly smaller radiation from the system.

An electronic system produces and uses various kinds of signals. These signals are generated by the active devices at the circuit inputs and by coupling from other circuits (crosstalk) or from the EMI environment (susceptibility). We will use the definition of an electrical signal as a time-varying magnitude and/or phase of a voltage, current, or field intensity, conveying information.

One should realize that the term information in the above definition means not just intentionally transmitted ("useful") information but also EMI information. The latter is "useful" by virtue of containing the parameters of the EMI environment.

The signal parameters of concern with regard to EMI are *amplitudes, time variations,* and *frequency spectra*.

◆ In the following discussion we address both *analog* and *digital* signals. Under the first we will understand signals *continuous* in time and in amplitude, while the second can be *discrete* in time and in amplitude. This permits the analysis in a large number of very diverse applications. Of course, the signal analysis itself is beyond the subject of this book, and we will as a rule use "ready" results, which are available in a large number of excellent texts, e.g., see [1.9-11]. ◆

Electrical Signals in System Circuits: Time Domain

As our signal definition suggests, the signal variations are a function of time; e.g., they exist in the time domain. To illustrate the signal fundamental time domain properties that affect the system EMI potential, consider several simplified examples of typical signals in electronic systems (see Fig. 1.22). The waveform 1.22a is an infinite (in time) sine wave or "pure tone". This is a typical periodic signal, and it may represent a fundamental frequency in carrier systems, a test signal, a simple control device signal, and so on.

To characterize a sinusoidal signal, only three parameters are required: amplitude (A), period (T) or frequency (f = 1/T), and starting point (t_0—initial phase).

Observation of Fig. 1.22a leads to important definitions and conclusions:
- This signal is periodic, which means that its shape repeats each time unit, T.
- Each unit of time (e.g., one second), the

$$f = 1/T \qquad (1.3)$$

repetitions of the signal occur, which of course is the definition of frequency.

- The phase of a periodical function is equal to:

$$\theta = \frac{2\pi}{T}(t + t_0) \qquad (1.4)$$

where t_0 is the point in time from which the countdown starts. Thus, in Fig. 1.22a, the signal countdown could have started from any point between 0 and T (e.g., in Fig. 1.22a, it is shown as having started from the point of time t_0). Now, the sine wave in Fig. 1.22 a can be expressed as

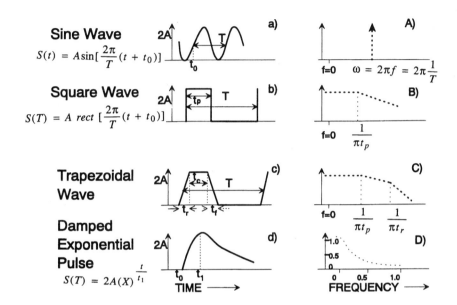

FIGURE 1.22 Signals in interconnection circuitw: time domain vs frequency domain

$$S(t) = A\sin\left[\frac{2\pi}{T}(t + t_0)\right] \quad (1.5)$$

Another important example of a signal in the system circuits is a clock signal, which time synchronizes the logic activity in digital devices. An ideal clock signal (Fig. 1.22b) presents an infinite periodic pulse train—a square wave in which the signal changes occur instantaneously. To characterize the square wave, some of the sine wave characteristics can be generalized, and additional parameters also must be used:

- amplitude, A—the maximum value reached by the signal
- period, T— the absolute value of the minimum interval after which the same characteristics of a periodic waveform repeat (Again, as was the case with sinusoidal signal, the frequency f = 1/T. Here, a generalized definition of the frequency can be given as the number of periods per unit time.)
- pulse duration, t_p—the absolute value of the interval during which a specified waveform exists
- duty cycle, α— the ratio of the pulse duration to the pulse period (In a square wave, the following relationship holds true: $\alpha = t_p/T$.)

Due to circuit inductance and capacitance, the pulse rise and fall actually take some finite amount of time, characterized as rise time, t_r, and fall time, t_f. In real circuits, very severe pulse distortions can occur, which may make it impossible or impractical to characterize and/or measure the "full duration" of the rise and fall time. Therefore, t_r and t_f are usually characterized only from 10 to 90 percent of the pulse amplitude. (Sometimes 20 to 80 percent, or 10 to 50 percent fractions of the pulse amplitudes are also used to characterize rise and/or fall times.) Along with the rise and fall time, three other parameters are used to characterize the pulse distortions: (1) pulse overshoot and (2) undershoot, describing the pulse amplitude variations over the nominal "high" and "low" levels, and (3) ringing, characterizing the oscillatory character of the pulse.

To account for the rise and fall time, a square wave is often modelled by a trapezoidal function (Fig. 1.22c). In a trapezoidal signal, $t_p = t_r/2 + t_f/2 + t_c$. Assuming for simplicity that $t_r = t_f$, the duty cycle of such a signal is $\alpha = (t_r + t_c)/T$.

In contrast to periodic signals, which infinitely repeat their shape, aperiodic signals are "one-time" events. As an example, Fig. 1.22d shows a damped exponential single-shot test pulse. In the time domain, this signal can be described by the expression $S(t) = 2A(N)^{t/t1}$, where $0 < N < 1$ is a coefficient. For example, a similar pulse shape is mandated by the specification IEC 801-5 for surge immunity testing.

The time-domain representation of electronic signals corresponds to actual signal variations that can be observed, for example, using an oscilloscope. These signal parameters vary in extremely broad ranges. In signal amplitudes, they run from microvolts (or microvolts per meter, when electromagnetic fields are evaluated) up to kilovolts (or kilovolts per meter). Rise and fall times range from picoseconds to seconds and minutes, and frequencies range from millihertz to terahertz to light and ionizing radiation (e.g., see Fig. 1.2). However, even if the period (in periodic signals), rise and fall times, and amplitudes provide important information to visualize these signals and come to certain conclusions with regard to their behavior, this information is often insufficient or inconvenient for practical applications—that is, for system engineering purposes, for a detailed comparison of different signals, for a study of the effects of the propagation media on these signals, or for evaluation and application of EMI mitigation techniques. Indeed, neither the crosstalk between circuits nor the radiated emissions from a cable transmitting these signals is readily deduced from the "appearance" of the pulse.

✦ It appears possible to simplify the analysis of the time-domain signal shapes using mathematical transformations of the signal functions. There are two main reasons behind this:
- bring all these diverse signals to a "common denominator", by describing them in the same terms
- Simplify mathematical processing. For instance, by using proper transformations such operations as differentiation and integration can be substituted by four elementary arithmetical operations: summation, deduction, multiplication, division.

Matematicians have developed many important transformations, which are available and can be used to performdigital signal analysis. Some of the most "popular" are Laplace transforms, Z-transforms, several types of and Fourier transforms: discrete (DFT), Fast (FFT), Discrete Time (DTFT). For our purposes, the two most important such transformations are Fourier series for periodic (infinite in time) functions (e.g., similar to those in Fig. 1.22a, b, and c) or Fourier transform for nonperiodic and/or periodic but finite in time functions (e.g., Fig. 1.22d). ✦

Electrical Signal Spectra: Frequency Domain

Using Fourier transformations, the signals can be presented as a sum or an integral of sine (cosine) functions with increasing frequencies. Each of these functions is a harmonic of the signal fundamental frequency, f_T, with the frequencies of harmonics $f_n = nf_T$ increasing with their number, n. This is the essence and the subject of harmonic analysis. Fourier series and integrals establish a correlation between the time domain G(T) and frequency domain H(ω) signal representations. Thus, a periodic function can be represented by a trigonometric polynomial:

$$S(t) = 0.5A_0 + \sum_{n=1}^{\infty} [A_n \cos(n\omega_t t) + B_n \sin(n\omega_t t)]$$

$$= 0.5A_0 + \sum_{n=1}^{\infty} [C_n \sin(n\omega_t t + \theta_n)] \tag{1.6}$$

where

$$A_n = \frac{1}{\pi} \int_0^{2\pi} S(t) \cos(n\omega_t t) dt$$

$$B_n = \frac{1}{\pi} \int_0^{2\pi} S(t) \sin(n\omega_t t) dt$$

$$C_n = \sqrt{A_n^2 + B_n^2}; \quad \theta = \text{atan}(A_n/B_n)$$

To characterize sinusoidal signals, it is convenient to use the angular measures of the sine (cosine) arguments. The angular fundamental frequency is

$$\omega_t = \frac{2\pi}{T} = 2\pi f_T \qquad (1.7a)$$

The angular current frequency is

$$\omega_n = n\omega_T = 2\pi f_n = 2\pi n f_T \qquad (1.7b)$$

As shown in Eq. (1.7), the angular frequency can be defined as an angular measure of progression of a periodic wave in time (or in space, for space coordinate functions) from a chosen instant (or position).

The angular frequency is measured in radians.

The angular phase, θ, of a periodical function is equal to:

$$\theta = \omega t + \theta_o \qquad (1.8)$$

where θ_o is initial phase; that is, the angle from which the time countdown starts. Like the angular frequency, the signal angular phase is also measured in radians. Thus, the signal countdown could start from any point between the 0 and 2π (e.g., the "starting" point in Fig. 1.22a is t_o, corresponding to $\theta_o = 2\pi(t_o/T)$. As a result, this "signal-harmonic" (Fig. 1.22a) can be written as

$$S(t) = A\sin(\omega t + \theta_o) = A\sin\left[\frac{2\pi}{T}(t + t_o)\right] \qquad (1.9)$$

✦ The basic property of periodicity of a time-function S(t) can be written as S(t + T) = S(t). Due to the signal repetitious nature, it is possible to introduce the concept of signal phase as *a fractional part of the period T, through which time t has advanced relative to an arbitrary origin.* Thus, if the point t_o at the sinusoid in Fig. 1.22a is considered the starting point of the signal, the initial phase is $2\pi t_o/T$. In general, the phase can vary from $-\infty$ to $+\infty$. But within this infinite interval, the phase changes that are equal to multiples of 2π bring about identical values of signal. This "granular" nature of phase parameter becomes obvious when it is described in angular terms. From trigonometry, one period of change of a trigonometric function corresponds to 360° or 2π radians rotation of the phasor vector, generating this function. Therefore, while the phase argument (in radians) changes monotonously, the function repeats itself infinite number of times, each 2π of its argument.

Even a signal that is periodic over some limited amount of time, but then does not repeat itself (e.g., is terminated) must be considered to be aperiodic. In this respect all practical signals, which have a beginning and an end, must be considered aperiodic. However, if the duration of the signal is long enough, the periodicity rules can be applied without major errors.✦

Introduction to System EMI and EMC 37

The distribution of the amplitudes (and/or phases) of the signal components as a function of frequency is called signal spectrum. The item A_o in Eq. (1.6) corresponds to the zero-order harmonic in the signal. It is easy to show that A_o is equal to the average value of the function S(t). The other components of the spectrum are sinusoidal (cosinusoidal) harmonics of voltages or currents corresponding to the transmitted information in the cable lines and circuits, or electric and magnetic field intensities in the EMI environment. The number n of any particular harmonic and the harmonic frequency are related as shown below:

$$f_n = \frac{n}{T} = nf_T \qquad (1.10)$$

From Eq. (1.10), $f_T = f_1$; i.e., the fundamental is equal to the first harmonic. The harmonic amplitudes are A_n, B_n, or C_n.

The theory of Fourier transformations is all too well known, and the mathematical background and the most important waveform expansions are given in numerous literature sources (e.g., see Ref. [1.12]). Therefore, we will limit this discussion to several examples only, illustrating the nature of the signals in electronic systems. For instance, by applying the formulas of Eq. (1.6), the same periodic signals presented in the time domain (e.g., in Fig. 1.22), can be mapped into the frequency domain.

Consider the sinusoidal signal in Fig. 1.22a:

$$S(t) = A \sin \frac{2\pi}{T} t$$

Substituting this function into Eq. (1.6), we obtain the coefficients $C_1 = B_1 = A$, all coefficients $A_n = 0$ (including A_o), and $B_{n>1} = 0$. On the frequency scale, such a signal is represented by a vertical line, with its height equal to the harmonic amplitude (Fig. 2.22A).

Applying Eq. (1.6) to the square wave in Fig. 1.22b, we obtain the modules of Fourier coefficients:

$$A_o = A_{av} = 2A\alpha$$

$$C_n = 2Af_T t_p \frac{\sin(t_p \pi n f_T)}{t_p \pi n f_T} = 2A\alpha \frac{\sin(\alpha \pi n)}{\alpha \pi n} \qquad (1.11)$$

where α is duty cycle and $f_T = 1/T$ is fundamental frequency. The fundamental frequency is the lowest frequency component in the Fourier transforms and series. In square wave expansion, the fundamental corresponds to the period of the square wave. It is also the first harmonic frequency, $f_1 = (n=1) \times f_T = f_T$. Thus its amplitude is obtained from Eq. (1.11) for $n = 1$. For example, in a square wave, when $\alpha = 0.5$, the $C_1 = 0.637$ A. Unlike a single-tone sine wave, the square wave spectrum consists of an infinite number of sinusoidal components. In accordance with Eq.

(1.11), these component frequencies are nf_T, so that the difference in frequency between any two adjacent components is f_T.

> ✦ The harmonic content of a signal depends on the properties of the signal function. If $S(t)$ is an odd function, i.e., $S(t) = -S(-t)$, then all the coefficients of the cosine terms $A_n = 0$ and Fourier series consist only of sine terms. If $S(t)$ is an even function, i.e., $S(t) = S(-t)$, then all the coefficients of the sine terms $B_n = 0$ and Fourier series consist only of sine terms (in general, including A_o).
> An odd or even function may contain odd or even harmonics. If $S(t) = S(t + T/2)$, then only even harmonics are present. If $S(t) = -S(t + T/2)$, then only odd harmonics are present.
> It should be mentioned here, that these properties may be a function of coordinate system, and by selecting an alternate coordinate system, an odd function will become an even one, and vice versa.✦

The analysis of expressions for C_n shows that the harmonic amplitudes oscillate while generally following a certain reduction trend with the increase of their number (i.e., rise in frequency). Since the EMC regulations usually specify the maximum emissions radiated from or affecting the electronic systems at specific frequencies, it is desirable to estimate the maximum bounds (MBs) of the signal frequency spectrum envelope. For instance, a simple observation of Eq. (1.9) yields the MB of a sine wave: a straight horizontal line plotted through the sinusoidal maximums.

Things are not that simple for a square wave, where the MB curve traces through the amplitudes of the harmonics in the frequency domain. Consider Eqs. (1.11). The "zero order" harmonic $A_0 = A_{av}$ exists at $f_0 = (n = 0) \times f_T = 0$ Hz; that is, at dc. Therefore, on the MB, it is this component that determines the starting point of the envelope at dc. Since the "zero harmonic" does not depend on frequency, the MB starts at dc and then retains the same value (i.e., goes parallel to the frequency axis) up until the frequency becomes equal to the first harmonic, when the second part of Eq. (1.11) must provide the next extreme points. In the second equation, with the increase of n (and therefore, the increase of $f_n = nf_T$), the maximum value of the sine function in the numerator can never exceed 1, while the denominator is monotonously increasing with n. Therefore, a conservative evaluation of the MB of the square wave expansion will monotonously decrease with the increase of n:

$$MB_{sq1} = \frac{A}{n\pi} = \frac{Af_T}{\pi f_n} \qquad (1.12)$$

This means that the amplitude evaluations given by the MB of a square wave are inversely proportional to the signal frequency.

Obviously, Eq. (1.12) is applicable only at frequencies larger than the fundamental ($n = 1$; $f_T = 1/T$) because frequencies below f_T do not exist in the spectrum. But at what particular frequency does the transition from the constant MB to the frequency-dependent MB [per Eq. (1.12)], occur? To answer this question, revisit the properties of the function

$$y = \frac{\sin x}{x}$$

At small x ($x \to 0$), the $\sin x \approx x$ and $y_0 \approx 1$. At large x, the MB of a sine function equals 1 (the sin function cannot be larger than 1), and $y_\infty \approx 1/x$. The MB breakdown point is located at the intersection of y_0 and y_∞:

$$1 = \frac{1}{x} \to x = 1 \tag{1.13}$$

Equating arguments $t_p \pi n f_T$ and $\alpha \pi n$ to 1, we obtain the breakdown point for an ideal square wave:

$$F_{T1} = n f_T = \frac{1}{\pi t_p}$$

$$n_T = \frac{1}{\alpha \pi} \tag{1.14}$$

If built in logarithmic scale, the spectrum envelope of a square wave look as shown in Fig. 2.22B. As we see, at the fundamental, the MB "breaks down." We will call this frequency the first breakdown frequency.

✦ Expression (1.12) describes the frequency dependence of the signal harmonic amplitudes in linear terms. At frequencies below the first breakdown frequency F_T (we will call it zone 0), the MB is independent of frequency ($A_0 = A_{av}$). At frequencies larger than F_{T1}, the MB is proportional to $1/f$, starting with the value A_0 at the F_{T1}. From this, the ratio of harmonic amplitudes (their MB) at any two different frequencies f_{01} and f_{02} within the zone 0 is $R_0 = 1$, and at frequencies f_{11} and f_{12} within the zone 1 is proportional to the ratio $R_1 = f_{11}/f_{12}$.

Since the emissions are often measured in logarithmic units, it makes sense to present frequency dependence as shown in Eq. (1.12) in these units also. The expressions and graphs become especially convenient when the decimal logarithm base is selected because then the values can be expressed in decibels (dB). The basic logarithmic units for voltage and field intensity are:

$$V, dB\mu V = 20 \log_{10} \left| \frac{V, \mu V}{1 \, \mu V} \right|$$

and

$$E, dB\mu V/m = 20 \log_{10} \left| \frac{E, \mu V/m}{1 \, \mu V/m} \right|$$

Substituting the f_{11}/f_{12} into these expressions, we obtain:

$$R_1 = 20\log_{10}\left|\frac{f_{11}}{f_{12}}\right|$$

These expressions are especially easy to interpret in the case when the ratio of the frequencies equals 10 (i.e., "decade"). Hence, the often used relationship for the harmonic amplitude reduction in zone 1 is *20 dB per decade*.◆

Applying the Eqs. (1.6) to the trapezoidal model of a square wave in Fig. 1.22C, and assuming symmetrical trapezoidal wave $t_r = t_f$, we obtain the module of coefficient

$$C_n = 2A\alpha \times \frac{\sin[(t_c + t_r)\pi n f_T]}{(t_c + t_r)\pi n f_T} \times \frac{\sin(t_r \pi n f_T)}{t_r \pi n f_T} \quad (1.15)$$

As in the case of a square wave, the item $A_0 = A_{av} = 2A\alpha$, the sines in the numerator of (1.15), cannot exceed 1, and

$$\lim_{x \to 0} \frac{\sin x}{x} = 1$$

However, there is an important difference in the envelopes of these two signal shapes. First, note that, as a rule, $t_r \ll t_c < T$. For instance, it is not unusual to have t_r in the nanosecond range, and t_c in microseconds. For this reason, the argument in the second $\sin(x)/x$ item in Eq. (1.15) is much smaller than in the first item. At very small arguments, when $\sin x/x$ approaches 1, the MB of the expansion in Eq. (1.15) is determined mainly by the first item, which is identical to the case of a square wave [see Eq. (1.11)]. Thus as in Eq. (1.12), at lower frequencies, the MB can be described as:

$$MB_{trapez1} = \frac{A}{\pi n} = \frac{Af_{T1}}{\pi}\frac{1}{f} \quad (1.16)$$

Again, the equation (1.16) is applicable starting with the fundamental (n = 1; f_T = 1/T), because the frequencies below f_T do not exist in the spectrum. So, the first breakdown frequency can be calculated similar to the ideal square wave. Similar to the case of the ideal square wave, the MB is an independent on the frequency straight line between the DC and F_{T1}, after which the MB reduces according to the 1/f law.

At higher frequencies, both sinx/x items in (1.15) will be reaching 1, so a conservative evaluation of the envelope at these frequencies can be obtained by equating both sine functions in (1.15) to 1, which results in an envelope of the expansion:

$$MB_{trapez2} = A \times \frac{1}{\pi^2 t} \times \frac{1}{f^2} \quad (1.17)$$

This means that, at the higher frequencies, the signal amplitudes (or rather their evaluation given by the MB) of a trapezoidal wave are inversely proportional to the square of the signal frequency, $1/f^2$. Similar to the previous derivation, the second breakdown frequency at which the Eq. (1.17) becomes true can be found by equating the arguments $t_r \pi n f_T$ to 1:

$$F_{T2} = n f_T = \frac{1}{\pi t_r} \quad (1.18)$$

Respectively, in logarithmic scale, the "trapezoidal wave" MB looks as shown in Fig. 1.22C.

The previous three examples illustrated the mapping of time-domain periodical signals into the frequency domain using Fourier series. Applying the Fourier transform, aperiodic signals can be mapped into the frequency domain (e.g., see Fig. 1.22D). Now, a correlation can be obtained between the rise (fall) time and frequency bandwidth occupied by the signal.

◆ Similar to Eq. (1.12), expressions (1.16) and (1.17) describe the frequency dependence of the signal harmonic amplitudes in linear terms. Thus, the MB frequency zone 0 below the first breakdown frequency F_{T1} is independent of frequency: $A_o = A_{av}$ = constant. In frequency zone 1, between the first F_{T1} and second F_{T2} breakdown points, the MB is proportional to 1/f, starting with the value A_o at F_{T1}. In the frequency zone 2, with frequencies larger than the second breakdown point, the MB is proportional to $1/f^2$. From this, the ratio of harmonic amplitudes (that is, their MB) at any two different frequencies f_{01} and f_{02} within the zone 0 is $R_0 = 1$; at frequencies f_{11} and f_{12} within the zone 1, it is proportional to the ratio $R_1 = f_{11}/f_{12}$; at two different frequencies f_{21} and f_{22} in the zone 2, it is proportional to the ratio $R_2 = (f_{21}/f_{22})^2$.

Again, as was true in the case of an ideal square wave, the expressions and graphs become especially convenient when the decimal logarithmic scales are selected for both the frequency and the amplitude (i.e., MB). Then, the values can be expressed in decibels (dB) and, when plotted, the dependencies look like straight lines with different slopes. Substituting the f_{11}/f_{12} and $(f_{21}/f_{22})^2$ into the voltage and field intensity unit definitions (dBµV and dBµV/m, respectively), we obtain: $R_1 = 20\log_{10}(f_{11}/f_{12})$ dB, and $R_2 = 40\log_{10}(f_{21}/f_{22})$ dB. When the ratio of the frequencies equals 10 (i.e., "decade"), we obtain the often used relationships for the harmonic amplitude reduction: 20 dB per decade in the first zone, and 40 dB per decade in the second zone.◆

Effect of Spectrum Bandwidth on System Emissions

At this point, an important question from the EMC standpoint is, "What is the effect of the signal bandwidth on electronic cable emissions?" Let us start with the following definition of spectrum bandwidth:

42 Electromagnetic Shielding Handbook

Spectrum bandwidth = the least frequency interval outside of which the power spectrum of a time-varying quantity (e.g., voltage, or field intensity) is everywhere less than some specified fraction of its value at the reference frequency (as a rule, from 0.5% to 1%, but other numbers can also be used); this reference frequency is usually that point at which the spectrum has its maximum value.

As demonstrated earlier, the spectrum bandwidth depends on the signal parameters: periodicity, rise/fall times, and transmission media distortions. As an example, Fig. 1.23 presents a plot of the radiated emission spectrum of a computer clock line.

a) Typical clock line emission spectrum

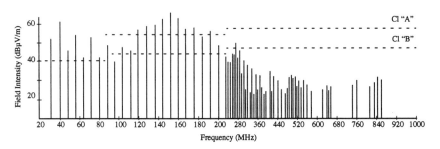

b) Effect of signal ringing on clock line emissions

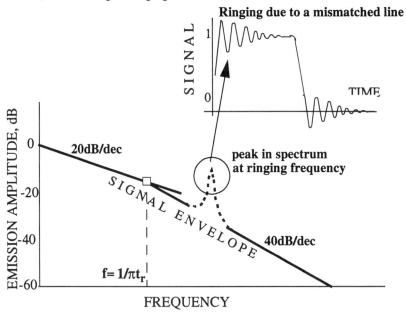

FIGURE 1.23 Clock line emissions

Also plotted are the radiated emission compliance limits The line was 20 cm long, unmatched, and powered by an unbuffered clock oscillator with a 8.064 MHz funda-

Introduction to System EMI and EMC 43

mental frequency. The field measurements were done at an EMC test facility at 3 m distance. Per Eq. (1.18), the signal bandwidth is easily found from the pulse rise time. As shown, due to the pulse distortions in the clock line, the harmonics of very high order (n > 100) have significant amplitudes.

✦ One may wonder, why the spectrum envelop in Fig. 1.23 looks to a certain degree different from that predicted by theory. In real life, there may be various reasons for such discrepancy, like the mismatch in the line or the signal distortion. Fig. 1.23 illustrates the effect of pulse ringing on the spectrum shape.✦

Figure 1.24. presents a frequency characteristic of radiated emissions from a 3 m flat cable, horizontally stretched over the ground plane at an open-area test site (OATS). Emissions were measured at 3 m distance from the cable with a horizontally polarized broadband antenna. The cable was powered by symmetrical trapezoidal pulses with 5.3 V amplitude, 10 MHz repetition rate, and three different rise times, t_r = 50, 5, and 1 ns. Equation (1.18) translates these rise time values into F_{T2} ~ 6.2, 64, and 320 MHz, respectively.

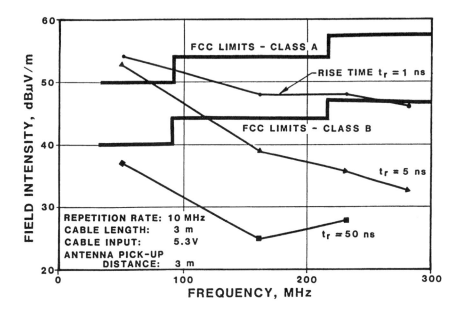

FIGURE 1.24 Radiated emission from unshielded flat cable carrying digital signal

Now, *if* this were our system ("if" is emphasized because the EMC regulations do not apply to cables as components but only to systems of which the cables are a part), both 1 ns and 5 ns rise time signals would have led to failing the FCC regulations.

1.3 DEFINING AND MODELING EMC PERFORMANCE PARAMETERS

1.3.1 Basic Emissions and Immunity Units

The general principles of system element EMC performance were formulated in Section 1.1 as a measure of the element input to the undesirable emissions and susceptibility of the whole system. A fundamental issue is what to specify as electronic cable EMC performance parameters. We saw that electronic cables contribute to system radiated EMI and EMC as radiators (i.e., emitters radiating part of transmitted signal) and/or sensors (i.e., receptors susceptible to ambient electromagnetic fields)

In this respect, the cables are no different from other system elements that act as radiators/sensors and together determine the system emissions and immunity. It follows, then, that to characterize the cable contribution into the system EMI, it is possible and expedient to use the basic system EMC parameters: radiated emissions and susceptibility.

A set of EMC performance parameters may consist of:

- for radiated emissions: field intensity (microvolts per meter or dB microvolts per meter) radiated by the given radiator under certain specified conditions
- for radiated immunity: in general case, the specified performance of the system or system elements under the impact of a given field; a "simplified" figure of merit of immunity evaluation may consist of voltage (dB microvolts or dB microvolts per meter), or current (microamps per meter or dB microamps per meter) induced in the sensor by a given ambient field intensity under certain specified conditions
- for conducted emissions: voltage (dB microvolts or dB microvolts per meter), or current (microamps per meter or dB microamps per meter), as measured at specified equipment ports
- for conducted immunity - similar to the radiated immunity, except that the EMI is introduced to given system ports as voltage or current
- for direct evaluations of coupling between interacting cables belonging to different systems or subsystems: crosstalk (in dB or percent)

✦ There are two related issues:
  Although the interconnections in general are a necessary element in the path of conducted EMI, their role is usually limited to that of transmission channels. Therefore, even if the interconnection parameters, in principle, do affect the conducted emissions (susceptibility) to a degree, this effect is secondary to the conducted emissions generators (receptors). In contrast, in radiated emissions, the effects of generators (receptors) and radiators (sensors) are both primary. ✦

Introduction to System EMI and EMC 45

◆ The use of electronic system element (e.g., cable) EMC performance parameters similar to the EMC parameters of the whole system should not obscure the fact that these two parameter sets are *different in principle*. As indicated earlier, the limits in EMC regulations are applicable *only* to the specially configured systems and not to individual system elements.

For this reason, *a separate cable cannot be compliant with EMC regulations!* This does not mean that there cannot be proper *cable EMI regulations and limits*. One good example of such cable EMI regulations are the limits to the shielding effectiveness of Ethernet™ cable.◆

1.3.2 Typical EMC Performance Models

From the EMI standpoint, electronic system *radiators / sensors* act as a means for radiating or receiving radio waves—that is, as *antennas*. Therefore, an "antenna" approach to the evaluation of the EMC performance of system elements appears quite natural. Such an approach also gives access to a well developed antenna theory and specific test methods.

It is useful to single out the elementary sources (ES) corresponding to the prevailing radiator/sensor shapes and radiation mechanisms and model them with a limited number of simple antenna geometries. In the following text, we will consider the modeling of four main electronic system element types: I/O and power interconnections, PWB and backplane level interconnect circuits, flat metallic surfaces (ground planes and shielding), and metallic 3D elements (shielding enclosure boxes, including boxed-in slots).

Several of the most important corresponding elementary geometries and their radiation patterns in free space are presented in Fig.1.25. They cover such example models like PCB traces and electronic cables - as wire antennas, shielding plates as "patch" antennas, shielding boxes - as broadband antennas . Analyses of these patterns and related formulas are widely available in antenna literature [1.12–17]. We will also return to their review in chapter 4, when discussing the radiating field electromagnetic coupling models.

Of course, the suggested antenna simulations of real system ES are to a certain degree arbitrary, and other models can and should be used as well, especially when modeling the near-zone fields. Nevertheless, they provide insight into the electronic system component radiation mechanisms and immediately lead to important general conclusions:

- Independent of configuration, a small ES (e.g., less than $\lambda/50$) can be modeled by an infinitesimal dipole with two symmetrical lobes in the E-plane.

- As a rule, a large ES exhibits multi-lobe patterns that can be symmetrical or asymmetrical, depending on the ES geometry.
- Improperly referenced metallic surfaces (e.g., ground planes) and shielding enclosures (e.g., shelves, racks, and cabinets) can be EMI sources with an infinitesimal dipole-type pattern for electrically small ($\lambda/50$) radiators, and with a multi-lobe pattern for electrically large radiators.

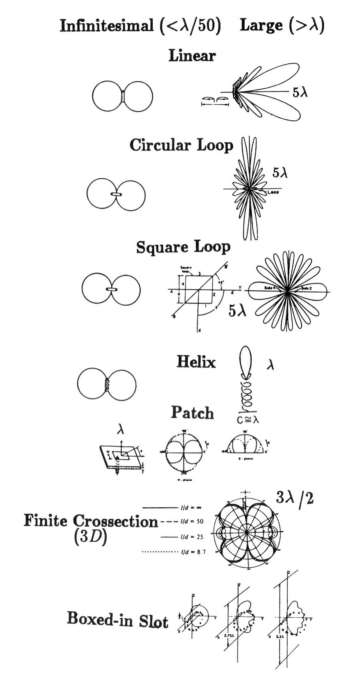

FIGURE 1.25 Elementary EMI source models and their radiating patterns

Introduction to System EMI and EMC 47

- Two- and three-dimensional radiators exhibit "smoothed" radiation patterns and can be efficient in a broad frequency range (not a very "good news", as far as EMC is concerned!).

Now, we will briefly review some of the characteristic features of several specific radiator / sensor models.

Linear Radiator / Sensor Models

Linear radiators / sensors are formed mainly by power supply and interconnect cable conductors and shields, PWB traces, and "one-dimensional" metallic elements which develop an RF potential with regard to the ground plane. Electronic *interconnections and cables* are *designed* so as to minimize the loops formed by direct and return current paths in the line conductors. These conductors are placed as close as insulation thickness permits, and in the cables the pairs are often twisted and/or shielded. Therefore, the cable emissions in a differential mode are not as prominent as in the *common mode* transmission (unless improper grounding and shield termination techniques are used—see the following chapters). On the other hand, the *traces on PWBs and backplanes,* especially when poorly designed, often have much smaller linear dimensions than cables but may produce extremely large area *loops*. That is why, in the higher frequency range (i.e., relatively short wavelength), *differential mode* radiation often accounts for the bulk of electronic system emissions. This is confirmed by experimental statistics [1.18] and illustrated by Fig. 1.26.

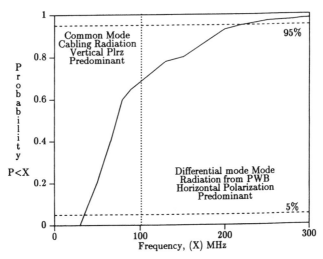

FIGURE 1.26 Cumulative distribution of electronic product top emissions frequencies

However, one must keep in mind, that both large loops and large common mode currents may be created unintentionally, if not sufficient attention is dedicated to the EMC design of the electronic or electrical system. Familiar examples of such unintentional radiators / sensors are common and differential mode currents appearing as

48 Electromagnetic Shielding Handbook

a result of common impedance coupling, ground loops, and crosstalk in the system circuits.

✦ By physical nature, the linear radiators / sensors are electrical dipoles (or monopoles) and loops. When the radiator / sensor length, L, is much smaller than the wavelength (e.g., $L < \lambda/50$), the elementary source can be considered infinitesimal. If an infinitesimal radiator carries current I_{com}, then the electric field intensity in the far-field zone at distance r from the radiator is equal to:

$$E_{\theta e} = \frac{j60\pi I_{com} \sin\theta}{r} \frac{L}{\lambda} \tag{1.19}$$

where θ is the elevation angle. As shown in Fig. 1.25, the pattern has two almost circular symmetrical lobes in the plane of the dipole (E-plane) at $\theta = \pi/2$. In the H-plane the electric field pattern is represented by correct circle. According to Eq. (1.19), maximum emissions of this kind will be produced by the longest radiators in the system (e.g., cables).

When the radiator (e.g., cable) length exceeds the wavelength, the resonant phenomena in the radiator drastically change the emission patterns and result in a multi-lobe patterns, depending on the L/λ ratio. In this case, the field intensity is:

$$E_{\theta e} = \frac{j60I_{com}}{r} \times \frac{\cos(\beta L\cos\theta/2) - \cos((\beta L)/2)}{\sin\theta} \tag{1.20}$$

where β is the free-space phase constant:

$$\beta = 2\pi/\lambda \tag{1.21}$$

It follows from Eq. (1.20) that the maximum field intensity is achieved when the radiator length equals half the wavelength. As a result, depending on the electrical length, long cable radiators emphasize different frequencies. Figure 1.27 shows the frequency characteristics of radiated emissions from three coaxial cable samples of different lengths. The cable lengths, together with respective first- and second-order resonant frequencies, are shown in the box at the top of the graph. The cables were powered by a CW sweep signal of identical amplitude, and the emissions were measured in the range from 20 to 200 MHz. Although the test results are affected by the effects of the test site ground plane, by receive antenna gain change with frequency, and by the equipment setup specifics, the emission peaks occur at frequencies close to theoretical predictions. However, the emission maximums for different cables have approximately the identical values—an important fact indeed! The explanation follows from Eq. (1.20), which contains the cable length only in the "oscillatory" part of the expression. This is quite different from the electrically small cable behavior as described by Eq. (1.19).

The resonant nature of emission dependence on cable length is also supported by statistical data (bottom box in Fig. 1.27). As shown, each statistical cable length produced emissions at a number of resonant signal harmonics, up to the frequencies at which the harmonic amplitudes dropped below the dominant radiators in the system. ✦

✦ It can be shown that a small loop is equivalent to an infinitesimal magnetic dipole whose axis is perpendicular to the plane of the loop. The simplest geometry is produced by circular loop. In the far field, the maximum electric field intensity occurs in the plane of the loop:

Introduction to System EMI and EMC 49

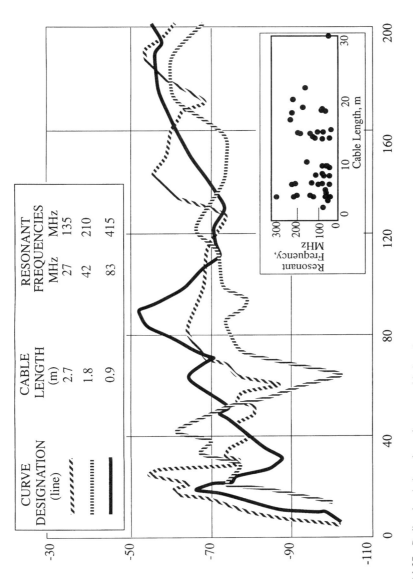

FIGURE 1.27 Radiated emissions dependence on cable length

$$E_{\phi e} = \frac{120\pi^2 I_{dif}\sin\theta}{r} \frac{A}{\lambda^2} \qquad (1.22)$$

where ϕ is the azimuthal angle and A is the loop area. Owing to the specifics of cable design (twisted pairs, coaxial pairs), their loop area, as a rule, is extremely small, so that the main sources of current loop type emissions are PWB and backplane-level lines (of course, in the process of EMC design, the loops created by these lines must be minimized!). Because the field intensity is quickly diminishing as the wavelength increases ($1/\lambda^2$), these current loops emphasize radiated emissions in the relatively high frequency range (e.g., available experimental data suggest a frequency range, as a rule, >100 MHz). As in the case of linear dipole, for large (relative to wavelength) size loops the multi-lobe patterns are typical. In general case, the calculation of field intensity produced by large loops is quite involved. In the case of a large loop with a uniform in-phase current:

$$E_{\phi e} = \frac{60\pi a I_{dif}\sin\theta}{r} J_1(\beta a \sin\theta) \qquad (1.23)$$

where a is the loop radius and J_1 is the Bessel function of the first order and of argument $\beta a \sin\theta$.

The radiating loops are usually assumed to be circular. A square loop pattern is very much similar to the circular loop—especially for small loops. The implication for the problem at hand is that the current loop emissions may be considered independent of the radiator shape and are determined only by the loop area. However, the radiation patterns of electrically large loops have multiple lobes. ✦

✦ Helix radiators may present a good approximation of inductive cable bundles (ideally, noninductive bundles can be approximated by a linear radiator). There are two working modes of helix antenna: the axial and the normal. In the normal mode, the field is maximum in the direction normal to the helix axis, while in the axial mode the field is maximum in the direction of the helix axis. The field is either circularly or elliptically polarized. Since a helix radiator features both axial and circular current components, it exhibits radiation properties of both dipole and loop, which radiated electric fields are described by the E_θ and E_ϕ components, respectively.

It can be shown [1.13-14] that, in the case of an infinitesimal helix in normal mode, the axial ratio is:

$$AR = \frac{|E_\theta|}{|E_\phi|} = \frac{2S\lambda}{\pi^2 D^2} \qquad (1.24)$$

where S is the helix spiral pitch and D is its diameter. However, while the circular current component in a "regular" loop is generated by differential mode currents, in the helix, both linear and tangential current components are generated by common mode currents. The axial radiation mode is obtained when the helix circumference is of the wavelength order. It can be treated in a manner similar to the normal mode way. ✦

Surface and Volume Radiator / Sensor Models

Two- and three-dimensional EMI sources are formed by ground planes, racks, and shielding enclosures (cabinets, shelves) which have a potential with regard to the site ground plane or which carry common mode currents. The "dimensions" here are understood in electrical sense, that is compared to the wavelength. While the theory of two-dimensional and three-dimensional antennas and radiators at large is quite complex, some general conclusions can be drawn even without a detailed analysis.

Isolated electrically small-dimension radiators behave as point sources and thus their models are no different from infinitesimal dipoles. With regard to larger dimension radiators / sensors the increased ratio of the antenna "width" to its "length" leads in general to the increase of the antenna bandwidth. Whereas for "real" antennas this may sometimes be desirable (when a broadband antenna is needed), it can quite negatively affect the electromagnetic compatibility of the system. Thus, the PCB ground planes, large metallic constructions (e.g., equipment racks), and shielding enclosures constitute especially harmful radiator / sensors.

In general, only approximate expressions are available for these antennas. Practically accurate radiating patterns can be calculated using numerical methods (e.g., the Moment Method). As predicted by theory, substantial second and third dimensions of a radiator result in very complex characteristics: gain, input impedance, and directivity. In particular, such radiators generate "smoother" radiation patterns because the intensity of minor lobes is reduced, and the nulls are "filled" with low-level radiation, while the intensity of major lobes remains essentially unaffected. This is illustrated in Fig. 1.25 by the patterns of a cylindrical radiator with a different length-to-diameter (l/d) ratio.

To model *arrays* of larger-dimension radiating metallic planes and solid elements, the concepts of surface patch and finite cross section antennas can be utilized. There exist numerous sources dedicated to such antenna analysis (for example, a quite detailed up- to-date review of such sources can be found, for example, in [1.19]). Fundamental to the concepts of patch antenna and patch antenna array are the interaction of the "patch" with the ground plane, or multiple "patches" - in patch arrays. In particular, the antenna radiation / sensing bandwidth becomes wider with the increase of the distance (or thickness of substrate in "real" antenna) and / or the reduction of the dielectric constant ε of the media between the "patch" and respective "ground plane". At the same time, the larger distance and smaller dielectric constant give rise to the surface waves, resonances and higher order modes, and in general, almost unpredictable distortions of the radiated pattern. Also, the behavior of such antennas is critical to the number, mutual location, and shape of interacting patches, number and location of conducting paths ("shorting pins") between them, "fed" and "parasitic" mode of their "operation", presence of slots and aperture coupling.

The pattern of a slot-type ES depends on the slot shape and size. According to Babinet's principle, when the slot is formed in an infinite conductive surface *(ground plane),* the radiation pattern of the slot is assumed to be identical to a complementary dipole consisting of a perfectly conducting flat strip of the same polarization. To model aperture effects in shielded enclosures (vents, shelf openings, doors, and so forth), a boxed-in slot antenna model can be used, formed in a finite size ground plane. As shown in Fig. 1.25, the radiation pattern of such an antenna depends on the ground plane size.

The problems quickly pile up, especially if *many* radiator non-uniformities combine into an *active array* fed by *one* or *multiple* generators. This results in complex radiated emission or susceptibility patterns of electronic systems, with the number of lobes at the pattern diagram increasing with the frequency rise. As an example, Fig. 1.28 presents a radiating pattern of complex shielded system (2.1 m high equipment

52 Electromagnetic Shielding Handbook

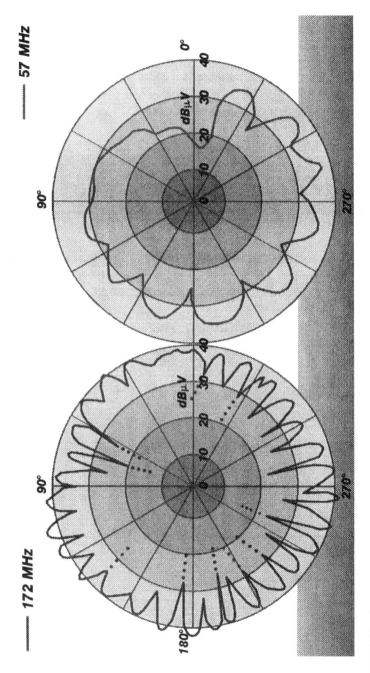

FIGURE 1.28   Complex electronic system radiating patterns: azimuthal plane, horizontal polarization

Introduction to System EMI and EMC 53

metallic frame, populated with shielded shelves containingmultiple electronic backplanes and "daughter"-cards), with many seams and slots in it shielding enclosures.

Of course, with all these problems and complications, we should not lose sight of our objective: we are not designing antennas, but using the existing theory to model respective EMC phenomena.

1.3.3 Elementary Source = Radiator + Generator

The expression in the title can be viewed as an equation of an elementary EMI source. This equation combines the effects of the generator and radiator performance to obtain ES parameters. As a rule, it is relatively easy to resolve with regard to electrically small radiators and simplified generators.

Consider first the radiator / sensor. We have shown that electrically small radiators behave as infinitesimal electrical dipoles (monopoles) when powered in a common mode regime, and as magnetic loops when powered in a differential mode regime. To investigate the "equation" in the headline, consider a radiating linear radiator of electrically short length L (e.g., $L < \lambda/50$), which carries common mode current, I_{com}, of frequency f. Then the radiated far zone field at distance from the radiator r can be approximated as:

$$E_e \approx K_L \times f \times I_{com} \times L \times \frac{1}{r} \quad (1.25)$$

where K_L is the coefficient. Thus, the radiated field intensity is proportional to the radiator length and current in it, proportional to the frequency, and inversely proportional to the distance from the radiator.

Equation (1.25) holds for electrically small radiators (i.e., $L \ll \lambda$) where the radiated field is proportional to their length. That is why, in the low frequency range (i.e., relatively long wavelength), the cables that often have by far the biggest length in the system develop large common mode currents and are capable of supporting strong emissions. The experimental statistical data [1.18] confirm the largest cable emissions below 100 MHz (e.g., see Fig. 1.26). In long cables, a radiation regime can be created that leads to multi-lobe radiation pattern shapes, as shown in Fig. 1.25.

Current loops are generated by the differential mode signals on PWB and backplanes: clock circuits, address lines, and data and power lines carrying high-frequency currents. If the radiating loop of area A is electrically small (e.g., $L < \lambda/50$) and carries differential mode current, I_{dif}, of frequency f, then the radiated far zone field at distance from the radiator r can be approximated as:

$$E_e \approx K_D \times f^2 \times I_{dif} \times A \times \frac{1}{r} \quad (1.26)$$

where K_D is the coefficient. Thus, the radiated field intensity is proportional to the radiator area and current in it, proportional to the frequency square, and inversely proportional to the distance from the radiator.

♦ Unfortunately, it is impossible to derive similar simple expressions for arbitrary signals and radiator configurations, and especially for arrays of such radiators. Graphs in Figs. 1.21 - 1.25 illustrate some practically important cases.

On the other hand, a typical generator (e.g., "trapezoidal" clock pulse train) can be described by Eq. (1.16) at frequencies between the first and the second breakdown (zone 1), and by equation [Eq. (1.17)] at higher frequencies (zone 2). Because, in principle, the MB_{trapez} in Eqs. (1.16) and (1.17) can be viewed as proportional to the current in the radiator, these expressions can be substituted into Eqs. (1.25) and (1.26). Then, with the accuracy to some constant coefficient (ζ in the case of common mode and ξ in case of differential mode), four important relationships can be obtained as shown in Table 1.3.

Table 1.3 Radiated Field Intensity Models of Electrically Small Cable EMI Sources

Generators	Radiators	
	Electric Dipole (Common Mode) $E_e \approx K_L \times f \times I_{com} \times L \times \frac{1}{r}$	Magnetic Loop (Differential Mode) $E_e \approx K_D \times f^2 \times I_{dif} \times LA \times \frac{1}{r}$
Zone 1 $\left(\frac{1}{\pi t_p} < f < \frac{1}{\pi t_r}\right)$ $I_1 \to \frac{A f_T}{\pi} \frac{1}{f}$	$E_{com1} = \zeta_1 A$ (0 dB/decade)	$E_{dif1} = \xi_1 A f$ (+10 dB/decade)
Zone 2 $\left(f > \frac{1}{\pi t_r}\right)$ $I_2 \to A \times \frac{1}{\pi t_r^2} \times \frac{1}{f^2}$	$E_{com2} = \zeta_2 A \frac{1}{f}$ (−10 dB/decade)	$E_{dif2} = \xi_1 A$ (0 dB/decade)

Respective plots shown in Fig. 1.29 provide an excellent illustration as well as a practical tool to determine the emissions from small size radiators. These plots are built in logarithmic frequency scale and dB units of field intensity but can also be built using linear units.

♦ How did it happen? First, realize that the radiated field intensity is the result of a sum (in logarithmic units) of the signal level at the model antenna terminals (generator parameter) and the

Introduction to System EMI and EMC 55

antenna effectiveness, e.g., as expressed by the antenna factor, - radiator parameter. On the generator side, with the raise of frequency the signal *falls down*:

a) at the rate 20 dB signal level / frequency decade below the breakdown frequency ($f_b = 1/(\pi \cdot t_r)$)

b) at the rate 40 dB signal level / frequency decade below the breakdown frequency

On the other hand, on the radiator side, with the raise of frequency the radiator effectiveness *increases* monotonously over the whole frequency band (bur within the constraints of electrically small radiator):

c) at the rate 40 dB signal level / frequency decade for differential (i.e., loop) radiators (which corresponds to the f^2)

d) at the rate 20 dB signal level / frequency decade for common mode (i.e., monopole / dipole) radiators

Combining a), b), c), and d) for respective ES, obtain the curve shapes in Fig. 1.29.

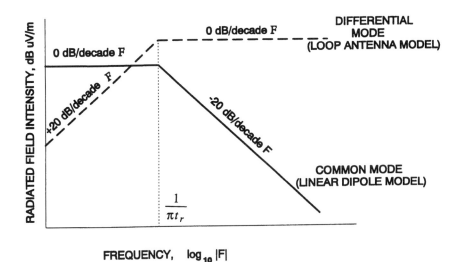

FIGURE 1.29 Envelop of radiated emission extreme boundaries for small size radiators powered by trapezoidal pulse train

1.3.4 Electronic System as an Active Antenna Array

It is typical for electronic systems to have many active and/or passive elements working simultaneously, which must be modeled with different ES. Based on the electronic system representation as an active antenna array, the electronic system EMC performance parameters within the system can be synthesized from the elementary EMI sources. Physically, such synthesis will combine separate elementary sources in arrays.

Mathematically, to determine the total system radiated field E_Σ, all the ES contributions must be superimposed with consideration of mutual coupling effects:

$$E_\Sigma = \sum_{n=1}^{N} \sum_{q=1}^{N} E_{en} \Phi_q K_{nq} \qquad (1.27)$$

Here, E_{en} is the field intensity generated by the nth ES in the absence of other sources; K_{nq} is the coupling coefficient reflecting the field contribution of the qth ES due to the signal coupled into it from the nth ES (obviously, $K_{n=q} = 1$); Φ_q is the phase coefficient reflecting relative field phase shift of the qth ES due to its position (distance) with regard to the observation point, and original excitation phase; and N is the number of EMI sources in the system. All of the variables in Eq. (1.27) are vectors.

◆ Combining separate ES radiation patterns into a system pattern is known in antenna theory as *pattern multiplication*. In the general case, the system field synthesis presents a serious technical and mathematical challenge. However, depending on actual system specifics, the field often can be modeled with simplified antenna array combinations. The pattern multiplication is especially expedited in arrays consisting of identical elements:

$$E_\Sigma = E_e \times AF \qquad (1.28)$$

where E_e is the field generated by a single element at a selected reference point, and AF is the array factor, which is a function of the number and character of elements, their spacing, and relative generator amplitudes and phases. For instance, in case of a two-element array with linear elements of constant amplitude positioned along the z-axis with spacing d:

$$AF = 2\cos\left[\frac{1}{2}(kd\cos\theta + \beta)\right] \qquad (1.29)$$

In actual product designs, the loop-shaped ES can simultaneously carry a differential mode current, I_{dif}, and a common mode current, I_{com}, which generates a radiated field of its own. Fig. 1.30a illustrates the effect of two identical vertical elements spacing and current phase difference on the H-plane array pattern. As was indicated, the radiation pattern of a single element is a circle also shown in the figure. Even such a simple array produces complex radiation patterns of specific shapes.

Passive system elements that do not carry current can be modeled by parasites complementing the active elements in the array. An example presented in Fig.1.30 b reveals a specific pear-like array pattern shape.

Another element combination often encountered in electronic system radiation pattern analysis consists of a loop and dipole elements (see Fig.1.30 c). Such a combination corresponds to a simultaneous action of differential mode and common mode generated emissions. One practical case when such an array is formed is when emissions are generated by a power supply cord or I/O cable (common mode emissions) and a PWB-located current loop (differential mode emissions). Another important case occurs when both common and differential mode currents flow in the same circuit.

Introduction to System EMI and EMC 57

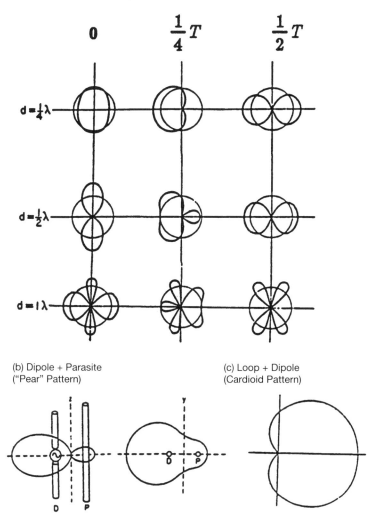

FIGURE 1.30 Models of elementary EMI source arrays

Even in the case of elementary radiators, when their dimensions exceed the wavelength, the patterns become extremely sensitive to frequency. In real-life situations, these ideal patterns are distorted by the cable and interconnection layout irregularities, proximity effects, ground planes, and absorber-lined surfaces in the test sites and absorber-lined rooms. Figure 1.31 illustrates the effects of the radiation pattern change with frequency variations for the same size radiator (or with the radiator size variations at the same frequency). Here, the emission patterns are shown for a simple PWB containing an 8 MHz oscillator driving a clock signal along a 0.2 m long trace line (the radiated spectrum of this setup is presented in Fig. 1.23). The patterns were taken at 57, 210, and 518 MHz (all frequencies are rounded up), which correspond to approximately $1/25\lambda$, $1/8\lambda$, and $1/3\lambda$, respectively, with regard to the 0.2 m long trace. While the 57 MHz pattern appears almost as a classical infinitesimal dipole (see Fig.1.25), at two higher frequencies, the electrically small size assumption does not hold, even with PWB radiators. ✦ .

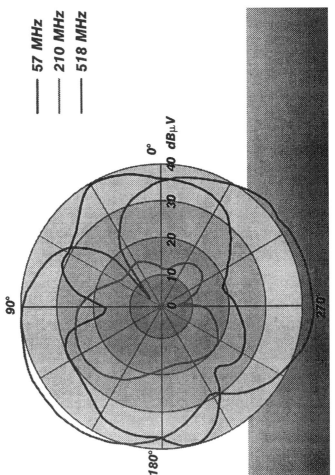

FIGURE 1.31 Radiated emission patterns of a PCB clock line

Introduction to System EMI and EMC 59

To summarize the preceding discussion, separate EMI sources within an active antenna array modeling an electronic system can be combined to yield a system EMI profile. This depends on ES electrical properties, different ES interaction, and system geometry. Although any particular system exhibits specific radiation patterns and frequency behavior, an analysis at the most general level provides extremely useful insight into generic system EMC performance. It was shown that, in a system, the cables are generally the source of common mode emissions, while the PWB and backplanes generate mainly differential mode emissions. Modeling the cable as a monopole and the PWB as a loop, a combined global system pattern can be generated .The theoretical patterns shown in Fig. 1. 32 (similar to Fig. 1.30 c) are almost identical to experimental data obtained by measuring a digital device with vertically polarized I/O cabling and horizontally polarized lines at the system backplane (see Fig. 1.33)

By changing the cable polarization to horizontal and conducting pattern measurements in horizontal plane, a "pear"-like pattern was measured (Fig. 1.34). Fig. 1.35 provides a possible theoretical explanation (it is similar to Fig. 1.30 b).

1.4 EMI SUPPRESSION: THE IMPORTANT OBJECTIVE OF EMC

1.4.1 EMI Mitigation Techniques

Presumably, if the electromagnetic compatibility definition and compliance evaluation are important EMC opbjectives, then the EMI control to meet the EMC limits - is another such objective. Table 1.4 lists some of the most important of such techniques. The EMI mitigation techniques are shown with regard to their applicability to system interconnection levels, and they cover both EMC design and EMC retrofit.

Different EMI mitigation techniques do have their specific areas of maximum technical and economic efficiency. Consider this: the selection of an IC family determines the signal pulse rise and fall times, which are intimately related to the transmitted bandwidth (i.e., EMI generator properties). But, as we know, the smaller bandwidth leads to EMI reduction. The grounding system has a direct effect on the transmission mode and coupling, and so do the layout, balancing, and decoupling. Even software used in the system is a factor, since it may define which signals in what combinations will be transmitted via certain cable circuits. To make things even more complicated, the system EMC performance depends not just on separate EMC parameters, including cable EMC performance parameters, but also on their interaction. For example, Fig. 1.29 illustrated how the generator and radiator properties combine to produce the specific frequency dependence of cable emissions.

Another complication stems from the differences in the selection of EMI mitigation techniques and their application to an electronic product for immunity and emissions purposes. Figs.1.36 and 1.37 illustrate some typical applications

Also, consider one of the most important distinctions in the EMI mitigation application: EMC design versus EMI retrofit. As a rule, EMC design provides much

60 Electromagnetic Shielding Handbook

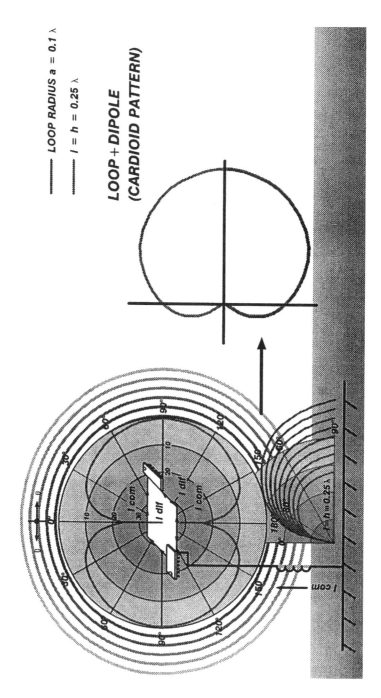

FIGURE 1.32 Generation of a combined differential+common mode radiation pattern

Introduction to System EMI and EMC 61

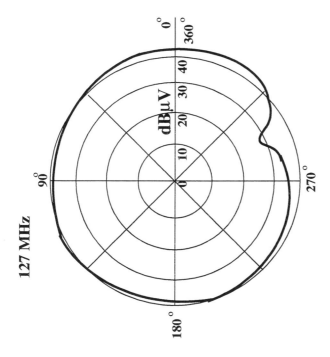

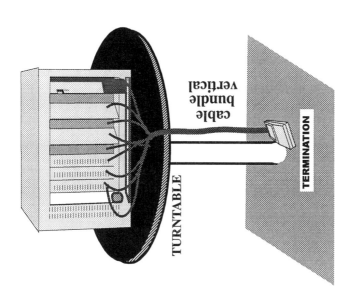

FIGURE 1.33 Complex system radiating pattern: azimuthal plane, vertical polarization

Table 1.4 EMC mitigation techniques and applications

Technique	Application*	Applicable level				
		Chip	PWB	Chassis	Device	Network
IC family/signal	D	✔	✔			
Grounding system	D	✔	✔	✔	✔	✔
Layout system	D	✔	✔	✔		
Balancing	D	✔	✔	✔	✔	✔
Filtering	D, R		✔	✔	✔	
Decoupling	D, R	✔	✔	✔	✔	✔
Cabling	D, R		✔	✔	✔	✔
Shielding	D, R	✔	✔	✔	✔	✔
Operating mode/software	D, R	✔	✔	✔	✔	✔

*D = design, R = retrofit

wider capabilities and many more choices than the retrofit, although some of the techniques can be applied in both situations. For example, electromagnetic shielding is one such technique, which can be built in at the earliest stages of product EMC planning. But it can also be used as a retrofit to compensate for the EMC design deficiencies after the product has been developed and even mass-produced.

Second, experience proves that neither of these techniques alone is usually sufficient: in mitigating EMI we need all the help we can get! Thus, as a rule, we will apply several techniques to the same product. These techniques should be considered not as isolated entities, but in their interaction, as well as with regard to the system objectives and the EMI environment.

For this reason, a certain strategy must be developed with regard to priorities and schedules of EMI mitigation technique applications.

In particular, similar considerations will determine when and whether shielding should be used instead of other EMI mitigation techniques.

Finally, since each elementary source (ES) in an active antenna array (representing the electronic system) consists of a generator (receptor) and radiator (sensor), in principle, EMI suppression can be addressed by affecting either the EMI generator/receptor or the radiator/sensor, or both. However, there is an important difference between these two EMI suppression levels. Since each EMI generator/receptor can "feed" multiple EMI radiators/sensors, it is much more efficient to work at the first level, that is, to cure the decease instead of the symptoms.

Introduction to System EMI and EMC 63

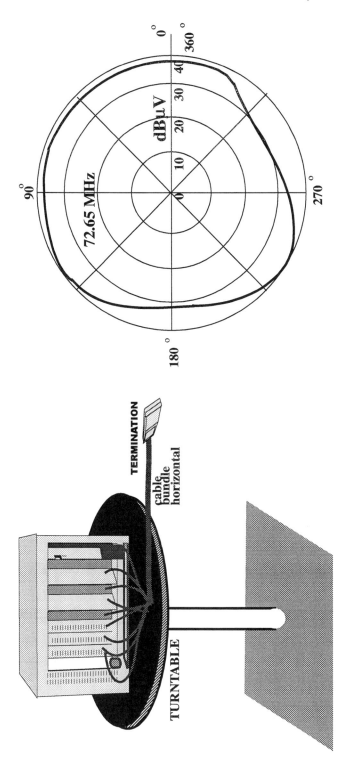

FIGURE 1.34 Complex system radiating pattern: azimuthal plane, horizontal polarization

1.4.2 Comparing Different EMI Suppression Techniques: Cable Crosstalk Example

Since abundant literature is available on the EMC mitigation techniques (e.g., see [1.20-23]), there is no need for a detailed analysis here. For this reason we will limit the discussion to an illustrative example. Consider the experimental results of crosstalk measurements. The crosstalk measurement diagram is presented in Fig. 1.38. Also marked on the diagram are the voltage, V, current, I, and power, P, at the near and far ends of the disturbing and disturbed circuits; the loads, Z, at the circuit ends (shown matched to the circuit characteristic impedance Z_c); and the propagation constants of both circuits, $\gamma = \alpha + j\beta$ (where α and β are the circuit loss and phase constant, respectively). The near end (NEXT) and far end (FEXT) crosstalk were measured in the same cable between coaxial pairs, balanced shielded pairs, twisted pairs, and parallel pairs in flat cables. In the last case, two circuit configurations were tested: one with alternating ground and signal conductors (designated GS, for ground-signal configuration), and the other where each two signal conductors are separated by two ground conductors (designated GSG, for ground-signal-ground configuration). Each cable was 3 m long.

The measurements were performed in the time domain—i.e., pulse crosstalk was measured. The obtained test results are illustrated in Fig. 1.39 by an example of pulse crosstalk between cable circuits in a flat cable with GS configuration. The disturbing circuit is driven by pulses with rise/fall time t_r. These pulses have the slope of curve 1 at the beginning and the slope of curve 2 at the end of the line. The distance between curves 1 and 2 corresponds to the propagation delay, and the difference be-

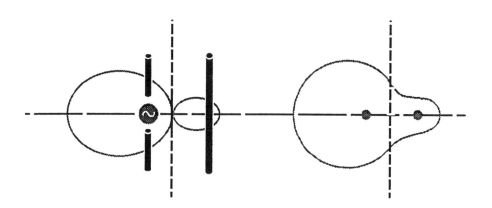

FIGURE 1.35 Combined radiating pattern: dipole + parasite

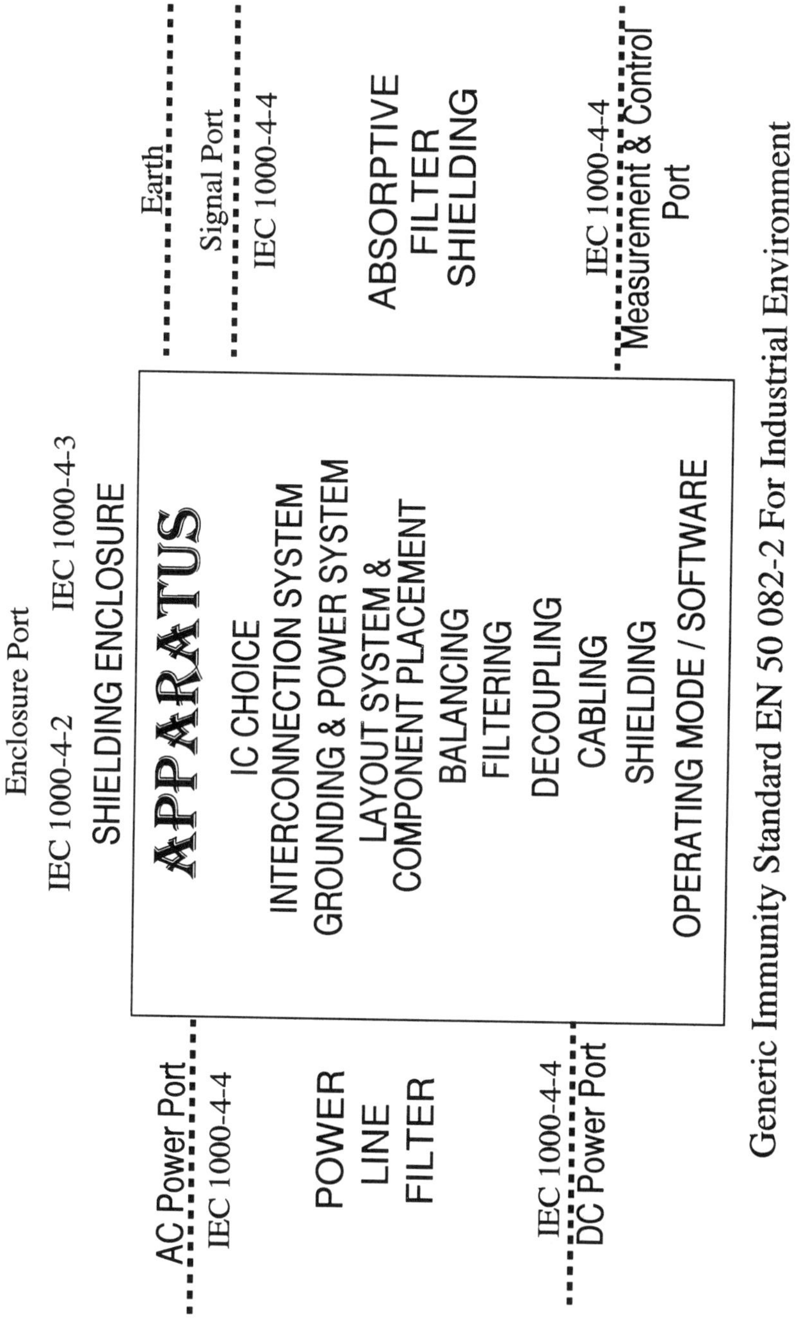

Enclosure Port
IEC 1000-4-2   IEC 1000-4-3
SHIELDING ENCLOSURE

Earth
Signal Port
IEC 1000-4-4

ABSORPTIVE
FILTER
SHIELDING

IEC 1000-4-4
Measurement & Control
Port

APPARATUS

IC CHOICE
INTERCONNECTION SYSTEM
GROUNDING & POWER SYSTEM
LAYOUT SYSTEM &
COMPONENT PLACEMENT
BALANCING
FILTERING
DECOUPLING
CABLING
SHIELDING
OPERATING MODE / SOFTWARE

AC Power Port
IEC 1000-4-4

POWER
LINE
FILTER

IEC 1000-4-4
DC Power Port

Generic Immunity Standard EN 50 082-2 For Industrial Environment

FIGURE 1.36   Product immunity test ports and mitigation techniques

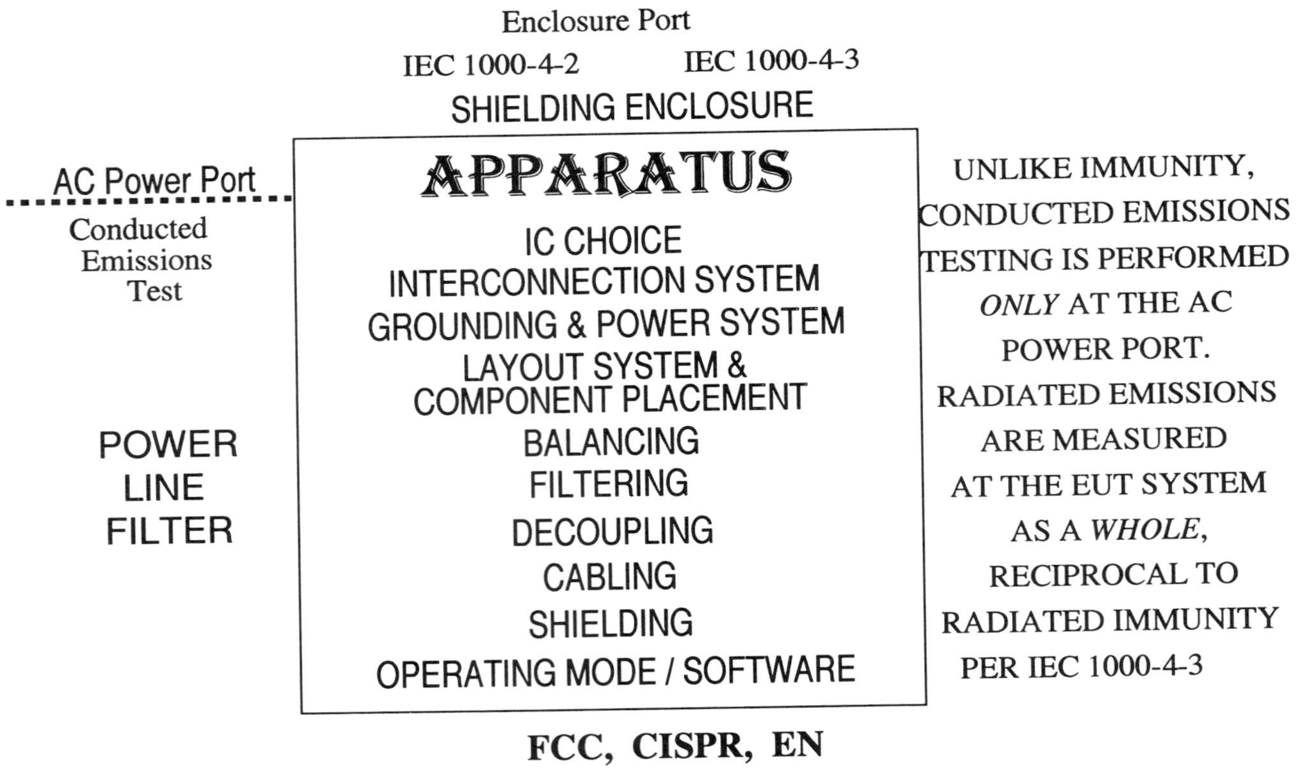

Enclosure Port

IEC 1000-4-2   IEC 1000-4-3

SHIELDING ENCLOSURE

APPARATUS

- IC CHOICE
- INTERCONNECTION SYSTEM
- GROUNDING & POWER SYSTEM
- LAYOUT SYSTEM & COMPONENT PLACEMENT
- BALANCING
- FILTERING
- DECOUPLING
- CABLING
- SHIELDING
- OPERATING MODE / SOFTWARE

AC Power Port
Conducted Emissions Test

POWER LINE FILTER

UNLIKE IMMUNITY, CONDUCTED EMISSIONS TESTING IS PERFORMED *ONLY* AT THE AC POWER PORT. RADIATED EMISSIONS ARE MEASURED AT THE EUT SYSTEM *AS A WHOLE*, RECIPROCAL TO RADIATED IMMUNITY PER IEC 1000-4-3

FCC, CISPR, EN

FIGURE 1.37 Product emissions test ports and mitigation techniques

tween the slopes is defined by cable length and propagation constant γ. As we know, a larger t_r corresponds to a smaller signal bandwidth, and vice versa.

The near end (fast) crosstalk at the beginning of the disturbed line (curve 3) is a miniature replica of the pulse signal in the disturbing line. In Fig.1.39, the NEXT crosstalk pulse has a flat top whose rise and fall times equal the original signal and whose width equals twice the propagation time down the circuit (the signal propagation speed in both the disturbing and disturbed circuits is assumed identical). It reaches maximum amplitude when twice the propagation time on the line equals the signal rise time. When longer lines are employed, the magnitude of the fast crosstalk will not increase. The far-end crosstalk (curve 4) appears at the far end of the disturbed circuit. It has opposite polarity to the signal, and its magnitude depends on the line length and the t_r of the signal. As indicated before, the pulse crosstalk is evaluated as a percentage of the ratio between the induced voltage and the driving voltage.

Similar time-domain measurements were performed with all the selected cable and pair designs. In Fig. 1.40, the measured data is plotted versus the circuit type and configuration, along with a "token" zero crosstalk between fiber optic cables. As we have expected, cable EMI is affected by circuit configuration—balanced, unbalanced, twisted, parallel conductors, and coaxial. It is also affected by shielding. The GS configuration exhibited very high levels of crosstalk, whereas the shielded balanced pairs have the least. Positive results were also obtained by using coaxial cable. As we can see, shielding really works!

It is another thing to explain these results. The explanation does not appear to be simple, if based on just the obtained experimental data. Consider crosstalk between coaxial pairs. A coaxial pair is also a shielded circuit. Therefore, it seems only logical that the fast crosstalk between coaxial circuits would be at the same level as for shielded pairs. However, it is surprising that the far-end crosstalk between coaxial cables is not that good. Also, twisted pair performance looks quite disappointing: the crosstalk between the twisted pairs was larger than between pairs in non-twisted and nonshielded GSG configuration, and no better than in GS configuration. After all, what's the use of a twisted pair? It is obvious that without understanding the mechanisms of and differences between the crosstalk in shielded balanced and coaxial cables, these behaviors cannot be explained, predicted, and mitigated.

◆ It so happens, that in this particular case in Fig. 1.40 the twisted pair was configured as a GS line (by virtue of using *unbalanced* test generator and receiver). Naturally, connecting one of the pair conductors to the ground resulted in an *externally* introduced unbalance, while galvanic connection of the interfering pairs via their grounded conductors created large common impedance coupling. Thus, all the positive effects of pair balancing and twisting were negated. Alternatively, when the same line was configured as balanced (signal-signal - "SS"), the crosstalk between the pairs *within* the cable was drastically reduced (for crosstalk comparison between balanced vs. non-balanced pairs, see chapter 4, e.g., Fig. 4.16). That's what behind the current trend to use unshielded twisted pair (UTP) for high speed data transmission! Nevertheless, this still leaves open the questions of the feasibility of even higher frequencies on UTP, as well as the UTP line immunity from common mode EMI and ESD, re-radiation by a UTP of externally picked up common mode signals, and effects of the environment factors on circuit balance. Thus, selection between UTP and shielded pair (STP) becomes a complex engineering problem (see chapter 7 for shielding engineering discussion.) ◆

68 Electromagnetic Shielding Handbook

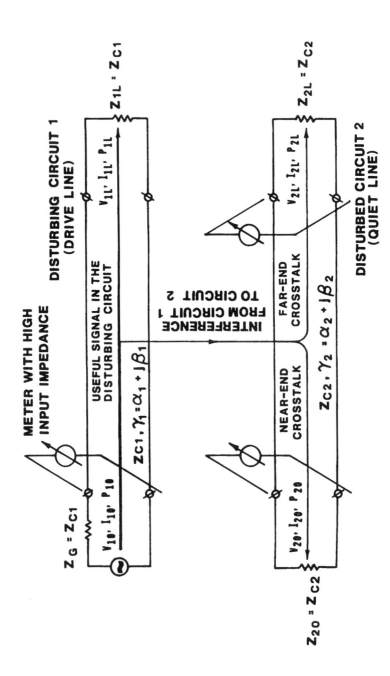

FIGURE 1.38 Near and far end crosstalk measurement diagram

Introduction to System EMI and EMC 69

3 ns pulse

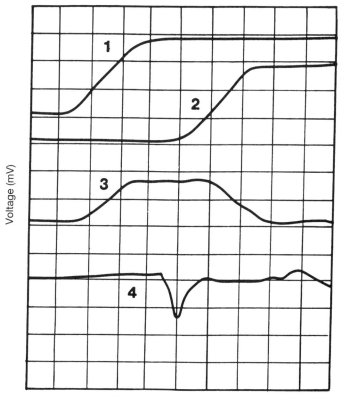

Time (ns), 2 ns per Division

1. Drive pulse of 300 mV at the input of the disturbing circuit
2. Drive pulse at the output of the disturbing circuit, vertical = 100 mV/division
3. Near-end fast crosstalk of 2.7%, vertical = 5 mV/division
4. Far-end crosstalk of 2.5%, vertical = 5 mV/division

FIGURE 1.39 Pulse crosstalk

70 Electromagnetic Shielding Handbook

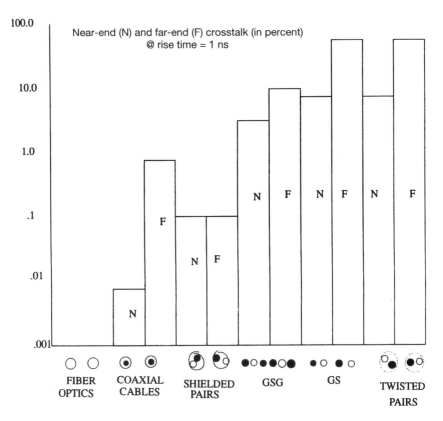

FIGURE 1.40 Pulse crosstalk between cable lines

2

Understanding Electromagnetic Shielding

2.1 SHIELDING AT THE EMC "FRONT LINE"

Consider the Table 1.4. It follows that for an EMC engineer electromagnetic shielding is just one method in a quite impressive "arsenal" of EMI mitigation techniques. Where should we place electromagnetic shielding within the general EMC mitigation hierarchy? How important it is, if at all? As always, "questions breed questions", and pretty "tough" questions they are:

- In what applications shieding is "relevant"?
- Can the shielded product do the required job?
- What about other EMI suppression methods? How do they fare in comparison with shielding? It may happen that these methods can offer a better solution to our EMI problems (whatever "better" means—we will address this subject later in the book).
- If we still insist on shielding, then what kind of shield do we need?
- How can and should the shield be grounded and terminated?
- What will be the effect of electromagnetic shielding on the whole system performance, both in terms of EMC and functionality?

To start with, an important insight can be obtained from the most general understanding of the shielding role. As discussed in the previous chapter, system elements contribute to field-coupled EMI and EMC problems as (1) emitters that radiate part of the conducted signal and (2) receptors that are susceptible to ambient electromagnetic fields. In this respect, the shielding objective is to confine EMI signals within (to solve emission problems) or outside (to solve susceptibility problems) of the shielded enclosure. In other words, shielding "works" by limiting the EMI field penetration into the protected environment - whether "inside" or "outside" the shield confines. While deferring the discussion of just how this is achieved, we will first try to illustrate the effectiveness of this shield's function.

The available experience proves that shielding can be an excellent technique to mitigate EMI generated by radiation and induction fields. For example, the data in Fig.1.13 and in Fig. 1.40 provides a "glimpse" at shielding capabilities with regard to cable line emissions and crosstalk, respectively. But these are just cable application, and many more exist: as a mitigation technique, "shielding" is applicable to a large diversity of interconnects and enclosures. In particular, shielding can be applied at *any* or *all* of the interconnection levels shown in Fig. 1.11. The range of "shielding-eligible" enclosures boast a similar diversity: from IC and discrete device packages, equipment shelves and racks, and up to quite large-scale products.

Following is a (quite incomplete) list of potential shielding applications and related "hardware":

- architectural shielding — shielded rooms (including anechoic), other shielding facilities
- "mobile" enclosures — cars, airplanes, ships, satellites, etc.
- equipment enclosures ("boxes") — cabinets, shelves, product housings
- any metallic structures of arbitrary shape, e.g., equipment racks, cable carriers, etc.
- cable shields
- partition shields — PCB power and ground planes used as separation shields, partial enclosures over separate system elements like chips, shelves, etc., partially shielded cabinet, closet, and room walls, floors, separation screens (solid, or mesh)
- parallel shields — PCB and backplane grounded traces and ground planes running parallel to important lines: clocks, I/O, etc.
- shielding gaskets — electromagnetically "sealing" components designed to reduce the "imperfections" at the interface of different elements of the shielding system.

We will see in the following examples that in many of these applications shielding can be extremely effective. In principle, when properly "administered," shielding can *always* provide the electromagnetic field-related EMI protection - to a degree, which is limited mainly by economical constraints. It can compensate for the "sins committed" by subassembly/component manufacturers and by system designers and assemblers. It may even be *the only* possible solution - the "last resort" in fighting EMI, when all other techniques fail or aren't practical to use. Therefore, while there also exist *other* techniques to fight cable EMI, electromagnetic shielding can be viewed as a potential "ultimate" EMI protection means — kind of an "EMC front line".

But then again (now "opposite" to our original thesis at the beginning of this section), does that mean that these other techniques are not necessary? Not at all—the other techniques fit certain applications, where they can be extremely effective! There are three main "causes" why shielding may "lose" in competition with other mitigation techniques.

1. The first cause is of the most general nature, involving the comparative merits of EMI mitigation at the radiator/sensor level, as opposed to generator /receptor level. Because shielding works at the radiator/sensor level (as opposed to generator receptor level) of EMI mitigation, it "cures" "symptoms", rather than desease. Thus, to shield only one or a limited number of possible "field escapes" may not solve the problem, because the EMI may "find other escapes". Only "total", "solid", and continuous shielding guarantees 100% success, but such shielding is practically impossible since "real life" electronic products must contain apertures, penetrations, and nonuniformities, and often, a large number of long cables which also have to be shielded.
2. The second cause is specific to a class of EMI mitigation techniques, which action is "distributed" at the radiator/sensor level. In particular, shielding as a rule involves *incremental* expenses: the larger is an enclosure surface or a cable length, the higher is the expense to provide the shielding. While expenses on shielding hardware can be quite significant to start with, they quickly "pile up" with the electronic product increase in size or the cable length. On the other hand, some of the alternative mitigation techniques may involve only "one-time" expenses, which sometimes may be quite small (e.g., compare shielding vs filtering - see Fig. 7.2.)
3. The third cause proceeds from the fact that application of shielding may result in undesirable change of the shielded circuit parameters (see section 7.5), affecting the transmitted signal. This is not to say that other EMI mitigation techniques are always free from this problem, but in any case, it should be investigated and accounted for.

For these reasons, a comprehensive analysis of physical, mechanical, and electrical shield performance, economic considerations, and shielding optimization must be integral parts of shielding applications. And to achieve optimal results, all these issues must be considered *simultaneously*, based on system analysis principles. As our discussion progresses in the following chapters, we will address the topics of shielding analysis, economics, and optimization. As for now, we will limit the discussion mainly to the effects and problems of selected tasks in electromagnetic shielding.

2.2 A SHIELD IS A SHIELD IS A SHIELD ...

Or is it? Well, there is more to it, than that. As you will see, not every shield provides enough shielding, if any, and in certain cases the use of the shield may be *counter-productive*, meaning that the shield application may result in *increased emissions* or *reduced immunity*!

In this section, we will review several practical applications of electromagnetic shielding. These examples prove important uses of shielding and also raise important issues: from semantics to physics to mathematics to the design/performance tie-in. At this time, we will not often be in a position to theoretically analyze these problems. We will do it in the following chapters of the book. Here, instead, we will use examples of "real-life" experimental studies to illustrate important phenomena accompanying these problems and to provide additional insight.

74 Electromagnetic Shielding Handbook

2.2.1 Shielding Enclosures for Electronic Products

Shielding Enclosure "Trivia"

The idea of using a shielding enclosure for electronic equipment EMI protection looks very attractive: let's "electrically" hide the product and stop worrying about electromagnetic compatibility - both, emissions and immunity. Indeed, such an idea is not without merits. Consider an important practical example. In Fig. 2.1 the radiated emissions are presented from a notebook digital assistant with a large LCD display (designed for such applications like medical, military, etc.). The emissions in Fig. 2.1a are from the product in plastic housing, while in Fig. 2.1b - the housing was *completely* sealed, including the display, with aluminum foil. As we see, the shield performance is impressive. Granted, the product was originally designed without any consideration given to EMI (otherwise, the emissions from the non-shielded version wouldn't have exceeded so much the FCC limits - 25 and 15 dB for classes 'B' and 'A', respectively). But this only emphasizes the "front line" capabilities of shielding, which application permitted to meet the regulations (well, judging from the Fig. 2.1b - "just about" in case of class 'B') with the product "as is".

Isn't this great? No wonder the industry has developed and widely uses all kind of shielding enclosures (see Fig. 2.2): from an integrated circuit "chip" to cable, connector, and termination shielding, to small and large cabinets and frames, to all kinds of civil and military riding, "creeping", "crawling", and flying vehicles, navy ships, and submarines, to large shielding rooms which can accomodate a tank or an aircraft, to whole buildings and building complexes hardened against external EMI, EMP (nuclear electromagnetic pulse), or eavesdropping.

Formed aluminum or extruded frame cases, copper, brass, steel, and / or aluminum cabinet panels, steel frames, multi-layer shielding room confines - a large diversity of shielding options is open to physical, electrical / electronics, and system designers.

Other metals are also used in applications: zinc, stainless steel, nickel, high magnetic permeability alloys, conductive plastics and composites, etc. Plastic-clad, nickel, chrome, or conductive paint covered surfaces serve diverse conducting, isolating, structural, environmental protection, and aesthetic look purposes.

The assembly of shielding enclosures can be done using snap-on holds, screws, crimps, soldering and welding. To provide for specific product needs - readout displays, locks, mechanical moving parts, cable penetrations, ventilation and thermal requirements, - the shielding enclosures may have apertures, vents, screw holes. Often, the shield continuity and electrical integrity is ensured with resilient, high contact conductivity gaskets placed at the interface between different parts of the *shielding system*: between shielding enclosures and equipment frames and cable shields, or between parts of the same enclosure, if assemblied.

To see some of the effects of "shielding hardware"design, let's return to the previous example. Indeed, as a matter of practicality, won't it be "useful" to take a look at the notebook's display? That's what this gadget is for, anyhow! So, we cut the foil over the display (leaving the shielding wrap untouched elsewhere), and learn the first harsh lesson in shielding (see Fig. 2.1.c): while some positive effect of shielding is still present, all in all, not too much "shielding effectiveness" was left!

Understanding Electromagnetic Shielding 75

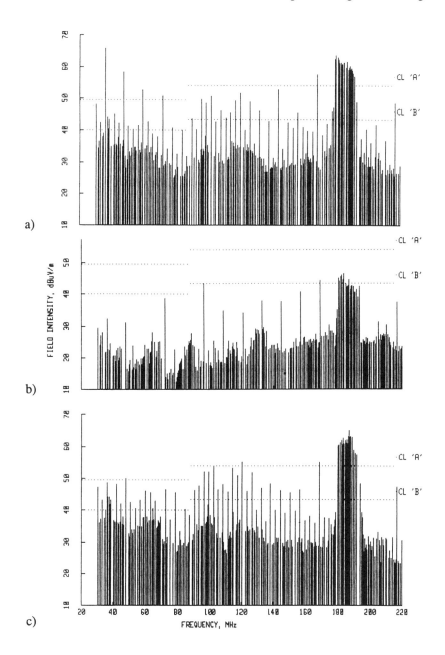

FIGURE 2.1 Notebook computer shielding enclosure effectiveness

76 Electromagnetic Shielding Handbook

FIGURE 2.2   The hardware world of shielding enclosures

Understanding Electromagnetic Shielding 77

Another example confirms that this problem is not "unique" to this particular application: problems exist also with other types of enclosures. The diagram in Fig. 2.3 presents quite surprising shielding effectiveness test results for a telecommunications office cabinet. The data represents the worst case (that is, the largest emissions) in the frequency band 30 -300 MHz. Indeed, with the open door, the cabinet *minimum* shielding effectiveness is *negative*, which means that *maximum radiated emissions* from a source placed inside the cabinet (i.e., a "shielded product") are *larger* then the emissions from the same product, but *out of* shielded enclosure (in "free space").

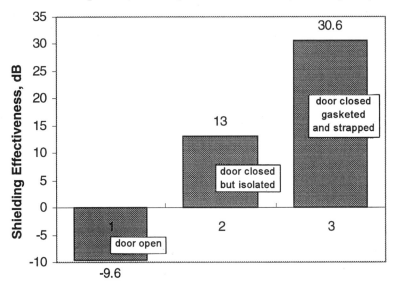

FIGURE 2.3 Shielding effectiveness of the telecommunications cabinet

Moreover, just closing the cabinet door, without providing a reliable electrical contact between the door and the cabinet body produces a meagre 13 dB shielding effectiveness (in the experiment, to emphasize the problem, the electrical contact between the cabinet and its door was completely eliminated by pasting an isolating paper gasket around the door perimeter). Only when the cabinet door was provided with low electrical resistance resilient fingerstock gasketing and, additionally, strapped to the cabinet body ("ground"), then closing the door resulted in over 30 dB shielding effectiveness.

◆ In the above text, pay attention to the italics. First, we have related the minimum shielding effectiveness to maximum radiated emissions. A more accurate formulation of the minimum shielding effectiveness should have related minimum emissions of an unshielded system to the maximum emissions of a shielded system in a given location, selected direction, and at a given frequency or frequency range of interest. *All such definitions are valid, as long as they correspond to the objectives of system applications.* Does that mean that there may be "different kinds" of shielding effectiveness figure of merit for the same product? We will return to this important subject later, when discussing the approaches to defining shielding effectiveness. Second, we have just introduced the concept of "negative" shielding effectiveness. As surprizing the negative shielding effectiveness may seem, in principle there is nothing controversial in it. Indeed, how would you qualify the effect of the shield, if its application results in the increase of the emissions or susceptibility? As we see, such "shields" do exist or can be "created" by non-careful, or non-competent shielding design and implementation.✦

How Much is Enough and What to Do About It

Another question is whether this is enough or not . To be sure, the 30 dB shielding effectiveness is not exactly "breathtaking". But then, just how much can the system emissions be reduced (or immunity improved) by using shielding enclosures in the electronic office? Again, let's call upon the experience. Fig.2.4 demonstrates the shielding effect of a 48×183 cm shielding cabinet with front and rear access doors. The walls of the cabinet are manufactured from 1.5 mm thick aluminum panels joined with screws 75 mm apart. Approximately 10 mm in diameter holes are used for ventilation. The cabinet also contains apertures of different size and shape (round, D-sub, etc.) for cable and connector penetrations. The cabinet specification calls for over 60 dB shielding effectiveness in the frequency range from 10 kHz to 1000 MHz. The shielding effectiveness testing was performed per MIL STD 285 (see chapter 6). After the respective "shielding hardening", the shielding effectiveness was measured as the attenuation loss between the transmit antenna outside the shielding enclosure and the receive antenna inside the enclosure was measured inside a shielding room.

As presented in Fig. 2.4, the shielding effectiveness requirements are met. At this, the effectiveness *falls off* in an *oscillatory* manner from the maximum over 130 dB @ 10 kHz down to around 60 dB @ 800 MHz. To obtain such high enough shielding effectiveness, the door was provided with conductive EMI gasketing, coaxial cables connecting the receive antenna to the spectrum analyzer were pulled through metallic conduit and had their shield terminated at the enclosure *proper surface*, other apertures were sealed with metallic overlays.

Problems and Questions ... More Problems and Questions

To summarize the just reviewed experimental data: looks like while we did achieve certain success with shielding, we've got more questions than answers. However, the "good news" is that now we can at least formulate some of these questions:
- Why the best obtained shielding effectiveness was not large enough?
- What effect, if any, have the shield material, thickness, and shape?
- What role play the size, shape, and location of the shield opening?
- What is the effect of mechanical contact and surface qualities of different parts of the shielding enclosure: at seams, gaps, overlaps?
- If there are cables penetrating the shielding enclosure, how will this affect the shielding effectiveness?
- Should or shouldn't the shielding enclosure be grounded, and if "yes", then where and how?
- Is there any difference between the shielding effectiveness performed using a "point system" (e.g., transmit or receive antenna) vs a "distributed system" (represented by an "active antenna array" - as discussed in chapter 1)?

This kind of questions is almost a pattern in the examples we consider next. As we forwarned you, the identified issues and problems as a rule will be left open until "bet-

Understanding Electromagnetic Shielding 79

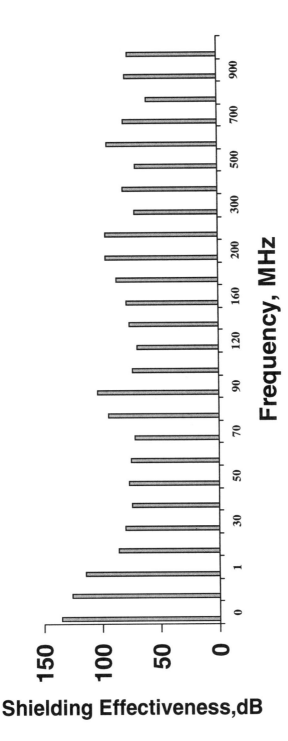

FIGURE 2.4 Shielding effectiveness of an EMI gasketed office cabinet

ter times", since they can be properly addressed only as we progress through the understanding of shielding theory.

2.2.2 Cable Shielding

Reducing Cable Crosstalk: Effect of Shield Construction

While discussing the effectiveness of shielding, we did not concern with the shield design. However, shield construction is one of the main factors that determine the amount of crosstalk between the shielded cables. In Fig. 2.5, the test results are presented of the far-end crosstalk between coaxial pairs in a 3-meter long ribbon cable.

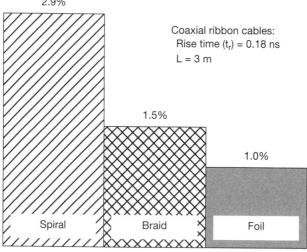

FIGURE 2.5 Shield design effect on pulse crosstalk

The crosstalk was measured in time domain, with 0.18 ns rise time pulses ($1/\pi t_r \approx 1.8$ GHz—see Chapter 1). Three different single-layer shield designs (i.e., coaxial pair outer conductors) were compared: spiral (serve), braid, and foil. Note that the foil shield provided for a lower crosstalk than the serve, or even braid. Of course, the performance of any particular shield depends on its design specifics: spiral and braid strand material and diameter, foil thickness, shield application angles, and so on. But a certain shield behavior trend is definitely there!

Reducing Radiated Emissions: Multilayer Cable Shield

A coaxial cable with a two-layer foil-braid shield was stretched over the ground plane at the open area test site. The cable was powered by a continuous wave (CW) sweep generator up to the maximum frequency of 500 MHz. A receive antenna was located at 3 m distance from the radiating line, picking up the electromagnetic emissions from the cable. At the beginning, the cable was tested with both shielding layers on the core. The tests were then repeated with the shield layers consequently peeled off; first the external shield layer (braid), and then the internal shield layer (foil). This technique ensured that the effects of the cable design and manufacturing tolerances on the measurement results were minimized. The last configuration was equivalent to a single-ended line (unshielded), with the return current path provided by the ground plane.

Figure 2.6 presents the results as displayed on the spectrum analyzer screen. As shown, the addition of each shielding layer reduces the maximum emissions by about 20 to 25 dB, with the total emission reduction achieved by a two-layer shield reaching 50 dB. Overall, while the emission amplitudes were strongly dependent on the shield design, the shape of the emitted signal did not significantly change. In other words, the EMI signal integrity was not strongly violated by shielding.

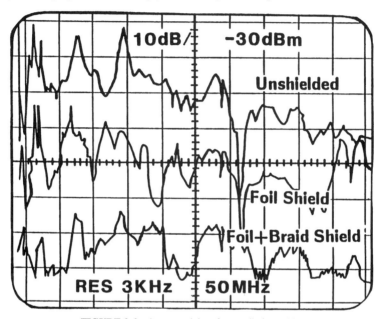

FIGURE 2.6 Antenna pickup from radiating cables

2.2.3 Architectural Shielding

Shielding rooms are used to eliminate ambient noise and / or protect the environment from unnecessary electromagnetic pollution. This is often desirable, or even necessary, to provide for normal work of telecommunications offices, computer rooms, medical facilities, for security reasons, and/or EMC testing (emissions and immunity).

As an example, Fig.2.7 presents a comparison between the ambient environment at the open area test site and inside a shielding room at the same location. As shown, separate spikes at the OATS reach 80 dBµV/m, while the ambients inside a shielding room were below the sensitivity threshold of the test setup (10dBµV/m) in the whole investigated frequency band (and could not be detected). Thus, a *reasonable* assumption can be made that the shielding effectiveness of *this particular room* is *over* 70 dB. The word reasonable was italicized because we are yet to come up the definition of the shielding effectiveness of a shielded room - see the author's note in the section 2.2.1.

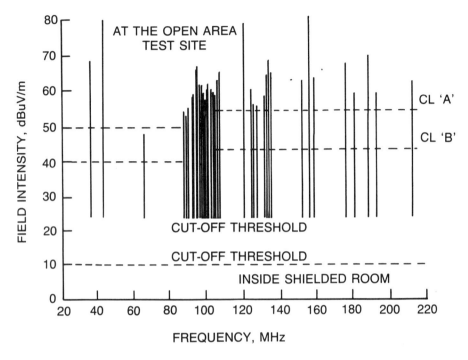

FIGURE 2.7 Ambient environment

Practically, it is not unusual to have requirements to shielding room effectiveness in excess of 100 dB in an extremely large frequency band: from Hz (e.g., hi fi audio testing) to hundreds of GHz (microwave). Can this be achieved?

Of course, to answer this question, we must first define, what is it we mean under shielding effectiveness. But for now, let us assume that MIL STD 285 is correct (as we will prove later, this assumption does not not always holds), and use the data measured per this standard. Fig. 2.8 presents catalog data (Lindgren, Inc.) for several standard modular single- and double-layer enclosures manufactured of different shielding materials.

In order to achieve that high field attenuation, the shielded room design must account for many "details": doors, cable penetrations, power and high frequency signal filtering, ventilation, fire protection, safety, and many other. Thus, designing a good shielding room is a "serious" business, with the cost in hundreds of thousands, or even millions dollars. The details of such design can be found in specialized literature (e.g., see [2.1, 2.2]) and in the abundant manufacturer catalogs. Another problem is the internal field reflection from the shielded room confines: walls, ceiling, floor. To mitigate these effects, the walls and the ceiling, and sometimes the floor of the shielding room is lined with RF absorber (e.g., see [2.1, 2.3]).

Understanding Electromagnetic Shielding 83

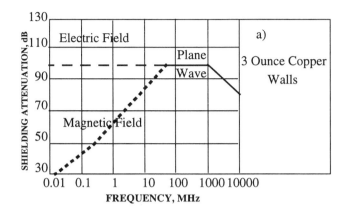

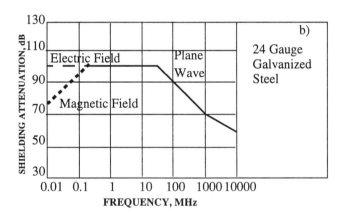

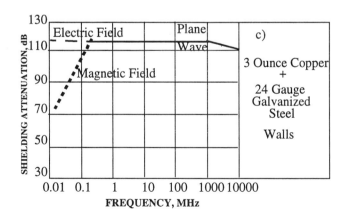

FIGURE 2.8 Shielded room effectiveness

2.2.4 "Mobile" Shielding: How Good Shield is a Taxicab

Another type of shielding enclosure is a shielded room "on wheels": cars, ships, airplanes, satellites, etc. Today, these vehicles are literally "stuffed" with electronics. In principle, here the problem is no different from other, non-mobile applications: protect the system electronics from the ambients and crosstalk within the same system, and contain the EMI generated by the vehicle.

However, don't forget that these objects have to communicate with the outside world. For example, take an application which may be pretty well described as the "sign of the time". We are talking "cellular" or PCS. Indeed, more and more people are using their mobile phones to communicate directly from the vehicle. As far as the subject of this book is concerned, the question is: how much the mobile enclosure (as a rule, metallic) attenuates the communications signal.

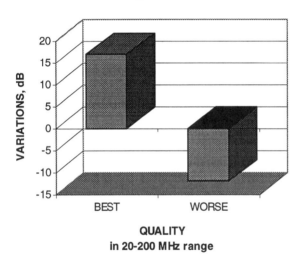

FIGURE 2.9 Shielding Effectiveness Variations in Passenger Vehicle

Fop example, a car or truck cabin is usually a pretty bad shielding enclosure, since they are *not* designed for electromagnetic shielding. However, this "bad news" may also be a "good news", because this gives the car riders an opportunity to use their cellular. As an example, Fig. 2.9 presents a diagram of a passenger vehicle shielding effectiveness variation extremes (per data in [2.4]) As we can see, many of the problems inherent to "bad" shielding enclosures, are present here', in particular, very low shielding attenuation and "negative" shielding.

✦ There are also many more problems in dealing with the RF /microwave signal propagation: multipath, Doppler effect, slow and fast fading, etc.While their analyses are definitely out of scope of this book, nevertheless, they are a part of the EMI environment which affect the shielding performance. The reader is referred to many excellent books available on the subjects. ✦

2.2.5 System Shielding: Limitations of System EMI Analysis

Can the shielding effectiveness of separate system components be determined, based on the *whole* system testing and evaluation? The answer is: only to a limited degree. Revisit Fig. 1.13. As we saw, the effect of cable shielding was dramatic below ≈100 MHz. However, two important questions remained unanswered.

1. How efficient was the shield? One way to determine the shielding effectiveness could have been to deduct the values in Fig. 1.13c from the values in Fig. 1.13a. Such data would represent the shielding effectiveness only if the cables remained the dominant radiators in both cases—with cable unshielded and shielded. However, the data itself does not give any direct indication of whether that was the case.
2. What was the shield performance in the "high" frequency range (i.e., over 100 MHz)? Here, the cable shield either did not perform at all, or the shielded cable was not a dominant radiator—otherwise, there should have been a difference between Fig. 1.13a, b, and c. In the second case, the only conclusion that can be made is that the shielded cable performance was at least as good (or bad) as to match the next highest secondary radiator in the system.

In chapter VI, we will return to the analysis of this particular shield evaluation strategy, based on the measurement of the whole system.

2.2.6 Shielding, Grounding, and All That 'Jazz'

The test results in the next example (see Fig. 2.10) emphasize the interrelated problems of shield design, grounding, circuit balance, and circuit-to-shield coupling, when dealing with cable EMC. A 3 m long flat (ribbon) cable with a matched load at the end was stretched parallel (i.e., horizontally) over a ground plane at a height of 1.0 m. A circuit in the cable was powered GSG ("ground-signal-ground" regime, when a signal conductor is "sandwiched" between two ground conductors in the cable plane). The signal was swept from 30 to 200 MHz with the cable input level 0 dBm. The radiated field intensity was measured at 3 m distance from the cable with a horizontally polarized antenna, parallel to the cable. Three different cable designs were tested (again, as in the previous example the same cable was tested with "peeled-off" shielding layers):

- cable 1, unshielded
- cable 2, with a one-layer longitudinal foil shield
- cable 3, with a two-layer longitudinal foil shield

In the shielded cables, the shield is connected to the ground conductors at both ends of the cable. At the near end (i.e., at the generator), the ground conductors (together with the shield—in shielded cables) are connected to the ground plane. With regard to the grounding conditions at the far end (i.e., at the load), the radiated emissions from each cable are measured for two cable working modes:

- The ground conductors (together with the shield, in shielded cables) are connected to the ground plane.
- The ground conductors (together with the shield, in shielded cables) are not connected to the ground plane.

86 Electromagnetic Shielding Handbook

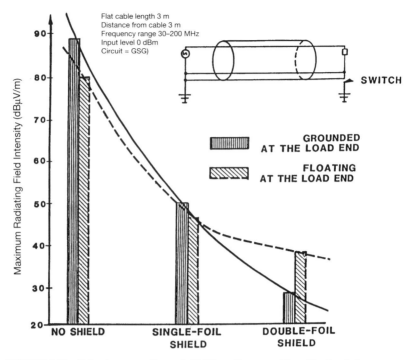

FIGURE 2.10 Balancing, grounding and shielding effects on cable radiated emissions

The maximum radiated field intensity (emissions) from the cables is plotted in the diagram in Fig. 2.10. As complicated these results may seem, they actually yield to quite simple and logical explanation. Before proceeding with the analysis, let us note several important considerations.

1. In unshielded cable, the GSG circuit configuration provides for small enough differential mode emissions. Therefore, it can be expected that the emissions from the cable are determined mainly by the common-mode currents.
2. In general, the emissions from the cable are determined by a combined effect of the cable circuit and the shield.
3. The circuit and the shield may respond differently to the grounding of the far end. Also, since the near end is grounded permanently (practically, this is done in the generator, which is unbalanced), grounding of the far end creates a large loop with dimensions 1 × 3 × 1 × 3 meters. The current in this ground loop is determined by the common-mode regime in the cable.
4. As we have shown(e.g., see Fig. 2.6 and 2.8), the double-layer shield can be expected to perform better than a single-layer shield.

Keeping this in mind, return to the diagrams in Fig. 2.10. The double diagram at the left just confirms our suspicion that a large ground loop (created by grounding the load end) is bad news, resulting in a 10 dB increase in radiated emissions. (Didn't they teach you to avoid ground loops?) The double diagram on the right is also quite logical: a "good" (i.e., double-layer) shield reduced emissions from the cable circuit by about 45 dB. This also corresponds to the data in Fig. 2.6 and 2.8. But there is more to this diagram. Comparing it to the unshielded cable diagram at the left, we see that the effect of grounding the load end was reversed! This means that the grounding of the load end of the shielded cable actually results in lower emissions than floating it (with regard to the ground). But didn't we see just opposite in unshielded cable? So as a "matter of fact," it means that the transmission line radiates more but the shield performs better, when the line is grounded at both ends.

Now, we can explain the middle double diagram for a single-layer shielded cable. First, we see that single-layer shield provides a more modest protection than a two-layer shield, although it is still quite significant—about 25 dB. But now, with regard to grounding of the load end, the achieved improvement in shield performance is just enough to offset the emission increase from the cable circuit *per se.* Thus, there is almost no difference between the emissions with grounded and nongrounded load ends.

A detailed analysis of the above example requires a solid background in shielding theory to be given in the following chapters. Then, using methodology based on antenna theory, a close enough correlation with experimental data can be obtained (this problem is often presented as exercise at the educational courses taught by the author).

2.2.7 Connectors: Don't Miss the Link

Shielded Power Supply Cord: A Case of a "Bad" Connector

This example illustrates the effect of connectors on radiated emissions from a power supply cable assembly. A shielded power cord was first tested with its original three-prong male and three-socket female connectors (see Fig. 2.11). During the test, the "hot," the "neutral," and the "green wire" were first connected together and formed the "signal line," while the shield was connected to the ground. Then, the test was repeated with the original cord connectors cut off and UHF connectors (they are coaxial-type) mounted instead at both sides of cable assembly. The test results shown in Fig. 2.11 prove that the connector change resulted in 20-to-40 dB reduction of radiated emissions. This means that the EMC performance of the power cable was determined not by its shield but by the radiation from the connector (which acted as a dominant radiator in cable/connector system). The inferior performance of the original cable terminated on power line plugs can be explained by large radiating loops at the cable ends. Thus, the effect of heavy and expensive cordage shielding (this was double foil + 60 percent braid) is almost brought to naught by bad cable termination. This radiation was eliminated by using UHF connectors.

Of course, nobody is going to use a UHF connector instead of regular electrical plugs. But to make proper use of the cable shielding, if necessary, for example, this plug could be located inside a shielded enclosure, with the cable shield terminated to the enclosure before entering it. We will discuss shield termination techniques in the following chapters of the book.

What About "Good" Shielded Connectors?

Data in Fig.2.11 can also be used to evaluate the *lower* boundaries of the UHF connector shielding effectiveness: at least, about 20 dB at frequencies below 100 MHz, and about 50 dB - at higher frequencies. The exact value of the shielding effectiveness

a) Shielded power cord: z-fold double foil + 60% braid

b) Radiated emissions from a shielded power cord

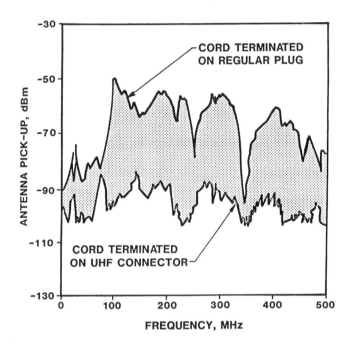

FIGURE 2.11 Effect of shield terminations on shielded power cord emissions

cannot be determined from the graph, because it is not clear "how much" of the emitted energy in both upper and lower curves can be attributed to the cable and to the connectors.

As a rule the performance of a UHF connector is limited to the UHF frequencies (indeed, what's in a name?). A much better performance is often expected from special connectors for microwave applications (e.g., type "N", or SMA) - well over a 100 dB! Such connectors are often used in antenna feeders and test equipment, as well as in other applications.

With the increased needs for broadband transmission and introduction of mandatory EMC regulations, the shielded connectors become more common in computer and telecommunications I/O circuits, e.g., shielded D connectors ("D-sub"), shielded connectors for telecommunication product backplanes, telecommunications plugs, etc.

Unfortunately, in many applications the physical constraints do not permit to create and "electrically tight" connector shielding system, so that connector shielding effectiveness 5 - 30 dB (throughout the frequency band of the application) is not uncommon. In chapter 7, we will review the effectiveness of some of the elements used to improve connector shielding performance: connector backshells, front shells, contact springs (see section 7.6.3).

2.2.8 Importance of Shielding System Integrity: Shielding System for Electrostatic Discharge Protection

System shield integrity appears to be one of the most important topics in the shielding applications. While the violations of the system shield integrity may occur for different reasons - inherent shield design, built-in vents and other "technical" apertures, physical damage, etc. - one of the most probable problems is created by inadequate shield terminations at the interface between the different system elements. Especially troublesome can be the interface between the electronic product shielding enclosure and cable shield at the location of the cable penetrations inside the enclosure. The following example proves this, along with emphasizing the importance of shielding for electrostatic discharge (ESD) protection.

> ✦ ESD is a major source of EMI with effects ranging from data errors to firmware memory loss to physical destruction of hardware. The sources of ESD are numerous: human body, furniture, friction between moving dielectric surfaces. In this way, voltages up to several tens of kilovolts can develop between the carrier of the electric charge and different circuit points and ground. Application of such high voltage leads to the dielectric (air, enclosure paint, cable jacket and conductor insulation, and PWB dielectric) breakdown and energy discharge. This process is accompanied by the flow of current through the discharge channel and conducting circuit elements, and generation of high electric and magnetic fields. The generated fields and currents can vary with frequencies up to hundreds megahertz. Accordingly, the ESD effects manifest themselves in various types of coupling: galvanic, near-field electric and magnetic, and far-field coupling.
>
> Electrostatic discharge is an extremely dangerous and even "insidious" phenomenon, Indeed, there are three groups of ESD effects which should be considered: electrical safety, equipment physical damage, and EMI. The case histories are abundant. Following are several examples of the "ill-effects" of the ESD, based on published materials:
> - A metallic chair rolled into a computer console causes a memory disc error,
> - A human body discharge causes a telephone electronic system reset and a burst of bit errors in a high speed system,
> - EMI from ESD results in the failure of a hospital nursery warmer,
> - EMI from a high voltage part of an automatic external defibrillator (around 4 kV), which is used to restart the heart of a person in cardiac arrest, disturbed the digital control circuitry of the same defibrillator
>
> And very often shielding from ESD effects is about the most efficient way to address these problems. While there are certain important specifics in shielding applications addressing each of these effects, the basic principles, as outlined in this book, are the same. ✦

To experimentally model the effect of system shielding integrity on the electrostatic discharge, a multiconductor shielded cable was used in a single-ended circuit mode, with the cable conductors forming the signal line and the cable shield together with the drain wire forming the current return conductor. At one end, the cable is connected to a metallic box, used as "dummy" product housing. The following four cases of cable shield and drain wire connection to the housing were investigated:

1. A 360° cable shield termination was created by soldering it to the box at the cable entrance location. The drain wire was connected to the shield.
2. The cable shield (drain wire) was soldered to a D-sub connector (forming a 1/4-inch pigtail) plugged into the mating connector, which is 360° soldered to the box.
3. Only the cable shield wire was connected to the box using a 2-inch long pigtail.
4. Neither the cable shield nor its drain wire were connected to the box.

At the other end, the cable shield is connected to the drain wire and grounded. Inside the box, the line (formed by the signal conductor and return conductor) is loaded at a high resistance load—20 Ω. At the other end, the line is connected to a measuring device—a spectrum analyzer and/or oscilloscope.

An electrostatic discharge was generated and applied to the metallic box with return via ground, and the induced voltage in the line was measured. As the test showed, the induced voltages depended significantly on the type of the shield, kind of termination, and specific grounding conditions. For a single-layer braid shield and using about 0.5 m long pigtails for ground connection, the maximum voltages induced in the four test cases were presented as a diagram in Fig. 2.12.

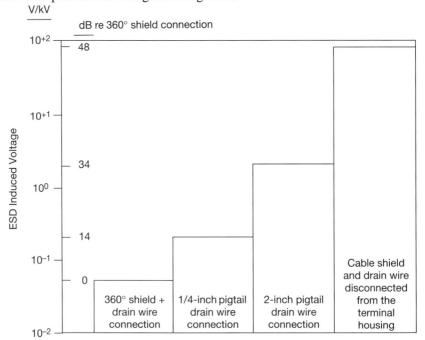

FIGURE 2.12 Shield grounding effect on cable ESD induced voltage

The vertical axis represents two scales: an absolute measure of induced voltage in V per 1 kV of the discharge source, and a relative (in dB) measure with regard to the reference induced voltage at 360° cable shield termination directly to the housing. When the cable shield is 360° directly terminated to the housing, the induced voltage is only about 60 mV/kV, while the introduction of a connector to the shield termination led to the three-fold increase of this voltage. Only a drain wire shield connection exhibited fairly poor performance, while no shield grounding at all produced almost 50 V/kV—enough to kill many types of IC.

Now, the "traditional" questions:

- What is the mechanism, or mechanisms, of the ESD voltage generation in a shielded cable circuit? One obvious cause of "ESD EMI" is galvanic coupling, i.e., the discharge current flowing in the shield. This current produces a voltage drop in the shield which, being the return conductor of the shielded single-ended circuit, applies this voltage to the circuit.
- Then the follow-up question is, how would the shield design affect this process and what could be done to reduce the generated voltage? Another cause of EMI is direct coupling to the shield of the electromagnetic field generated in the course of electrostatic discharge. This coupling depends on many factors, including the character of the discharge, field configuration, shield geometry, electrical characteristics, and position in space with regard to the incident electromagnetic fields. Again, if properly constructed, grounded, and terminated, the cable shield will act to reduce the effect of this field on the shielded circuit.

2.2.9 Ground Plane in a Printed Circuit Board: Parallel Shielding Effect

The positive effects of ground planes in multi-layer printed circuit boards (PCB) are numerous and well publicized: enhanced circuit parameter stability and signal integrity, reduced crosstak, reduced radiated and conducted emissions, better immunity - you name it. As a result, in modern electronics the use of ground planes in PCB design and layout is almost a rule. As far as shielding is concerned, we can relate the effects of the ground plane on PCB EMC performance in two ways — shielding modes, which we will call separation shielding and parallel shielding, respectively. In the first mode, the ground plane is placed *between* the interfering systems and acts in a "traditional" shielding role of electromagnetic separation. In the second mode, the ground plane is placed *parallel* to one, or several, or all the interfering systems. Although in this case, the "shielding" role of a ground plane may not be as obvious as in the separation shielding, nevertheless it does exhibits "shielding effects".

While we are not yet ready to analyze these, as well as other, shielding modes (we will do it later in the book), here we will illustrate the effects of parallel shielding using experimental data. Two PCBs were manufactured, one single-layer, second - double layer, one of them - ground plane. Except for the extra ground plane layer in the second PCB, both boards were identical in all the details. On the component layer, the boards contained a circuit, containing a 10.752 MHz oscillator which acted as a comb generator. The oscillator was powered by a battery and connected to a (6×16) cm trace loop with a 50 Ohm load — simulating a clock distribution line.

The boards were tested in identical conditions at the EMC test site. The radiated emissions measured @ 3 m from the boards are presented in Fig. 2.13..

As shown, the addition of a ground plane results in dramatic reduction of radiated emissions from a PCB — a typical example of parallel shielding effectiveness.

2.3 THIS CONFUSING WORLD OF ELECTROMAGNETIC SHIELDING

Hopefully, the reviewed specific examples have generated a useful insight into the capabilities and difficulties of shielding. The next step is to come up with general models and definitions, which summarize and explain this experimental data and capitalize on the obtained results to develop respective shielding recommendations. Surprisingly, such generalization appears much more difficult than expected. In this section, we will review some of the most wide-spread approaches to this problem. As you will see, this "common knowledge" wisdom is of little help, and the main result to draw from this exercise will be the realization of the fact that we need more understanding.

2.3.1 What Is Electromagnetic Shielding, Anyhow?

Let us start by attempting to define electromagnetic shielding. Simple? Try to define it yourself, or look it up in the literature. It appears that the problem is not in the lack of such a definition, but just opposite—in the existence of too many definitions of electromagnetic shielding and shielding effectiveness. And numerous and confusing they are! For example, the IEEE Standard Dictionary of Electrical and Electronics Terms (IEEE Std. 100) contains about 2 dozens (TWO DOZENS!) shield- and shielding-related definitions. Here's an example:

> **"shielded-type cable.** A cable in which each insulated conductor is enclosed in a conducting envelope so constructed that substantially every point on the surface of the insulation is at ground potential with respect to ground under normal operating conditions".

While such a definition may do "some good" when it is possible to define and assess the ground potential at the insulation surface (e.g., in case of a coaxial cable with an ideal shield and/or at low enough frequencies), it does not address such important real-life problems as single or multiple unshielded pairs within a common shielding enclosure; the practical nonuniform shields, as well as shields of finite conductivity and complex geometry; the skin effect in the shields; and so on. Then again, what are the "normal" operating conditions, and how much and in what sense can these conditions be "abnormal?" For example, are the conditions aboard a satellite normal with regard to the "ground?" Moreover, at high enough frequencies, the

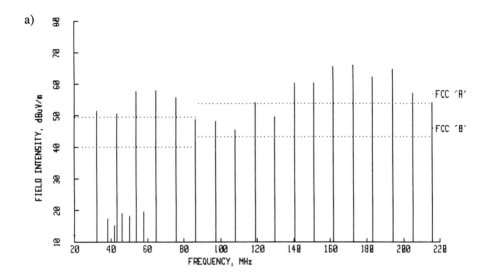
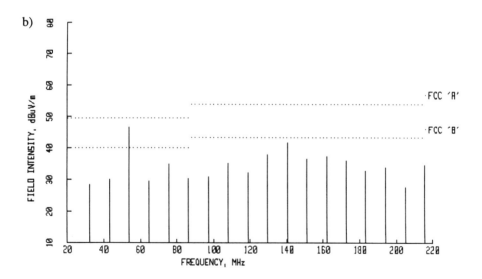

FIGURE 2.13 Radiated emissions from PCB containing a clock line
  a) without ground plane
  b) with ground plane

different points of the "ground" can be at various potentials, which effectively "removes the ground" under this definition.

Similar "claims" can be made also with regard to other definitions

In general (not just in the IEEE Std. 100), a dozen other parameters often used as definitions of shielding effectiveness might include the following:

- ratios of incident on and transmitted through a specified shield sample electric and magnetic field intensities, measured in free space environment
- ratios of incident on and transmitted through a specified shield electric and magnetic field intensities, measured in a particular type of shielding enclosure, e.g., shielded room, TEM/GTEM cell, dual TEM cell, or reverberating chamber
- measure of the field energy loss in the shield
- measure of electromagnetic energy reflection from and/or absorption in the shield
- ratio of the voltages induced in a shielded and nonshielded line by some specified field
- ratio of the voltage induced in the shielded circuit to the current in the shield (termed transfer impedance)
- voltage induced in a sample probe (e.g., wire loop or a monopole) by the shielded cable under test, powered with a specified signal
- ratio of crosstalk measured between shielded and nonshielded disturbing and disturbed cable circuits
- shield material resistivity, e.g., ohms per square, for foil shields
- manufactured shield dc resistance per cable unit length (e.g., ohms per meter)
- certain shield design features, e.g., optical density in percent for braid, or foil thickness, or a number of shielding layers
- magnetic permeability of shields made from high-μ metals

This list can be expanded ad infinitum by varying both general approaches to and specific details of each shielding parameter and/or evaluation method, as well as the characteristics of the applied test signal (e.g., time domain or frequency domain). No wonder that, to many practitioners, the term electromagnetic shielding (or screening, as it is more commonly known in Europe) may mean different things, and about the only thing everybody agrees upon, is its association with "black magic."

2.3.2 Do You Speak "Shieldinese" ?

Each of the quoted definitions, as well as related "jargon", reflect a certain philosophy of understanding and evaluating the shielding phenomena. Which is correct? As we will see, the answer could be anything between "all" and "neither"! While in some applications any of these definitions and approaches may be useful, in other situations they can be either incomplete, or make little physical sense, or fully disregard system EMI specifics and be inaccurate, or result in almost insurmountable mathematical and logistics difficulties.

Consider one of the most "popular" shielding effectiveness definitions:

Understanding Electromagnetic Shielding 95

A ratio of electric and magnetic field intensities that are incident (E_e and H_e) and transmitted through the shield (E_i and H_i)

Respectively, in terms of linear units,

$$S_E = \frac{E_i}{E_e}; \quad S_H = \frac{H_i}{H_e} \tag{2.1a}$$

and in terms of shielding loss (logarithmic units),

$$A_E = 20 \log_{10}\left|\frac{1}{S_E}\right|; \quad A_H = 20 \log_{10}\left|\frac{1}{S_H}\right| \tag{2.1b}$$

It is also often (and not always with sufficient explanation) suggested that

$$S_E = S_H, \text{ or } A_E = A_H \tag{2.1c}$$

Under certain conditions, there are no objections to definitions (2.1a through 2.1c); when the fields are uniform and the shield and the propagation media adjacent to the shield are homogeneous, the effect of the fields on the source and receptor of EMI can be disregarded. However, their application in general, and specifically to electromagnetic shielding, is not always simple or justified. Just what do we do if the incident field is not uniformly distributed in space, or the shield properties are not homogeneous, or both—which is very often the case in real life? For example, would it be right to use the field peaks or the averages over some area, and if so, over what area (for example, consider Fig. 2.9)? Also, statement (2.1c) is correct only if the shield transfer properties for the electric and magnetic field components and the impedances of the propagation media adjacent to both sides of the shield are identical. However, if the incident and exit waves encounter different losses in and impedances of the media, then the distribution of the energy related to these components changes, E and H ratios in (2.1a) are not equal, and (2.1c) is not correct. This is a typical situation for near zone fields (see Chapter 4). It is true for any kind of shields, including the "box"-like shielding enclosures and cable shields. In reality, one cannot expect the electric and magnetic field propagation conditions inside and outside the shielded enclosure to be identical. But if $S_E \neq S_H$, then should we use S_E or S_H, or some combination of the two?

✦ For example, what happens in the case of a high-μ shield with apertures, placed in the induction electromagnetic field? The shield's effects on the electric and magnetic components of the incident field and field sources will be different, resulting in different shielding effectiveness values of these components for the protected environment. And these differences will vary as the EMI signal frequency, incident field configuration, and the shield position with regard to the incident field change.

Let's face it, "surprising" you with such an example at this stage of our discussion is almost like "serving a blow below the belt," but you don't have to answer the question yet. We just want to emphasize the seriousness of the situation. Besides, we hope to be able to make more sense out of this as we proceed, don't we?✦

Even if there weren't basic obstacles to using definition (2.1), the specifics of electromagnetic shielding often make them impractical. One of the reasons is that *system EMI effects* depend not only on the field intensities in the parts of space separated by the shield, but also on the mode and degree of coupling of the fields to the shield (on one hand) and to the shielded circuits (on the other). But these considerations may have little or nothing to do with the shield design *per se*, but rather with the specifics (e.g., subsystem interactions, grounding system, etc.) of the electronic system, of which the shielded circuit is a part. Surely, there is much more to electromagnetic shielding than a simple model can yield. We will defer a detailed discussion of the subject until a necessary theoretical background has been laid out. By the meantime, in the next section we will use the concepts of one very widespread approach to illustrate some of the dangerous "reefs in the troubled waters" of electromagnetic shielding. Hopefully, this example will deter the reader from indiscriminate use of even the most popular and respected theories and formulas.

2.3.3 Transmission Theory of Shielding: User Alert

Consider an important example: an infinitely large flat conducting shield illuminated by an incident electromagnetic wave. The incident field may be a far zone plane wave field from an unspecified source. Alternatively, it could be the field of a pair of closely spaced parallel and opposite-direction current filaments enclosed in an infinitely long thin (comparing to the radius) cylindrical shield. While this later model reminds you of a shielded cable, its utility is much wider: any set of conductors running parallel an infinitely large shield. In either case, the shield attenuates the wave propagating through it to the "other side". We will use the shielding effectiveness definition as formulated in (2.1). Using the transmission line analogy, it was shown [2.5] that such shield attenuation can be related to two physical mechanisms: energy reflection from the shield surfaces and absorption in the shield body (see Fig. 2.14). Thus, the shielding performance (e.g., shielding loss, A_s) can be formulated in terms of these two processes (A_a and A_r, respectively) and their interaction (A_m):

$$A_s = A_a + A_r + A_m \qquad (2.2)$$

Corresponding formulas were derived many years ago by S.A. Schelkunoff, based on what he called the transmission theory of shielding, "since it dealt with waves passing through the shield" [2.5].

This is an easy understandable and even elegant (if you will!) model. It is often used as a first approximation of shielding problems. Our purpose is to determine how well such a model matches *practical* needs of shielding evaluation. In the case of plane waves in free space, incident on an infinite length homogeneous shield with high conductivity, flat or with thickness much smaller than the shield radius and the wavelength (whoo!), the expressions for items in (2.2) are especially simple. We will give the ex-

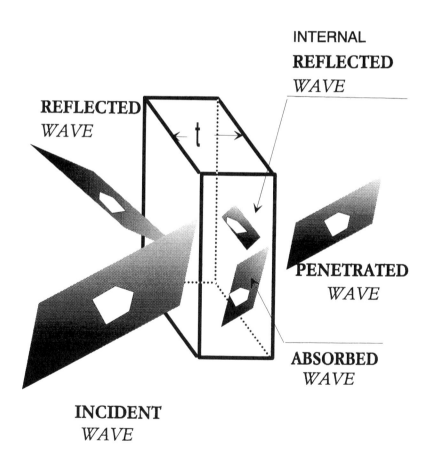

FIGURE 2.14 Plane wave incident on a flat shield

pressions here without derivation, with the further analysis provided later (see Section 2.4.1). The absorption losses in the shield are

$$A_a \approx 8.686\left(\frac{t}{\delta}\right), \text{ dB} \tag{2.3}$$

The reflection shielding losses depend on the mismatch between the propagation media intrinsic wave impedances, which can be calculated as follows:

$$Z = \sqrt{\frac{j\omega\mu}{\sigma + j\omega\varepsilon}} \tag{2.4}$$

In a metallic shield, the item with ε can be disregarded:

$$Z_m = \sqrt{\frac{j\omega\mu}{\sigma}} \tag{2.5a}$$

If the space adjacent to both sides of the shield (i.e., at the wave "entrance into" and "exit from" the shield) is dielectric, then $\sigma = 0$ and

$$Z_d = \sqrt{\frac{\mu}{\varepsilon}} \tag{2.5b}$$

Across two interfaces between the shield and two identical media at its sides (e.g., at the wave "entrance in" and "exit from" the shield), the reflection attenuation is

$$r = 20 \log_{10}\left|\frac{(\xi+1)^2}{4\xi}\right|, \text{ dB} \tag{2.6a}$$

where the impedance ratio $\xi = \dfrac{Z_d}{Z_m}$

Assuming that the incident on the shield wave impedance in free space is

$$d = Z_0 = 120\pi \approx 377 \text{ }\Omega \gg Z_m$$

Schelkunoff's reflection loss formula for two air/metal interfaces can be presented as

$$A_r = 20 \log_{10}\left|\frac{377}{4Z_m}\right|, \text{ dB} \tag{2.6b}$$

Similar formulas also apply to the interface between two metals or two dielectrics that have significantly different values of Z. These values can be calculated per Eqs. (2.4) and (2.5) and substituted in Eq. (2.6).

A_m is a multiple reflection factor within (that is, "inside") the shield. It is determined by "bouncing" reflections of the signal between the shield two boundaries with the dielectrics. These inter-reflections are accompanied by energy absorption in the shield:

$$A_r = 20 \log_{10} \left| 1 - \frac{(\xi - 1)^2}{(\xi + 1)^2} e^{-2\gamma_c} \right|, \text{ dB} \qquad (2.7)$$

In Eqs. (2.3) through (2.7), t and r are the shield thickness and radius, respectively (in meters), and the current penetration depth (skin depth) of the shield is

$$\delta = \sqrt{\frac{2}{\omega \mu \sigma}}, \text{ m} \qquad (2.8)$$

The intrinsic propagation constant of the shield material is

$$\gamma_c = \sqrt{j\omega\mu\sigma}, \text{ 1/m} \qquad (2.9)$$

which is intimately related to the skin depth,

$$\delta = \frac{\sqrt{2}}{|\gamma_c|}; \quad \gamma_c = \frac{1+j}{\delta} \qquad (2.10)$$

Other variables and parameters are as follows:

$\omega = 2\pi f$, the clock frequency

μ = magnetic permeability of the shield

σ = conductivity of the shield

ε = dielectric constant of the wave propagation media outside the shield

For convenience, the values of δ and Z for different shield metals were tabulated (see Table 2.1)*, and available just about from any reference on shielding — which makes their use especially easy.

---

*You can find the conductor and dielectric basic properties (σ, ε, μ) in special literature, e.g., [A.25]

The described approach is one of the most familiar to and widely used by electronics engineers. Unfortunately, the users often forget the major limitations of it. We will discuss these limitations in details later in this chapter, after gaining the necessary theoretical background. Here instead, we will use an illustration to see how close the theory gets to experimental evaluation of several practical shields.

To see the point, let us use this approach to calculate the shielding effectiveness of several designs of two shield types:
- an aluminum foil cable shields with thicknesses $t = 8.5 \times 10^{-3}$ mm (i.e., 1/3 mil), $t = 25.4 \times 10^{-3}$ mm (1 mil), and $t = 76.2 \times 10^{-3}$ mm (3 mil),
- a "hinged" steel 1 mm thick "box" with dimensions $50 \times 20 \times 15$ mm, consisting of two halves with the seam at the junction made mechanically "tight", and "enforced" with a pair of latches (do you recognize your tool box?).

The investigated shields are "regular" designs, like many others often used in electronic systems.

The results of shielding effectiveness calculations (A_s) are plotted in Fig. 2.15. The calculated values of shielding effectiveness for the three considered cable aluminum shields with different thicknesses at *low frequencies* do not differ too much, and exceed 100 dB. The calculations show that, at these frequencies, the theoretical shielding effectiveness is determined mainly by the *reflection losses*, which do not depend on the shield thickness. At *higher frequencies*, the *absorption losses* become significant (especially for thicker shields), and this leads to the increase in shielding effectiveness. This is especially manifested in thicker shields where thickness exceeds the skin depth. Also plotted in Fig. 2.15 are theoretical values of A_s for a steel shield (second considered shield type - "box"). Note, the difference between the A_s characteristics of steel and aluminum shields.

Table 2.1 Skin Depth, δ, and Wave Impedance, Z

Metals: $\delta = \sqrt{\dfrac{2}{\omega \mu \sigma}}$ mm ; $Z = \left|\sqrt{\dfrac{j\omega\mu}{\sigma}}\right|$ ohm

Air: $\delta = \infty$; $Z_{planewave} = \dfrac{120\pi}{\sqrt{\varepsilon}} = 377$ ohm

Frequency, f, in Hz	Copper		Steel		Aluminum	
	$Z(\cdot 10^{-3})$	δ	$Z(\cdot 10^{-3})$	δ	$Z(\cdot 10^{-3})$	δ
10^4	0.037	0.667	1.044	0.187	0.048	0.864
10^5	0.12	0.21	3.3	0.059	0.15	0.27
10^6	0.37	0.067	10.44	0.019	0.48	0.086
10^7	1.18	0.021	33.030	0.006	1.53	0.027
10^8	3.72	0.0067	104	0.0019	4.833	0.0086
10^9	11.8	0.0021	330	0.0006	15.3	0.0027
Formula	$0.372 \cdot 10^{-6} \sqrt{f}$	$66.68 \dfrac{1}{\sqrt{f}}$	$10.44 \cdot 10^{-6} \sqrt{f}$	$18.7 \dfrac{1}{\sqrt{f}}$	$0.483 \cdot 10^{-6} \sqrt{f}$	$86.44 \dfrac{1}{\sqrt{f}}$

Understanding Electromagnetic Shielding 101

So far, so good. Now, let us compare the theoretical values we just calculated to experimental data. For example, also plotted in Fig. 2.15 are cable shielding data based on Fig. 2.6. The foil shield thickness of the cable in Fig. 2.6 was 1/3 mil. The experimental curves are approximate and rounded, having been obtained by deducting the "foil shield" and "foil + braid shield" data from "unshielded" data. But what a miserable failure! At no frequency did the calculated values come in reasonably close agreement with the experimental data!

Similar discrepancy between theory and experiment is true for a steel box. Fig. 2.16 presents experimental data obtained by first measuring emissions from a self-contained source in "free space" and then repeating the measurements with the source placed inside our shielding enclosure. Owing to the self-contained nature of the source, no cables, lines, wires, or "whatsoever" were penetrating the shielding enclosure. As shown in Fig. 2.16, the theoretical vs measurement comparison of a "box" shield is nothing better: an average about 7 dB measured shielding effectiveness, with "negative" shielding at some frequencies, vs well over 1000 dB calculated values at higher frequencies!

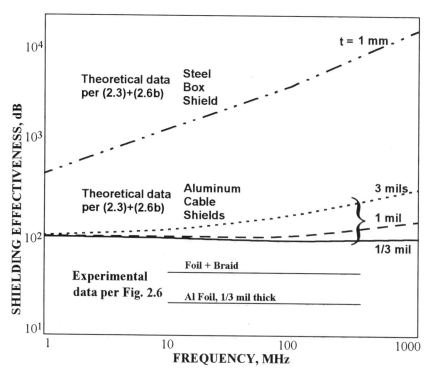

FIGURE 2.15 Shielding effectiveness of aluminum foil cable shield

Honestly, can you live with the discrepancy often exceeding several hundred decibels? (Well, 100 dB more, 100 dB less—what's 100 dB between friends!). Unfortunately, we see that our first attempt to apply the theory "where black magic belongs", leaves us with quite troubling conclusion:

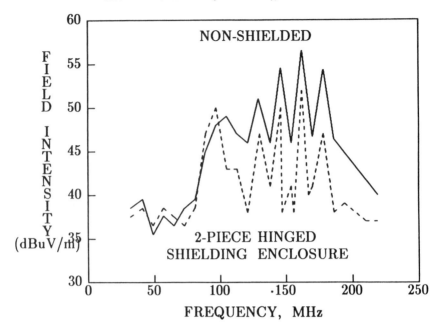

FIGURE 2.16 Shielding enclosure effect on electronic product radiated emissions

the *indiscriminate* use of transmission theory of shielding, as well as the formulas (2.2) through (2.4), can lead to very serious errors and *as a rule cannot be recommended* without the analysis of the application specifics.

✦ If you feel as if you are lost in some kind of a electromagnetic shielding jungle, you are not alone there. Back in the early 1960s, two coaxial cables had to be connected to transmit a signal while producing low radiated emissions. In the absence of "real" connectors, a young technician installed a "home-made" substitute, which resulted in much higher emissions than could be tolerated. As a remedy, over the cable joint he installed an impressively large cylindrical metal can with holes at the center of the butt-ends to pull the cable through, and connected the can to the cable external conductors using aluminum foil. This resulted in an emissions... increase. Pondering, that a *larger* and *thicker* can would do *better* (don't you think so?), the technician used a larger can, which resulted in a... further increase in emissions. But wasn't this can supposed to provide shielding effectiveness of several hundred decibels at these frequencies? That's when an engineer was called upon (guess who?). It was a real eye-opener: instead of reducing emissions, "improved" shielding led to their increase. And the "more shielding," the worse the result! At that time, it took about a day to solve the problem by using the *smallest possible* sheet of copper, *tightly* wrapped around both cables and 360° *soldered* to the their external conductors and soldered at the *longitudinal seam*. Then it took many years to figure out the causes of and remedies to many similar "inverse shielding" practices. To be specific, in this particular case it appeared that a larger and thicker shielding can facilitated the EMI coupling and propagation, while its improved *transfer function* (that is, larger absorption loss) was "bypassed" by the imperfections of the shield and its interface with the rest of the shielding system (that is, the cable outer conductors). The most prominent of these imperfections were the holes at the can butt-ends, poor contact resistance between the foil and the cable outer conductors, and spiral application of the foil.✦

2.3.4 What We Have Learned and What's Missing

In summary, the previous discussions positively proved the great potential of electromagnetic shielding as an EMI mitigation technique. We accumulated a significant body of "real life" experimental data which illustrate the role of shielding as an "EMC front line" in a large variety of important applications. However, the "first encounter" of the facts and theory ended in a failure (for the theory, to be sure!).

Does all of this mean that the "transmission theory of shielding," or any other of the named approaches to electromagnetic shielding, are wrong? No, but... a theory is correct only within the realm of its assumptions and constraints. Unfortunately, as a rule for any particular theory, those fit only a limited number of specific applications and may not be met in others. What's even worse, these assumptions often are not explicitly spelled out, nor are they alw*ays understood*. Therefore, *each theory must be used only in conjunction with* the proper analysis. As you will see, the direct penetration of the EMI signal through the solid shielding wall, addressed by transmission shielding theory is often *the least* of the many shielding concerns (later, we will call this process *diffusion* and incorporate it into the general shielding model). There is much more to shielding than meets the eye "at the first sight".

Next, in all the considered applications, did we *really* understand the role of our shield? Did we want to just reduce the field at the "other side", or improve immunity of *some* system, or reduce crosstalk between two lines or systems, or was it making any difference, what is it we were trying to achieve? The problem was that while literally dozens of shielding effect definitions and shield evaluation techniques were available, we were lost in the "shieldinese", lacking their conceptual comprehension. It appears that the definition of shielding is a *relative* concept, very much dependent not only on the shield *per se, but also* on the considered system configuration and general EMI protection goals (start seeing the light?)!

We have also failed to achieve the degree of understanding of shielding phenomena and their dependence on the shielding design and application, which would permit us to properly *engineer* the shields and *predict* their performance, as well as compare different shielding evaluation philosophies and measurement techniques. As a result, the considered popular approaches don't have neither the plan nor the tools to evaluate and compare the shield and shielding system designs, evaluation techniques, and performance data. That's why those approaches were helpless in answering often confusing questions we posed during our previous discussions of electromagnetic shielding definitions and examples of applications.

"Black magic" notwithstanding, there are objective reasons for our problems. But how does one account for these? And what can be done if the results of such accounting become technically and/or mathematically unmanageable? Should we simplify the models, relax limitations and accuracy requirements, or resort to empirical methods? As a first step to answer these questions, we need to develop realistic physical and theoretical concepts and models of electromagnetic shielding. What should such models be based on? With the knowledge we accumulated so far, we can identify at least three most fundamental sources of such models:

- Diverse experimental data. By now, we have reviewed a number of important practical examples, and more will be given as the discussion progresses.

- Theoretical analysis based on system analysis and electromagnetics and circuit theories. That should provide an insight into the available theories' assumptions and constraints and help to avoid possible "mismatch" with reality.

  ✦ While the theoretical background *per se* is not the subject of this book, it is absolutely necessary not only to solve, but just to understand the problems at hand: we will need a clear understanding of the physical processes taking place, as well as corresponding mathematical description. As you will see, the development of such an understanding is based on at least an introductory level of electromagnetics and circuit theory, years of experience, and lots of engineering common sense. Unless you are willing to make your way through the rigors of theory and experience, we may as well abandon the whole effort to understand electromagnetic shielding and risk depending on the mercy of some "smart" salespeople who want to sell their "stuff."* ✦

- Basic *EMC system* concepts of EMI synthesis and analysis. These were developed in chapter I.

  ✦ You really must consider "the whole" system. Just to prove the point, look at the following "real life" example. At the educational courses on shielding, I often exhibit an official certificate issued by a well-known company (the name of the company is withheld to protect the innocent; that is, myself). In the certificate, the company solemnly swears that the shielding cables it produces meet the requirements of FCC Regulations, Part 15. Do you need a shielded cable that meets Part 15 Regulations? Caveat emptor!✦

Hopefully, by now you are a strong advocate of using theory to the benefit of electromagnetic compatibility, and shielding in particular. Thus, it looks like a good opportunity to take a detour to the Appendix, at the end of the book, where we placed a very brief refresher of your old "favorites" - electromagnetics and circuit theory. It is certainly worth to do this *before* you proceed to the next sections, to refresh the knowledge of these important tools. Even if you are an expert in these disciplines, it is still worth to familiarize at least with the "architecture" of the chapter 8, because we will continuously refer to its equations throughout the remainder of the book.

In the remainder of this chapter we will first analyze the basic assumptions behind the transmission theory of shielding (the following section) and then develop a model of electromagnetic shielding based on the system approach to the EMC.

2.4 SHIELDING — THIS IS NOT VERY SIMPLE

So, why did the transmission theory of shielding failed in our examples? In this section, we will use the powerful "arsenal" of electromagnetics and circuit theories to analyze the results of the transmission theory of shielding. Using this particular problem, we will try to gain an understanding of the advantages and limitations of the two theories, as well as their abilities to resolve physical shield design and EMC system issues.

---

*But, of course, most salespeople are not like that.—A.T.

2.4.1 A Fresh Look at a Familiar Problem

First, we will revisit the problem of an infinite plane homogeneous linear metallic shield illuminated by a plane wave. As we know, this shielding problem was solved (see Ref. [2.5]) by applying the *transmission line approach*. We presented the resulting formulas without derivation (those interested in this derivation are referred to the original source (Ref. [2.5]); also, an excellent review of such solution is given in Ref. [2.6]. However, it can be shown (see Appendix) that, under certain conditions, both the field and the circuit approaches lead to identical differential equations of the second order in full derivatives. Thus, we could expect identical results using both approaches. We will test this statement by deriving the same transmission shielding theory equations using the field approach.

As you remember, our experience with matching transmission shielding theory with experimental data was far from satisfactory. So, why go through this exercise again? There are several reasons. First, we will check the validity of our theoretical tools and gain some experience in using them. As a matter of fact, we will derive more general formulas than are available in the literature. Our formulas will be able to account for different impedances of the dielectric inside and outside the shield, as it is the case with the cable shields. Second, having been through the process of the derivations from the most general positions (i.e., using field theory), we will be better equipped to clarify the reasons for the inaccuracy of transmission shielding theory and to find the limitations of the respective approaches. Finally, we will use the results of this theoretical analysis to lay the groundwork for a system-level approach to shielding. In the suggested later system shielding model, we will "find a place" for at least a part of the obtained result by identifying it with just one (of many!) shielding mechanisms - diffusion. Those not interested in the derivations, can skip directly to section 2.4.2 (however, they will miss on important insights!).

✦ Our shield is a metallic sheet of thickness, t, and infinite dimensions, with the dielectric media adjacent to both its surfaces. Select a Cartesian coordinate system with the axis z perpendicular to the shield, and the 0 coordinate point at the left surface (see Fig. 2.17). The shield is illuminated at z = 0 (in Fig. 2.17, from the left) by a plane wave with electric $E_x = E_1$ and magnetic $H_y = H_1$ field vectors, orthogonal and situated in the plane perpendicular to the direction z of the wave propagation. The *energy flux* associated with the propagating wave is determined by the Poynting vector, P_z. We will adopt the shielding effectiveness definition, similar to Eq. (2.1), as a ratio of the electric field intensities measured, respectively, at the surfaces of the shield opposite to the wave incidence and facing the wave (which does not mean that we endorse this definition). Thus, we will work to find these fields.

Assuming harmonic signals, we will use the Maxwell's equations as presented in Eq. (A.16). For the described problem, this system leads to differential Eqs. (A.49) or (A.25), with the actual variables substituted per Table A.1. Thus, from Eq. (A.49) we obtain:

$$\frac{d^2}{dz^2}\dot{E} = \gamma^2 \dot{E} \qquad (2.11)$$

The solution to (2.11) is (A.51):

$$\dot{E} = \zeta_1 e^{-\gamma z} + \zeta_2 e^{\gamma z} \qquad (2.12)$$

106 Electromagnetic Shielding Handbook

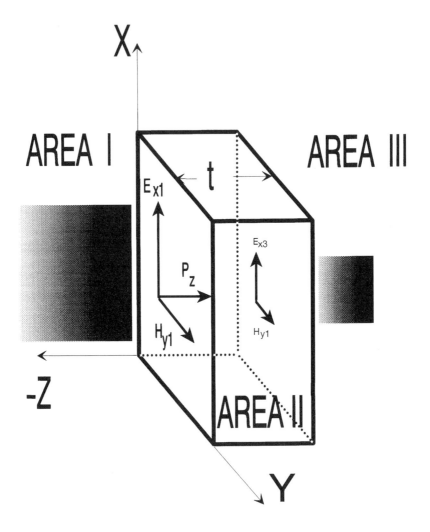

FIGURE 2.17 Electromagnetic wave incident on a plane metallic sheet

The metallic sheet separates the space in three areas:
- Area I
  $-\infty < z \geq 0$: dielectric area in front of the shield with regard to the incident wave (that is, on the left in Fig. 2.17). The parameters of the medium are:

$$\varepsilon_1; \mu_1; \sigma_1 = 0, \gamma_{d1} = j\omega\sqrt{\mu_1\varepsilon_1}, z_{d1} = \sqrt{\frac{\mu_1}{\varepsilon_1}}$$

- Area II
  $0 \leq z \geq t$: conducting area within the shield. The parameters of the medium are:

$$\varepsilon_2 = 0; \mu_2; \sigma_2; \gamma_c = \sqrt{j\omega\mu_c\sigma_c}; Z_m = \sqrt{j\omega\frac{\mu_c}{\sigma_c}}$$

- Area III

  $0 \leq z < \infty$: dielectric space area "behind" the shield (i.e., the protected space). The parameters of the medium are:

$$\varepsilon_3; \mu_3; \sigma_3 = 0; \gamma_{d3} = j\omega\sqrt{\mu_3\varepsilon_3}; z_{d3} = \sqrt{\frac{\mu_3}{\varepsilon_3}}$$

In the dielectric areas (I and III), $\sigma = 0$ and substitution into Eq. (2.12) yields:

$$\dot{E}_{1d} = \dot{E}_{1d}^{inc} + \dot{E}_{1d}^{rfl} = \zeta_{1d1}e^{-\gamma_{d1}z} + \zeta_{2d1}e^{\gamma_{d1}z} \qquad (2.13)$$

$$\dot{E}_{3d} = \dot{E}_{3d}^{inc} + \dot{E}_{3d}^{rfl} = \zeta_{1d3}e^{-\gamma_{d3}z} + \zeta_{2d3}e^{\gamma_{d3}z} \qquad (2.14)$$

In the shield body (II), $\sigma_c \gg j\omega\varepsilon$, and substitution into Eq. (2.12) yields:

$$\dot{E}_c = \dot{E}_c^{inc} + \dot{E}_c^{rfl} = \zeta_{1c}e^{-\gamma_c z} + \zeta_{2c}e^{\gamma_c z} \qquad (2.15)$$

Three other equations are provided by the expressions for magnetic fields. In general, these can be obtained by substituting Eqs. (2.13) through (2.15) into Eq. (A.47) per Table A.1, with the end result corresponding to Eq. (A.52):

$$\dot{H}_{1d} = \dot{H}_{1d}^{inc} + \dot{H}_{1d}^{rfl} = \frac{1}{Z_{d1}}[-\zeta_{1d1}e^{-\gamma_{d1}z} + \zeta_{2d1}e^{\gamma_{d1}z}] \qquad (2.16)$$

$$\dot{H}_{3d} = \dot{H}_{3d}^{inc} + \dot{H}_{3d}^{rfl} = \frac{1}{Z_{d3}}[-\zeta_{1d3}e^{-\gamma_{d3}z} + \zeta_{2d3}e^{\gamma_{d3}z}] \qquad (2.17)$$

$$\dot{H}_c = \dot{H}_c^{inc} + \dot{H}_c^{rfl} = \frac{1}{Z_m}[-\zeta_{1c}e^{-\gamma_c z} + \zeta_{2c}e^{\gamma_c z}] \qquad (2.18)$$

Equations (2.13) through (2.18) show that, in principle, in each of the areas I, II, and III, direct and reflected waves can propagate. These correspond to the "−" and "+" signs in the power items of e and are designated respectively with the E^{inc}, H^{inc} for the direct (incident) waves, and E^{rfl}, H^{rfl} for the reflected waves. However, this does not mean that all these waves always exist in real life. The mathematical solutions just indicate that such waves are *potentially possible*. For example, in area III, there is no reflected wave. This is because the area is considered infinite, and there is no reflective boundary at the "far end" (that is, at $z \to \infty$); thus, it should be assumed $\zeta_{2d3} = 0$.

In general, Eqs. (2.13) through (2.18) can be used to determine the six integration constants: ζ_{1d1}, ζ_{2d1}, ζ_{1d3}, ζ_{2d3}, ζ_{1c}, and ζ_{2c}. The usual technique to determine the integration constants is to utilize boundary conditions (A.17) through (A.20) as well as certain "common sense" considerations, e.g., the "symmetry" of the problem, impossibility of infinitely strong fields, and so on. Because, in our problem, all field intensities are tangential to the shield surface, only conditions (A.17) and (A.18) suffice. The boundaries are located at $z = 0$ and $z = t$. Thus, at the wave entrance into the shield,

$$\dot{E}_{1d}(z=0) = \dot{E}_c(z=0) \tag{2.19}$$

$$\dot{H}_{1d}(z=0) = \dot{H}_c(z=0) \tag{2.20}$$

and at the wave exit from the shield,

$$\dot{E}_{3d}(z=t) = \dot{E}_c(z=t) \tag{2.21}$$

$$\dot{H}_{3d}(z=0) = \dot{H}_c(z=t) \tag{2.22}$$

Using "common sense," we can reduce the number of variables without resorting to the formal mathematical transformations. Keep in mind that the fields incident at the shield surface at $z = 0$ are the given source fields: E_{x1} and H_{y1}. This, along with the fact that there are no reflections in the area III, results in the relations $\zeta_{1d1} = E_{x1}$, $\zeta_{3d1} = E_{x3}$, and $\zeta_{2d3} = 0$. As a result, we obtain four equations linking the four integration constants:

- at $z = 0$,

$$\dot{E}_{x1} + \zeta_{2d1} = \zeta_{1c} + \zeta_{2c} \tag{2.23}$$

$$-\dot{E}_{x1} + \zeta_{2d1} = \frac{Z_{d1}}{Z_m}[-\zeta_{1c} + \zeta_{2c}] \tag{2.24}$$

- at $z = t$

$$\dot{E}_{x3} = \zeta_{1c}e^{-\gamma_c t} + \zeta_{2c}e^{\gamma_c t} \tag{2.25}$$

$$-\dot{E}_{x3}\frac{Z_m}{Z_{d3}} = -\zeta_{1c}e^{-\gamma_c t} + \zeta_{2c}e^{\gamma_c t} \tag{2.26}$$

We can now use Eqs. (2.23) through (2.26) to find the shielding effectiveness coefficient in the form:

$$S_E = \frac{E_{3x}}{E_{1x}}, \text{ or } A_E = 20\log_{10}\left|\frac{1}{S_E}\right| \tag{2.27}$$

Denoting

$$\xi_{1c} = \frac{Z_{d1}}{Z_c} \text{ and } \xi_{3c} = \frac{Z_{d2}}{Z_c}$$

we obtain the following expressions:
- from Eq. (2.23) minus (2.24)

$$2\dot{E}_{x1} = \zeta_{1c}(1 + \xi_{1c}) + \zeta_{2c}(1 - \xi_{1c}) \tag{2.28}$$

- from Eq. (2.26)

$$-\frac{1}{\xi_{3c}}\dot{E}_{x3} = -\zeta_{1c}e^{-\gamma_c t} + \zeta_{2c}e^{\gamma_c t} \tag{2.29}$$

- from Eq. (2.25) minus (2.29)

$$E_{x3}\left(1 + \frac{1}{\xi_{3c}}\right) = 2\zeta_{1c}e^{\gamma_c t}; \quad \zeta_{1c} = E_{x3}\frac{1 + \frac{1}{\xi_{3c}}}{2}e^{\gamma_c t} \tag{2.30}$$

- from (2.25) plus (2.29)

$$E_{x3}\left(1 - \frac{1}{\xi_{3c}}\right) = 2\zeta_{2c}e^{-\gamma_c t}; \quad \zeta_{2c} = E_{x3}\frac{1 - \frac{1}{\xi_{3c}}}{2}e^{-\gamma_c t} \tag{2.31}$$

Substituting Eqs. (2.30) and (2.31) into Eq. (2.28), after simple (although somewhat cumbersome) algebraic transformation, we obtain:

$$S_E = \frac{2}{\left(1 + \frac{\xi_{1c}}{\xi_{3c}}\right)\cosh\gamma_c t + \left(\xi_{1c} + \frac{1}{\xi_{3c}}\right)\sinh\gamma_c t} \tag{2.32}$$

Expression (2.32) permits theoretical determination of the shielding effectiveness for different types of dielectric at both sides of the shield. Assuming $\xi_{1c} = \xi_{3c} = \xi$ (that is, identical dielectrics at both sides of the shield), from Eq. (2.32) we obtain a formula identical to that derived by Kaden [P2] and by Grodnev [P3]:

$$S = \frac{1}{\cosh\gamma_c t + \frac{1}{2}\left(\xi + \frac{1}{\xi}\right)\sinh\gamma_c t} \tag{2.33}$$

Further approximations can be obtained by assuming that $\xi \gg 1$ (which is almost always the case) and $RE(\gamma_c t) = t/\delta \gg 1$. This takes place at high enough frequencies, when the skin effect significantly reduces the effective thickness of the conductor. Given

$$\left|\sinh \gamma_c t\right| \approx \frac{e^{t/\delta}}{2}$$

then

$$\left|S_E\right| \approx \frac{4}{\xi} \frac{1}{e^{t/\delta}}, \text{ or } A_E \approx 20\log_{10}\left|\frac{Z_d}{4Z_m}\right| + 8.67 \text{ dB} \qquad (2.34)$$

which is identical to Eqs. (2.3) and (2.6), suggested by Schelkunoff.

For simplicity, we conducted this derivation for a plane shield. We also could have done this for a cylindrical shield, using a cylindrical coordinate system. Such derivations lead to more complex differential equations, the solutions to which are expressed in terms of Bessel, Hankel, and Neumann functions. However, for a shield with a thickness much smaller than its radius, and for high enough frequencies, these solutions reduce to equations identical to what we have obtained in a much simpler way. Those interested in exact solutions for cylindrical shields are referred to the literature sources (e.g., see Refs. [P1–P3]). These sources also provide expressions for the reflected fields. If desired, the derived expressions also permit the determination of the reflected fields using the coefficients ζ_{2d} and ζ_{2c}. ◆

2.4.2 Transmission Line vs. Plane Wave Shielding Models

The success of matching the transmission line theory and field theory to yield identical formulas should not obscure the fact that these theoretical results differ from experimental data. We have already indicated this "sad" fact before. But unlike in the previous discussions, now we possess the necessary background to clarify the causes of this difference.

Let us start by comparing the transmission line theory with the plane wave theory of shielding. Why and under what conditions are these two approaches equivalent in the first place? The answer is in the field configuration associated with these two ways of wave propagation: both are transverse electromagnetic (TEM) fields. We saw that in the plane wave the vectors of electric and magnetic fields are orthogonal (e.g., E_x and H_y) and result in the wave energy propagating in the direction perpendicular to the plane defined by the field vectors (and signified by the appropriate Poynting vector components). A similar situation exists in a transmission line. While very different from a plane wave by nature, generation means, and propagation mechanisms, under certain conditions the electromagnetic field in a transmission line has a similar transverse electromagnetic structure.

For example, consider a coaxial cable (Fig. 2.18) with homogeneous and electrically linear center and outer conductors, between which a voltage is applied. The center and outer conductors are separated by a dielectric material. The current $+\dot{I}$ in the center conductor returns $(-\dot{I})$ in the outer conductor.

Understanding Electromagnetic Shielding 111

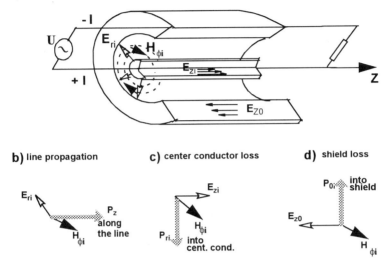

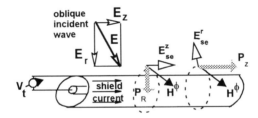

FIGURE 2.18 Electromagnetic fields and energy fluxes in coaxial cable

Select a cylindrical coordinate system with the coordinate axis z coinciding with the cable center conductor axis. If the energy losses in the cable conductors and dielectric can be neglected, only radial electric E_r and tangential magnetic H_ϕ field vectors exist, and both are located in the perpendicular to the axis cross-section plane of the cable. In Fig. 2.18a, the electric field lines start at the surface of the center conductor and end at the internal surface of the outer conductor. If the voltage between the line conductors is U, then from Eq. (A.44) it follows that

$$E_r = \frac{dU}{dr} \qquad (2.35)$$

where r is the outer conductor internal radius. The magnetic field lines are concentric circles around the center conductor. Applying Ampere's law [Eq. (A.21, II)] yields the magnetic field intensity between the center and outer conductors:

$$H_\phi = \frac{I}{2\pi r_i} \tag{2.36}$$

where r_i is the magnetic line radius.

In the cable, the Poynting vector associated with the transverse field is directed along the z axis (Fig. 2.18b), thus indicating that the wave energy propagates along the cable. The analogy in the electromagnetic field TEM distribution leads to similar formulas for the transmission line and plane wave propagation, as shown in the Table A.1. However, this analogy is only formal and is based on the assumption of a lossless propagation media. Indeed, the plane wave can propagate wholly in a dielectric, including free space. In the transmission line, the wave also propagates mainly in the dielectric (e.g., cable insulation) between the line conductors, but along these conductors, which act as wave directors. And in the process, part of the signal energy enters the line conductors. When the line conductors possess a finite conductivity and form a closed circuit, the current in them produces electric field intensity in the same direction, which results in energy losses in these conductors: the electromagnetic energy of the field transforms into heat.

In a coaxial cable, the longitudinal electric field intensities on the external surface of the center conductor E_{zci} and the internal surface of the outer conductor E_{zoi} have opposite directions. However, the magnetic field $H_{\phi i}$ inside the cable is determined only by the current in the center conductor per Eq. (2.36). (Remember that we are discussing a homogeneous outer conductor—for nonhomogeneous conductors this may be not true, as discussed in the next chapter.) As we can easily see, the Poynting vector components P_{rci} and P_{roi} are directed inside the respective conductors and characterize the losses in these conductors (see Fig. 2.18c and d). The vector P_{roi} also indicates at the energy flux that enters the outer conductor and propagates through it, reaching the space outside the cable. In fact, if the coaxial cable outer conductor is viewed also as a shield, the losses of this energy flux constitute the losses in a shield.

In developing the circuit analog of the plane wave field theory of shielding, we used only longitudinal (i.e., along the cable) energy flux generated by the transverse field components and did not account for the losses in the conductors, R, nor in the dielectric, G. However, it actually is just these losses that account for at least the absorption part of the shielding effectiveness! It is clear now that the circuit/plane wave analogy is formal and holds only as long as the losses in the circuit conductors can be disregarded. Obviously, such a deficiency could have been fixed by introducing the conductor and dielectric losses into the wave propagation models. But these losses (as well as the inductance, L, in magnetic conductive materials) are frequency dependent. Which brings us to skin effect, which brings us to the Maxwell's equations to be used to determine the skin effect parameters, which brings us to the concepts of fields, instead of the concepts of currents and voltages. We have come back to where we started!

In summary, whereas the circuit laws can always be expressed in terms of electromagnetic field theory (at least in principle), the reverse is not always true. This is one of the main reasons why many "nice" theories of electromagnetic shielding based on circuit theory have never "worked." This does not mean, however, that we should abandon circuit theory altogether. In many cases, we still can enjoy its relative simplicity and vivid physical interpretations—as long as we check the problem formulations and the obtained solutions versus the objectives and constraints of particular problems being solved.

2.4.3 Plane Shield / Plane Wave Model Constraints

So far, we compared the circuit theory to the plane wave theory, as applied to a plane infinite shield. But the plane wave theory is in itself only a specific case of general electromagnetic theory, described by the Maxwell's equations. Together with the idealizations of the shield design (that is, homogeneous infinite plane), the plane wave conditions impose severe limitations on the applicability of the respective theories to the practice of electromagnetic shielding. Let's summarize the most important field configuration and shield design constraints that we used so far in theoretical analysis:

1. The shield is a "flat" plane and must have infinite dimensions.
2. The shield is considered uniform and homogeneous; no holes, seems, gaps, or other nonuniformities are present.
3. A transverse electromagnetic field structure is assumed.
4. The incident field propagates in a direction perpendicular to the shield plane.
5. There is no "back" effect of the shield on the field source.

As we can see, these limitations (and others, not mentioned here) can be separated in two groups: those related to the field configuration, and those related to the shield design. As a rule, the limitations related to both these groups are too strict to meet the practical needs of shielding. The plane wave field configuration is typical for the so-called far field zone, at a distance from the source of radiation much larger than the wavelength. Although this situation does take place in some shielding problems, there are also many problems dealing with configurations that differ from the plane wave field forms. Two important and typical problems of this kind are (1) near field emissions/susceptibility and (2) crosstalk between the lines in the same or neighboring cables (in principle, cable crosstalk as well can be considered a near-field phenomenon). In the near field, the distribution of the EMI energy between the electric and magnetic field components depends on the physical nature and mutual arrangement of the EMI source, EMI receiver, and the shield. This results in the specific mechanisms of electromagnetic coupling between the shield and the field source on the one hand, and between the shield and the EMI receiver on the other. It also results in the shield reaction back at the field source (whether this reaction actually affects the source is a different matter).

✦ As an illustration of complications arising from the near-field effects, we will consider a study performed on two coaxial loops axially perpendicular to and separated by an infinite plane (Fig. 2.19). One loop is the EMI source, and the other is the receiver. The loop radii and the distance between the loops and the plane are much smaller than the wavelength, and the applicable frequency range was limited to 50 kHz. The shielding effectiveness was defined as the flux density or voltage *insertion loss*.

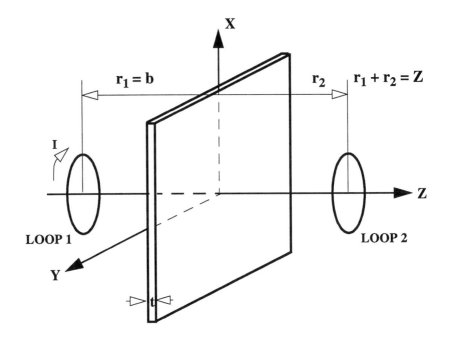

FIGURE 2.19 Coaxial loop antennas separated by an infinite plate (loop radius = 3.5 cm)

This is a typical near-field problem, although simplified by the assumption of the plane being an infinite shield. The solutions to this problem were reviewed in Ref. [2.7] and revisited by the same author in Ref. [2.8]. Without going into derivations, we will briefly summarize the main findings of this work. Three possible approaches were considered:

1. Theoretical determination of the induced eddy currents in the shield. This approach was first suggested in Ref. [2.9]. Its main restriction is an assumption that the induced potential on both sides of the shield would be identical. However, such a condition takes place only when the skin effect in the shield can be disregarded. This essentially limits this theory to very *low frequencies* (for "usual" cable shields, below ~10 kHz) and thin nonmagnetic shields. These considerations were confirmed by experimental data.
2. The second approach was based on the solution of the vector wave equation for the vector potential $\vec{A}$:

$$\nabla^2 \vec{A} + \gamma_d^2 \vec{A} = 0 \tag{2.37}$$

where

$$\vec{B} = \nabla \times \vec{A} \tag{2.38}$$

The solution to the problem looks as follows:

$$S = 20\log_{10}\frac{1}{4}\mu_r \left| \frac{\int_0^\infty \frac{\lambda^2}{\tau_o} J_1(\lambda a) e^{-\tau_o z} d\lambda}{\int_0^\infty \frac{C\lambda^2 \tau}{\tau_o^2} J_1(\lambda a) e^{-\tau_o z - t(\tau - \tau_o)} d\lambda} \right| \qquad (2.39)$$

By far, this is a much better solution than the previous one. The calculations according to this expression were shown to agree quite well with the experimental data (within the stated low frequency range and dimension limitations). But look at the mathematical "monster" that was created! So far, the researchers were able to obtain only numerical solutions of Eq. (2.39). While this is useful when evaluating manufactured shields, it's sure not too convenient for general analysis and for design purposes.

3. The third considered approach was the Schelkunoff's transmission theory. It was shown that a relatively good agreement can be achieved with the experimental data *only* under certain dimension limitations in theoretical analysis (in particular, when for calculation purposes the shield is placed at 1/3 of the distance from the source to the point of observation). Using numerical analysis, it was also shown [2.10] that under these constraints Eq. (2.39) becomes equivalent to Schelkunoff's formulas.

Thus we see that, even at relatively low frequencies (i.e., below 50 to 100 kHz), for even an infinite size plane shield, and even with all these quite severe constraints on the source-shield-receiver spacing, only the solution of electromagnetic field equations yields an accurate agreement with the experimental data. But we must pay the price of serious mathematical complications.◆

But the field configuration complexities constitute only one part of the problem. The second part of the problem is related to the shield homogeneity conditions, which are met only in a limited number of cable designs; practical cable shields are not usually solid homogeneous shields. Why is this an issue? Because the formulation of the EMI problem and consequent application of the boundary conditions are very much dependent on these factors. Just think of it. In a plane homogeneous shield, the field vectors in the shield had the same direction as the incident field. Applying the boundary conditions of Eqs. (A.17) through (A.20) at the interface between different media, we could determine the current in the shield, which had the same direction as the tangential component of the electric field vector on the shield surface. Suppose the shield homogeneity is violated. Then, the shield may be considered as having an anisotropic conductivity and anisotropic magnetic permeability (the latter being for high-μ shields). Now, the current and magnetic fields in the shield are determined by conductivity and magnetic permeability tensors. Instead of simple relationships of Eq. (A.3a) and (A.4a), we have to deal with Eqs. (A.3b) and (A.4b).

The third part of the shielding problem stems from the fact that within an electronic system the shield is simultaneously a part of the system circuitry, connecting different elements within the system. Thus a cable shield has a "twofold relationship" with the system EMC performance. On one hand, the EMI signals induced in the shielded line are applied to the system elements interfacing the line. In this respect, the better is the shield, the less EMI is induced in the circuit, and the more immune the system is (don't forget, we are still discussing the susceptibility case).

On the other hand, the shield itself is connected to different elements of the circuit; e.g., hardware, logic, or analog grounds, power supply common points, and overstress protection devices. These elements have "their own" circuital parameters (R, L, G, C), serve as return paths for system currents and receptors of EMI, and can be at different potentials with regard to the reference grounds and each other. When the cable shield is connected to these elements, it becomes a part of the system circuitry, providing a path for the system currents, including ground loops. In this case, as far as the system is concerned, the shield is just a regular conductor, with its own resistance, inductance, and so on. This function of a cable shield in a system does not depend on the cable shielding properties. But the reverse is not correct: the system does affect the shielding performance of a cable shield. Thus, the currents in the shield will depend not only on the source field but also on the characteristic and terminal impedance of the circuits in which those currents flow, including the losses in the direct and return paths of the currents, as well as on the mutual inductances and capacitances between the shield and other system elements.

2.5 A SYSTEM VIEW ON ELECTROMAGNETIC SHIELDING

It is quite obvious that the transmission theory of shielding cannot accomodate the identified "real life" constraints, without serious modifications (if at all!). But now, learning from the "past mistakes", we are ready to suggest more general, system EMC-based models, which will be flexible enough to cope with these factors, and on the other hand, will incorporate the best features of many existing theories, including the transmission theory of shielding.

2.5.1 The Roads We Take

Let us begin with summarizing the problems we face. The electromagnetic shielding issues can be grouped in two classes dealing with (1) the shielding problem formulation and (2) the shielding problem solution.

1. The shielding problem formulation complexities are related to three major factors:
   a. shield design (e.g., non-homogeneity)
   b. EMI environment (e.g., incident field nature and configuration in susceptibility cases, and nonuniform field distribution)
   c. electronic system specifics (e.g., system effect on shield performance)
   These complexities are not easy to overcome. In studying shielding, the usual trend is to oversimplify the conditions, which results in significant inconsistencies with actual performance. We have identified, evaluated, and illustrated some of the problems of this kind—e.g., see the analysis of the transmission theory of shielding.

2. The picture is just as confusing with regard to available solutions. Here is what we have been able to find out so far:
   a. The field theory can provide (at least in principle) accurate solutions to shielding problems. However, these solutions may lead to unmanageable mathematical difficulties, especially with regard to the whole system EMC performance.
   b. The circuit theory can provide "elegant" physical models leading to relatively simple solutions to shielding problems. But these solutions may be quite inaccurate in important practical applications.

Where do we go from here? Although the electromagnetic shielding problems look formidable and available solutions may not be quite adequate, we did make significant strides in the right direction. As a matter of fact, the problem identification is in itself a first step to a solution. Indeed, even a "superficial" look at the very "spelling" of these problems indicates at possible strategies for tackling them. In this respect, two considerations are especially important. First, it is clear that the problem formulations should somehow refer to the whole system EMC performance. Second, although both approaches to the solution (based on field theory and circuit theory) do possess their shortcomings, the deficiencies inherent to each of the approaches may be orthogonal, i.e., are not always interdependent. Thus, by properly combining the techniques based on field and circuit theories, the advantages of both of them can be realized with many of their individual problems bypassed or "neutralized." In a way, these techniques together may possess the "synergy" we need to address the solution problems. It is from these general system positions that we will attempt to define the shielding and its performance parameters.

We will revisit now the main physical "events" associated with shielding, making the emphasis on their system EMC-related nature. At this time, we will not try to specify the "details" so as not to "hide the forest behind the trees." But we will return to these details in the following chapters.

Consider a shielded cable illuminated by an electromagnetic field (Fig.2.18e). From the EMC point of view, this is a typical system susceptibility case. Obviously, the first task is to define the incident EMI field distribution. In general case, the ambient field has an arbitrary configuration. Depending on the specific problem and the available information, the distribution of the field incident on the shield may be determined in different ways:

1. The ambient field may be specified.
2. If the EMI-generating systems are known, their EMI signatures can be synthesized by treating them as active antenna arrays (see Chapter 1).
3. If the field is not given and the potential EMI sources are unknown, then the field distribution can be assumed, based on:
   - expected maximum or average field
   - built-in system immunity
   - maximum permissible field, e.g., per certain standards and regulations

118 Electromagnetic Shielding Handbook

Very often, the EMI field is given, specified, or assumed to be uniform, which simplifies both theoretical analysis and testing. If the EMI field is nonuniform, the general technique is to subdivide the space around the shield on small enough elements and consider the effect of each element on the shield separately.

The next task is to determine the coupling of the incident field to the shield. As a rule, the elementary fields are incident on the shield surface at some oblique angle. By decomposing the field vectors (e.g., vector E in Fig. 2.18 e) on longitudinal (that is, parallel to the shield surface — also called *tangential*) and *normal* components and applying the boundary conditions at the shield surface, the longitudinal $E^z{}_{se}$ and normal $E^r{}_{se}$ field components at the shield external surface can be determined using Maxwell's equations. The shield reactions to the tangential and normal components are different and can be described using boundary conditions.

We have already discussed the phenomena associated with the longitudinal components of electric and magnetic fields at the boundaries: they are continuous [see Eqs. (A.17) and (A.18)]. The behavior of the normal components follows from Eqs. (A.19) and (A.20). The boundary conditions of the normal magnetic field components are determined by Eq. (A.20), which results in

$$\frac{H_1}{H_2} = \frac{\mu_2}{\mu_1}$$

When the shield and the dielectric adjacent to it are nonmagnetic, the magnetic permeability of the shield and dielectric can be assumed to be $\mu_s = \mu_d = \mu_0$, and therefore at the shield surface, $H_{n1} = H_{n2}$, which means that there is no reflection of the magnetic field. However, if the shield is ferromagnetic (i.e., $\mu_s \gg 1$), then $H_2 \ll H_1$; that is, a large part of the magnetic field is reflected back.

The boundary conditions of the normal electric field components are determined by Eq. (A.19). Although, in general, the relationship can get pretty complicated, in practice the problem can be simplified. Thus, assuming a thin shield with thickness, t, much smaller than its radius, r, ($t \ll r \ll \lambda$), the current across the shield thickness can be neglected, which results in a zero normal component of electric field E_{ntot} at the shield surface. But the electric field at the shield surface is a sum of the incident and reflected fields, so $E_{nto} = E_{ni} + E_{nr} = 0$, and $E_{ni} = -E_{nr}$. This means that the normal component of the electric field is fully reflected from the shield.

Now, using Maxwell's equations, the expressions for the field components can be formulated and solved (at least, in principle), with substituting boundary conditions to find the integration constants. Thus we see that this part of the electromagnetic shielding problem requires the application of the field theory.

2.5.2 "Parallel Bypass" Model of Energy Transfer Through the Shield

The real-life shields are not isolated homogeneous infinite metal planes, "boxes", or tubes. The violations of the shield integrity come from gaps, seams, holes, spiral structures. The enclosures may have multiple points of contact with each other and "ground", incorporate cables penetrating the shield. In their turn, these "other" enclosures and cables can be shielded or unshielded, balanced or non-balanced, grounded or

Understanding Electromagnetic Shielding 119

non-grounded (whatever it means - see later section 7.6), etc. The presence of these factors may seriously affect the shield performance.

> ✦ To get "an idea" of the contributions of some of these factors, consider a series of test on a shielded frame for telecommunications office. A standard frame is populated with shelves containing "electronics". The system is configured to transmit, receive, and switch T1 signals (1.544 MHz clock frequency). It contains multiple DSPs, other digital and analog circuitry, round and ribbon I/O electronic cables and DC power supply cables. Each shelf can be shielded with a snap-on box enclosure. Originally, the radiated emissions from the system are sweep-tested in an anechoic room in the frequency range 30-220 MHz with the I/O and power cables freely dangling outside the frame, the shelving shields snapped on without paying attention to the contact resistance between the shield and the frame, and ribbon cables inside the shelving shielded boxes are sometimes passing adjacent to the vents in the shielding. The respective "EMI signature" is presented in Fig. 2.20a (top graph).
>
> Fig. 2.20b (next below) corresponds to the EMI emissions obtained when the I/O round cables are looped and placed inside the shielded enclosure. As shown, this results in a significant emission reduction, especially at higher frequencies. The emissions are further reduced (although not as much — Fig.2.20 c) when the power supply cables are also placed inside the shielded enclosure, in addition to the I/O cables. Now, let's eliminate the majority of remaining shielding system imperfections by sealing with copper foil the seams between the shelf shielding enclosures and the frame, and sealing the vents in the shielding enclosures, located adjacent to the ribbon I/O cables inside the enclosures. The dramatic emission reduction (Fig. 2.20d — bottom graph) confirms the effectiveness of the taken measures.
>
> Not only the emission amplitude, but also the radiating system field configuration is drastically affected by the introduction of additional elementary sources in the active antenna array model, generated by the shield imperfections. One example of the frequency-dependent complex radiating patterns resulting from imperfections in shielding enclosure is shown in Fig. 1.28. We will return to this subject chapter 4. ✦

Using the terms of "court proceedings", we will now address the question of "right or wrong" for transmission theory of shielding.

- The *"Prosecution case"*: The theory gives wrong results — forget it!
- The *"Defense case"*: This theory *was not* designed to account for *"physical deficiencies"* of the shield and shielding system. With no shielding "deficiencies and anomalies" and in a certain setup, the theory matches experimental data. Thus, "use it with discretion"! It may be used, for example, *to compare* various shield materials or shielding enclosures assumed homogeneous and "infinite".
- The *"Judge"*: Theory cannot be "blamed" for "wrong results". It was *the user's* failure to conceptualize the shielding in terms of the electronic system EMC performance objectives and to use the theory in the "right" context.
- The *"Verdict"*: Develop "technically correct" *shielding performance model* which will *incorporate* the transmission theory of shielding in the "right context".

To generalize this conclusion (the "Verdict"), let's assume, that "originally" a homogeneous "perfect" (no apertures, seams, etc) and infinite size shield presents the best possible electromagnetic barrier, with certain "transparency" which can be achieved with respective material of given thickness. For certain, quite simplified, system configurations, this effect can be calculated using the transmission theory of shielding. Then, the shield imperfections will result in "bypassing" this "maximum" performance, as illustrated in Fig. 2.21.

Also, another important consideration follows from this model: not only the shield *per se* imperfections, but also *other* elements of the sytem may have negative effects on the shield performance. Two examples shown, are the ground loop "bypass" path

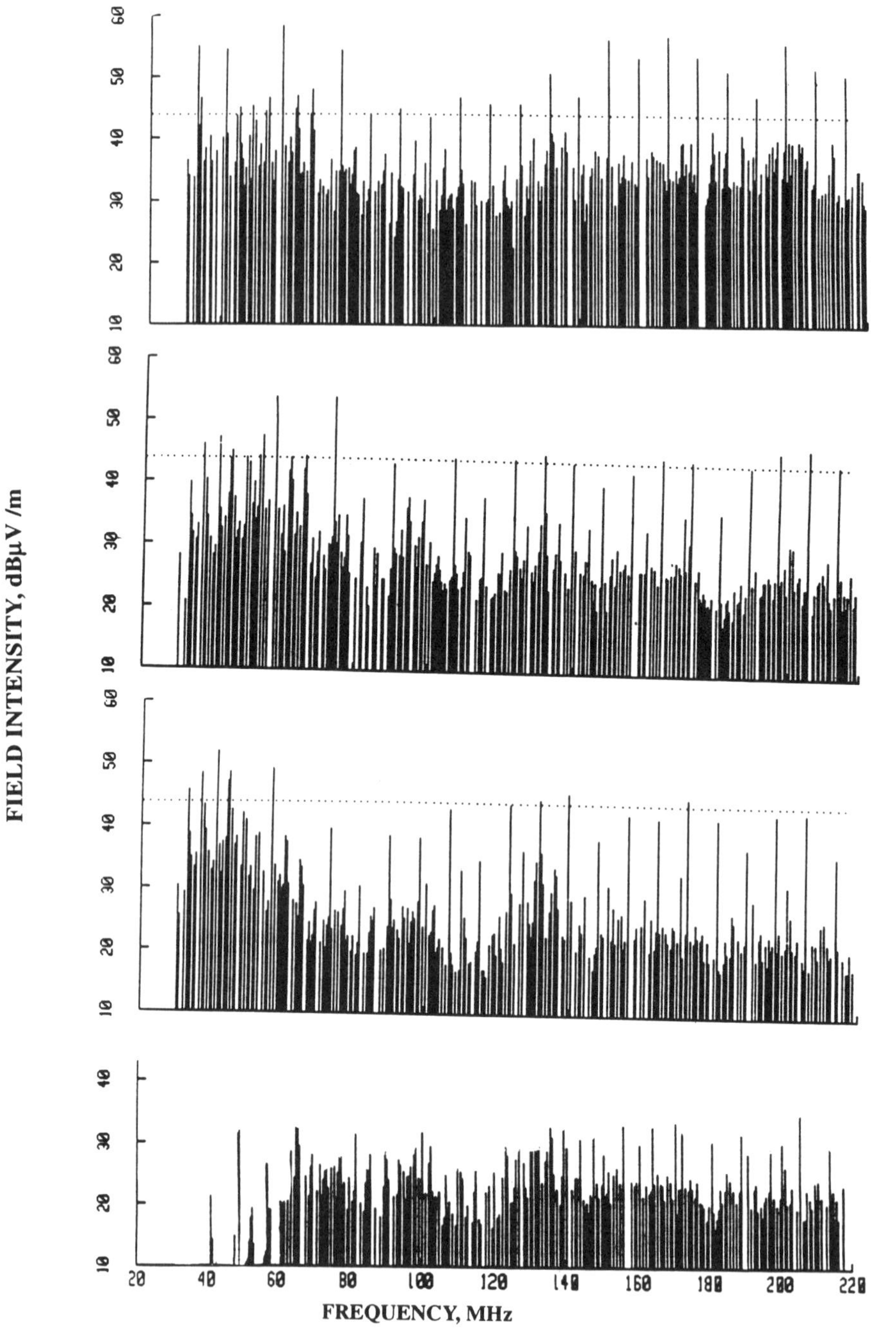

120 Electromagnetic Shielding Handbook

FIGURE 2.20 Shielding performance of a telecommunications shielded frame

and cable-"carried" interference, which can be brought in the protected by the shield area.

As indicated in Fig. 2.21, the negative effects of shield "imperfections" can be mitigated using gaskets, filtering, and "EMI hardened" system elements, e.g., shielded cables. As we saw, the transmission theory of shielding is helpless to explain, let alone to evaluate, these effects. In the following chapter III, the necessary means will be developed and used to evaluate the shield imperfections both qualitatively and quantitatively.

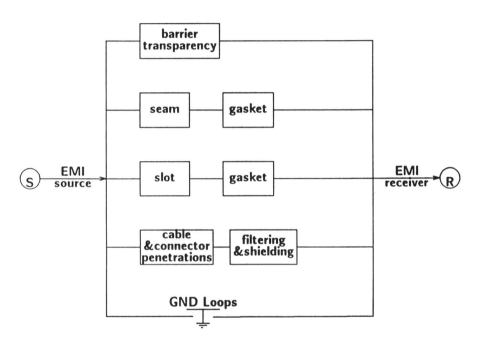

FIGURE 2.21 "Parallel bypass" model of energy transfer through a shield

2.5.3 "Series" Model of System Electromagnetic Shielding

It Takes Five to Shielding

Even if shield imperfection factors were somehow accounted for and respective corrections were made, this approach still may fall short of real-life shielding *system complexities*. As important as the "shield-proper" performance is, it is only one of the issues among the electromagnetic shielding problems. Still other issues are generated by the specifics of shield coupling to the incident EMI field and to the protected (i.e., shielded) circuits, to the effect of the rest of the system on transmission mode and mutual coupling between the system elements, to the effect of shield and circuit terminations, and to the grounding techniques. All these parameters should be controlled, or at least specified identically for all the compared shields. Just how to do this is not a trivial problem but may be a matter of specific system conditions, or a standard (we will return to this subject in chapters 3, 4 and 5).

> ✦ For example, Fig. 2.10 can serve as an illustration of the effect of the system grounding scheme, shield termination, and the return path topography on the performance of the shields with identical designs.
>
> Another example is presented in Fig. 2.22, where crosstalk between the ECL integrated circuit outputs was reduced in half, just by adding four grounding pins to the metallic cover of the chip carrier socket. ✦

Return now to Fig. 2.18. In a homogeneous cylindrical shield with the axis coinciding with the coordinate z, the field component E^z_{se} generates on the shield surface an interference current I_{se} with an external return path—outside the shield and associated with this current magnetic field H^ϕ_{se}. In general, Ohm's law as in Eq. (A.4 a) in differential form can be used to determine the current I_{se}. However, if the involved electric circuits are "identifiable" and possess a relatively simple geometry, the problem can be greatly simplified by the application of the network theorems of circuit theory to determine the field components and currents at the shield surface. For example, in the case of crosstalk between circuits, both direct and return current paths and voltages can be identified. As a rule, the respective phenomena present excellent opportunities to apply the principles of circuit theory. Even when the incident field is nonhomogeneous, the directions of the currents are determined by the combined characteristics of the incident field and shield design. Then, the elementary currents coupled into the shield from the ambient field should be integrated as indicated previously.

The electric field vector E_z^{se} at the shield surface is the source of the current in the circuits formed by the shield and external with regard to its conductors. But if the return path of this current is situated outside of the shield, there is a component of electric field E_r^{se} that is associated with the voltage between the shield acting as a circuit conductor and the conductor(s) forming the return path for the current in this circuit. At the shield external surface, this component is present along with the E_z^{se}. This current is also the source of the magnetic field H_ϕ, which can be determined from the Ampere's law as shown in Eq. (A.21, II). Calculating now the Pointing vector, we see that along with the energy flux across the cable entering the shield and producing absorption losses in it (which is determined by Poynting vector P_r), there is also an energy flux along the cable (which is determined by Poynting vector P_z):

$$P_r = \int_0^{2\pi} E^z_{se} H^{\phi *}_{se}; \quad P_z = \int_0^{2\pi} E^r_{se} H^{\phi *}_{se} \qquad (2.40)$$

Vector P_z describes the energy propagating along the shield external surface. The shield then acts as an antenna, which radiates energy back into the outer environment, or as a circuit conductor (coupled with other conductors) if crosstalk is considered. When the shield is a perfect conductor, then, of course, $E^z_{se} = 0$, the shield

a) IC connector with 4 grounding pins for cover grounding

Noise on unswitched output when other outputs are active, switching from Low to High state

b) Connector without grounding pins

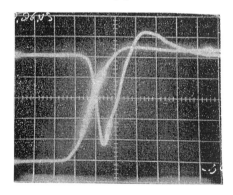

c) Connector with four grounding pins

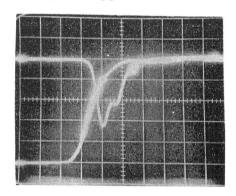

FIGURE 2.22 Effect of grounding on crosstalk in IC connector

will present a short for current I_{es}, and all the energy is reflected from the shield. If the original source and other conductors outside of the shield are close enough, the reflected energy couples into them, changing their parameters and causing them to reradiate.

The next phenomena are associated with the energy penetration through the shield. Vector P_r describes the energy flux directed inside the shield. Just how much energy actually penetrates through the shield, depends on the shield quality. If the shield conductivity is poor, then the $E^z{}_{se}$ is large and so is P_r: the bulk of the EMI energy diffuses through the shield. The shield quality also depends on the presence of apertures which facilitate the energy penetration through the shield and also provide for direct coupling of the ambient field to the shielded circuits. Another shield design specific has to do with the spiral shield application. The "spiralities" facilitate generation of tangential electric E^ϕ and longitudinal magnetic H^z field components which contribution to the energy transfer is proportional to the frequency (see next chapter). In general case, the quality of the shield can be described using the concept of a shield transfer function, which takes into account all relevant phenomena: diffusion, aperture penetration, spirality effects.

Again, evaluating the phenomena associated with the Poynting vector components P_z and P_r, we see that the first phenomenon is a typical subject of circuit theory, while the transfer function definition may require the application of both methods, e.g., field theory to determine diffusion processes, and circuit theory to determine spirality effects. We will discuss these processes in the next chapter.

Return now to the shielded cable exposed to the impact of ambient electromagnetic field (see Fig. 2.23). As a result of the interaction of the shield with the field, an interference current I_e is induced in the shield, with external return path—outside the shield. Due to the shield reaction, the original electromagnetic field in the vicinity of the shield distorts, and the incident energy flux splits: one part of it is reflected from the shield, and the second part attenuates while propagating through the shield inside the cable. These energy fluxes are associated with the absorption and reflection shielding coefficients. This is a "traditional" model of shielding. However, for electromagnetic shielding and from the point of view of the whole system, such a model is incomplete. Indeed, in the system, the shield carrying the induced current I_{se}, together with this current return path via other system elements, forms a closed circuit. Consider then secondary electric and magnetic fields associated with this circuit (we will call them secondary because they were generated by the currents induced in the shield by the primary incident field). In general, this circuit can have an arbitrary type and configuration, to which Maxwell's equations in integral form [Eq. (2.31)] apply. These equations can be used to determine the secondary fields, depending on the configuration of the circuit.

The energy penetrating through the shield couples into the shielded conductors by inducing an electromotive force applied to the shield's inner surface and generating current I_i in the shield with an internal return path—conductors inside the shield. This results in the interference voltage V_i between the shielded conductor and the shield. At the cable or connector terminal, this voltage is measured as voltage V_t (see Figs. 2.22 and 2.23).

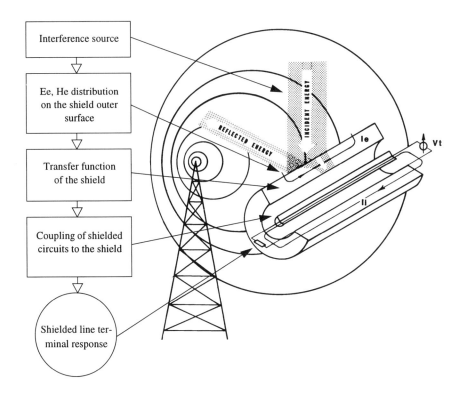

FIGURE 2.23 Physical process in electronic shielding

In a system, the cable outputs are connected to certain system elements. In this way, the induced EMI voltage V_t is applied to the system in accordance with the circuit theory laws as described in the preceding section. For example, the cable/system joint can be considered as a port of a lumped or distributed circuit. As a rule, such connection is made via specially designed connectors. Through this connection (connectorized or otherwise), the cable circuits and cable shield are galvanically connected (that is, coupled) to other system elements, components, and subassemblies, and to the ground. Thus, the working regime of the cable circuits and the shield, is determined by these system elements. Another mechanism affecting the system is the electromagnetic coupling of EMI which is reradiated from the affected cable (or cable shield, for that matter).

It follows that the electromagnetic shielding actually incorporates *five* different phenomena, which are lined up *reciprocally* (don't confuse with *reciprocity principle* - see next section) for the emissions and immunity, as illustrated in Fig. 2.24.

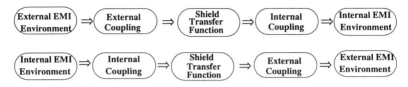

FIGURE 2.24 Electromagnetic Shielding Model

Blueprint for Shielding Study

Therefore, the study of shielding involves five steps associated with shielding phenomena and respective coupling mechanisms and requiring the use of field theory, or circuit theory, or both. Consider first the case of immunity.

Step 1. *Study and specification of the EMI field*
This includes understanding and modeling of the external EMI environment and particular electronic system and subassembly specifics. The main goal of this work is to obtain the EMI field distribution in space around the system containing the shielded cable.

Step 2. *External field coupling to the shield*
The coupling mechanisms depend on the configuration of the EMI field, and the shield electricals and geometry (e.g., length, layout), and the working mode, physical configuration, and the grounding system specifics of the electronic system of which the cable/connector of interest is a part, and the presence and configuration of extraneous physical objects, including ground planes. As a result, the analysis of external coupling should be done on a system level, as general as possible. If such general level is not achievable, then the lower system levels should be considered—cable/connector assembly, shield itself, shield material—with substituting the absent system elements by a set of respective constraints. Useful techniques for the analysis of external coupling can be based on the concepts of antenna or crosstalk theory.

Step 3. *Energy transfer through the shield*
This can be characterized by the shield transfer function. Although there can be different definitions of the shield transfer function, in general it is expedient to evaluate it on a fairly small shield element. This is because at larger electrical dimensions the shield behaves as a transmission line, and the measured shield characteristics become dependent on the cable/connector dimensions and circuit working mode; that is, on assembly-level and system-level specifics. If, additionally, the field and current distributions on the shield surface are specified and fixed, then the "elementary" shield transfer function may be assumed as dependent on the shield design only (e.g., transfer impedance). It then presents an excellent figure of merit to compare different shield designs, as opposed to shielding evaluation as a system parameter.

Step 4. *Internal coupling from the shield to the shielded conductors*
This mechanism depends on the shield design and shielded circuit specifics: configuration, grounding, line length. If the "elementary" shield transfer function is known, then the effects of all the shield elements can be integrated along the shielded line, and the internal coupling may be considered at the assembly level only.

Step 5. *EMI energy transfer from the affected shielded elements and the shield to other system elements, components, and subassemblies*
As indicated earlier, this can be done via galvanic or electromagnetic coupling.

Reciprocally, similar five ingredients/steps of electromagnetic shielding phenomena can be formulated for the system emissions case. Compared to susceptibility, the same steps will go in reverse order:

Step 1. EMI energy generation and transfer from the system elements, components, and subassemblies to the affected shielded line

Step 2. Internal coupling from the shielded conductors to the shield

Step 3. Energy transfer through the shield

Step 4. Identification of the shield as a circuit element within the system and as a system radiator

Step 5. Study and specification of the external EMI field generated by the shield

We will return to the electromagnetic shielding models

2.5.4 Reciprocity Principle in Electromagnetic Shielding

In the course of preceding discussions, we illustrated electromagnetic shielding phenomena by either emissions or immunity EMC problems, without addressing the applicability of the deduced conclusions to both. It would have been very tempting to extend the results achieved by analyzing the system emissions to system immunity, and vice versa. On first sight, this problem is easily solved on the basis of the reciprocity principle. Unfortunately, this is often not the case.

In the first part of Section A.4.2, we briefly discuss the reciprocity principle as applicable to the lumped electric networks. There we state that whereas, in general, the reciprocity principle is one of the most basic laws of nature, it is applicable only to passive and linear circuits. We also discussed reciprocity in Chapter 1 and found that the application of the reciprocity principle to EMC should be done in the context of the system EMI. Comparing the cases of system immunity and emissions, we showed that the dominant radiator in the system (emission case) was not necessarily the dominant receptor in the same system (immunity case). For instance, one of the reasons that the radiator is dominant may be a high level of signal within it which, on the other hand, can make this particular part of the system more immune to EMI. Thus unfortunately, we cannot apply the reciprocity principle to the most general system level EMC. It is not that the reciprocity principle fails—it is just that this particular problem does not meet the necessary requirements, which renders the reciprocity inapplicable in this case.

However, this does not mean that the reciprocity principle should be excluded when considering EMC at lower levels. A general principle in reciprocity requires us to consider only those parts of the system that meet the criteria of "passiveness" and linearity. In this sense, within an electronic system, the properties of radiators/sensors can be often assumed to be reciprocal, just as is the case with antennas. There is abundant literature available on the reciprocity. Each of the general sources on circuit theory, antennas, and electromagnetism addresses this subject. Therefore, we will limit this discussion only to several specific topics that are important in shielding applications.

Another important concept is related to the process reversibility in the system. Indeed, in the reciprocity theorem in the form Eq. (A.36), the coefficients $Z_{12} = Z_{21}$ and $Y_{12} = Y_{21}$ relate the current at one port to the voltage at the other port, and vice versa (but—very importantly—not the voltages to voltages or the currents to currents at different ports). What happens if, say, conductivity is one-directional? Suppose we inserted a diode in the circuit. It is easy to see that the reciprocity relations are no longer true. But then, we "deserve" it, because we violated the reversibility principle. In shielding practice, a somewhat similar phenomenon is often caused by the "rusty bolt" effect, which results in nonlinear and nonreversible conductivity. Another characteristic example from shielding practice, is a multilayer shield, with one of the layers having nonlinear magnetic permeability. Suppose such nonlinear magnetic layer is external, and the nonmagnetic linear layer is internal, facing the shielded conductor. If the EMI field comes "from inside" the cable (i.e., emissions case), it first reaches the nonmagnetic layer, and then after being reduced by this layer, it reaches the magnetic layer. If an identical EMI field comes from outside the cable (susceptibility case), it first reaches the magnetic layer while being "full strength" and then reaches the nonmagnetic layer. Then, obviously, if the magnetic layer is nonlinear, its shielding properties will be different for emissions and susceptibility. Of course, if both layers are linear, the effect will be identical in both cases.

Often, the reciprocity is associated with symmetry. But for the reciprocity principle to be true, symmetry is not necessary. Note, however, that if the circuit symmetry between the ports does exist, then not only is Eq. (A.36) is true, but also $Z_{11} = Z_{22}$ and $Y_{11} = Y_{22}$. The errors most frequently made by EMC engineers are extending this later relationship to nonsymmetrical circuits and trying to apply reciprocity to nonlinear circuits. Also, we saw that because electronic systems in general contain active elements and may be nonlinear, it is difficult to apply the reciprocity principle at system levels. It can be applied only to some parts of the system, i.e., those that do not contain active and nonlinear elements.

A shield can be viewed as a circuit with the inputs at one side of the shield and outputs at the other. As we know from the circuit theory, the ratio of the voltage at one port of such circuit to the current at the other port is Z_{21}. But, as we will see, this is the definition of the transfer impedance! If the shield is linear, the reciprocity principle requires that $Z_{21} = Z_{12}$. The same requirements for passive, linear, circuits hold. As long as the shield is a passive network (yes, there are also active shields—see the last chapter) and its material exhibits linear electric and magnetic properties, the reciprocity theorem holds, and the same shielding parameters are true for emission and susceptibility. However, remember that this is true only at the levels of the shield transfer function, but not at the system level shielding effectiveness.

2.6 SO, WHAT IS ELECTROMAGNETIC SHIELDING, AFTER ALL?

At long last, we are ready to formulate the system definition of shielding phenomena for EMC purposes and give it a most general intepretation. (It is strongly recommended that you agree with this statement, or else ... we will have to return to the very beginning of the chapter 2 and repeat the "whole" journey; anyway, may be it's not such a bad idea?). In this section, we will first consider one "innocent" fallacy about shielding, but then proceed with the "serious stuff".

2.6.1 System Definitions of Electromagnetic Shielding

What is a "good" (or a "bad", for that mater) shield? The analysis we have just conducted leads to a conclusion that the answer to this question depends on two factors:

1. The objectives of the system, of which the shielded element is a part
2. The generality (or specificity) of the selected system model

Keeping in mind these factors and based on the listed ingredients/steps of system shielding model, we can relate the shielding quality to the electronic system EMC performance. As a rule, there is not and cannot be a single criterion of the shield performance, applicable to all possible system applications.

This situation results in many levels of electromagnetic shielding definitions. In the most specific case, we define the shielding quality by the ratio of EMI signal obtained at certain points of the system when using shielded and nonshielded system elements. As we discussed in Chapter 1, by signal evaluation we will understand electric field, magnetic field, voltage, current, power. The point in the system where the signal is to be evaluated depends on the specific requirements to the system EMC. Such a point can be situated anywhere within the system (in the susceptibility case) or anywhere outside the system (in the emissions case). The choice of the signal nature and the location where the signal is evaluated are determined by the problem specifics, i.e., whatever is relevant in the particular application. It follows that from the system EMC performance point of view, the shielding effectiveness in susceptibility and emission cases may be different. Using the "five step" series model of the system shielding phenomena, the shielding effectiveness can be modeled by structurally simple mathematical expressions. Thus, in the immunity case, the corresponding generalized expression of shielding effectiveness can be written as follows:

$$S_i = \frac{N_{si}}{N_{ni}} = F_{ei} \times C_{ei} \times T_{si} \times C_{ii} \times L_{ii} \qquad (2.41)$$

where F_{ei} is the ambient field distribution function, C_{ei} and C_{ii} are the field-to-shield and shield-to-shielded circuit coupling functions, respectively, and T_{si} and L_{ii}

are the shield transfer function and shielded line-to-system coupling function. A similar expression can be written for emissions case:

$$S_e = \frac{N_{se}}{N_{ne}} = L_{ie} \times C_{ie} \times T_{se} \times C_{ee} \times F_{ee} \qquad (2.42)$$

Here, the sign $\times$ should be understood in a rather "symbolic" way. Variables N_s and N_n are the EMI signals associated with shielded and nonshielded system elements under consideration, respectively. Since the "ultimate" objective of shielding is to protect the electronic system from interference, the shielding effectiveness definition, S, is thus system-specific with regard to N.

2.6.2 Shielding Model Generality Vs Relevancy

Expressions (2.41) and (2.42) can relate the shielding performance of any system element (subsystem) to the whole system EMC performance. For example in case of "immunity-oriented shielding effectiveness," it compares the noise power, P, voltage, V, or current, I, at the electronic system output for two cases: when the evaluated element (e.g., product enclosure, connector, or cable-connector assembly) is used with and without the shield:

$$S_p = \frac{P_s}{P_n}; \text{ or } S_v = \frac{V_s^2 Z_n}{V_n^2 Z_s} \text{ and } S_i = \frac{I_s^2 Z_s}{I_n^2 Z_n} \qquad (2.43)$$

where Z_s and Z_n are the impedances at the respective system points. Also, logarithmic measures can be used, expressed in dB.

In these definitions, the subscripts s and n correspond to the shielded and nonshielded system elements, respectively. Again, it is worth repeating that the whole system output does not necessarily coincide with that of the evaluated element. Though such a definition may seem too far-fetched, it is logical, because without stating the shielding objectives and EMI environment specifics, the shield performance figure of merit may be *ambiguous* or make *no sense* at all. Indeed, the shield with "good" performance in one system may yield inordinate system noise in another system, and vice versa. In this way, the whole concept of a shield "goodness" becomes irrelevant, unless it is related to the "whole" system performance at an appropriate level (e.g., remember the "inverse shielding" examples in Section 2.2?)

Shielding evaluation or test per (2.41-2.43) can be easily visualized. One realization may consist of a "regular" EMC test per applicable regulation if the shield is used to reduce system emissions or susceptibility, or a system noise test if the shield is designed to prevent intra-system crosstalk. For instance, by comparing system emission field intensity with shielded E_s and nonshielded E_n, V/m cable (see Fig. 1.13), the cable under test shielded effectiveness can be determined as $S = 20\log|E_s/E_n|$ dB (see also section 6.1.3). Such test can be performed at the open area test site (OATS), or in an anechoic room. Another alternative is to directly measure the current on the cable (e.g., using absorbing clamp), while the system is functional (see chapter 6)

However, it is difficult to imagine calculating such a shielding effectiveness parameter. Also, the same system elements may be (and as a rule are) used with multiple systems. Thus, the shielding designers often do not know the systems in which their products will be implemented. Therefore, while being unquestionably "correct," the system-specific definition and the "absolute" test may also be impractical, because different electronic systems can yield a different shielding effectiveness figure of merit for the same system element.

It is still possible to provide a comparison of different shield designs on a more general basis (as opposed to the system-specific absolute test) by evaluating only those parts of the system that are important for given applications and then conducting an evaluation or test corresponding to the required system EMI evaluation and test. The respective evaluation topologies can be modeled by "cutting out" and considering the desired parts of the system (containing the shielded element in question) and substituting the "severed" ties with the rest of the system by some kind of boundary constraints. Referring to general expressions (2.41) - (2.43), this corresponds to assuming certain items from the formulas to be equal to unity or constant coefficients, instead of variables. Of course, the so obtained figures of merit are applicable only for the assumed conditions.

The problem is especially simplified if instead of evaluating the noise at the system output, the shielded element output is considered. A logical step in this direction is to limit the test just to the shielded element or containig this element assembly (subsystem), which in this case must be provided with terminations simulating some "generic" circuit working mode and grounding conditions.

Obviously, shielded element-level evaluation, or even containing this element assembly (subsystem) - level evaluation is simpler than the whole system-specific evaluation, and provides a more *general* insight into the shield performance. This *generality* is understood in a sense, that the obtained results can be applied to a wider class of systems using the element in question, than the results obtained in a more system-specific context. However, in this case, it is hardly possible to simulate the *actual* field distribution at the shield external surface and the effect on the system performance of the penetrated through the shield EMI signal. This results in only a limited degree of general shield performance evaluation *relevancy* to specific applications. Any departure from the absolute evaluation may distort the actual external coupling mechanisms and shield working mode. On the other hand, the less system-specific the evaluation, the wider the applicability of the obtained results. In general terms, the closer the evaluation routine is to the absolute one, the higher the shield evaluation relevance will be (with regard to its performance in particular system and given environment), but the more system-specific and less general will be the results (see Fig. 2.25).

Of course, it is just about time to ask an "insidious" question: so, what should we do in any particular case? You may not like the answer: learn the shielding theory, study your specific system, do some probing (theoretical, numerical, and / or experimental), apply some common sense based on available experience (yours and otherwise) and then "enjoy" the consequences ("good" or "bad"). In short - there is no easy, cook-book like answer!

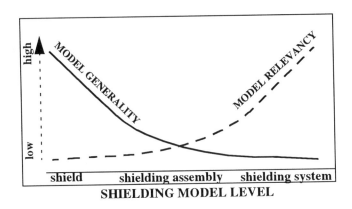

FIGURE 2.25 Shield evaluation model generality and relevancy vs system levels

2.6.3 What's Next?

As we saw, the system shielding effectiveness presents quite a formidable problem. To facilitate its understanding and practical applications, our approach will be to separate the task into a number of "smaller" and easier problems according to the formulated shielding process phases. Then we will address each "subtask" in sufficient detail in the respective chapters of the book: the shield transfer function, electromagnetic coupling, shielding performance evaluation and measurements, shield and shielding design and optimization.

3

Transfer Parameters of Electromagnetic Shields and Enclosures

3.1 FIGURE OF MERIT OF ELECTROMAGNETIC SHIELD PERFORMANCE

3.1.1 Transfer Impedance - Shield Transfer Parameter of Choice

In the preceding chapter, we have developed a system model of electromagnetic shielding and introduced a fundamental relationship between this model generality and relevancy. Central to this model (both conceptually and graphically—see Eqs. (2.41) and (2.42)—is the shield transfer function, which is intimately related to the shield design, largely without regard to particular electronic system applications. In a way, the shield transfer function leads to the *shield*-oriented characteristics, as opposed to the *system shielding* characteristics, which are derived from system-specific models. Inspite of the imposed by the model high generality limitations (actually, just because of them!), the analysis provided by the transfer function may be a highly *desirable* feature in certain applications:

- The shield transfer function can be extremely useful in comparing and optimizing shield designs on the most general basis.
- If little or no information about the specific shield applications is known, it may be *the only* way to evaluate and compare different shields.

134 Electromagnetic Shielding Handbook

- If the application data is available, the transfer function may serve as a cornerstone by adding the necessary details to develop system-specific shielding effectiveness parameters with the required relevancy to the task.

To be of practical use, any characteristic must be capable of evaluation and measurement, even if only qualitatively, and the shield transfer function is no exception. But what does it mean, "capable of evaluation and measurement"? We suggest the following interpretation of this "thesis":

2. The evaluation and measurement *parameters* must constitute a *true representation* of the shield performance
3. The evaluation and measurement *results* should be *repeatable*
4. Both the evaluation and measurement parameters and results need to *match* the specific sytem objectives and design, as well as to the technical, economical, and regulatory environment.

Meeting these requirements looks like a pretty "formidable" task, especially with regard to the shielding system as a whole. A way to facilitate the task is to consider, one by one, the shielding effectiveness system model elements, by "stripping" it off all its elements, *except* for the *one* under consideration. In this chapter, we will concetrate on the *transfer function*. We will suggest and then investigate the general concepts of this parameter, as well as its specific applications to the variety of shield constructions: homogeneous solid wall shields, shields with apertures, seams, gaps, mesh shields, spirals, and braids.

Fortunately, in developing quantitative attributes and parameters of the shield transfer function, we don't have to start from "scratch". Several characteristics which would belong to this category have been known and widely used for many years. By far, one of the most familiar and often used, is the *transfer impedance*, first introduced by Schelkunoff [P.1] for cylindrical (cable) shields. The concept of a *cylindrical shield* transfer impedance, Z_t, is illustrated in Fig. 3.1.

Consider a small element of a shielded cable illuminated by an ambient electromagnetic field. As a result of coupling of this field to the shield, an EMI current, I_e, is induced in the shield with a return path outside the shield. At this time, we don't have to be specific about this return current path; suffice it to note that for the current to flow, the circuit *must* be closed, which implies *some* sort of return path. This current in the shield is the source of a secondary field which interacts with the original field. As a result of this interaction, the incident at the shield energy flux is partly reflected from the shield and partly penetrates through the shield inside the cable, where it generates an interference voltage, dV_t, at the terminals of the shielded circuit.

So far, this is a "repeat" of Fig. 2.23 and corresponding explanations. But consider this: While propagating through the shield, the EMI energy is attenuated. The greater the attenuation of the EMI energy passing through the shield, the smaller the dV_t generated by the same field and corresponding to the same magnitude of induced current, I_e, and the better the shield. Therefore, if we "forget" about that part of the shielding process, which is associated with the ambient field configuration and distribution, its coupling to the shield and reflection, and mutual influence of different-

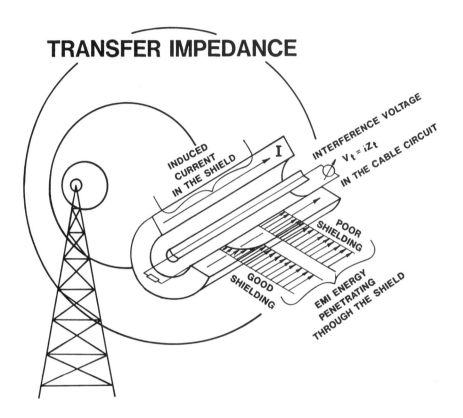

FIGURE 3.1 Concept of transfer impedance

parts of the same shield including the conditions at its ends, edges, and borders, the ratio of voltage dV_t to the current I_e may serve as a figure of merit of the shield performance:

$$Z_t = \frac{1}{I_e} \frac{dV_t}{dl} \tag{3.1}$$

Being the result of voltage-by-current division, parameter Z_t presents an impedance per unit length. But unlike, for example, the characteristic impedance, which determines the signal propagation properties *along* the cable, Z_t characterizes the energy propagation *across* the cable—through the shield. So the name transfer impedance is very much justified.*

Transfer impedance parameter is an inherently *differential* characteristic applicable to a *small, isolated,* and *terminated in a special way* shield element. Thus, it excludes or limits the effects of the interaction of signals along the line - such as difference in phases of the currents and voltages associated with different line elements, the actual loads at the ends of these circuits, and the system functions and interactions of both the shield and the shielded line with other system elements. Because there are no other system elements affecting the test results, such a model provides the deepest insight into the shield-"proper" performance, which is not obscured by many factors unrelated to the shield itself. However, for the same reason, it is also quite detached from the specific system applications— remember the phenomena that we so conveniently "forgot about" when coming up with the Eq. (3.1)? Therefore, to represent the shielding performance within a system, the transfer impedance parameter must be integrated along the shield length and complemented with system-specific parameters as expressed by the models of Eqs. (2.41) and (2.42) and corresponding to the model Fig. 2.24.

As you see, the question of the degree of relevancy of the transfer impedance to the particular problems of electromagnetic shielding (see Fig. 2.21) is not a trivial one. Let us repeat (how many times?): the transfer function models ignore the actual, within a system, field distribution at the shield surface, as well as circuit functions of the shield and shielded line, and substitutes these with some predetermined conditions that are independent of specific applications.

Not less important is also the origin of current I_e at the shield. Indeed, the ratio in Eq. (3.1) might imply that this current is the source of the EMI. But more often than not, this current is itself generated by the ambient field. In this case, I_e can be considered a *secondary* EMI source, which it is not necessarily identical to the actual EMI source (although they are related). Only when within the considered system this current is directly injected into the shield can it be viewed as truly representative of this original EMI source with regard to the shielded circuit. The "indirect" *shield performance* measure of transfer impedance may not readily relate to the specific *system performance* (that is, in high relevancy models) - but such is the "nature of the beast".

All these limitations are the "price we paid" to achieve the maximum generality of our model. Yet, was it worth it? Indeed, the popularity of transfer impedance for cable shield characterization is well deserved, as it is a practical tool for shield performance evaluation. The transfer impedance is an extremely convenient *shield effectiveness* parameter (in contrast to *system shielding effectiveness*) for "regular" metallic shields. Just because the transfer impedances of all shields are evaluated under similar conditions (the same attribute that prevents the direct use of this parameter to evaluate shielding performance at a system level), this technique is especially useful for shield design and comparison on the most general basis. Owing to the well defined test circuit configuration, this characteristic usually can be calculated in closed form, or at least numerically evaluated. It can also be physically interpreted: the smaller is the transfer impedance, the better is the shield, resulting (but only po-

---

*Other terms which are also used in literature to identify the transfer impedance parameter, especially overseas, are mutual impedance or coupling impedance.

tentially, because it does not account for all the factors involved in system-level shielding) in lower emissions from the cable and/or less induced noise in the cable circuits. That's why this parameter is a favorite in EMC applications of electromagnetic shielding.

Even more important is the transfer impedance methodological value. As we will show in the following sections, this parameter can be of great value not just for cylindrical (e.g., cable) shields. Its expansion on the separation shielding planes and surfaces and product enclosures - from a chip to a desktop product enclosure to a large chambers - presents a valuable tool to evaluate and design such shields.

Another important point: the suggested in this book transfer function is *not limited* to the transfer impedance. Depending on the particular task, other related parameters will be developed by expanding the underlying concepts, as required for different levels of system modeling. In fact, we will develop a *philosophy of shield performance* evaluation based on a set of shield transfer parameters: transfer impedance, transfer admittance, loss impedance, and transfer effectiveness. We will give practical definitions of these parameters, develop the mathematical apparatus to calculate them, and then provide a rigorous analysis of the most popular cable shield designs, in conjunction with experimental data.

> ✦ Previously, we have elaborated on the *operational* setbacks of the transfer impedance parameter generality and "indirect" relation to the EMI source. There is also an important *conceptual* limitation, due to the fact that this parameter is based on the very *existence* of current in the shield. To comprehend the seriousness of this fact, consider a shield composed of a material that is nonconducting but absorbent of electromagnetic energy (e.g., ferrite). Since the shield is nonconducting, no current can be induced in it. Thus, the transfer impedance concept seems irrelevant with regard to such a shield. On the other hand, such nonconducting shields may perform pretty well in high-frequency applications as well as in magnetostatic fields, if the shield possesses high magnetic permeability. Now, how would you relate this shield performance to its transfer impedance?
>
> While the evaluation of magnetostatic performance of non-conducting shields can be done similar to conductive shields, as discussed in chapter 6, their high frequency performance is definitely a problem. One way to "by-pass" this limitation is to *combine* a non-conducting shield with a conducting shield (one conducting layer will suffice!) with a known transfer function. Then, by injecting the current into and evaluating performance of this composite shield and "deducting" the known layer characteristics from the "total" shield performance, some evaluation of the effect of the non-conducting layer can be determined. ✦

3.1.2 Universal Set of Shield Transfer Parameters: from Cables to Small and Large Products to Architectural Constructions

So far, we have dedicated considerable attention to the study of the fields and energy fluxes associated with electromagnetic wave propagation through a shield. Considering an electromagnetic shield illuminated by an incident electromagnetic field, we arrived at a useful practical concept of the shield transfer function. We also have seen that a similar approach can be used to contain the EMI sources within the shielding enclosure.

In other words, we created an *electromagnetic field model* of the transfer function, as illustrated in chapter 2 and Fig. 3.1. However, we also invoked an important *circuit representation* of the process: the incident field generated an EMI current in the shield, which acted as a secondary EMI source with regard to the protected conductors. Or alternatively, the current could be directly injected into the shield. We utilized this circuit model to formulate one of the transfer function parameters - the transfer impedance .

We will now develop a more *general approach* to the shield transfer function and use it to define a *universal* set of transfer parameters. Note, that as long as the current in the shield was defined, further analysis was done with regard only to this current, without reference to its origin. Therefore, we could expect that to define and evaluate the transfer parameters, it is unnecessary to immerse the shield in an *actual* field, or consider an *actual* current injection circuitry, for that matter. Indeed, the

EMI current can be injected into the shield using any appropriate conducted or radiated method.

It is difficult to overestimate the importance of this conclusion! By selecting the most convenient methods to generate the EMI current in the shield, the theoretical derivations and practical evaluation techniques can be streamlined and simplified, as well as the interpretation of the obtained results . As we will see, this also facilitates the introduction of other (with regard to transfer impedance) transfer parameters. Such a philosophy opens wide opportunities to study the shielding phenomena using powerful methods of circuit theory, or its combination with field theory.

In particular, one of the most obvious methods of injecting the EMI current into the shield is galvanic coupling (see more about electromagnetic coupling in the next chapter). Corresponding circuits for immunity mode and emission mode shielding are presented, respectively, in Fig. 3.2 a and 3.2b. As shown in Fig. 3.2a, a current generator, I_e, with and *external* return path, is connected to an elementary length, dl, of electromagnetic shield. In this external circuit, current I_e causes the source voltage drop dV_{es} at the shield surface facing the current return path. Checking with Fig. 2.18 confirms our "suspicion" that this voltage drop is associated with the energy losses in the shield. Due to current I_e in the shield, transfer voltage dV_{it} is generated at the shield surface facing the internal current return path. This voltage is a result of the transfer through the shield of the part of the energy associated with the EMI current I_e.

Reviewing the Figs. 3.2 a and b, one may get an impression that we are limiting the discussion to cylindrical shields, e.g., as exemplified by cable shielding. And indeed, that was the original intent of the creators and users of the transfer impedance concept from the times of S. A. Schelkunoff and H. Kaden, to the most recent developments. But this doesn't not have to be so! Really, the Figs. 3.2 a and b are actually dealing with *crosstalk* between two circuits which have *one common conductor* - the shield- and current return pathes external and internal with regard to the shield. (O.K., this crosstalk is expressed in somewhat peculiar form as a ratio of the disturbing current to induced voltage, but there is nothing wrong with it, isn't it?).

Are crosstalk phenomena limited only to cables? - You bet, not! Then, nothing prevents us from building a similar model of crosstalk for other shield applications and shapes. As an example, Fig.3.2d presents the concept similar to that in Fig. 3.2 a,

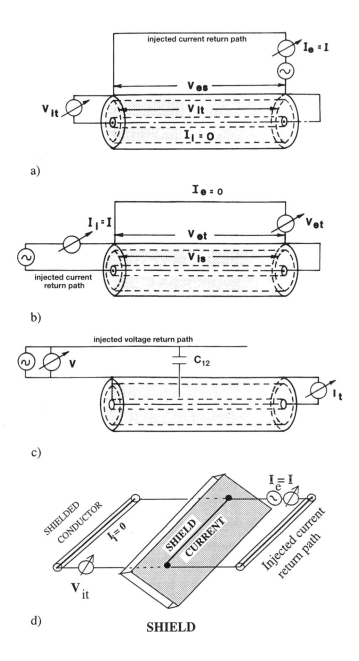

FIGURE 3.2 Defining shield transfer function

140 Electromagnetic Shielding Handbook

but for a plane flat shield. In principle *any other* conducting surface, solid and homogeneous, or perforated and non-uniform, and / or exhibiting anistropic conductivities - can be described by the same model.

In other words, this model is *universal*. But that means, that the *shielding parameters* based on such a model are also *universal*. Of course, you should understand that the analogy is mainly *methodological*, related to the philosophy of derivation, interpretation, and implementation of the transfer parameters. Thus in the process of the study, we will have constantly check our models: the current and field distribution in the shield, the transmission modes and applicability of different transmission regimes, electromagnetic field configuration and propagation conditions, and evaluation and test setups. However, as you will see in the next several sections of this book, very often the effect of existing discrepancies can be relatively simply accomodated, or reduced to manageable assumptions. A practical example of such "control" will be presented in the sections 3.2.1 and 3.2.2, where a comparison is given of the field and current distribution in a cylindrical shield, vs a flat shield, vs a shield of arbitrary shape.

With the above considerations and comments in mind, we will proceed to further derivations on the most general basis, using the representations in Fig. 3.2. The transfer voltage, dV_{it}, measurement circuit can be arranged by providing a respective return path inside the shield. In Fig. 3.2 a, this is achieved by measuring the voltage between the shielded conductor and the shield at one end of the cable element while shorting the shielded conductor to the shield at the opposite end. However, voltage V_{ie} at the output of this circuit will be equal to dV_{it} only in the absence of the current in this circuit. The absense of current prevents resistive voltage drop at the circuit elements (other than the shield) and the generation of magnetic fields crossing the voltage measurement circuit, both of which contribute to the measured voltage (in principle, magnetic flux through the circuit contour can also be generated by external sources, with the same effect of adding to the voltage V_{ie} —see later equation (3.10)). The two necessary conditions for the absence of current in this circuit are a high impedance of the voltmeter and a short length of the evaluated cable element (to avert capacitive currents). In this case, voltage dV_{it} can be thought of as an electromotive force (emf) generated at the shield in the internal circuit (inside the shielding enclosure).

Figure 3.2b presents an alternative arrangement corresponding to the cable emission case. Now, the source voltage dV_{is} is generated at the shield internal surface by the EMI current I_i in a circuit formed by the shield and the shielded conductor. As before, this voltage drop is associated with the energy losses in the shield. The transfer voltage dV_{et} corresponds to the energy transfer from inside the shield to the outside space, and acts as an emf in the circuit outside the shielding enclosure. Again, the cable element must be small, and the voltages dV_{et} and dV_{is} should be measured with a high input impedance voltmeter.

Voltages dV_{et} and dV_{it} are associated with the EMI energy that impacts the circuits protected by the shield, respectively, outside and inside the shield. The magnitude of each of these voltages is determined by two factors: the magnitude of the respective EMI currents, I_i and I_e, and the losses of the energy propagating through the shield (i.e., the shield effectiveness). Thus, when normalized with regard to the current, these voltages can serve as figures of merit of the shield effectiveness:

$$Z_{it} = \frac{1}{I_e}\frac{dV_{it}}{dl} \tag{3.2}$$

$$Z_{et} = \frac{1}{I_e}\frac{dV_{et}}{dl} \tag{3.3}$$

Comparing Eqs. (3.2) and (3.3) with (3.1) reveals that the first two formulas yield the transfer impedance of the shield. Hence, they can and usually are used as definitions of the transfer impedance parameters. Also, in principle, each shield has two transfer impedance parameters for two EMI propagation modes: into the cable and out of the cable. With regard to the shielded conductors, these two modes correspond to the immunity and emissions. In further discussions, we will use all these terms interchangeably, depending on the descriptive needs.

In a similar way, the voltages dV_{is} and dV_{es} also define impedance parameters, which we will call respectively internal loss impedance and external loss impedance:

$$Z_{is} = \frac{1}{I_i}\frac{dV_{is}}{dl} \tag{3.4}$$

$$Z_{es} = \frac{1}{I_e}\frac{dV_{es}}{dl} \tag{3.5}$$

The transfer and loss impedances are measured in ohms per meter (Ohm/m).

Since the interference current is related to the magnetic field (Ampere's law), the transfer impedance can serve as a measure of magnetic (inductive) coupling between the internal and external (with regard to the shield) surfaces. The last statement is true for solid shields in which only diffusion processes are manifested and for spiral and perforated shields in which magnetic field penetration via additional mechanisms also takes place[*]. Also, if the shield contains holes, gaps, and other nonconducting elements, the normal component of the electric field acts directly through these perforations, resulting in straight electric (capacitive) coupling, C_{12}, between the conductors at the opposite sides of the shield. To evaluate this process, the transfer admittance parameter, Y_t, was introduced (see Fig. 3.2c):

$$Y_t = j\omega C_{12} = \frac{1}{Vdl} = \frac{Z_f}{Z_{c1}Z_{c2}} \tag{3.6}$$

Equation (3.6) defines the immunity mode and, as usual, the derivations are similar for the emissions mode. Here, voltage V is the source of the penetrating electric field, which generates interference current I_t.

---

[*]As you will see later, this is a fundamental departure from the Schelkunoff's transfer impedance concept and definition

To facilitate the simultaneous evaluation of both electric and magnetic (that is, caused by voltage and by current, as shown in Fig. 3.2) coupling, the International Electrotechnical Commission (IEC) in 1974 introduced the capacitive coupling impedance parameter, Z_f, which is a fictitious transfer impedance producing the same result as Y_t interference. This parameter is introduced by the last equality in Eq. (3.6). Variables Z_{c1} and Z_{c2} are the characteristic impedances of the cable (that is, internal circuit) and external circuit, respectively.

Since the impedances in Eq. (3.6) are capacitive in nature, sometimes the so-called "normalized through elastance" parameter, K_t, is introduced, which is independent of the outside circuit geometry [unlike the impedances in Eq. (3.6)]:

$$K_t = \frac{C_{12}}{C_1 C_2} \sim \frac{\sqrt{\varepsilon_1 \varepsilon_2}}{\varepsilon_1 + \varepsilon_2}$$

As shown, K_t is related to the permittivities of the dielectric in external and internal circuits but does not depend on their dimensions.

In certain applications, it is more convenient to use dimensionless transfer parameters, and we now have all the necessary "building blocks" to define such parameters. Indeed, instead of normalizing the voltage at one side of the shield with regard to the current with return path on the other side, we can compare the voltages at the opposite shield sides directly by taking their ratio:

$$T_e = \frac{dV_{it}}{dV_{es}} = \frac{Z_{it}}{Z_{es}}, \quad T_i = \frac{dV_{et}}{dV_{is}} = \frac{Z_{et}}{Z_{is}} \quad (3.7)$$

We will call parameters T_e and T_i the voltage transfer effectiveness parameters for susceptibility (or immunity, depending on the measure of one's optimism) and emissions, respectively.

✦ Interestingly, we actually achieved more than expected. Not only did we create *dimensionless* transfer parameters, but also ones that "pull" together both transfer impedance and loss impedance parameters! To appreciate this fact, recall that the transfer impedance relates the "transferred" EMI voltage in the shielded circuit *directly* to the EMI source current (albeit at the opposite surfaces of the shield). Alternatively, the transfer effectiveness parameters introduce an additional factor, because, instead of the EMI source current, they use the current *product* by the loss impedance (which is the EMI source voltage). You may argue that this additional parameter can obscure the properties of the transfer function, and this could be very well the case in some applications. Then you would want to use the transfer impedance as the shield characteristic. But isn't this additional parameter—the loss impedance—one of the factors that determine the magnitude of the induced current in many "real life" applications? For example, assuming the same (and low enough) impedance of the EMI current-generating circuit, the larger is the shield loss impedance, the smaller will be the current induced in the shield. Such a situation takes place in many important applications, e.g., crosstalk between the shielded lines (we will have an extended discussion of this subject in Chapters 4 and 5).

Therefore the ratios of Eq. (3.7) not only express a measure of the transfer function but also give an indication (though only to a degree) of the shield *interaction* with the environment (external or internal, depending on the shielding mode). In this respect, the transfer effectiveness can be viewed as a kind of low-level shielding effectiveness: the model is limited to a small section of an isolated shielded cable, which is evaluated under certain generic conditions. To emphasize the shield-oriented nature of such definition, we will name it *shield effectiveness*, as opposed to the system oriented term - *shielding effectiveness*. As a matter of fact, we can trace down to identical or similar parameters many of the "shielding effectiveness" definitions used by practitioners, e.g., see the "dozen" definitions in section 2.3.1 of Chapter 2. Of course, there is nothing wrong with this, as long as the applicability limits and constraints of this parameter are realized, as we have just discussed.◆

3.1.3 Finding EMI Voltage: A Union of Field and Circuit Theories

We have developed the universal set of the shield transfer parameters considering the corresponding voltages in external and internal circuits, normalized to the current in the shield. The next step is to define these voltages as functions of the current. This can be done using the diagrams in Fig. 3.2.

We start with the loss impedance. Each voltage drop dV_{is} and dV_{es} at the shield surface is just one of the contributing voltages in the measuring circuit. For example, consider a small length of a coaxial cable in an internal transmission mode (per Fig. 3.2 b): the current I_i return path is fully inside the shield through the center conductor, and no EMI energy arrives from the outside EMI environment. Then, applying the Kirchhoff's voltage law [Eq. (A.27)] to the closed circuit between the coaxial cable outer conductor (i.e., the "shield") and the center conductor, we obtain:

$$\Sigma V_{ci} = V_{ig} + V_{is} + V_{ci} + V_{Li} = I_i(Z_g + Z_{is} + Z_{sc} + j\omega L_{ei}) = 0 \qquad (3.8)$$

where ΣV_{ci} is the sum of all voltage drops at all the elements of the closed circuit inside the shield, including voltage drop V_{ig} at the generator; V_{is} and Z_{is} are the voltage drop and loss impedance at the shield internal surface, and Z_{cs} is the center conductor loss impedance. L_{ei} is what is called the external inductance of the coaxial circuit. This inductance is determined by the magnetic flux crossing an area inside the cable which is bounded by the contour formed by the external surface of the center conductor and the internal surface of the outer conductor. It is a commonly used parameter of the coaxial cable. If a current generator, I_i, is used its internal impedance can be considered zero, so no respective voltage drop occurs. If a voltage generator with internal impedance, Z_g, is used instead, the corresponding voltage drop item, $I_i Z_g$, must be accounted for in Eq. (3.8). The equation can be used to find the voltage at any of the elements of the contour, which in that case may be viewed as input or output of the circuit. We will take full advantage of this in the following discussions.

A similar equation can be written for the external circuit, in which voltage drop $V_{es} = I_e Z_{es}$ acts. However, very often, the identification of the external circuit elements presents a more complicated problem than in internal circuit. To avoid such complications in the course of the EMI problem analyses, it is important to properly select the circuit models of the studied processes. Such considerations as geometric and electrical symmetry and simple patterns of electric and magnetic field distribution become paramount to the success of the analysis. For instance, one important example of a popular structure used in the cable shield analysis is triaxial line, which will be discussed later.

What are the ingredient voltages in Eq. (3.8)? By the definition of Eq. (A.44), these voltages can be expressed via the electric field intensity:

$$V_r = \vec{E} \cdot \vec{dl} \tag{3.9}$$

Compared to Eq. (A.44), the "minus" sign in Eq. (3.9) is omitted for simplicity. We may do this by properly selecting the direction in which the voltage is measured. Since this direction is arbitrary anyway, the only thing we should watch for is that we observed the same direction for all other ingredient voltages. As indicated in Section A.5, the voltage definition applies differently to an isolated length of current-carrying conductor as opposed to a closed circuit. In the first case, the voltage drop is determined only by resistive effects associated with the currents I_i or I_e. The voltage drop can be defined by Ohm's law [see Eq. (A.4)]. In a closed loop, Maxwell's equations identify two potential sources of the electric field $\vec{E}$, acting together: the resistive effects (as above) and magnetic induction effects. The later are associated with the alternating magnetic flux crossing the area bounded by a closed contour formed by respective conductors (and defined by the Faraday's law [see Eq. (A.21, I)]. To summarize: voltage V_i in a closed circuit between the current-carrying shield and the shielded conductor depends on both electric and magnetic field configurations and intensities. Applying the Maxwell's equations in integral form (for phasors), the magnetic and electric field-related ingredients of this voltage can be written as follows:

$$V_c = j\omega\mu_s \int_S \vec{H}_n \cdot d\vec{s} + \oint_l \vec{E}_l \cdot \vec{dl} \tag{3.10}$$

The double integral is taken over the area S bounded by the segment of cable radius between the shield inner surface and the shielded conductor outer surface and the two current filaments through the ends of the segment. $\vec{H}_n$ is the magnetic field component normal to the surface S. The double integral defines the part of the voltage produced by magnetic flux inside the shielded space, while the field $\vec{H}_n$ can be originated by the outside field penetrating through the apertures and slots, as well as be produced just inside the shielding enclosure. The linear integral is to be taken around the contour l of the surface S; $\vec{E}_l$ is the electric field component along this contour. Since $\vec{E}_l$ depends on energy losses in conductors, this term is related to diffusion processes. This expression is applicable to both internal and external circuit configurations. (We didn't put forward any conditions, did we?)

We will now investigate whether the same expression, Eq. (3.10), could also be applied to the transfer impedance. If our shielded cable is immersed into the EMI environment, and no energy arrives from the inside environment (Fig. 3.2 a), then by definition $V_{it} = I_e Z_{it}$. There, V_{it} is the voltage drop on the shield as perceived by an internal circuit, but which is due to the current I_e in external circuit. This voltage acts as an emf in the same circuit for which we defined Eqs. (3.8) through (3.10), associated with the loss impedances. And in loss impedance definition and evaluation, the current in the circuit acted as a source of both electric field (resulting in resistive components of the voltage) and magnetic field (resulting in inductive components of the voltage). But when defining the transfer impedance, we required that no current flow in this internal circuit! If so, then what are the sources of the electric and magnetic fields in the internal circuit?

Because no energy is generated inside the shield, the only EMI energy in the internal circuit is that which penetrates through the shield from the outside (that is from the external circuit). Thus, the shield is the single source of EMI energy insideits confines. Then, given the current in the shield with external return path, our objective will be to determine the electric field intensity at the internal surface of the shield (that is, at the surface facing the internal return path) and the magnetic field generated by the shield in this closed internal circuit. Whether such electric and magnetic fields are actually generated (and, if yes, then to what extent) depends on the particular shield design and its reaction at the given signal parameters (e.g., the frequencies and the amplitudes of the interference current). Only after these fields are found, can Eq. (3.9) be used to determine the transfer parameters. Then, substituting Eq. (3.10) into Eqs. (3.1) through (3.7), the appropriate transfer and loss impedances can be determined, as well as the transfer coefficients.

◆ What's so "special" about our universal set of transfer parameters? Aren't they identical to the existing ones—say, transfer impedance and transfer admittance? If we look hard enough, we may even find some equivalents to the transfer effectiveness parameters!

The answer is not so much in the formal presentation of these parameters as in their contents, physical interpretation, and application scope. To start with, we have expanded the respective transfer function concepts onto *ALL* shields, and not just cable shielding. The transfer impedance parameter was first introduced by S.A. Schelkunoff into the shielding theory for *cylindrical* shields and coaxial transmission lines. He studied the transfer impedance and surface impedance parameters for a homogeneous tube model and related them to the axial electromotive intensities on the shield surfaces. The two theorems suggested by Schelkunoff (see below) were formulated in these terms. When considering losses in a homogeneous shield, the voltage at the respective shield surfaces is proportional to the longitudinal electric field intensity, per Eq. (A.44). There, looking at the longitudinal components of the electric field at the shield surfaces yields a very logical and graphic interpretation, linking the impedances with the appropriate shield surfaces with regard to the return current path.

In a more general case of nonhomogeneous shields, cylindrical or otherwise, the longitudinal electric field component presents only a fraction of the voltage associated with the current propagation along the shield [another part is due to the magnetic field, per Eq. (3.10)]. An excellent example is a spiral shield, which generates both longitudinal and tangential electric field components as well as the longitudinal magnetic field component inside the shielded space (that is, in the internal circuit) when the source current is flowing in the external circuit. In this case, depending on the shield design specifics, the other components of the electric field, as well as the effects of magnetic field, must be also considered in evaluating the voltage drop associated with the shield. In such shields, it can be misleading to limit the consideration to only the longitudinal electric field component. This task was first addressed by H. Kaden. However, although H. Kaden has expanded the transfer and loss impedance concepts to non-homogeneous shields, the utility of these parameters was still limited to cylindrical shields and cables. On the other hand, our universal parameter set is not limited to cylindrical shields and coaxial cables, but is expanded onto shielding surfaces and enclosures of any shape and size.◆

146 Electromagnetic Shielding Handbook

We will conclude this section with several clarifying comments.

✦ Formula (3.10) was first suggested by Kaden[P.2]. However, at that time, it was not treated as expressing the combination current in the shield, with partly external and partly internal returns. This very important case takes place in many practical applications (e.g., in triaxial line—see Fig. 3.3) or when the impedance of the circuit formed by the shielded conductor and the shield is not large enough, which allows a current to flow in the transfer voltage measurement circuit. A similar situation also occurs when a shielded (or coaxial) cable which is being used to transmit a "useful" signal is immersed into the EMI environment. Such shield (or cable outer conductor) simultaneously will carry two currents belonging to two different circuits: with internal and external return current paths.

We can still use Expression (3.10) if we complement it with appropriate network considerations. Note that current I is now the total current in the shield, i.e., the current with internal return path, plus the current with external path (e.g., as illustrated in Fig. 3.3). However, each of the two ingredients of this total current will affect, in a different way, the voltages at the shield surfaces associated with the internal and external returns. If, for example, the return path is internal, voltage V_{is} associated with the internal surface is determined by the internal circuit (current I_i) energy flux entering through this surface, and the external circuit (current I_e) energy flux which exits through this surface. This latter energy flux, of course, entered the shield at its external surface and as a rule was attenuated before reaching the internal surface (e.g., attenuation associated with energy diffusion). The surface crossed by the total magnetic flux is still limited by the contour formed by the outer conductor on one hand, and, on the other hand, by the center conductor for internal circuit or external return current path (whatever it may be) for the external circuit. ✦

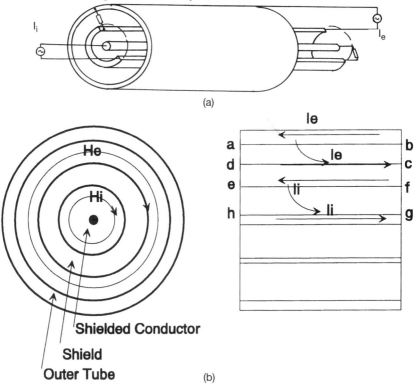

FIGURE 3.3 Total current in the shield

✦ Intuitively, we expect the voltages associated with the transfer and loss impedances to be proportional to the current in the shield and its shielding effectiveness. Thus, the adopted definitions seem to imply that whenever the EMI current in the circuit is increased or reduced by some amount, these voltages are changed proportionally. But this is true only if the shield impedances are linear. Then, of course, the proportionality takes place. However, in real-life applications, this may be not the fact. A case in point is a magnetic shield. In such a shield, the magnetic permeability can change with an increase in the magnetic field intensity and, in the limiting case, the saturation occurs at a certain field level, and the shield magnetic permeability is abruptly changed. Thus, in general, the ratios in Eqs. (3.1) through (3.9) may also include the current dependence of the transfer and loss impedances.✦

✦ Why do we associate the voltages dV_{es} and dV_{it} with the shield surfaces? The answer is based on Poynting theorem, which is briefly reviewed in Appendix (section A.3). We know that the signal energy (whether "useful" or EMI) is concentrated within and propagates along the circuits, formed on one hand by the shield, and on the other hand by the current return path. For instance, in a case of immunity shielding (Fig. 3.2 a), the EMI current return path is outside the shield ("external" return current path), and the EMI energy is concentrated outside the cable. There it propagates in two directions: along the external circuit (i.e., "regular" transmission) and across the external circuit (leading to EMI). In this second case, it enters the shield at the outside surface and then propagates through the shield body, "emerging" (hopefully, weakened) at the inside surface. Thus, the voltages associated with the energy entering into and emerging from the shield in the circuits with external and internal circuits *can* be related to the respective shield surfaces. Now the question is just the opposite: whether it is necessary to reference these voltage drops to the respective shield surfaces. Although it may come as a surprise to many readers, the answer is *no*. In principle, voltages dV_{es} and dV_{it} can be referenced to *any* level across the shield thickness, or even outside the shield body. That's why we used the word "can" instead of "must." We will prove and illustrate this statement in the last author's note of section 3.2.4, while looking at the effects of attaching the interconnecting wires to different points "along the shield thickness." In the meantime, we offer this problem as "food for thought."✦

✦ Substituting electric field intensities with proper voltages, we obtained more general definitions of Eqs. (3.1) through (3.9), which are applicable to nonhomogeneous as well a homogeneous shields. As we will see later, when discussing nonhomogeneous shields, this change has a very profound effect on the understanding of the shielding phenomena. In particular, the use of the voltage to define the transfer and loss impedance parameters permits us to account not just for longitudinal (axial) components of the electric field, but also for other components of electric and magnetic fields, if present. Referring to the same spiral shield, we will see that the degradation of the shielding properties is due mainly to the effect of the longitudinal component of the magnetic field inside the shielding enclosure. For the same reasons, we have changed the Schelkunoff's term *surface impedance* to *loss impedance,* which more accurately describes the role of this parameter, although it is less graphic.

With such broad understanding of the transfer impedance and loss impedance parameters, we can summarize the preceding discussion by modifying the two important theorems, formulated by Schelkunoff, using respective changes which summarize the preceding discussion (also, we have tried to retain the original wording as much as possible):

Theorem 1: When a current flows along a tubular conductor and the return path is wholly external ($I_i = 0$) or wholly internal ($I_e = 0$), the voltage between the tubular conductor and the conductor(s) forming the return path equals the loss impedance per unit length, associated with the surface of the tubular conductor facing the return path, multiplied by the total current flowing in the conductor; and the voltage associated with the other surface equals the transfer impedance per unit length multiplied by the total current

148 Electromagnetic Shielding Handbook

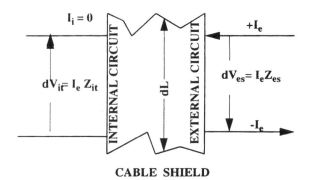

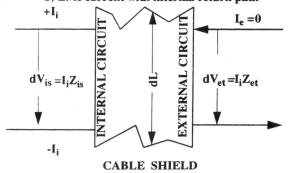

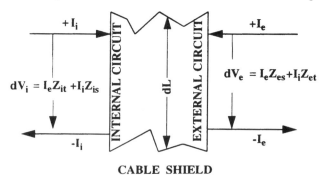

FIGURE 3.4 Transmission and transfer impedance in shielded conductors with internal and/or external return current paths

Theorem 2: If the return path is partly external and partly internal, the separate components of the voltage due to the two parts of the total current are calculated by the above theorem and then added to obtain the total voltages.

Both theorems follow directly from Fig. 3.4. We will see in the next section that when the tubular conductor is homogeneous, the voltages defined by these theorems are proportional to the longitudinal electric filed intensities (*longitudinal electromotive intensities,* as per Schelkunoff) at the respective surfaces of the tubular conductor, which are facing or opposite to the return paths, respectively. Then our formulations become identical to Schelkunoff's original formulations.

Later, in section 3.3.1 we will briefly review the corresponding theoretical expressions, Eqs. (3.79) through (3.81)). Then, in section 3.4.2 on the spiral shield performance, these two theorems will be complemented with the ***"Theorem 3"*** *("solenoid" theorem)*, which will provide even broader perspective at the relation between the current return path and shield transfer function. ✦

3.2 THIN HOMOGENEOUS SOLID SHIELD PERFORMANCE

In this section, we will first prove that current and field distribution in a thin homogeneous field is independent on its curvature, like for plane sheet, cylindrical surface, or any other shape enclosure. We will then derive and analyze the transfer parameters of such shield. It follows that a flat cable shield, or a large shielding room with arbitrary shape confines, will have *identical transfer impedance,* with a cylindrical shield (as long as their thickness is much smaller than the enclosure curvature)! W*hy* does the shield behave in a certain way and what are the trends of such behavior? W*hy* did such different shielding enclosures as a "box", cylindrical tube, and sphere behave almost identically with regard to the transfer parameters? Hopefully, the following will provide enough insight and understanding.

3.2.1 Current and Field Distribution in Homogeneous Metallic Tube - Coaxial Cable Shield / Outer Conductor

We start the analysis of field distribution in a solid homogeneous tube for several reasons. First, some products do have solid single- or multi-layer cylindrical shields, so this is a matter of practicality. Second, under certain conditions, many practical shields can be *approximated* with homogeneous tubes. Third, because of inherent symmetry and geometry, we expect solid tubes of a regular shape to be "less trouble" in theoretical analysis than other shields and thus yield faster results which are better adaptable to interpretation. And last but not least, we plan to use the obtained characteristics as theoretical "building blocks" to develop more complex models of different electromagnetic shield designs. While deferring the discussion of the physical and mechanical properties, as well as design and optimization principles to Chapter 7, we will concentrate here on the transfer parameters.

Consider an infinitely long homogeneous metallic tube conducting current, I, uniformly distributed along the tube circumference (but not necessarily uniform across the tube thickness). Of course, for the current to flow, a closed circuit must exist to provide the return current path. We saw that two different ways exist to form the re-

150 Electromagnetic Shielding Handbook

turn current path, resulting in two different currents in the shield: current I_i, with an internal return path (inside the shielded space) and current I_e, with an external return path (outside the shield). Correspondingly, the shield performance can be modeled by the circuit networks in Fig. 3.4a and b. On the other hand, there is no reason why the two currents, internal and external, wouldn't be present simultaneously, as shown in Fig. 3.3. The respective network model is presented in Fig. 3.4c. Now, the total current in the shield is the sum of internal and external currents:

$$I = I_i + I_e \qquad (3.11)$$

The currents I_i and I_e may have any direction, depending on particular sources. And for analysis purposes, it makes no difference that one or both these currents can be either "useful transmission" and/or EMI—they are "just signals."

Our goal is to determine the universal set of the shield transfer parameters as functions of signal characteristics, as well as the shield geometry and material properties. In the previous section, we introduced and formulated transfer parameters as functions of each respective current and voltage in the internal and external circuits. Although the definitions of Eqs. (3.2) through (3.7) were derived for only one of these currents present (expressions (3.79) through (3.81) prove that, as long as the transfer parameters are known, the presence of both currents can be easily accounted for). We also developed the basic Expression (3.10) linking these voltages to the electric and magnetic fields within the respective closed contours and surfaces. This expression defines all four voltages of interest: V_{it}, V_{is}, V_{et}, and V_{es}.

The "small" thing left is to determine the electric and magnetic field distribution inside or outside of the tube for internal and external return currents. Although this task is not an easy one, the symmetry of the circuit helps. Indeed, we assume that the shield is cylindrical, infinitely long, and homogeneous; the shielded conductor is situated in the shield geometrical center. (Isn't that what an "ideal" coaxial cable is all about?) Thus the internal circuit is symmetrical both "around" and "along" the cable; that is, we have the longitudinal and angular symmetry. To enhance the symmetry even more, we also make the external circuit symmetrical by selecting a concentric external current return path via another homogeneous solid tube (in later discussion, we will call it "external tube"). What we have built as a result, is a triaxial line* (Fig. 3.5).

Consider first the case presented in Fig. 3.5. The EMI current is applied between the shielded conductor and the shield (or between the center conductor and the outer conductor of the coaxial cable, which is the same) and shorted at the other end. Also, no energy is generated in the external circuit formed by the shield and the external tube. Because the line is long, we can disregard the edge effects at its ends (along with the shield homogeneity, this provides the longitudinal symmetry) and the signal mismatch reflections produced by the shorts at the line loads.

---

*We will find the triaxial line to be a remarkable tool with many uses: as a multipurpose practical shield, an idealized model of important real life problems, and as a test and measurement instrument. We will return to the triaxial line concept many times

FIGURE 3.5 A triaxial line

On the other hand, although the line is long, we consider it only as a small element. In this way we can simultaneously use both the assumptions of an infinitely long and infinitely small line ("have our cake and eat it, too"). We have selected the current generator (as opposed to the voltage generator) to make sure that its input impedance is "zero" (although in principle this is not necessary—this was done just for convenience, to simplify the analysis). Together with the "short" at the other end of the circuit and the absence of energy arriving from the external circuit, this fits Eq. (3.8).

The current I_i generates three voltages in the internal circuit: at the shield, V_{is}; at the center conductor, V_{sc}; and inductance voltage, V_{Li}, generated in the closed contour formed by the internal circuit. Actually, we could "care less" about the voltage V_{sc}, because under our assumptions it has nothing to do with the shield effectiveness. We are interested only in the voltages V_{is} and V_{Li} associated with the shield. However, we had to incorporate the center conductor into Eq. (3.8) in order to obtain a closed contour—only so we could consider the inductive part, V_{Li}. Therefore, in further discussion, we "demonstratively" will pay little attention to this voltage (another "cake," isn't it?). By the way, this does not mean that the voltage drop at the center conductor of a coaxial cable is not important in other problems. As a matter of fact, it is responsible for the most of the signal losses in coaxial cable transmission and thus may be a part of the system shielding effectiveness by affecting the EMI energy exchange between the cable and the system.

In the external circuit in Fig. 3.5, the input impedance of the voltmeter is by definition infinitely large, so no current flows in the circuit ($I_e = 0$). In general, the voltages associated with this circuit are resistive voltage drop V_{et} and inductive voltage drop V_{Le}:

$$V_{ee} = V_{et} + V_{Le} \qquad (3.12)$$

To use Eq. (3.10), we need to find the distribution of the electric and magnetic fields generated by current I_i in three areas:

- Area I: free space between the shield and the external tube
- Area II: the shield body
- Area III: free space between the shield and the center conductor

152 Electromagnetic Shielding Handbook

As always, these fields are found by integrating the Maxwell's equations and finding the integration constants by substituting the boundary conditions at the shield surfaces into the general solutions. Due to the symmetry, it is possible in this case to eliminate from consideration all the components of the fields except two—one electric and one magnetic. In the process of this analysis, we will find important relationships between the electric and magnetic field components, which will lead us to the transfer parameters. Although this material is important, it's somewhat "dry." Those who don't "enjoy" the derivations may skip directly to the end results in the next section. (The book's format permits this "superficial" approach, which is why we selected this format to begin with.)

✦ We start by writing down the first and second of Maxwell's equations. Generally, both the vector notation, e.g., Eq. (A.16a), or its expansion in cylindrical coordinates as in Eq. (A.16c) can be used—they are absolutely equivalent. Operating with the vector notation facilitates a more general analysis, while dealing at the field component level invokes more detail. In our case it is possible to simplify the beginning stages of the analysis using vector notation and only afterward proceed at the field component level. We simplify the equations right from the start by writing them separately for the dielectric ("free space" is assumed inside and outside the shield) and for the metal (i.e., shield "body'). Assuming quasistationary regime (that is, disregarding the displacement currents in the dielectric—see Chapter 4) and expressing the vectors of magnetic induction and electric displacement via the respective components of magnetic and electric field intensities and magnetic permeability and dielectric permittivity, we obtain:

In dielectric,

$$\nabla \times \vec{E} = -j\omega\mu\vec{H} \qquad (3.13)$$

$$\nabla \times \vec{H} = j\omega\varepsilon\vec{E} \qquad (3.14)$$

In metal,

$$\nabla \times \vec{E} = -j\omega\mu\vec{H} \qquad (3.15)$$

$$\nabla \times \vec{H} = \sigma\vec{E} \qquad (3.16)$$

Fortunately, the geometrical symmetry goes a long way toward simplifying the task. In a straight thin tubular conductor with small enough radius and/or linear return current parallel to this conductor (lots of "caveats," but pay attention), there is only one component of current—along the conductor. If we situate the cable along the axis Z, this current component is $I_i = I_z$. In such a conductor, only one component of electric field, E_z, is present, while the two other electric field components are not generated. Per Ampere's law (rather, per an also experimental equivalent to that, the Biot-Savart law), there is only one magnetic field component, H_ϕ, present. That's all we need to know to apply Maxwell's equations.

Consider first the fields in the shield body (i.e., in metal). Finding $\vec{E}$ from (3.16) and substituting it into (3.15), we obtain

$$\nabla \times \nabla \times \vec{H} = (-j\omega\mu\sigma)\vec{H} \qquad (3.17)$$

Using the well known vector algebra relationship, in particular,

$$\nabla \times \nabla \times \vec{H} = \nabla(\nabla \cdot \vec{H}) - \nabla^2 \vec{H}$$

and for the stated conditions, this can be reduced to:

$$\nabla^2 H_\phi = j\omega\mu\sigma H_\phi \qquad (3.18)$$

Now we switch to the component level. The expression in the left side of Eq. (3.18) is the Laplacian (see Appendix). Noting that because of the symmetry the field component H_ϕ does not vary with the coordinates Z (longitudinal symmetry) and ϕ (angular symmetry), and using the expansion of the Laplacian in cylindrical coordinates [see Eq. (2.23)], we obtain the final differential equation:

$$\frac{d^2 H_\phi}{dr^2} = j\omega\mu\sigma H_\phi = \gamma^2 H_\phi \qquad (3.19)$$

The solution of this equation is well known:

$$H_\phi = \varsigma_1 e^{\gamma r} + \varsigma_2 e^{-\gamma r} \qquad (3.20)$$

Then from (3.16) and (3.20),

$$E_z = \frac{\gamma}{\sigma}[\varsigma_1 e^{\gamma r} - \varsigma_2 e^{-\gamma r}] \qquad (3.21)$$

The integration constants ς_1 and ς_2 are found by applying the boundary conditions at the shield surfaces. To find the field intensity at the shield surfaces, there is no need to resort to the differential equations; due to the simple geometry, the magnetic fields can be determined from Ampere's law. According to Ampere's law, a line integral of magnetic field $\vec{H}$, taken over any closed path L enclosing a current $\vec{I}$, is equal to that current:

$$\oint_L \vec{H} \cdot d\vec{l} = \sum I \qquad (3.22)$$

Here, the *dot* symbol stands for scalar product of the multiplied vectors. Assuming the conductor is placed along the axis Z, the current component may be denoted I_z. Applying the Ampere's law to the space inside of the shield (i.e., in Area III), we see that any path inside the shield encloses *all the current* I_i returning via the center conductor. (However, no "forward" current flowing in the shield is enclosed!) Moreover, because of the symmetry, the field H_ϕ is constant when the coordinates Z and ϕ change, and it varies only with r. Therefore, applying Eq. (3.22) to a circle with an arbitrary radius r inside the shield, determine that this current produces circular magnetic field lines in the planes perpendicular to the conductor axis; that is, there is only one component H_ϕ of the magnetic field:

$$H_{ii} = \frac{I}{2\pi r}, \text{ A/m} \quad (3.23)$$

We omitted the component index ϕ because no other components exist (so there is nothing to confuse). For all other directions, the current component is zero, and thus the dot product yields no field. On the other hand, outside the shield, any contour will enclose both the direct and return current I_i (which are, of course, equal in amplitude and opposite in phase), and the total current enclosed will be 0. Therefore, the magnetic field outside of the shield will be equal to 0:

$$H_{ie} = 0 \quad (3.24)$$

In a similar way, if there is only a external return current path, then applying Eq. (3.22) to a circle with the radius r outside the shield, determine that any path outside the shield encloses all the current I_e returning via the shield. Again, because of the symmetry, the field H_ϕ is constant with the changing coordinates z and ϕ and varies only with r. This current produces circular magnetic field lines in the planes perpendicular to the conductor axis; that is, there is only one component of the magnetic field:

$$H_{ee} = \frac{I}{2\pi r}, \text{ A/m} \quad (3.25)$$

For all other directions, the current component is zero, and thus the dot product yields no field. On the other hand, inside the shield, any contour will enclose no current at all, and the magnetic field outside of the shield will be equal to 0:

$$H_{ei} = 0 \quad (3.26)$$

Now that we know the magnetic field intensity H_ϕ at both surfaces of the shield with either external or internal return current path, we can proceed to determine the integration constants ς_1 and ς_2. Let us express the magnetic fields at the internal H_i and external H_e surfaces of the shield (again, the index ϕ is omitted because only this field component is present), as functions of shield characteristics, using (3.20). The component H_i will be obtained at $r = r_i$, and the component H_e will be obtained at $r = r_e$. As a result, we obtain two equations that can be written in matrix form. The general notation is:

$$[H] = [e^{\gamma r}] \cdot [\varsigma] \quad (3.27)$$

Or expanded to the field component level:

$$\begin{bmatrix} H_e \\ H_i \end{bmatrix} = \begin{bmatrix} e^{\gamma r_e} & e^{-\gamma r_e} \\ e^{\gamma r_i} & e^{-\gamma r_i} \end{bmatrix} \cdot \begin{bmatrix} \varsigma_1 \\ \varsigma_2 \end{bmatrix} \quad (3.28)$$

With regard to ς_1 and ς_2, this a simple system of two linear equations. Applying the Kramer's rule, we obtain:

$$\varsigma_1 = \frac{\Delta_1}{\Delta}, \quad \varsigma_2 = \frac{\Delta_2}{\Delta} \tag{3.29}$$

where Δ is the determinant of the middle matrix in (3.28), and Δ_1 and Δ_2 are obtained by substituting the first and second columns, respectively, in this middle matrix by the left matrix in Eq. (3.31). As a result, we obtain:

$$\varsigma_1 = \frac{1}{2\sinh\gamma t}(H_e e^{-\gamma r_i} - H_i e^{-\gamma r_e}) \tag{3.30}$$

$$\varsigma_2 = \frac{1}{2\sinh\gamma t}(H_i e^{\gamma r_e} - H_e e^{\gamma r_i}) \tag{3.31}$$

where the shield thickness $t = r_e - r_i$.

Substituting Eqs. (3.30) and (3.31) into Eqs. (3.20) and (3.21), we obtain the expressions to calculate the magnetic and electric field intensity within the shield body ($r_i < r < r_e$):

$$H_\phi = \frac{1}{\sinh\gamma t}\{H_e \sinh[\gamma(r - r_i)] - H_i \sinh[\gamma(r_e - r)]\} \tag{3.32}$$

$$E_z = \frac{\gamma}{\sigma}\sinh\gamma t\{H_e \cosh[\gamma(r - r_i)] - H_i \cosh[\gamma(r_e - r)]\} \tag{3.33}$$

◆

3.2.2 Thin Homogeneous Solid Current-Carrying Plane or "Other" Shield Shapes Have Identical to Cylindrical Shield Field Distribution

It can be shown, that the same expressions which describe the field distribution in a tube, are applicable to a plane sheet.

◆ There is really no need to "re-invent the wheel": all the derivations were done about 50 years ago by Kaden [P.2], and they are almost identical to what we did above for cylindrical tube. Kaden considered two parallel metallic walls of a "large enough" size and thickness t, connected with transverse metallic plates at the top and the bottom, but open in the front and back. You have probably already guessed why we need "large enough" dimensions - to be able to disregard the "end efect" and the "edge effect" of the currents and fields at the bottom, top, front, and back of our "shielding enclosure".

The structure is illuminated by a uniform magnetic field H_y^e from the sides. The magnetic field vector is parallel to the side walls and to the axis y - as follows from the variable designation in Carthesian coordinates (that is, only one component along the axis y is present). When the shield is illuminated by magnetic fie d H_y, an electric field E_z is generated in the shield, which results in the current I_z in the shield. This is a typical immunity shielding problem, where the field H_y^i inside the shielding enclosure is expected to be smaller than H_y^e. This way, we can arrive at the shielding effectiveness of only the side walls.· (And, by the way, since no mention is made about the incident electric field, we can assume that this is a near field problem - see chapter 4).

Easy to see, that with the spelled out conditions in place, the same equations (3.13) through (3.17) are applicable and valid, since they are written in vector notation which is invariant to the coordinate system. Then, reducing the vector equations to the respective differential equations, we arrive at solutions, which are identical to (3.18)-(3.21) and (3.30)-(3.33), as long as we substitute H_ϕ for H_y. ✦

For example, in case of Cartesian coordinates equation (3.17) is reduced:

$$\frac{d^2 H}{dx^2} = \gamma^2 H$$

with solutions

$$H_y = A_1 e^{\gamma x} + A_2 e^{-\gamma x}$$

$$E_z = \frac{\gamma}{\sigma}[A_1 e^{\gamma x} - A_2 e^{-\gamma x}]$$

which are identical to (3.20) and (3.21).

Conducting similar derivations for other shield shapes (preferably, in respective coordinate systems), it can be shown that as long as the shield thickness is infinitesimal (comparing to its curvature radius) and the shield dimensions are infinitely large to disregard the edge effects, the same transfer impedance equations will hold for a large majority of practical shield shapes. For instance, Kaden [P.2], developed formulas for spherical shields, which yield the values not too much different from cylindrical and "plane" shields. Another matter, what happens when the shield infinitely large size condition isn't met. This problem is more of a shielding system character and, as a rule, cannot be resolved within the transfer function domain. We will address it in chapter 4.

3.2.3 Transfer Parameters of Thin Solid Homogeneous Shields

Now we can apply the developed theory to thin homogeneous solid shields. As we just saw, all the considered shield types have identical field and current distribution. But if the current and field distribution in the shield body are the same, so are the transfer impedance parameters. So, any of the shield configurations can be used to derive transfer parameters applicable to the shields *at large*. Here, we consider tubular shield in a setup per Fig. 3.2a and corresponding definitions. To determine the voltages V_{es} and V_{et}, we need to know the electric field intensities at the shield surfaces: external, at $r = r_e$; and internal, at $r = r_i$. We also must operate with the magnetic fields in the external and internal circuits. Naturally, when $r = r_i$, then $H = H_i$ and $E = E_i$, and when $r = r_e$, then $H = H_e$ and $E = E_e$. As far as magnetic fields are concerned, just $H = H_i$ and $H = H_e$ are obtained by substitution of $r = r_i$ and $r = r_e$ into Eq. (3.32). [What else should we expect, when expressing magnetic field components via "themselves?" This exercise merely serves as a simple "check of validity" for Eq. (3.32)]. On the other hand, substituting the $r = r_i$ and $r = r_e$ in Eq. (3.33), we obtain the expressions to calculate the electric field intensity via magnetic field intensity at the respective shield surfaces:

$$E_{zi} = \frac{\gamma}{\sigma}\left\{H_e\frac{1}{\sinh\gamma t} - H_i\coth\gamma t\right\} \qquad (3.34)$$

$$E_{ze} = \frac{\gamma}{\sigma}\left\{H_e\coth\gamma t - H_i\frac{1}{\sinh\gamma t}\right\} \qquad (3.35)$$

Those not interested in further derivations can proceed directly to the final expressions for transfer impedance later in this section [starting with Eq. (3.47)].

✦ Substituting Eqs. (3.23) through (3.26) into Eqs. (3.34) and (3.35), we obtain the electric field intensity at the shield surfaces as functions of the current in the shield and shield design characteristics γ, r, and t:
• External return current path (Fig. 3.2a):

$$H_i = 0;\ H_e = \frac{I_e}{2\pi r} \qquad (3.36)$$

$$E_i = \frac{I_e}{2\pi r\sigma}\frac{\gamma}{\sinh\gamma t}\frac{1}{} \qquad (3.37)$$

$$E_e = \frac{I_e}{2\pi r\sigma}\gamma\coth\gamma t \qquad (3.38)$$

• Internal return current path (Fig. 3.2b):

$$H_e = 0;\ H_i = \frac{I_i}{2\pi r} \qquad (3.39)$$

$$E_i = \frac{I_i}{2\pi r\sigma}\gamma \coth\gamma t \qquad (3.40)$$

$$E_e = \frac{I_i}{2\pi r\sigma}\gamma \frac{1}{\sinh\gamma t} \qquad (3.41)$$

When writing Eqs. (3.36) through (3.41) we took advantage of the fact that $r_i \sim r_e = r$, and $r_e - r_i = t$. Also, to reduce the error even more, it is often adopted that

$$r = \sqrt{r_i r_e}$$

The components of electric field intensity can also be related to the shield direct current (dc) resistance. For a *thin* shield ($t \ll r$), the dc resistance of a homogeneous tube can be approximated as:

$$R_0 = \frac{1}{2\pi r\sigma t} \quad \text{Ohm/m} \qquad (3.42)$$

Substituting Eq. (3.42) into Eqs. (3.37) through (3.38) and Eqs. (3.40) through (3.41), we obtain the "generic" formulas:

$$E_{(ii)/(ee)} = I_{i/e} R_0 \gamma t \coth\gamma t \qquad (3.43)$$

$$E_{(ie)/(ei)} = I_{i/e} R_0 \frac{\gamma t}{\sinh\gamma t} \qquad (3.44)$$

In Eqs. (3.43) and (3.44), the indices ii and ee correspond to those sides of the shield which face the return path for the currents I_i and I_e, respectively, while the indices ie and ei correspond to the opposite surfaces.

At last we are ready to "plug in" the electric and magnetic fields into the expression (3.10). We will do this job in three moves:

1. Define the linear contours, and the surfaces delimited by them, for the integrals.
2. Calculate the double integrals.
3. Calculate the linear integrals.

Refer to the definition of the integration contour and surface, which accompanied Eq. (3.10), and to Fig. 3.3b (remember, we are using a triaxial structure). We have already stated that, in the case of a homogeneous solid tube, there is only one current component, I_z, present in the shield and the shielded conductor. (This is always true only when the radii of these conductors are small compared to the signal wavelength.) Therefore, internal contour l_i is a rectangle in the cable axial section. The contour is limited by two straight segments (hg and fe, in Fig. 3.3) running at the elementary length dl on the external surface of the shielded (or center) conductor of the cable elementary length dl and internal surface of the shield, and by two shield radius segments (gf and eh) between these lines. Similarly, the external contour l_e is also a rectangle (badc) limited by two segments of external tube radius at both ends of the elementary length dl between the internal surface of external tube and external surface of the shield and two straight segments running at these surfaces.

Transfer Parameters of Electromagnetic Shields and Enclosures 159

First notice, that when the magnetic field is absent ($H_f = 0$), the double integral in Eq. (3.10) is also zero, the corresponding (inductive) voltage is zero, and V_t is a function of only the electric field intensity at the respective shield surface, which is opposite to the current return path. Thus, for external return path, this voltage is determined along the internal contour efgh (as we stated, the direction is arbitrary but once selected, should be fixed) and according to Eq. (3.10), the voltage V_{it} is:

$$V_{it} = \oint_{l_i} \hat{E}_z \cdot d\hat{l} = \int_e^f E_z dl = E_{it}^z dl \qquad (3.45)$$

The closed integral in Eq. (3.45) yields 0 at all path segments except fe, because the field component E_z is perpendicular to the respective path segments which results in

$$\cos(\hat{E}_z, d\hat{l}) = \cos(90°) = 0$$

(See the definition of a scalar product in Chapter 1.) In a similar way, for the internal return path, this voltage is determined along the external contour abcd, and the voltage V_{et} is:

$$dV_{et} = E_{et}^z dl \qquad (3.46)$$

In Eqs. (3.45) and (3.46), we changed the notation slightly, to avoid congestion with indices.◆

Now substituting Eqs. (3.37) and (3.41) into Eqs. (3.45) and (3.46) and then into Eqs. (3.2) and (3.3), respectively, we finally obtain the transfer impedance for external return path, that is immunity transfer impedance:

$$Z_{it} = \frac{\gamma}{2\pi r \sigma \sinh \gamma t} \qquad (3.47)$$

In a similar way, we can obtain emissions transfer impedance:

$$Z_{et} = \frac{\gamma}{2\pi r \sigma \sinh \gamma t} \qquad (3.48)$$

Expressions (3.47) and (3.48) are identical, which proves the reciprocity of transfer impedance immunity and emissions! Thus we can write

$$Z_{(et)/(it)} = Z_t = \frac{\gamma}{2\pi r \sigma \sinh \gamma t} \qquad (3.49)$$

But let's not forget the assumptions on which such conclusion hinges: homogeneous tube of small thickness. Also, expressing the transfer impedance via the dc resistance of the tube per Eq. (3.44), we obtain

$$Z_{(ei)/(ie)} = R_0 \frac{\gamma t}{\sinh \gamma t} = Z_t \qquad (3.50)$$

In a similar way, a homogeneous thin tube loss impedance can be obtained. In this case, however, along with the field intensity at the respective shield surfaces facing the current return path, the voltage associated with magnetic field must be accounted for. Fortunately, the expressions for magnetic field are very simple [Eqs. (3.23) and (3.25)]. As you remember, in both external and internal circuits, it was only one component of the field Hf, whose direction is parallel to the surface bounded by the rectangle contour (that is, parallel to the normal to the surface). Thus, from Eq. (3.10) we obtain:

$$V_s = j\omega\mu_s \int\int_{S_e} H_s^\phi ds + \oint_{l_e} E_s^z dl = j\omega\mu H_s^\phi S + E_s^z dl$$

The left part of the final equality constitutes the circuit external inductive impedance Z_{sL}. It is defined by what is called the circuit external inductance (that is, "regular" inductance due to the magnetic flux through the dielectric associated with internal or external circuit, whichever is being evaluated (don't get confused with such terms as "external inductance of external circuit" and "external inductance of internal circuit"—they are all "legal"). The external inductance of a coaxial circuit is well known:

$$L_s = \frac{\mu_r \mu_0}{2\pi} \ln \left| \frac{r_i}{r_c} \right| \qquad (3.51)$$

where r_e and r_c are the internal radius of the outer conductor, and external radius of the internal (e.g., center) conductor. The relative magnetic permeability of the cable dielectric as a rule is $\mu_r = 1$, although for the external circuits this is not always true: high magnetic permeability shielding layers and/or magneto-dielectric jackets (see the previous chapter) may result in much larger μ_r. In free space, the magnetic permeability $\mu_0 = 4\pi \times 10^{-7}$. The value of L_s is pretty small; for example a 75 Ohm coaxial cable with a "free space dielectric" yields $L_s \sim 0.25 \times 10^{-6}$ H/m.

The right part of the final equality with substitution of (3.38) and (3.40) yields the conductor loss impedance Z_{ss} of the respective circuit. As a result,

$$Z_s = Z_{ss} + Z_{sL} = \frac{1}{2\pi r \sigma} \gamma \coth \gamma t + j\omega L_s = R_0 \gamma t \coth \gamma t + j\omega L_s \qquad (3.52)$$

The voltage transfer effectiveness parameter

$$T_{e/i} = \frac{Z_t}{Z_s} = \frac{R_0 \frac{\gamma t}{\sinh \gamma t}}{(R_0 \gamma t \coth \gamma t + j\omega L_s)} = \frac{1}{\cosh \gamma t + \eta \sinh \gamma t} \qquad (3.53)$$

where coefficient

$$\eta = \frac{1}{\gamma t} \frac{j\omega L_s}{R_0} \qquad (3.54)$$

◆ It is time to "set the record straight" with one problem that we may have. In deriving our formulas for cylindrical shields (tubes), we made a basic assumption that, although the shield is cylindrical, we could disregard its curvature and treat it as a plane (this is *why* we could claim similar field distribution in homogeneous solid current-carrying metallic plane sheet and cylindrical shield!). Essentially, we substituted a cylindrical wave in the shield body with a plane wave. Return to the Eqs. (3.18) and (3.19), which is where our "sentimental journey" into the "good old" world of homogeneous tube transfer parameters started.

Have you noticed that when transferring from Eq. (3.18) to (3.19) we "cheated?" Per Eq. (A.13), the Laplacian in cylindrical coordinates in Eq. (3.21) should have been written as Ξ_{cyl}, but instead we approximated it in Eq. (3.19) with a "Cartesian" Ξcart:

$$\Xi_{cyl} = \frac{1}{r} \frac{\partial}{\partial r} \left(r \frac{\partial H_\phi}{\partial r} \right) \qquad (3.55)$$

$$\Xi_{cart} = \frac{d^2 H_\phi}{dr^2} \qquad (3.56)$$

There is a fundamental difference between Eqs. (3.55) and (3.56). The solution of Eq. (3.19) is (3.20), which led us to exponential and hyperbolic functions in the expressions for transfer parameters of Eq. (3.47) through (3.53). However, the solution of the cylindrical equations would have led us instead to the modified Bessel functions of the first kind, $I_\nu(\Upsilon r)$, and the second kind, $K_\nu(\Upsilon r)$, of order ν. Then, with the cylindrical expression for the Laplacian, the solutions of Eq. (3.19) for the loss impedance and transfer impedance would have been:

$$Z_{se} = \frac{\gamma}{2\pi r_e \sigma D} [I_0(\gamma r_e) K_1(\gamma r_i) + K_0(\gamma r_e) I_1(\gamma r_i)] \qquad (3.57)$$

$$Z_{si} = \frac{\gamma}{2\pi r_i \sigma D} [I_0(\gamma r_i) K_1(\gamma r_e) + K_0(\gamma r_i) I_1(\gamma r_e)] \qquad (3.58)$$

$$Z_{it} = Z_{et} = \frac{\gamma}{2\pi \sigma D \sqrt{r_i r_e}} \qquad (3.59)$$

$$D = I_1(\gamma r_e)K_1(\gamma r_i) - K_1(\gamma r_e)I_1(\gamma r_i) \tag{3.60}$$

Along with the difference between the behavior of the exponential and Bessel functions, Eqs. (3.57) through (3.60) include both internal and external radii of the shield and also indicate that there may be a difference between the loss impedances on the inner and outer surface of the shield. Thus, theoretically, the approximation we assumed is an undoubtedly important, and it is worthwhile to investigate this question in more detail. Fortunately, this has been already done (and long ago) by Schelkunoff [P.1], who originally derived Eqs. (3.57) through (3.60). To be sure, Schelkunoff was the first to reduce Eqs. (3.57) through (3.60) to formulas that are similar (although not fully identical) to our Eqs. (3.49) and (3.52), using the asymptotic approximations of the Bessel functions at large arguments. On the other hand, in practical cable shielding applications, the need for Bessel function-based rigorous solutions is rare because of the relatively thin shields and high frequencies used. For this reason, we concentrated on expressing the shield transfer parameters via hyperbolic functions. Such representation is very convenient for engineering use and physical interpretation of the results, and thus is often used in the engineering literature. For example, many of these expressions were originally derived and analyzed by Kaden [P.2].

Note that, mathematically, the cylindrical-to-Cartesian approximation of the Laplacian can be adopted if in the cylindrical expression r can be "pulled out" from under the partial derivative symbol. This is possible without significant error when the change of the radius within the shield body is negligible. Then, it can be assumed that the tube radius does not depend on the coordinate r—which is valid if the shield thickness is much smaller than its radius. But that's exactly the assumption that we made. On the other hand, the difference between the Laplacians written in cylindrical and Cartesian coordinates for any particular shield design may serve as a measure of the error if an approximate solution is adopted.✦

✦ The transfer parameters above were derived under conditions that the current flows in *only* one circuit (external or internal), while the opposite (with regard to the shield) circuit was loaded on infinitely large impedance (e.g., of a voltmeter) and did not carry any current. You may object, that this situation is not exactly what happens in the majority of applications. However the assumed conditions are of consequence only for nonlinear shields. As long as the shield material is linear, that is, its performance does not depend on the amplitudes of the currents and fields, we can easily combine Eqs. (3.49) through (3.52) to obtain expressions for the shield with both currents (with external I_e and internal I_i returns) existing *simultaneously* in the shield. In this case, each of the voltages in the external V_e and internal V_i circuits consists of two parts generated by the internal and external return path currents:

$$V_e = I_e Z_s + I_i Z_t = I_e(R_0 \gamma t \coth \gamma t + j\omega L_s) + I_i\left(R_0 \frac{\gamma t}{\sinh \gamma t}\right) \tag{3.61}$$

$$V_i = I_i Z_s + I_e Z_t = I_i(R_0 \gamma t \coth \gamma t + j\omega L_s) + I_e\left(R_0 \frac{\gamma t}{\sinh \gamma t}\right) \tag{3.62}$$

Now, as you will see in section 3.3.1, Eqs. (3.61) and (3.62) are identical to the generalized network equations of the shield (3.79) through (3.81) The latter set will be written without invoking the physical processes in the shield, just as a hypothesis stating that the shield might be viewed as a two-port network. The equations (3.61) and (3.62) can be considered as a proof of that hypothesis.✦

3.2.4 Thin Homogeneous Solid Shield Performance Analysis

To gain further insight into the thin homogeneous shield problem, we will evaluate the variation bounds of the transfer parameters Z_t, Z_s, and T, at some critical points of their argument γt. To do this, we will apply the same approach we used in Chapter 1 to estimate the maximum bounds (MB) of the signal frequency spectrum (similar modus operandi was also used by Schelkunoff and Kaden to investigate the shielding performance trends and develop engineering formulas). Subsequently, we will write down the end results of such analysis.

- At $\gamma t \to 0$,

$$MB(Z_t) = R_0 \tag{3.63}$$

$$MB(Z_s) = R_0 + j\omega L_s \tag{3.64}$$

$$MB(T) = \frac{1}{1 + \dfrac{j\omega L_s}{R_0}} \tag{3.65}$$

Obviously, if the circuit inductance is not accounted for,[*] or at low frequencies, where "low" is determined by the ratio $(j\omega L_s)/R_0 \to 0$,

$$MB(Z_t) = EX(Z_s) = R_0 \tag{3.66}$$

$$MB(T_1) = 1 \tag{3.67}$$

- At $\gamma t \to \infty$,

$$MB(Z_t) = 2\sqrt{2}R_0 \frac{t}{\delta} e^{-\frac{t}{\delta}} = 2R_0 \frac{t\sqrt{\mu\sigma}\sqrt{\omega}}{e^{t\sqrt{\mu\sigma}\sqrt{\omega}}} \tag{3.68}$$

$$MB(Z_s) = \sqrt{2}R_0 \frac{t}{\delta} + j\omega L_s = R_0 t\sqrt{\mu\sigma}\sqrt{\omega} + j\omega L_s \tag{3.69}$$

---

[*]We may want to do that to "distill," for general purposes, just the shield performance from the characteristics of the circuits in which the shield is to be used.

$$MB(T_1) = \frac{2e^{-\frac{t}{\delta}}}{1 + \frac{1}{\sqrt{2}\frac{t}{\delta}}\frac{j\omega L_s}{R_0}} \tag{3.70}$$

or if the circuit inductance is not accounted for,

$$MB(Z_s) = \sqrt{2}R_0\frac{t}{\delta} = R_0 t\sqrt{\mu\sigma}\sqrt{\omega} \tag{3.71}$$

$$MB(T_1) = 2e^{-\frac{t}{\delta}} = 2e^{-t\sqrt{\omega}\sqrt{\frac{\mu\sigma}{2}}} \tag{3.72}$$

A quite cognitive exercise is also to present these formulas in the logarithmic form ($20 \log_{10} X$, dB), which results in much simpler notation that is easier to interpret. Thus, from Eqs. (3.71) and (3.72), we obtain relative evaluation of the transfer function parameters for homogeneous tube:

- At $\gamma t \to 0$,

$$Z_{tr} = Z_{sr} = 20\log_{10}\left|MB\left(\frac{Z_t}{R_0}\right)\right| = 20\log_{10}\left|MB\left(\frac{Z_s}{R_0}\right)\right| = 0, \text{dB RE } R_0 \tag{3.73}$$

$$T_{1r} = 20\log_{10}\left|MB\left(\frac{T_1}{T_0}\right)\right| = 0, \text{dB RE } T_0 \tag{3.74}$$

- At $\gamma t \to \infty$,

$$Z_{tr} = 20\log_{10}\left|MB\left(\frac{Z_t}{R_0}\right)\right| \approx 8.7\left(1 - \frac{t}{\delta}\right) + 20\log_{10}\left|\frac{t}{\delta}\right|$$

$$= 8.7\left(1 - \frac{t\sqrt{2}}{\sqrt{\omega\mu\sigma}}\right) + 20\log_{10}\left|\frac{t\sqrt{2}}{\sqrt{\omega\mu\sigma}}\right|, \text{dB RE } R_0 \tag{3.75}$$

$$Z_{rs} \approx 20\log_{10}\left|\frac{t\sqrt{2}}{\delta}\right| = 20\log_{10}(t\sqrt{\omega\mu\sigma}) \tag{3.76}$$

$$T_r = 8.7\left(1 - \frac{t}{\delta}\right) = 8.7\left(1 - \frac{t\sqrt{2}}{\sqrt{\omega\mu\sigma}}\right) \tag{3.77}$$

The following are the basic relationships between the hyperbolic functions that we have used to obtain the bounds (MB) of the transfer parameters of homogeneous cylindrical tubes.
- When the argument of a hyperbolic function $\gamma t \to 0$, then

$$\left|\frac{\gamma t}{\sinh \gamma t}\right| \to 1, \ |\coth \gamma t| \to \infty, \ |\gamma t \coth \gamma t| \to 1, \ \left|\frac{\gamma t}{\sinh \gamma t} \cosh \gamma t\right| \to 1$$

- When the argument of a hyperbolic function $\gamma t \to \infty$, then

$$\left|\frac{\gamma t}{\sinh \gamma t}\right| \to \left|\gamma t e^{-\gamma t}\right| = \sqrt{2}\frac{t}{\delta}e^{-\frac{t}{\delta}}, \ |\coth \gamma t| \to 1, \ |\gamma t \coth \gamma t| \to 1$$

- The relationship between the intrinsic propagation constant and skin depth is

$$\gamma = \frac{1}{\delta} + j\frac{1}{\delta}, \ |\gamma| = \frac{\sqrt{2}}{\delta}$$

It is time to "reap the fruits" of our labors. Consider a current-carrying homogeneous cylindrical shield. The current return can be external (Fig. 3.6 a) or internal (Fig. 3.6b). The field intensities at the surfaces of the shield facing the return are $E_s = IZ_s$, and the opposite surfaces $E_t = IZ_t$, where Z_s and Z_t can be determined by Eqs. (3.71), (3.73) through (3.74), or the equivalent logarithmic equations. It follows that at dc, the $E_s = E_t = IR_0$. The loss impedance and transfer impedance are identical and equal to the tube dc resistance. But the dc resistance of the tube depends on its design characteristics: conductivity, thickness, and radius [see Eq. (3.42). The larger their values, the smaller the transfer and loss impedances. Thus, for example, a copper shield will have smaller transfer and loss impedances than a lead one of the same thickness. (Sounds trivial, doesn't it?) The field is uniformly distributed within the shield thickness, so the field intensities at both surfaces of the shield are equal. Therefore, the transfer coefficient $T_a = 1$; no shield effectiveness is exhibited. But that does not mean that there isn't any shielding effectiveness. First, there are the mismatch losses (coefficient η) and, second, the shield can still provide protection from fields (even within the same cable) that are not associated with the currents in the shield, and in particular from the electric fields whose components are perpendicular to the shield surface. This second mechanism is relevant to both far fields and near-field regimes (e.g., protection from electric coupling between the "crosstalking" circuits—see Chapter 4. Another important example is the electrostatic field).

As the frequency of the signal increases, the E_s starts increasing with the law t/δ, which is proportional to the square root of frequency [$\sqrt{\omega}$—see Eq. (3.76)], while E_t quickly reduces with the law

166 Electromagnetic Shielding Handbook

$$\frac{t}{\delta} e^{-\frac{1}{\delta}}$$

which is proportional to $\sqrt{\omega} e^{-a\sqrt{\omega}}$ [see Eq. (3.73)]. The field distribution within the shield thickness changes, as illustrated in Fig. 3.6. This is the essence of energy diffusion in the shield.

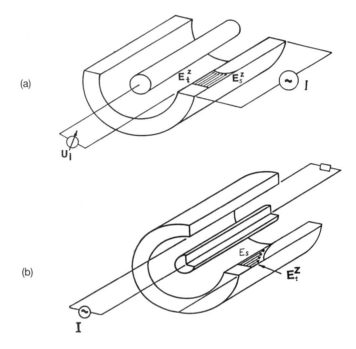

FIGURE 3.6 Energy diffusion through cable shield

The smaller the skin depth and the thicker the shield, the larger the diffusion losses in it. Therefore, the ratio t/δ is the determining factor of diffusion in the shield (see Fig. 3.7)This data is easy to translate into the frequency dependence for particular shield designs, as illustrated in Fig. 3.8 for copper shields of different thicknesses. The physics is as previously covered: while the current skin depth is of the same order as the shield thickness, no diffusion ("absorption") losses take place, and the shield performance is determined mainly by the energy reflection due to the mismatch at the shield interfaces with the dielectric media. However, with the frequency increase, the absorption losses in the shield quickly increases. Note that with the frequency rise, the transfer impedance falls down much faster than the loss impedance increases. The transfer and loss impedance curves in Figs. 3.7 and 3.8 are plotted as ratios of the transfer parameters to the shield dc resistance. However, the dc resistance also plays a role in the shield performance.

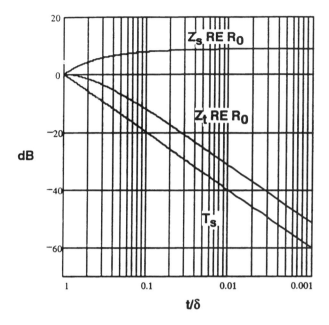

FIGURE 3.7 Transfer parameters of solid copper tubes as a function of skin depth

This problem is not as simple as it may seem, because the tube dc resistance depends on the material conductivity and tube thickness—the same parameters that also affect the diffusion processes—and on the tube diameter, which is related to the circuit external inductance. As an illustration, Fig. 3.9 includes experimental frequency characteristics of homogeneous tubes made from two metals with significantly different conductivity: copper and stainless steel. As shown, not only is the copper tube transfer impedance about two orders lower, but the phase of its "precipitous" decrease starts at much lower frequencies than for the stainless steel tube.

Figure 3.9 also contains the frequency characteristics of the transfer impedance of several practical shield designs other than homogeneous tubes. These will be discussed in the following sections.

In Fig. 3.8, the transfer parameters are presented as functions of frequency, with the shield thickness assumed as a parameter, accepting some discrete values. However, there is no reason why transfer parameters could not be treated as functions of two (or more) variables. Such treatment presents excellent opportunities in analyzing the behavior of the shields. An example of a "3-D" representation of transfer impedance as a function of the frequency and the shield thickness is given in Fig. 3.10. The geometry corresponds to the lower part of the graph in Fig. 3.8.

Being functions of a complex variable γ, the transfer parameters are themselves complex numbers. This fact can be used to obtain additional insight into the physics of the described phenomena. For example, a parallel analysis of the real and imaginary components of the shield transfer impedance and loss impedance was utilized

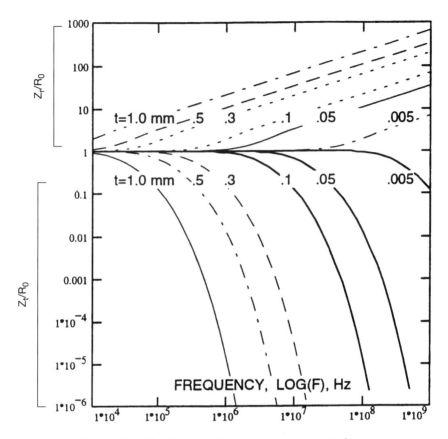

FIGURE 3.8 Transfer parameters of homogeneous copper tubes

by Kaden. Modern computing hardware and software permit even wider use of these techniques. As an illustration, Fig. 3.11 shows a vector diagram of the transfer and loss impedance frequency characteristics for a copper shield. Additionally, Fig. 3.12 presents separately the frequency characteristics of the real and imaginary part of the transfer impedance. All the curves are the ratios of the respective characteristics to the dc resistance R_0. In complete agreement with the physics, the curves start as real numbers with value 1 (which means that the transfer and loss impedance are equal to the dc resistance of the shield) and then vary in accordance with the discussed rules. Of course, for shield thicknesses other than shown, the dependencies will be different.

✦ Here's an interesting problem, and it can shed additional light on the physical processes in solid homogeneous shields. Consider a solid homogeneous shield within a triaxial structure. The shield thickness is relatively large (not large enough, however, to invalidate the use of a plane wave propagation model in the shield). In such a shield, the interconnecting conductors can be attached at different points across the shield thickness, from the internal surface (radius r_i) to the external surface (radius r_e). The question is: how will the transfer and loss impedance depend on such connection?

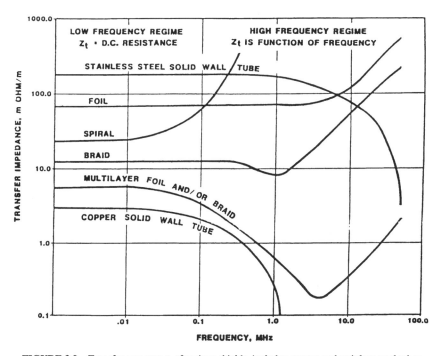

FIGURE 3.9 Transfer parameters of various shields, includng copper and stainless steel tubes

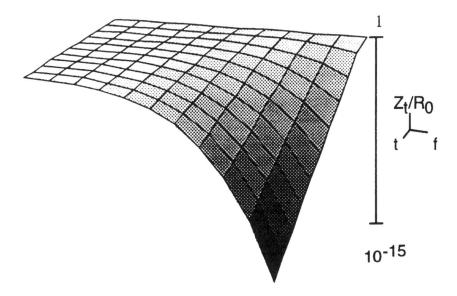

FIGURE 3.10 Transfer impedance of solid copper tubes as a function of frequency and thickness

170 Electromagnetic Shielding Handbook

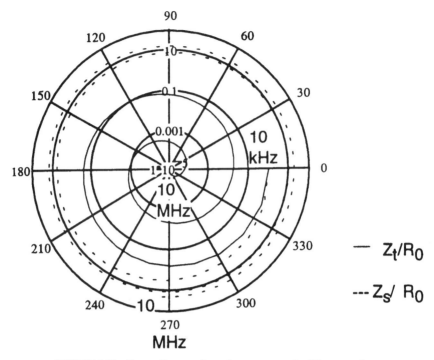

FIGURE 3.11 Vector diagram of transfer parameters of solid copper tubes s

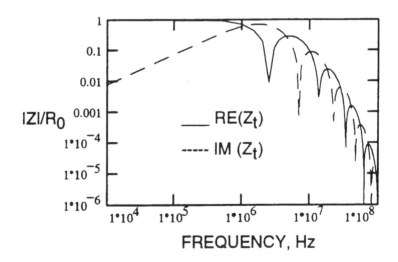

FIGURE 3.12 Real and imaginary parts of transfer impedance

Suppose, the shield (or the outer conductor of a coaxial cable) is powered at the internal circuit. In this case, the signal energy is concentrated inside the cable; that is, in the space between the direct and return current path. From there the energy diffuses into the shield thickness and partially penetrates through it. Although this energy "enters" the shield from its internal surface, we don't have to select this surface as a reference in order to evaluate the shield properties. Therefore, it *does not make any difference* at which level (determined by coordinate r *within the thickness*) the current enters the shield. Except, of course, for edge effect—but we assume an infinitely long cable, so the edge effect can be disregarded.

To prove this, let us investigate the "physics" of the phenomena involved. The shield loss impedance Z_{ss} is determined by the voltage drop at the shield internal surface. However, the voltage drop in question is not only the *resistive* voltage drop at the shield surface, but also the *inductive* voltage drop, which is determined by the magnetic flux through the surface formed by the contour bounded by the direct and return current filaments [see Eq. (3.10)]. We think in terms of taking this shield "current filament" at the shield's surface. But in principle, nothing precludes our selecting this "filament" at any distance within the shield body. The thing to remember, though, is that as a rule (except at dc and low frequencies), the farther we move from the shield internal surface (that is, the surface facing the return path) in the direction of the external surface, the smaller the electric field intensity E_z [e.g., see Eq. (3.33)]. If we change the reference surface within the shield thickness by moving it, for example, from the internal to the external surface (see Fig. 3.13), voltage drop $V_r = E_z dl$ changes from $V_{ir} = IZ_{is}$ to $V_{er} = IZ_{et}$. However, by moving "deeper" inside the shield, we also change the area of the contour that determines the inductive impedance, so that the inductive ingredient of the voltage changes. As a matter of fact, with the movement inside the shield, the enclosed area includes not just the dielectric enclosed by the shield, but also a part of the shield body that is enclosed by the contour.

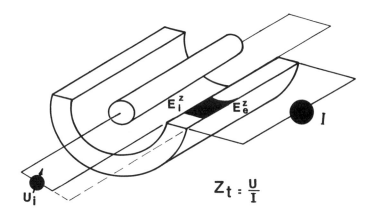

FIGURE 3.13 Energy diffusion through a cable shield

We now will show that, as we move the reference surface from one shield surface toward the opposite one, the *total voltage* determined by the *sum* of the diffusive and inductive ingredients stays *unchanged*. Let us prove this first for the shield internal loss parameter, Z_s. Note that the part of inductive voltage caused by the magnetic flux in the dielectric inside the cable stays the same, independent of what is happening in the shield body. (Per Ampere's law, it depends only on the current in the center conductor.) Therefore, all the changes due to our "action" will take place within the shield body.

When the generator is connected to the shield internal surface and this surface is selected as a reference, the voltage drop $V_{is} = E_{zi}dl + j\omega L_i$, where L_i is the inductance associated with the magnetic flux in the dielectric inside the cable. However, when the generator is connected to the shield at some distance $(r - r_i)$ from the internal surface and this level (i.e., $r - r_i$) is selected as a reference

172 Electromagnetic Shielding Handbook

surface, the voltage drop is $V_{isr} = E_{isr}dl + j\omega L_i + j\omega L_{sr}$, where L_i is the inductance due to the magnetic flux in the shield body. By substituting Eq. (3.32) and (3.33) into Eq. (3.10), with the proper values of magnetic fields H_e and H_i, we can prove that for any r,

$$V_{isr} - V_{is} = (E_{isr} - E_{is})dl - j\omega L_{sr} = 0 \qquad (3.78)$$

This means that, with the selection of the different *reference* plane, *nothing* changed in the shield performance. We can also prove the same for both external and internal return paths and when considering the loss or transfer impedances.✦

3.3 A "SECOND LOOK" AT TRANSFER PARAMETERS ...

3.3.1 Circuit Theory / Reactive Near Field Nature of Transfer Parameters

✦ The transfer parameters, as they are suggested in this book, are fundamentally based on the circuit theory representations. This means that the basic "tools" to define, evaluate, compute, and measure these characteristics are based on the concepts of *current*, *voltage* and / or corresponding *reactive near zone* electric and magnetic *fields*. However, at this time we have not yet classified the fields by the *wavelength-to-distance-to-shield size* criteria and thus do not yet see the *big* comprehensive picture of electromagnetic phenomena in a broad frequency range, as it applies to electromagnetic shielding. The subject as a whole will be addressed in necessary detail in chapter 4, and only certain specific considerations will be noted here.

The developed models make possible very useful circuit network representation. Revisit Fig. 3.2a and b. Here, I is the interference current in the shield; dV_t and dV_s are the voltages generated by the current I on the dl length of the shield and measured (that is, voltmeter circuit is arranged at), respectively, on the opposite and the same shield side with the current I return path. As we saw, the return can be either internal ($I = I_i$ and $I_e = 0$) or external ($I = I_e$ and $I_i = 0$), which corresponds to the diagrams of Fig. 3.2a and b. Referring to the Appendix (Section A.4), we will hypothesize (and in principle, can present certain proofs of this hypothesis) that this is a typical two-port network. Thus, even without knowing the actual processes inside the shield, we now can view this length of the shield as a "black box" two-port network, with the input on one side (internal, or external) and the output on the other side (opposite to the first), as shown in Fig. 3.4a and b. The respective relationships between the currents and voltages in such networks are described by Eqs. (A.27) through (A.41).

In general, the currents with both external and internal return paths will be present in the shield so that the total current in the shield $I = I_i + I_e$ (Fig. 3.4c). Then the currents are injected into and the voltages are measured at the terminals of the "black box" from both sides. For example, consider a coaxial cable placed in the EMI environment and carrying a signal. The current in the cable outer conductor (which is also the shield) is a sum of the " useful " signal I_i with the internal return path and interference signal I_e, with the external return path (e.g., similar to the currents in the center shield in triaxial line in Fig. 3.3). The voltage at the cable terminals (that is, between the center and outer conductors) is the sum of the " useful " voltage V_{is} produced by the transmission generator and interference voltage Vit, "transferred" from the ambient environment through the shield. Now, each of the voltages at the both sides of the shield is determined by both currents: with external and internal return paths. Using the theory of the two-port networks (see Appendix) or direct derivations (revisit (3.61) and (3.62)), the generalized expression relating the currents and voltages at both sides of the shield is

$$dV_i = I_e Z_{it} + I_i Z_{is} \tag{3.79}$$

$$dV_e = I_e Z_{es} + I_i Z_{et} \tag{3.80}$$

Or in matrix form,

$$\begin{bmatrix} dV_{it} \\ dV_{et} \end{bmatrix} = \begin{bmatrix} Z_{is} & Z_{it} \\ Z_{et} & Z_{es} \end{bmatrix} \cdot \begin{bmatrix} I_i \\ I_e \end{bmatrix} \tag{3.81}$$

This compares with Eq. (A.30) or (A.37) (which are equivalent). If our network is linear and reversible (remember reciprocity principle?), then $Z_{it} = Z_{et}$. Also, if the shield has a geometric and electrical symmetry and thickness much smaller than the radius and the wavelength ($t \ll r \ll \lambda$), then $Z_{is} = Z_{es}$.

The condition $t \ll r \ll \lambda$ implies that signal wavelength is *much* larger than the shield *transverse* geometry (since the shield length is a *longitudinal* measure of the shield). In other words, the signal frequency (which is reverse to λ) is relatively *small*. Of course, we should have expected this, while so "heavily" relying on the circuit theory.

All this means, that in our model there is *no place* for radiating, displacement, currents. Thus, our transfer parameters do not directly "participate" in the EMI radiation! Of course, this doesn't mean, that they do not affect radiated emissions and susceptibility of the shielded products. But as a rule these parameters are used to model the *generator* part of the radiating EMI sources, while the radiators are modelled in terms of antenna theory.✦

3.3.2 Shielding "from Currents" vs Shielding "from Fields"

Let's have another look at Eq. (3.53). Compare it to Eq. (2.33) in Chapter 2—they look identical. But Eq. (2.33) was obtained as shielding effectiveness under typical plane wave incident on a plane shield conditions, whereas Eq. (3.53) yields the transfer effectiveness (read, "simplified" shield effectiveness) parameter related to the current in the shield. Isn't it interesting that such basically different premises (shielding from fields versus shielding from currents) yield identical results? By the way, this "shielding from fields" versus "shielding from currents" terminology was introduced by Kaden. He attributed so much importance to this difference that he even separated his book [P.2] into two distinctive parts, literally titled so. However, as we can see, in fact the difference is not that great. We shouldn't be surprised by this coincidence, because these two approaches have actually many more commonalities than differences. First, the departing Eqs. (3.19) and (2.11), which were used to obtain Eqs. (3.53) and (2.33), respectively, are identical. Second, both Eqs. (3.53) and (2.33) relate the signal at the opposite sides of the shield: the first deals with the voltages developed in the internal and external circuits, and the second deals with the fields by definition (2.1) of the shielding effectiveness. So where is the "catch?"

To identify the difference between the "field versus current" approaches, let us look at the "physics" behind the items in the respective equations. In both Eqs. (3.53) and (2.33), the $\sinh \gamma t$ and $\cosh \gamma t$ relate the shield performance to its design: thickness t and material properties. [Intrinsic propagation constant γ is a function of the shield material magnetic permeability and conductivity—see Eq. (A.56)]. This part is identical for "fields" and "currents." However, the contents of the coefficient η in Eq. (3.53) and ξ in Eq. (2.33) are different. Per Eq. (3.54), along with the dependence on γt, coefficient η also incorporates the ratio of the inductive impedance to the dc resistance of the shield and is a function of the EMI-generating circuit characteristics (in the simplest case, just the tube radius). By comparison with Eq. (2.33), η in Eq. (3.53) can be also viewed as a mismatch loss which is due to the difference between the external inductive impedance of the circuit and conductor loss impedance of the tube. But per definitions in Eqs. (2.27) and (2.34), ξ represents the ratio of the interfacing media intrinsic wave impedances of Eq. (2.4), with associated reflection loss.

Now we really have arrived at "the heart" of it. Indeed, Eqs. (3.53) and (2.33) are based on inherently different physical models: near field based circuit theory for the first, and far field plane wave theory for the second. And they lead to identical results! However, even when the concepts of the far field zone and plane wave are invoked, it is done so by analogy with the near reactive fields, and / or transmission lines. This is possible because the electromagnetic field structure in both far field plane wave and near field within a cable is transverse electromagnetic (TEM), the basic initial equations and their solutions were identical in form. (Also, the respective physical mechanisms resulted in corresponding differences in reflection coefficient η. We will continue this discussion in Chapter 4, and you can also revisit Section 2.4).

To summarize:

- The "shielding from currents" and "shielding from fields" coefficients are described by formally identical expressions incorporating two parts: absorption losses in the shield and reflection losses at the interface between the shield and the surrounding media.
- The absorption losses are physically identical in both "current" and "field" shielding and are determined by the shield thickness and material properties. The absorption losses are representative of the shield transfer function and casn be related to the shield transfer impedance
- The reflection losses in "current" and "field" shielding are physically different and are determined by the mismatch in the near field in the first case, and the far field in the second case.

3.3.3 The Role of Coupling in Shield Transfer Function

Shield Transfer Parameters Still Depend on Coupling

In this section we will address certain coupling issues, even if this whole chapter is mainly dedicated to the shield transfer function. We will return to a more comprehensive "coupling" discussion in Chapters 4 and 5. As much as we wanted to make the concept of transfer function *independent* on coupling, we still had to "settle" for an at least simplified and "fixed" coupling item in the transfer parameter definitions!

To streamline the transfer function problem and ensure a high degree of model generality, we had to "forget" about that part of the shielding process, which is associated with the ambient field configuration and distribution, its coupling to the shield and reflection, and mutual influence of different parts of the same shield. We also assumed a symmetrical (coaxial) placement of the shielded conductor, the shield, and the common mode return — in short, we based our derivations on a triaxial configuration.

Of course, we understand now that just because of this simplification we were able to obtain shield transfer parameters which are *almost*, but *not fully* invariant with regard to the shielding system configuration! Here we should note, that while the transfer parameters do indeed depend on certain system parameters, this dependence is relatively limited and can be controlled and standardized, that is, made similar for different shields, at least as a requirement of the shield evaluation model and / or test procedure. Thus, the correlation and repeatability of the transfer parameter evaluations and test results can be ensured.

If not properly addressed, this "almost" can easily become the "fly in the ointment", because in reality the transfer parameters *do* contain certain system-related items. As a result, to avoid the pitfalls, it is *always wise* to start the shielding performance study with *identifying* the *philosophy* and *techniques* used to obtain the data, and then *account for* these specifics in further analysis.

"External" Coupling in Shield Loss Impedance and Transfer Effectiveness

First of all, let's "find the culprit". Consider in Eq. (3.52) *only* the voltage drop at the shield surface (internal or external, depending on the immunity or emissions case), and in Eq. (3.53) consider only the absorption shielding. This assumption has not only theoretical, but also a practical value, since a similar situation takes place in practice, when the inductance of the cable circuit is small and/or can be neglected in comparison with resistive impedance, and the reflection ingredient of the plane wave shielding can be neglected in comparison with the absorptive ingredient. This leads to quite simple expression for the voltage transfer effectiveness parameters which is identical (instead of just "looking" identical) for both circuit and plane wave:

$$T_a = \frac{E_t}{E_s} = \frac{1}{\cosh \gamma t} = S \tag{3.82}$$

Eq. (3.82) describes all too good familiar absorption shielding losses. And as we see, this parameter *does not* contain any system-related items and is identical for every shield!

Since the basic equations are identical for other solid homogeneous shield designs (e.g., plane sheet or sphere) we could have repeated the derivations with the same results (3.49), (3.50), (3.52), (3.53), except for the items related to the external inductance L_s and depending on this inductance coefficient η. In particular, the coefficient η presents a ratio of the external inductive impedance of a coaxial circuit to its

DC resistance R_0, and thus can be viewed as a reflection coefficient between the two media: shield metal and air (for instance, see equation (2.33) and related text).

But doesn't the reflection coefficient belong to the "family" of *coupling* parameters? And if "yes", how did we end up with it in the expressions for shield *transfer* parameters? The answer to this question can be obtained by reviewing equations (3.52)-(3.53). Indeed, these equations feature the shield *external* inductance L_s, which, by definition, depends on the configuration and characteristics of the *other* system elements, than the shield under study. Of course, we could have "stripped" off that inductance from our definition of the loss impedance (in other words, we could have defined the shield loss impedance without this external inductive item). But practically, it is just this item that is as a rule the largest and the most important part of the shield loss impedance parameter. So, we have to "live with it" in the definitions of the shield loss impedance and consequently of the shield transfer effectiveness.

Looks like we've "got ourself a problem": the external inductance is a part of the loss impedance and shield transfer effectiveness and *does depend on the shield shape*, as a function of the orientation and the area of the respective inductive contours. In particular, the expressions for the coefficient η were derived and are available in literature[P.2-3]: if we "plug in" the respective expressions for η in (3.53), the *ratio of the shield transfer effectiveness of the plane ("box"), cylindrical ("tube"), and sperical shield is $T_b:T_c:T_s = 1:2:3$.*

That's why we have *almost*, but not completely, achieved our goal of formulating system-independent set of shield transfer parameters. As hard as we tried, only the transfer impedance Z_t (e.g., see equations (3.49) and (3.50)) appears to be "free" of system" influence", while the loss impedance Zs and transfer effectiveness T are functions of the circuit external inductance L_s. Easy to understand, that L_s does depend on the configuration of the shield evaluation system "outside" of the shield itself. Of course, we also understand, that we had no choice: we did need a *physical circuitry* to inject the current into the shield at its "one side" and measured the generated EMI voltage - at the "other side". Even if we could come up with different a specific implementations of these circuits, we do need *a circuit* of some kind.

As any other inductance, the inductance L_s is a typical "circuit theory" concept, and it is generated by reactive magnetic flux through the respective contour. Therefore, it is implied that the distance between the direct and return path in the external circuit meets near field zone conditions. So, what happens if this distance is large (electrically), not meeting the near field conditions? Don't even ask, because this will lead to quite "unfortunate" complications! Because then we will have to take into account the radiation (and, most probable, in the near field radiation zone!), the scattering, and the field diffraction. Now the "things" will get so "out of hand" that we will lose any hope to make sense of the whole process. That's why in our model we separated the shield transfer function from the coupling functions within and outside the shielding enclosure - to be able to address the respective problems separately.

Transfer Impedance is a Function of Internal and External Coupling and Excentricity of the Shield

✦ Note that even the transfer impedance "independence" on coupling holds only for certain shield designs and/or test configurations, and may be not true for others. Indeed, consider cylindrical shield in Fig. 3.2. It follows from the most general equation (3.10), that only in the case of current-carrying homogeneous cylindrical solid shield the magnetic field is absent inside the shielded space. But, as we will see later, shields with apertures and spiral shields generate internal magnetic fields inside the shielded space. Then the *internal coupling* related to this field is a function of and can be characterized by internal inductance, which is dependent on the shield construction, as well as on the geometry and spacing of the shielded conductors. This is the case with any shields: cable (round and flat), plane sheets and surfaces, three-dimensional shields. As long as the respective inductance contours inside the shield meet the electrically infinitesimal size conditions, the concept of inductance can be sustained.

However, under certain conditions, even solid homogeneous cylindrical tubes will generate this internal inductance. But how could that be, if there is no magnetic field inside a cylindrical tube, generated by the current in this tube? Isn't this what a coaxial cable with solid homogeneous cylindrical external conductor is all about? Yes, but for this to be true, the cable *must* be *coaxial*. An excentricity in placement of the center conductor within a cylindrical tube may result in the generation of magnetic field *inside* the tube, which affects the transfer impedance values and frequency characteristics. In particular, when the center conductor is excentrically placed within a solid homogeneous tube, the high frequency transfer impedance becomes dependent on both the measure of excentricity and the way that the current is injected into the shield. This problem has been addressed before by Kaden [P.2]. However, it didn't lose its timeliness in modern times. For example, theoretical analysis and measurements were performed [3.17] on a shield (copper tube with dimensions l = 2.03 m, D = 28 mm, t = 1.39 mm) with the external current injection (return path) made with a wire in close proximity to the shield. When the center conductor moved within the tube from the center to periphery, the high frequency transfer impedance values varied up to 10 dB. Similar data were obtained for a braided shield. That's a persuasive manifestation of internal and external coupling association!

To appreciate the importance of the situation, note that practical multi-conductor shielded cables — round and flat, paired and single-ended — as a rule do *not* have a coaxial configuration. As an illustration, consider a flat shielded cable. In principle, the distribution of electric and magnetic fields within and outside the shield body changes because of the shield curvature variations and enclosed conductors' excentricity. However, when we derived our formulas, we neglected both the shield curvature and conductor excentricity, anyway. Nevertheless, though inductance L_s changes, the inductive part of the loss impedance does not have a significant effect on the shield transfer effectiveness until we reach fairly high frequencies. Also, at lower frequencies, the transfer effectiveness is determined mainly by the rate of reduction of the transfer impedance. Therefore, as a first approximation, we can still use the derived formulas. If necessary, the higher order effects can be accounted for by integrating Maxwell's equations and applying the specific boundary conditions, although the solutions are very often "quite not trivial" and, as a rule, require the utilization of numerical techniques. Although much has been written on the subject, there is still a great deal to be done.

Does that mean that the *single* transfer impedance / transfer admittance parameter is not sufficient to characterize shielding transfer function? Indeed, there were attempts by some authors to introduce *multiple* transfer impedance/admittance - like *parameters* [3.18-20], each of them would account for a *specific* type of field configurations and coupling. In fact, the authors of the referenced sources claim that, in principle, there should be *"an infinite"* (!) number of transfer impedance/ admittance parameters to account for *each* such combination. And, as approximation, they recommend the use of at least five such parameters.

Should that be the way to go? Not at all! Just think of it. While there are practical needs to account for different shield designs, field configurations, and electromagnetic coupling particulars, the *merits* of introducing *a large number of separate* Z_t / Y_t - *"type" parameters* for each possible combination of these factors are highly questionable. One thing is to solve a number of technical problems by using the single evaluation parameter "of choice" and determining its numerical value for any of these problems. Then, as complex as these problems and their solutions are, one can always compare the respective designs, because the *parameter of merit* is *the same*. Another thing is to introduce a separate parameter of merit for each such problem. Then, the problem is considered on a very *specific* level — the *generality* (or rather, *lack of* it) of each such parameter would all but make them useless (see Chapter 1), since the obtained results are *not applicable* to *any other* problem or design. Now, we *have created* an additional hurdle: how to compare this infinite number of "orthogonal" evaluations ("apples with oranges"?)! But such comparison *is* the main objective of the designers and users of any products. Thus, the "infinite parameter number" approach destroys the very rationale of introducing an evaluation figure of merit, which permits to compare different design solutions.

We can easily avoid these difficulties by "not creating them". It is really unnecessary to resort to such "desperate" measures as an infinite number of a shield (or any other product, for that matter) evaluation parameters. *ALL* practical applications can be accomodated within a single universal parameter. And this parameter of choice is the transfer impedance/transfer admittance, as formulated by Schelkunoff and further developed by Kaden and many, many other researchers. In particular, Kaden had successfully applied this philosophy to one of the problems at hand - effect of excentricity on transfer impedance.

Also, keep in mind the objectives and limitations of the transfer impedance parameter: it was conceived to evaluate the shield transfer function at a very general level, and employs simplified and standardized ("generic") coupling regimes. The rationale for shield *transfer parameters* of choice is not necessarily *exact* performance of the system but *true representation* of the *shield - "proper"* performance , *repeatability*, and matching the *system objectives* — see the very beginning of section 3.1.1 for the rationale. Such rationale minimizes the effects of coupling on the shield transfer function eand permits comparison of specific shields using a common figure of merit. Then the different specific field configurations, coupling mechanisms, as well as their interaction with the transfer function, are accounted by the shielding model as presented in Chapter 2 of this book. In the following chapters we will address this subject many times and in many details. ✦

Shielded Conductor-to-Shield Coupling in 1-, 2- and 3- Dimensional Shields

We will now concentrate on the role of *"internal" coupling* between the shielded conductor and the shield. Along with purely engineering aspects, this discussion of coupling has definite methodological implications. Consider first an electronic cable. As far as signal propagation is concerned, there is only one degree of freedom in a cable - along its axis. Therefore, an electronic cable shield (cylindrical tube) can be treated in the majority of applications as *electrically one-dimensional* object. Indeed, we have defined the transfer parameters on small line elements. But for any infinitesimal shield element the distance from the shielded conductor can be assumed constant. In a cable, this distance can as a rule be also considered *electrically small* — and thus present no "threat" of the differential mode radiation conditions. So, when evaluating cable shield transfer parameters, as formulated, we may not "bother" with this distance variations and with radiation. Also as a rule, the usual design of an electronic cable (cylindrical tube) resulted in the symmetry and "controlled nature" of the electromagnetic field within the shielded space — TEM field configuration — see chapter 4).

This is not the case for plane sheets, as well as product enclosures, electronic office cabinets, shielding rooms, etc., which possess 2- or 3-dimensional geometries. The two- and three-dimensional shields present a problem, because they provide additional "degrees of freedom" for signal current propagation and for the shielded conductors placement within and with regard to the shield confines. No doubt, the additional "degrees of freedom" affect the electromagnetic fields and circuit parameters (currents and voltages), resulting in certain complications in shielding performance. But how important is this in the context of electromagnetic shielding?

In our derivations we have always assumed that the shielded conductor (e.g., the center conductor of a coaxial cable) is parallel to the shield. While this was true in one-dimensional shields, in two- and three-dimensional shields it may be not the case. Now, we may have three problems:
- The field along the shielded conductor may be non-uniform
- The distance between the shielded conductors and the shield is variable (see Fig. 3.14)
- The distance between the shielded conductor and the shield may be *not* electrically small

So, how do we handle this situation when analyzing the shield transfer parameters?

Consider Fig. 3.14 for an "immunity" shielding. If a current I_s is induced or injected in the ABCD shield confine with an external return path, respective EMI voltages can be measured between the shield and current-carrying conductors at 2F and 1E. And this is true for each of the shield confines, e.g., AMND, as shown in Fig. 3.14. The magnitude of the induced in the shielded conductor 1-2 voltages (as well as the current I in the shielded conductor 1-2 with the internal to the shield return path) will depend on the transfer impedance of the shield.

Now, all the distances between the shielded conductor ends and the different surfaces of the shield (e.g., 1E, 2F, 1G, and 2H in Fig. 3.14), are as a rule different. But if an infinitesimal current element is considered, instead of the different distances 2F and 1E, for example, some average distance 3-4 can be used.

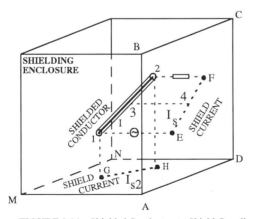

FIGURE 3.14 Shielded Conductor-to-Shield Coupling in a 3-dimensional shielding enclosure

Further, assume that the size of a surface ABCD is large enough to disregard the effects of the shield surface edges on the current I_s. Then, the treatment of a differential (i.e., infinitesimal) current element is no different from the analyzed above cylindrical, or "flat" shield. For instance, "shorting" (or loading at a corresponding impedance - as shown in the figure) the 2F and measuring the voltage 1E for a specified value of current I_s, the transfer impedance can be determined of this particular shield surface element. Of course, this should be done for every surface of the shield confine (e.g., current I_{s2} at the enclosure surface AMND), and the effects of all surfaces must be integrated to obtain the performance of the shielding enclosure as a whole in a given environment. Obviously, for a finite size shielded conductor or line, the effects of all small elements also must be integrated to obtain the performance of the shielding enclosure as a whole in a given environment.

When the 2- and 3-dimensional shields are large enough to disregard the reflection from their ends and edges, and the distances are electrically small, the difference with a 1-dimensional cylindrical shield is not that great. Indeed, as we just saw, this is true for voltage transfer effectiveness of a box, a cylindrical tube, and a sphere. Because the "large" shield size and inifnitesimal size of the shielded conductors were stipulated as preconditions to introduce the transfer parameters, we were able to assume the transfer parameters *independent* on specific shield shape! That's the "good news".

Now, several "bad news". First, in the two- and three-dimensional shields, *unlike in the one-dimensional cable shields*, we may often encounter a situation, when the distance between the shielded conductor and the shield is electrically large, resulting in near - and far field radiation conditions. Another problem develops, when the two- and three-dimensional shield sizes are finite (that is, not large enough). In the chapter 4, we will address this case as a radiation coupling problem. However, as "bad" as this may seem, even in these cases our transfer parameter formulations are extremely useful. In the two- and three-dimensional shields, as well as in the one-dimensional cable shields, such approach provides the shield's most general figure of merit, largely independent on the specifics of the system in which the shield is used. Also, this approach facilitates the physical understanding and interpretation of the underlying phenomena, and the ease of integration of transfer function in system-specific models of different levels.

3.3.4 High Frequency Current Return Path in the Shield Mainly Follows the Direct Path

✦ In Fig. 3.14 , as in Fig. 3.2.d, the current in the shield is shown to propagate *parallel* to the current in the shielded conductor. But haven't we noted that multi-dimensional shields possess in principle multiple degrees of freedom for this current to propagate? For example, what prevents the current in a homogeneous plane shield, as shown in these figures, to propagate any other way, or even at the "whole" surface of the shield? In fact, this is a valid question, and the answer is not that simple! By no means this problem is new, and the task here is how to select one out of a host of explanations (all true and correct!) found in the literature.

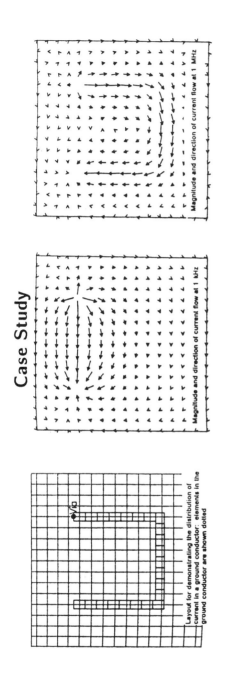

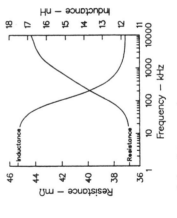

FIGURE 3.15 Current return in ground plane interconnections

182 Electromagnetic Shielding Handbook

As an alternative, we will review a practical example, presented in the literature [3.1]. Consider a U-shaped direct current trace over a ground plane, as shown in the figure Fig. 3.15 a. The signal generator is placed behind the ground plane and connected to the trace through a *via*[*] in the ground plane. At the other end of the line, the current-carrying trace is directly connected to the ground plane.

The ground plane is divided on elements, as shown. Obviously, the direct current path follows the trace - since it is a *one-dimensional* structure. But the ground plane under the direct path is *two-dimensional*. The question is, what is the current return path in the ground plane? This problem was simulated using the method of moments (see Appendix).

Here's the problem. Obviously, if you measure the DC resistance, the shortest *electrical distance* (that is, *smallest resistance*) between the ends of the direct trace will "top" the "U" of the direct path trace. Therefore, you could expect the bulk of the return current to take this shortest electrical distance path, with only insignificant current "branching out" around this shortest distance, their amount being inversely proportional to the increase in length and consequently, in resistance of these other, alternative paths. And this actually is the case at DC and lower frequencies (see Fig. 3.15 b). However, the situation is drastically different at high frequencies. As shown in Fig. 3.15 c, the return current "doesn't care" about the shortest trace and smaller resistance, and follows the direct path! Doesn't the Ohm's law apply any more?!

Fortunately, the sky didn't fall and the Ohm's law is still valid. Remember, that at high frequencies it is not DC resistance, but *AC impedance,* that matters. An AC impedance of a current contour (that is, the contour formed by the direct *and* return paths), consists of the AC resistance and inductive impedance of the contour: $Z_{AC} = R_{AC} + j\omega L$, where the inductive impedance ωL is proportional to the frequency and the area of the contour (it determines the inductance). It is clear, that the closer the return path is to the direct path, the smaller is the area of the inductive contour, and the smaller is the inductive impedance. Because with the frequency rise the contribution if the inductive impedance ingredient into the total impedance of the current contour quickly increases, the return current tends to keep closer to the direct current, thus reducing the contour area and inductive impedance. At high enough frequencies, almost all the return current will "collect" under the direct current path.

✦

3.4 PRACTICAL NON-UNIFORM SHIELDS

In this section, we will present analysis, as well as important quantitative evaluations, of transfer parameters for several important typical shield designs and applications. The data illustrates the fundamental phenomena associated with these shields, and simultaneously may be used as a reference for respective practical designs. We will explain the observed behavior in the next section. use the solid shield theory as a starting point to investigate the specific effects introduced by the shield nonuniformities.

3.4.1 The "Troubling World" of Shield Non-Uniformities

"Solid" homogeneous shields find application in shielded rooms, product enclosures, and in rigid and some designs of semi-rigid cables. However, even an extruded enclosure needs certain provisions for cable and connector penetrations, visual readouts, and ventilation. Similarly, the majority of cable applications require a certain

---

[*]Via is an isolated through-hole in a printed circuit layer, interconnecting circuitry in different layers.

degree of flexibility (see Chapter 6), which cannot be met with a solid homogeneous tube. Therefore, practical product enclosures and cable shields as rule contain different kinds of *seams, gaps,* and *perforations.* Additionally, to assure the shield flexibility, the cable designers must resort to *spirally* applied tapes and wires and corrugated tubes. Meanwhile, the realization of these *physical* features may have a drastic effect on the shield performance, as we have witnessed in the examples in chapter 2.

There are "non-uniformities" and "non-uniformities", meaning that there is a need for some classification. We start with the degrees of freedom for the shield current to flow. In non-uniform shields, the shield current path is usually *pre-determined* by the shield design and, as a rule, the number of *"degrees of freedom"* for the shield current to flow is be severely *limited*. The effect of these limitations depends on the "alignment". of the shield nonuniformities with the field vector. To simplify the solutions, it is expedient to consider three types of problems:

1. The shield consists of *narrow* enough elements. The currents in it always propagate in the direction of these elements, because there is only one degree of current flow freedom - *along* the narrow elements. Such designs the most often are used in cables, e.g., *serve, spiral tape, braid.*
2. There are still left multiple degrees of freedom, even if the non-uniformities impose the current flow limitations. Now, multiple current paths are possible, and it may be quite difficult to figure out the proper ones and what is the end effect.
3. The shield has finite dimensions, which results in the effects of the current reflections and propagation environment changes at the shield edges and interface with other system elements.

As an illustration, Fig. 3.16 shows the problem explanation, as presented by Kaden in Ref. [P.2]. As follows from basic laws of electromagnetism, the currents are generated at the shield surfaces perpendicular to the respective field vectors (in these models, Kaden considered magnetic field). While in a homogeneous solid shield of *infinite dimensions in all directions* the currents in the shields can propagate in any direction, the "obstacles" in the shield and the finite size of the shielding enclosure *may* or *may not* result in drastic changes to these 'regular" current flow patterns. Indeed, in the shield with a "horizontal" (at the figure) seam there will be almost no effect on the propagation of the currents generated by "vertical" field. In the same cable, the currents generated by the "horizontal" field, will have to "go through lots of troubles" to close the current loops (hopefully, you don't question the currents need to close?). Not less educational are the models of cable shields consisting of circular coils.

To limit the number of possibilities, Kaden usually considered two "extreme" cases: the first, where the shield non-uniformities *did not* present obstacles to the propagation of the induced (or injected) into the shield currents, and second, where the shield non-uniformities *did* present obstacles to the propagation of the shield currents. The "moral" was that, if possible, the shield nonuniformities must be placed so as not to prevent the induced current flow in the shield. This is certainly a good advice, where it can be followed, e.g., , when placeng a transformer in a seamed shielding enclosure. But, in general, when the "real-life" alignment of the shield with the field is unknown, the shielding must be evaluated at the worst alignment, producing

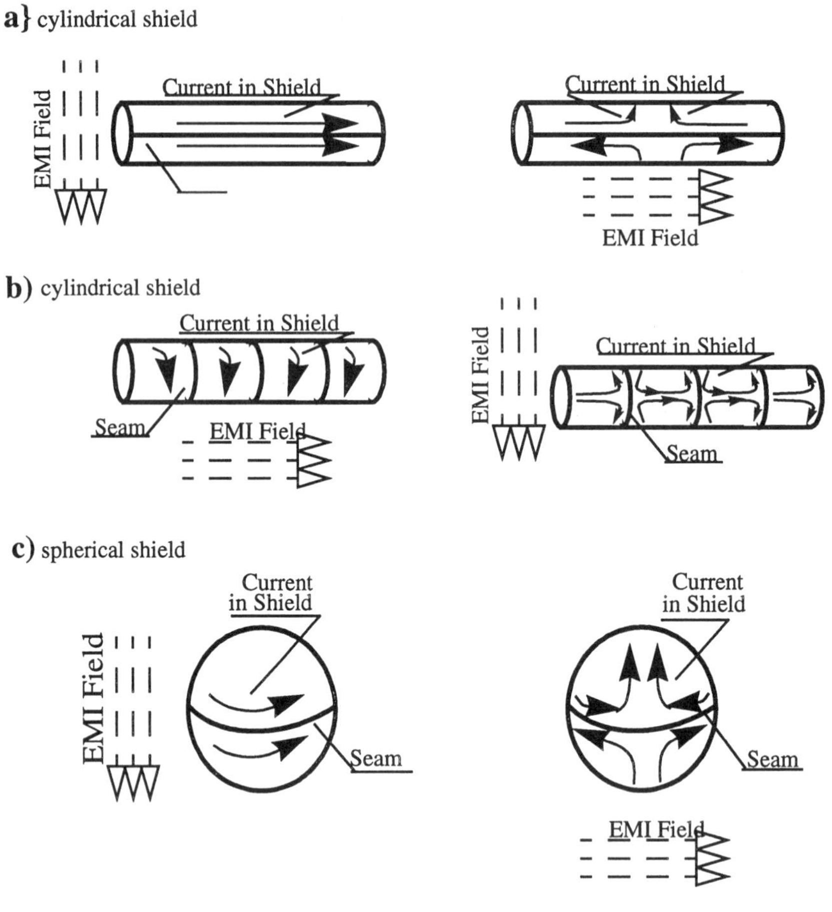

184 Electromagnetic Shielding Handbook

a) cylindrical shield

b) cylindrical shield

c) spherical shield

FIGURE 3.16 Effect of seams on shielding currents

the lowest possible shielding effectiveness (highest possible shielding coefficient). This follows directly from general EMC requirements:during the testing, EMI must be maximized.

The situations in Fig. 3.16 deal with seams in the shield. More problems arise, if the shield seams are accompanied by the gaps or apertures, which provide a bypass inside (or outside) the shielding enclosure (see the shield performance" parallel" model in chapter 2). For example, considering Fig. 3.2 c, we see, that if there were no gaps and apertures in the shield, there would be no capacitive coupling through the shield.

Depending on electrical size of the apertures and gaps, the electric or magnetic near or far zone fild phenomena will be emphasized in the shielding effects. In this chapter, we will limit the discussion to electrically small apertures. The reason is that the transfer parameters (as we have introduced them in this book) are inherently near

Transfer Parameters of Electromagnetic Shields and Enclosures 185

field / circuit characteristics. The electrically large gaps and apertures often involve complex coupling mechanisms and require far field modeling. They will be treated (at least, to a certain degree) in chapter 4.

As usually in shielding, the reviewed classification is incomplete. Of course, there may be suggested many more classification features of the shield non-uniformities. In any particular case, this can (and should) be done, based upon the specifics of the task and analysis objectives. In principle, any violation of the shield symmetry, or any local changes in conductivity, magnetic permeability, or the ambient environment (e.g., the dielectric constant), result in the additional "dimensions" in the non-uniformity classifications.

3.4.2 Spiral Shields

There are two main reasons behind the spiral shield proliferation: flexibility and manufacturability. Spiral shields are formed by the spiral application of tape or wires (in this latter case, it is called serve) as shown in Fig. 3.17. They can be manufactured from a variety of metals—copper, aluminum, steel, and so on. The winding can be laid down with gaps, seams, or overlaps. The spirality characteristics are the pitch (or weave) angle α and the lay length, h. We will define the weave angle with regard to the cable axis (it is arbitrary, and different authors and even industry standards may also define this angle with regard to a perpendicular to the cable axis).

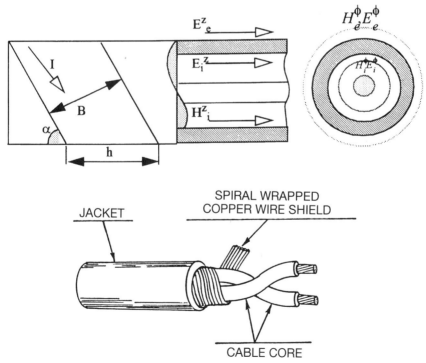

FIGURE 3.17 Spiral shield

Currents and Fields in Spiral Shields

Simple physical and geometrical considerations reveal the basic difference between solid homogeneous and spiral shields. Assuming that the isolation between the adjacent coils of the spiral is large (which is usually the case in practice, the current in the shield also follows a spiral path. The total current, I, in the shield can be decomposed on two components: axial I_z (along the cable axis) and circular I_ϕ (around the cable). The basic relationships between these two components follow the geometry:

$$I_z = I\cos\alpha, \; I_\phi = I\sin\alpha, \; \frac{\pi D}{h} = \tan\alpha, \; \frac{B}{h} = \sin\alpha \qquad (3.83)$$

In the last equality in Eq. (3.83), B is the spiral conductor diameter or tape width. Even if the shield tape is relatively wide, or the serve conductor diameter is large, the current is considered to consist of separate filaments shaped in the form of the shield spiral. The axial current component acts in the same manner as the current in a solid homogeneous tube. It originates the axial (longitudinal) components of electric fields E_i^z and E_e^z at the respective shield surfaces, and circular components of magnetic field: H_i^ϕ when the current return path is internal, and H_e^ϕ when the current return path is external. Similarly, the circular current component gives rise to the circular components of the electric field at the shield surfaces: E_i^ϕ and E_e^ϕ. Pretty simple, but here the "simple stuff" ends: the effects of the circular current component on electric and magnetic fields are dramatically different from those produced by the longitudinal component. Saving the explanations for the "fine print," we will express these effects in the form of a theorem:

Theorem 3: Solenoidal Field Theorem

In an infinitely long, continuous circular current sheet, a longitudinal (axial) component of magnetic field is always generated inside the shielding enclosure, and no magnetic field is generated outside the shielding enclosure. This is independent of whether the return current path is internal or external.

An "artist's view" at the solenoidal field is shown in Fig. 3.18.

FIGURE 3.18 Longitudinal magnetic field in spiral shield

The Mystery of the Return Path in Circular and Spiral Currents

✦ How about this? To see that this is correct, we apply Ampere's law to one spiral lay length, h, which contains one closed coil of the circular current. According to Ampere's law, a line integral of magnetic field $\vec{H}$, taken over any closed path L enclosing a current $\vec{I}$, is equal to that current:

$$\oint_L \vec{H} \cdot d\vec{l} = \sum I.$$

In a straight thin tubular conductor, there is only one longitudinal component of current—along the conductor. As we have already discussed, this current produces circular magnetic field lines H^ϕ in the plane perpendicular to the conductor axis. For all other directions, the current component is zero and the dot product yields no field. The picture is just opposite for ideal solenoidal fields, which have only circular component of the current I_ϕ and $I_z = 0$ (of course, in real-life spiral shields, both current components are present, but here we are looking at an idealized situation). This can be approached in different ways. To add "credibility", we selected the explanation given in Ref. [A.2]. Figure 3.19 shows a section through the axis of an infinite solenoid, which has n turns/m and carries current nI_ϕ A/m. Taking the line integral around the path ABDEA, we obtain

$$\oint_l \vec{H} d\vec{l} = \int_A^B \vec{H} d\vec{l} + \int_B^D \vec{H} d\vec{l} + \int_D^E \vec{H} d\vec{l} + \int_E^A \vec{H} d\vec{l} = nI$$

As was the case with the longitudinal current, inside the shield any radial component of magnetic field H_r is canceled by an opposite direction component produced by a symmetrically placed element. Therefore, $\vec{H} d\vec{l}$ along paths BD and AE is zero.

Also, we know that, at the axis of a solenoid, there exists an axial component of the magnetic field

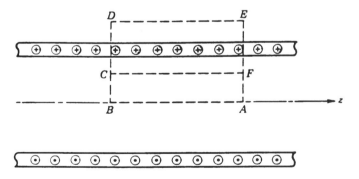

FIGURE 3.19 Section through the axis of an infinite solenoid

$H_z = nI$. Taking the line integral around path ABDEA and setting it equal to the enclosed current gives

188 Electromagnetic Shielding Handbook

$$\int \vec{H} d\vec{l} = nI + \int_D^E \vec{H} d\vec{l} = nI$$

This means that the integral from D to E is zero. Since path DE is arbitrary, external field H must be zero. Note that we never mentioned the return current path. We don't need it! Current I_ϕ is always "short-circuited on itself," providing its own return path. The field energy is concentrated inside the current coil and enters the shield from the inside surface. (You can easily prove this by building the respective Poynting vector.) For both internal and external returns, line contour l constitutes this circular path inside the shield.

If the path ED is equal to one lay length, h, then only one "turn" of the current is enclosed (n = 1), and

$$H_i^z = \frac{I}{h} \tag{3.84}$$

✦

Transfer Parameters of a Spiral Shield

Let us apply the knowledge we have gained.* The same basic Eq. (3.10) can be used to determine the spiral shield transfer parameters. Consider an external return (this is arbitrary). The contour l for a spiral shield can be chosen as follows: It runs along the inner surface of the spiral shield in the axial direction on the lay length, h, then describes a complete revolution with path length πD. It then passes to the surface of the inner conductor, travels axially in the opposite direction, then performs a complete revolution in the reverse direction to end up at its starting point.

Neglecting losses inside the inner conductor, we obtain the voltage per cable length h in the internal circuit:

$$V_{it} h = E_{it}^z h + \pi D E_{is}^\phi + j\omega \Phi_i^z \tag{3.85}$$

It is designated here as E_{it}^z with index "t" because, in the internal circuit the component, E^z is taken at the shield surface opposite to the current return, and E_{is}^ϕ is designated with the index "s" because the "return" path of current I^ϕ is always inside the shield, and the surface "i" is always facing it, as stated before. Magnetic flux due to the axial magnetic field inside the shield is

$$\Phi_i^z = j\omega \mu_0 \frac{\pi D^2}{4} H_i^z \tag{3.86}$$

Now, remember that the axial electric field is related to circular (it is also called tangential, and the terms are equivalent) magnetic field, and vice versa. Also in our case, $H_i^\phi = H_e^z = 0$; H_i^z is defined by Eq. (3.84) and H_e^ϕ is defined by Eq. (3.25). Then, finding the electric field components from Eq. (3.34) and the magnetic flux

---

*The following analysis was done first by Kaden [2].

from Eq. (3. 86), and substituting their values into Eq. (3.85), we obtain after dividing on current I and on lay length h:

$$Z_{t\alpha} = Z_{th} + \left(Z_{sh} + j\omega\frac{\mu_0}{4\pi}\right)\tan^2\alpha \quad (\Omega/m) \tag{3.87}$$

Here Z_{th} and Z_{sh} are the transfer and loss impedances of a homogeneous pipe having the same with the spiral shield parameters t, D, and σ.

For the current I internal return path, the same contour yields the loss impedance. For the external return there will be similar treatment of the components E_i^ϕ and H_i^z (see **Theorem 3**), while with the components E_i^z and H_i^ϕ, the same approach to defining the surfaces facing and opposite to the return path has to be used, as in homogeneous shell. Thus we obtain

$$Z_{s\alpha} = Z_{sh} + \left(Z_{sh} + j\omega\frac{\mu_0}{4\pi}\right)\tan^2\alpha = \frac{Z_{sh}}{\cos^2\alpha} + j\omega\frac{\mu_0}{4\pi}\tan^2\alpha \tag{3.88}$$

The dc resistance of a spiral shield is obtained from either Eq. (3.87) or (3.88), assuming $\omega = 0$.

$$R_{0\alpha} = Z_{0t\alpha} = Z_{0s\alpha} = \frac{R_{0h}}{\cos^2\alpha} \tag{3.89}$$

We see that the transfer impedance now contains two items: one decreasing with frequency per the law of the transfer impedance of a homogeneous tube, and the other increasing with frequency per the law of a homogeneous tube loss impedance. Both the transfer and loss impedances of a spiral shield also depend on the pitch angle. The larger the angle, the larger will be the high frequency values of both transfer and loss impedances. Figures 3.20 and 3.21 illustrate these dependencies.

✦ Compare transfer and loss impedance expressions for homogeneous solid shields and spiral shields. Though the latter are more complicated due to additional circular current component, they both are expressed in terms of homogeneous cylindrical tube parameters Z_{th} and Z_{sh}. This fact proves the identical nature of the diffusion processes originated by axial and circular current components. This could be expected, owing to the symmetry of the basic Eqs. (3.34) and (3.35) with regard to the axial and circular field components. This fact can be used to provide generalized expressions for magnetic and electric field intensities in homogeneous and spiral shields in terms of the transfer and loss impedances of homogeneous tubes of equivalent diameter and thickness.

$$E_i^{z,\phi} = \pi D(H_i^{\phi,z} Z_{sh} + H_e^{\phi,z} Z_{th}) \tag{3.90}$$

$$E_e^{z,\phi} = \pi D(H_i^{\phi,z} Z_{th} + H_e^{\phi,z} Z_{sh}) \tag{3.91}$$

190 Electromagnetic Shielding Handbook

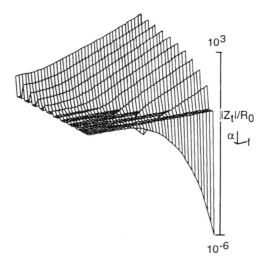

FIGURE 3.20 Spiral shield transfer impedance as a function of weave angle (0-72°) and frequency (10 kHz-10 MHz); rotation = 240°, tilt =20°

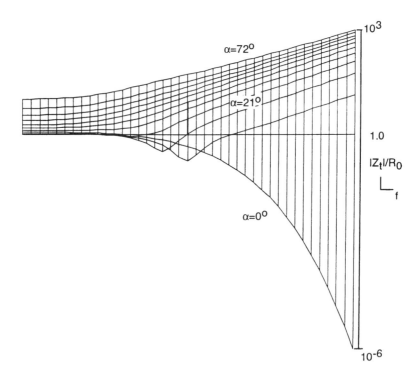

FIGURE 3.21 Spiral shield transfer impedance as a function of weave angle (0-72°) and frequency (10 kHz-10 MHz); rotation = 270°, tilt =0°

Here, the longitudinal components of electric field intensity are related to the circular magnetic field intensity, and the circular components of electric field intensity are related to the longitudinal magnetic field intensity; Z_{th} is calculated from Eq. (3.50), and Z_{sh} from Eq. (3.52) without the inductive item.✦

3.4.3 Shields with a "Longitudinal" Seam and / or Electromagnetic Gasket

Seams vs Electromagnetic Gaskets

The seam qualification as longitudinal may need additional clarifications. Probably, the most immediate association is that of a cylindrical cable shield formed by a metallic tape which is "longitudinally" (i.e., along the cable length) applied to the cable core. On the other hand, in a product shielding enclosure it may be just a matter of assigning the directions to different shield elements, including the seams. The point is, that in both cases it is implied that the seam is a "straigt" line, at least at some length. We have already "got a flavor" of what the seam does to the current flow in the shield - see Fig. 3.16.

Related to the seam issue is that of electromagnetic gaskets, which are used to bring separated conducting surfaces into electrical contact. By properly selecting the type, material, and geometry of a gasket, the seams can be effectively "bridged". As a result, the shield performance may approach that of a homogeneous solid shield.

There are numerous types of gaskets at the market: fingerstock, spiral, rubberized mesh, metallised and composite plastics, etc.The most often used material for metallic gaskets is berillium copper, which provide good contact resistqance, high conductivity and resiliency.

In this section, we will attempt an at least *qualitative* explanation of the involved physical mechanisms.

Physical Processes in a Seam

Consider Fig. 3.22 (which is based on Fig.3.2 d). If the seam is "perfect" - no current can cross it. Then the currents generated in the shield will "loop on themselves". A shown in Fig. 3.22, two "basic" loops are formed, with opposite direction current in them. We have already seen, in the previous section, the effect of current loops: they produced magnetic field vectors, perpendicular to the plane of the loop, with the magnitude proportional to the current in the loop.

In general case, the currents in the parts of the shield separated by the seam, will rotate in opposite directions and will have different magnitude, depending on the shape, size, and electrical/magnetic properties of these parts. So, the only thing which we know in general case, is that the currents are exactly opposite at both sides of the seam. Thus, the seam acts as a dipole. Now, we have to consider two processes:

1. The effect of the two current loops - two magnetic field vectors produced by the two parts of the shield separated by the seam
2. The effect of a magnetic dipole, associated with the seam.

Both problems were addressed by H.Kaden [P.2] at a most general level. Unfortunately in practice, there are two many unknown variables to obtain numerical solutions. Therefore so far, the main method to study these processes remains experimental.

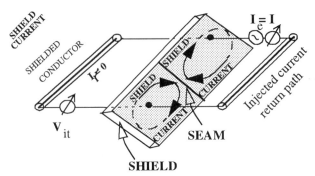

FIGURE 3.22 Current loops at the shield seam

◆ Remember, that we are discussing the transfer parameters, which we have defined as near field / circuit theory characteristics. Surprisingly, the far field (radiation) treatment is much more palatable. Not that it is easier, but the study method suggested in this book - application of antenna theory principles to the radiation shieldingis problems - results in quite logical physical interpretations, and permits to use the available mathematical appratus for practical numerical results. As you will see in chapter 4, a modification of the Babinet principle will easily yield final results without any derivations. ◆

We will consider experimental data two typical cases of a shields with seams: a cable and a small electronic product enclosure.

Coaxial Cable

Figure 3.23 presents typical transfer impedance modulus frequency characteristics of a coaxial cable with an outer conductor formed by a steel tape applied longitudinally with an overlap. The legend describes the shield design data.The plotted characteristics are measured (using the shorted triaxial line techniques described in Chapter 6 of this book and represented by the "bullet" line) and calculated [per Eq. (3.46) and represented by the smooth curve). The graph carries both "good news" and "bad news": the theory and experiment either quite accurately coincide or have nothing in common whatsoever. On the positive side, if you think of all the assumptions we made in deriving the formulas and, on the other hand, of the extremely small values to measure (the measurements were done on a 1 m cable length, so fractions of a milliohm had to be measured), the accuracy looks remarkable indeed! On the negative side, this accuracy is limited to a relatively low frequency band—below 500 kHz.

Shield Characteristics:
thickness $t = 0.18$ mm
diameter $D = 10$ mm (20% overlap)
magnetic permeability $\mu_r = 100$
electrical conductivity $\sigma = 10\ m\Omega \times cm$

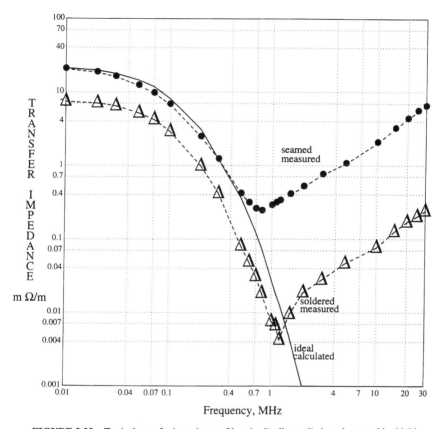

FIGURE 3.23 Typical transfer impedance of longitudinally applied steel tape cable shield

What is the problem? Two come to mind: the theoretical model limitations and the test technique limitations. Deferring the discussion on the merits of different transfer impedance test methods until Chapter 5, for now we will concentrate on the unaccounted-for shield construction "details." The first "suspect" is the longitudinal seam (or, possibly, gap) at the tape overlap. But this is easy to check: just eliminate this seam or gap. In practical terms, this was done by soldering a copper strip over the seam. Because the tape is made from steel, the soldering is also easy to do. It appears that the test results (i.e., the Δ curve in Fig. 3.23) confirm our hypothesis: now the coincidence between the measured and calculated data extends both to smaller values and higher frequencies. Still, at frequencies above 1 MHz, and with the values of several micro-ohms (!), the experimental data "takes off" again and, after some "turn-around," increases proportionally with the frequency. Now, the cause of the discrepancy can be either in measurement techniques or in some undetected faults of

194 Electromagnetic Shielding Handbook

the test assembly (which also can be treated as a test technique imperfection), but because of the extremely small measured values, this time further investigation becomes more of an "art" than something that can be based purely on engineering skill and hardware capabilities.

Several important conclusions follow from the presented data:

- Where applicable, the theory of solid homogeneous cylindrical shields is accurate enough, at least for engineering applications. The transfer impedance of such shields quickly diminishes with the frequency.
- Seams (and/or gaps), even longitudinal ones, violate the shield homogeneity and result in the transfer impedance frequency characteristic "turnaround" and increase starting at sufficiently high frequencies.
- After the transfer impedance frequency characteristic "turnaround," it grows in proportion to the frequency. This indirectly indicates at an inductive nature of the involved processes: indeed, such proportionality to frequency may be modeled (at least formally) by ωL.
- Available transfer impedance measurement techniques yield sensitivity down to extremely small values.

✦ By now, you have already probably guessed the reason why the "Δ" curve has lower values than the "bullet" curve. Yes, of course, this is because the high conductivity copper tape acts in parallel with much more resistive steel tape.✦

✦ Solid homogeneous tubes or longitudinally applied tapes are often corrugated. The corrugation enhances the shield flexibility, especially at larger diameter cables. A *low depth* corrugation is also often used to facilitate the contact between the overlapping surfaces. While the problem of corrugation effects on shielding and transmission properties of cable shields is exciting, we won't dwell on it here for two reasons: the effects of corrugation on the shield parameters are, as a rule of a second order, and they are quite difficult to account for.✦

Small Product Enclosure with a Seam or a Gasketed Lid

✦ The following data is based on a report presented in Ref. [3.2]. Fig. 3.24 shows the transfer impedance frequency characteristics measured on a $120 \times 90 \times 50$ mm tin-plated die-cast box, with a 3 mm thick brass lid covering a 16 mm wide flange, cut out of the bottom of the box. The flange can be attached to the box by 4 screws. We will consider here four test setups:

1. The brass lid is applied to the box, without any contact treatment
2. The brass lid is applied to the box, with insulating material (PTFE film) pasted to the contact surfaces
3. The brass lid is applied to the box, with electromagnetic gasket pasted to the contact surfaces; the gaskets are compressed leaving the lid spaced from the die-cast box at 2 mm
4. The brass lid is applied to the box, with electromagnetic gasket pasted to the contact surfaces; the gaskets are compressed more reducing the lid spacing from the die-cast box at 1mm

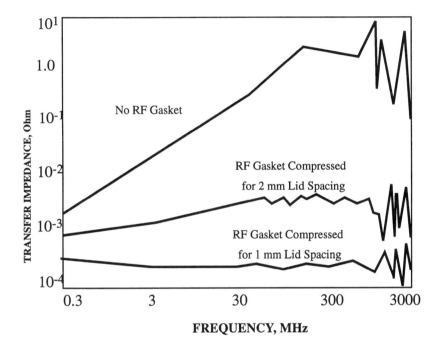

FIGURE 3.24   Transfer impedance of gasketed shielding box

The measurements were made using the line injection test method described in chapter 5. Several important conclusions:

- The first two tests yielded identical results, which proves that you cannot rely on the contact resistance to provide continuity.
- The gasketing can provide drastic improvement of the non-uniform shield performance.
- Especially dramatic improvement is observed at the higer frequencies
- The design, quality, and implementation of gasketing is of the greatest importance. ✦

3.4.4   Foil Shields: Effects of Seam and Overlap

Foil Shield Design and Applications

Foil shields are formed by metallic sheets applied to the structural frames or plastic enclosures or tapes wrapped longitudinally or spirally around the cable core (Fig. 3.25). Aluminum foil is usually used for this purpose, although copper, magnetic alloys, metal-clad, and various kinds of conductive plastic tapes are utilized in some applications.

196 Electromagnetic Shielding Handbook

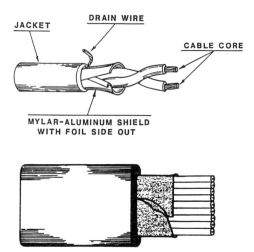

FIGURE 3.25 Foil shields

The foils so used can be several mm thick or quite thin (e.g., 0.35 mil (0.01 mm) foil thickness is a typical specification for cable shielding applications). Thin foil results in two technical challenges: high dc resistance and low mechanical strength. To provide the necessary mechanical strength, the foil is generally either bonded to the plastic enclosure or cable insulation or laminated to a polyester tape, or both. This approach is realized in several popular foil shield tape modifications developed for cable use, as shown in Fig. 3.26 . To reduce the foil dc resistance and facilitate shield termination, drain wires are used, which as a rule are manufactured from high enogh gauge copper and hove high conductivity. In laminated shields the drain wire is applied to the foil side of the shield.

Foil Shield Problems

If no gaps are left, the foil shield can provide up to 100 percent coverage, estimated as the ratio of the area of the shielded surface covered by metal to the total area of the product surface (enclosure, cable). It might be expected, therefore, that the foil shield transfer impedance would imitate the transfer impedance of a corresponding solid homogeneous shield of the identical thickness and shape. Yet experience proves that the "gapless" foil shield becomes less effective at higher frequencies, as suggested by the experimental data in Figs. 3.23, 3.24 and 3.9. Assuming "perfect" quality of the shield itself (that is, no foil perforations or manufacturing-related faults and gaps), one of the most obvious reasons for foil shield performance degradation is the slot at the shield overlap. Figure 3.27 illustrates the

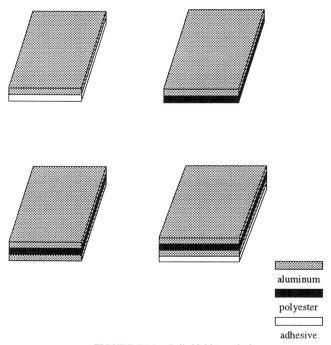

FIGURE 3.26 Foil shield tape designs

formation in a cable shield of a longitudinal slot with the width equal to the insulating tape thickness. As shown, the slot exists in both "foil-in" and "foil-out" shield application techniques, even if the tapes are applied with an overlap. Obviously, the same is rue for any type of enclosure.

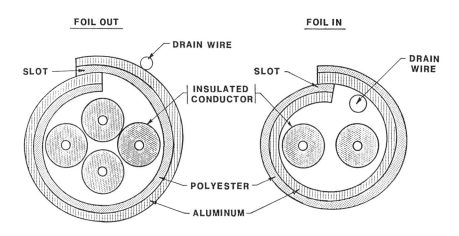

FIGURE 3.27 Slot in a film-foil shield with overlap

The effect of the slot on the shield performance is two-fold, consisting of (1) direct coupling and radiation through the slot and (2) generation of a circular component of the current in the shield, which produces effects similar to those in a shield with a seam.

Gap or Slot Width vs Shield Thickness

✦ Although a gap width may be very small, its impact on the shield performance at high frequencies can be dramatic. This impact is defined by two effects: the direct capacitive coupling between the conductors at the both sides of the shield, and direct radiation through the gap (as opposed to diffusion through the shield).

The theory of electric and magnetic coupling through the longitudinal shield gaps and circular apertures was first developed by Kaden [P.2]. Although later researchers keep improving on the accuracy of the formulas (albeit at the price of enormous mathematical complications and likely loss of easy physical interpretation), for engineering purposes the original theory will suffice. Therefore, herein we will give only the resulting formulas based on Kaden's theory.

Consider a triaxial structure in which our shield is the middle tube. The radii of the center conductor, the shield, and the external tube are ri, rs, and re respectively. The width of the gap is w, and the shield thickness is t. Also, $r_s \gg t \geq w$. Then the direct capacitive (i.e., electric field) coupling through the gap is

$$C_{ie} = \left(\frac{w}{\pi r_s}\right)^2 e^{-\left(\frac{\pi t}{w} + 2\right)} \frac{C_e C_i}{\pi \varepsilon_0} \tag{3.92}$$

where the external and internal circuit capacitances are

$$C_e = \frac{2\pi\varepsilon_0}{\ln\frac{r_e}{r_s}}, \quad C_i = \frac{2\pi\varepsilon_0}{\ln\frac{r_s}{r_i}} \tag{3.93}$$

Due to the direct radiation through the gaps or slots, the penetrating electromagnetic field induces in the considered coaxial system the voltage Ztg, in addition to voltage Zt due to diffusion, so that the total transfer impedance is

$$Z_{ttot} = Z_t + j\omega \frac{\mu_0 w^2}{\pi^3 r_s^2}\left[1 + (1-j)\left(2 + \pi + \frac{\pi t}{w}\right)\frac{\delta}{w}\right]e^{-\left(\frac{\pi t}{w} + 2\right)} \quad (3.94)$$

where Zt is the transfer impedance of the corresponding solid homogeneous tube.

The second item in Eq. (3.94) can be presented as $j\omega L_{gap}$, which indicates an inductive nature of the phenomenon. At low frequencies, the second item in Eq. (3.94) is small, and Zttot is decreasing due to the decrease of Zt. At some high enough frequency, the second item in Eq. (3.94) prevails, and Zttot starts to increase. With a further rise in frequency, the total transfer impedance is determined mainly by this second item and increases proportionally to the frequency. ✦

The second effect is similar to that discussed in the previous section. With modern thin foil shields, this second effect plays an extremely important part in EMI generation in foil-shielded products.

Shorting Fold Techniques

To improve the high-frequency performance of a monolayer foil shield, the gaps in the shield generally should be avoided (or if "designed-in," they should be "shielded" by other metallic layers in multilayer shields). Indeed, as shown in Fig. 3.28, even a modest 1.5 mm gap results in a tenfold increase of the transfer impedance, as compared to the 1.5 mm overlap.

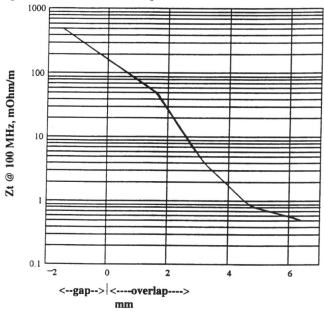

FIGURE 3.28 Effects of gap and overlap on shielding performance of foil shields (experimental data)

200 Electromagnetic Shielding Handbook

To deal with the overlap slot, two approaches are used: we can (1) increase the width of the overlap of the shield edges or (2) apply special shorting fold techniques designed to eliminate the slot. By further increasing the overlap four times (from 1.6 mm to 6.3 mm) the transfer impedance at high frequencies can be further reduced almost 80 times. As you might expect, the radiation from the slot also depends on its width (that is, on the thickness of the insulating tape). The data in Fig. 3.29 can be summarized as follows: a threefold increase of the insulating tape thickness, from 0.08 mm to 0.25 mm, results in a threefold increase in transfer impedance when the overlap is 1.6 mm, and only in 1.5-fold increase in transfer impedance when the overlap is three times larger (4.8 mm).

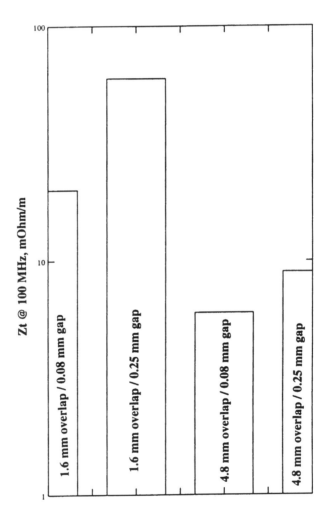

FIGURE 3.29 Effects of insulation thickness and overlap on shielding performance of foil shields (experimental data)

The shorting fold techniques are designed to eliminate the slot altogether. Figures 3.30 and 3.31 present two such techniques: the so-called "S-fold" and "Z-fold" methods (the terminology is not standardized).

Other types of shorting folds may be encountered. As you can see from the

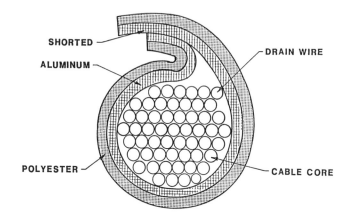

FIGURE 3.30 Shorting fold design

drawings, the idea is to fold the shield in such a way that "metal faces metal," thus eliminating the slot.

Now we surely can expect the shield to behave as a solid homogeneous tube. But

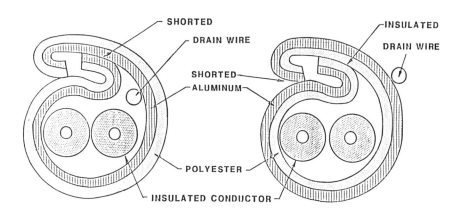

FIGURE 3.31 Shorting fold design

consider the transfer impedance measurements performed on flat cables with different shield designs (Fig. 3.32). Performance comparison of a "regular" foil shield design (it is sometimes called cigarette wrap) with the shorting fold design does not indicate any difference in the frequency range below 100 MHz. Only at higher frequencies do the advantages of the shorting fold become obvious, but even then the transfer impedance characteristic fall short of a solid homogeneous tube. Of course, the shield in Fig. 3.32 is not a very effective one, and the effect of the shorting fold can be more dramatic when used in better shields, including multilayer designs (see later), but the data in Fig. 3.32 certainly deserves an explanation.

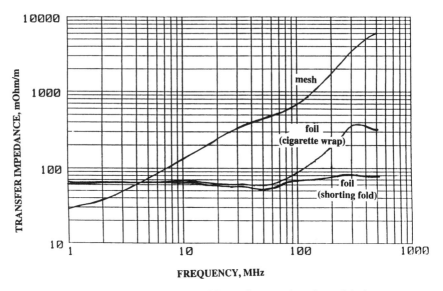

FIGURE 3.32 Flat cable shielding performance (experimental data)

The problem is that just applying two metallic surfaces "face-to-face" will not necessarily "do the trick," because the effectiveness of the shorting fold strongly depends on the contact resistance between the two interfacing surfaces—and this contact resistance is a function of the metal surface conditions, the area of the interface, and the pressure applied to these surfaces. It is also a function of the signal frequency. Some of these effects are illustrated in Fig. 3.33. The contact resistance of a 12×12 mm aluminum foil overlap was measured in the frequency range from 1 to 1000 MHz with and without applied pressure (0.5 kg). When no pressure is applied, the contact resistance is large and reduces with frequency, probably only because the capacitance between the two pieces of foil provides a parallel by-pass to the very few contacts between them. However, with the pressure applied, the resistance is determined mainly by the contacts. Even when a bronze spring was inserted between the two foil pieces, no changes were observed in the measured values. Actually, the measured values are comparable to the ohm per square (Ohm/sq) values measured on a continuous piece of foil.

It appears that the transfer impedance values are extremely sensitive to the contact resistance (or, rather, it should also be called impedance, since it has both real and imaginary components) between the overlapping surfaces at the shorting fold. To confirm this, experiments were done where the number and quality of contact be-

Transfer Parameters of Electromagnetic Shields and Enclosures 203

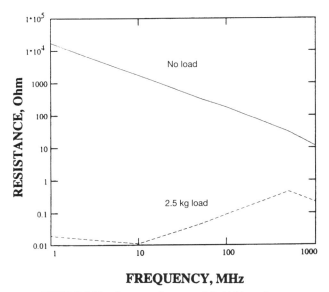

FIGURE 3.33 Contact resistance at foil shield overlap (experimental data on 12×12 mm overlap

tween the overlapping surfaces was strictly controlled. As shown in Fig. 3.34, the better the contact, the smaller the transfer impedance and the better the shield.

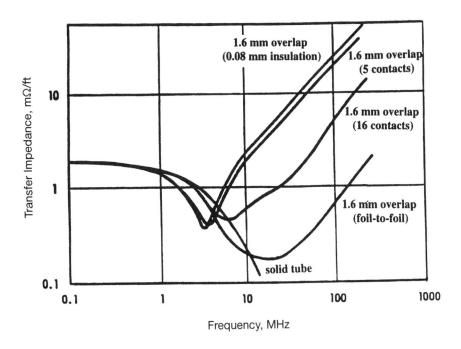

FIGURE 3.34 Transfer impedance of bonded shields (experimental data)

More Mysteries Solved: Longitudinal Foil Shield with Overlap

We have already mentioned the two mechanisms affecting the foil shield performance: direct coupling through the gap and generation of a circular current component in the shield. While we have addressed the first mechanism in a manner similar to the gap in the shield, we deferred the explanation of the second mechanism until this section. Using the concepts of a spiral shield, we can address this mechanism now.

Look at Fig. 3.27. Isn't such a shield a wide ("infinitely wide," by our assumptions) one-turn spiral? A specific feature of a spiral is that the external surface becomes internal. However, when longitudinal current flows in the shield, the electric field intensities at these surfaces are different: E_i^z, and E_e^z. Then, as shown in Fig. 3.35, a potential difference, $E = E_e^z - E_i^z$, is generated between the overlapping parts of the spiral. The corresponding equivalent circuit is presented in Fig. 3.35.

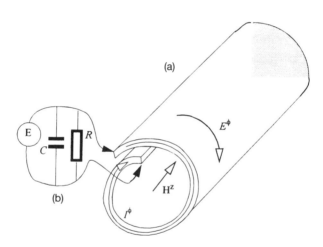

FIGURE 3.35 One-turn spiral shield (a) and equivalent circuit (b)

The "source" $E = E_e^z - E_i^z$ drives a circular current I^ϕ in the shield at the cable whole length as one "turn," which generates a longitudinal component of the magnetic field Hz. But we know that the field Hz results in the rising transfer impedance frequency characteristic. Capacitance C is proportional to the area of the overlap. The larger this capacitance is, the smaller the shorting impedance $1/(j\omega C)$ due to this capacitance will be, and the more "mitigated" will be this effect. When the shield is shorted (shorting fold), the potential difference is shunted by the contact resistance R = 0 of the shorting fold. This resistance acts in parallel with the capacitance C. If the short is not perfect, resistance R has some finite value, and a residual circular current will still flow in the shield.

Now, it presents no difficulty to quantify those effects using elementary concepts of circuit theory.

3.4.5 Mesh Shields

Expanded copper (as well as other metal) mesh is a popular electromagnetic shielding material. It is mainly used (often instead of foil) in flat cables as a cigarette wrap shield and as a ground plane. Also, for some "cheap" shielding rooms.

The advantages of the mesh compared to the solid metallic sheet consist of reduced material consumption, weight, and (eventually) cost. These qualities are especially important when the use of relatively expensive and heavy copper is desirable for some reason (e.g., for solderability). The specific "design" feature of mesh is the presence of perforations that are uniformly distributed over the material surface. Figure 3.36 presents a 40× view of a sample of a mesh screen. The main disadvantage of a mesh shield is the result of the same characteristic that secured its advantages: the perforations.

FIGURE 3.36 Mesh shield

As it was the case with the gaps and slots in "solid" foil shield, electric and magnetic fields can penetrate the perforations, causing direct electric and magnetic field coupling to the shielded conductors. This coupling acts in addition to "regular" diffusion in a solid homogeneous tube. For example, due to the field penetration through the apertures, the circular magnetic field component H_ϕ appears inside the shielded space even when the return current path is exclusively external, and outside the shielded space even when the return current path is exclusively internal.

The amount of coupling (i.e., mutual capacitance and mutual inductance) through small apertures depends on the number, size, and shape of the apertures, as well as their interaction. When the apertures are much smaller than the wavelength, their effect can be modeled by correspondingly polarized (electrically) infinitesimal dipoles situated in the center of the aperture (see Refs.[P.2, 3.16]; a historical review of the subject is also given in [8.2]). The dipole axis lays perpendicular to the shield surface for an electric dipole (electric field coupling) or the axis lays in the plane of the shield surface for magnetic dipole (magnetic coupling). Treating circular apertures as such dipoles Kaden [P.2] obtained:

206 Electromagnetic Shielding Handbook

$$M_{12} = \upsilon\mu_0 \frac{d_0^3}{6\pi^2 D^2}, \quad C_{12} = \upsilon \frac{C_1 C_2}{\varepsilon_1} \frac{d_0^3}{6\pi^2 D^2} \qquad (3.95)$$

where

M_{12}, C_{12} = inductive and capacitive coupling coefficients per unit cable length

C_1, C_2 = capacitances per unit length between the shield and (respectively) the internal (shielded) conductors and conductors providing current I external return path

d_0 = the aperture diameter;

D = shield diameter

ε_1 = conductor insulation dielectric constant

ν = number of apertures per unit cable length

In the shielding industry, it is customary to characterize mesh (or any other perforated shield, for that matter) by the amount of coverage density K (it is also called optical density), which is the percentage ratio of the metal covered surface to be shielded to the total area of the shield. However, for convenience in formula notation, we will use a fractional expression, instead of percentage:

$$K = \frac{S_m}{S_{tot}} \qquad (3.96)$$

Expressing the total area of the apertures per cable unit length, S_a, as a function of the coverage density from Eq. (3.96), and summing the longitudinal linear density of υ_1 apertures per meter and the circumferential linear density of υ_c apertures per meter, obtain the area of each aperture as

$$a_s = \frac{(1-K)\pi}{\upsilon_1 \upsilon_c} \qquad (3.97)$$

The total number of apertures at the shielded cable unit length is

$$\upsilon = \upsilon_1 \upsilon_c \pi D \qquad (3.98)$$

where D is the cable diameter. If the apertures are circular irises with diameter d_0, then $\upsilon_1 = \upsilon_c = \upsilon_0$, and equating the area of such an iris (which, of course, is $\pi d_0^2/4$) to Eq. (3.97), we obtain

$$d_0 = \frac{2}{v_0}\sqrt{1-K} \qquad (3.99)$$

Substituting the number of apertures and their diameter from Eqs. (3.98) and (3.99) into Eq. (3.95), we obtain

$$M_{12} = \frac{4\mu_0}{3}\frac{1}{\pi D}\frac{(1-K)^{\frac{3}{2}}}{v_0}, \quad C_{12} = \frac{4C_1C_2}{3\varepsilon_1}\frac{1}{\pi D}\frac{(1-K)^{\frac{3}{2}}}{v_0} \qquad (3.100)$$

In an electromagnetic shield, the aperture coupling acts simultaneously with diffusion processes. The same basic Eq. (3.10) can be applied, with the fields penetrating through the apertures accounted for. Thus, joining the expressions for these two physical mechanisms, we obtain

$$Z_t = Z_{td} + Z_{tp} = \frac{R_0 \gamma t}{K \sinh \gamma t} + j\omega \frac{4\mu_0}{3}\frac{1}{\pi D}\frac{(1-K)^{\frac{3}{2}}}{v_0} \qquad (3.101)$$

$$Z_s = \frac{R_0}{K}\gamma t \coth \gamma t + j\omega L_s + j\omega \frac{4\mu_0}{3}\frac{1}{\pi D}\frac{(1-K)^{\frac{3}{2}}}{v_0} \qquad (3.102)$$

$$Y_t = j\omega \frac{4C_1C_2}{3\varepsilon_1}\frac{1}{\pi D}\frac{(1-K)^{\frac{3}{2}}}{v_0} \qquad (3.103)$$

Here, the item Z_{td} reflects the diffusion processes in the shield and represents the transfer impedance of a homogeneous pipe having the same dc resistance and diameter as the mesh shield. However, it is still convenient to relate the transfer impedance to the dc resistance, R_0, of a respective solid tube; that's why R_0/K is entered in the formulas. The items $j\omega M_{12}$ and $j\omega C_{12}$ express the inductive and capacitive coupling effects due to the fields penetrating through the apertures. Similar considerations are also applicable to the loss impedance.

Some of the typical trends of a mesh shield transfer impedance frequency characteristics are plotted in Fig. 3.37 using Eq. (3.101). Curves 1 and 2 are plotted for solid copper shields with thicknesses of 0.1 and 0.3 mm, respectively. They exhibit behavior typical of a solid homogeneous tube, and their dc resistance is equal to "their dc resistance" (what else?). All other shields are perforated (with $K = 0.5$ or $K = 0.75$), and this causes a respective increase of their dc resistance with regard to respective solid tubes. Curves 3 and 4 correspond to the shield of the same thickness 0.1 mm and coverage density 50 percent, but with the different linear density of apertures—2500/m and 250/m, respectively. The trend is clear: it is better to have many small apertures then fewer large ones (with the same coverage). Curves 4 and 5 differ only with the shield thickness while having identical large apertures. As you can see, the effect of the shield thickness increase (with respective increase in weight and cost) is "not worth it" when the large apertures are used. And finally, curve 6 has large thickness (0.3 mm) and large aperture size, and $K = 0.75$. First, the large thickness leads to the reduction of the transfer impedance, but then the negative effect of the apertures overtakes it—although the relatively high coverage keeps its performance at a pretty decent level.

As a matter of fact, curve 5 corresponds to the "real life" mesh shown in Fig. 3.36, and the results of its measurements are presented in Fig. 3.37. The experimental and theoretical curves are almost identical.

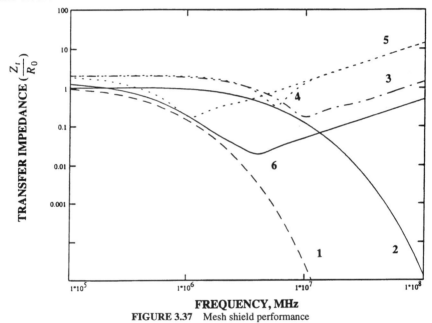

FIGURE 3.37 Mesh shield performance

✦ As widespread and convenient the coverage density parameter is, it is insufficient to characterize the shield EMI-protection potential, and we should caution against relying solely on it. The problem is that the coverage density depends on the number, area, and shape of the apertures. For example, the sample in Fig. 3.36 has two rhombic shape apertures per 1 mm of shield length and provides about 50 percent coverage. If we model the rhombuses with circles, then the "visible" size of the aperture is proportional to its linear dimension, the area of the aperture is proportional to the square of that dimension, and the coupling (that is, EMI potential) is proportional to the 3-D power of this dimension per Eq. (3.95). That can become very deceiving. For instance, a twofold increase in iris diameter leads to an eightfold increase in coupling! ✦

3.5 BRAIDED SHIELD

For many years, braid has been one of the most popular types of electronic cable shields. It also finds limited applications as a electrical conductor (especially "popular" in bonding and grounding). Braided shield usually consists of two groups of copper or aluminum strand *interleaving* carriers woven in *opposite directions*; one set is applied in a left-hand lay, and the other in a right-hand lay.

If the cable core is round (braided shield is also used in flat cables), then the strands form a circular spiral.

As a matter of shield design classification, braid can be considered either one- or two-layer shield, with specific mechanical and electrical performance properties: braid flexibility approaches that of a spiral shield, while the solenoidal fields of both unidirectional layers are compensating (or so we think—see below). This brings quite satisfactory shielding performance.

3.5.1 Geometry and Design Parameters of Braided Shield

A braid is formed by crossing several belts of conductor sets so that each belt passes alternately over and under the other belts. The main construction elements and definitions of a braided shield are as follows (see Fig. 3.38):

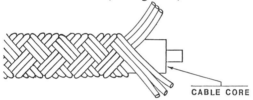

FIGURE 3.38 Braid shield

- Cable core diameter (D). If no other shielding or isolating layers are applied to the core, D is also the braided shield internal diameter.
- Strand (d). Usually, strands are single round conductors of diameter d. In large diameter cables, the strands can also have rectangular or special shape. In flat cables, narrow foil strips are sometimes used.
- Carrier or ends (C). A set of N conductors that run in parallel. The most popular values for N are from 3 to 7, although as few as one and as many as a dozen can be used. The carrier width

$$C = Nd$$

- The number of carriers (n) in the shield.
- Weave angle α. Similar to the spiral shield, we will count this angle from the cable axis. Obviously, if one set of carriers is applied at the angle α, then the opposite set is applied at the angle $-\alpha$.
- Pick (P). The distance between two adjacent crossover points (that is, the "width" of a carrier along the cable axis)

210 Electromagnetic Shielding Handbook

$$P = C/\cos\alpha$$

- The lay length.

$$h = \pi(D+d)\tan\alpha$$

- The length of one turn of the spiral.

$$L = \pi(D+d)/(\cos\alpha) = h/(\sin\alpha)$$

- The fill of the braid.

$$F = \frac{Nnd}{2\pi D\cos\alpha} \tag{3.104}$$

- Optical coverage (we will use the fractional measure instead of percentage).

$$K = 2F - F^2 \tag{3.105}$$

- The dc resistance of a braided shield.

$$R_{0\alpha} = \frac{4}{\pi d^2 \sigma Nn\cos\alpha} = \frac{2}{\pi^2 Dd\sigma F\cos^2\alpha} \tag{3.106}$$

3.5.2 Physical Processes in Braided Shield: Noah's Ark of Shielding Problems

Braid is a real "Noah's ark" of cable shielding problems. Many phenomena are involved in the basic shield designs we have reviewed so far, and we will meet them all again in analyzing the braided shield. Diffusion into solid metal, field penetration through the apertures, effects of a spiral wrap—you name it: all the main electromagnetic field coupling and radiation mechanisms are applicable to modeling braided shield, as well as some specific problems that are not encountered in other types of shields. We will review now, one by one, the physical mechanisms as applicable to the braid. We will use the same familiar Eq. (3.10). Fig. 3.39 schematically depicts the voltage contours inherent to these mechanisms.

We start with diffusion. Everything we discussed in the section on homogeneous solid tubes is applicable. The corresponding voltage contour is shown in Fig. 3.39a. The surface S and the contour L are formed by straight segments. For the external return path of the current, only axial electric field components E^z_{es} and E^z_{it} are present, correspondingly, on the shield's outer and inner surfaces, with circular magnetic field H^z_e outside the shield, and no magnetic field inside the shield.

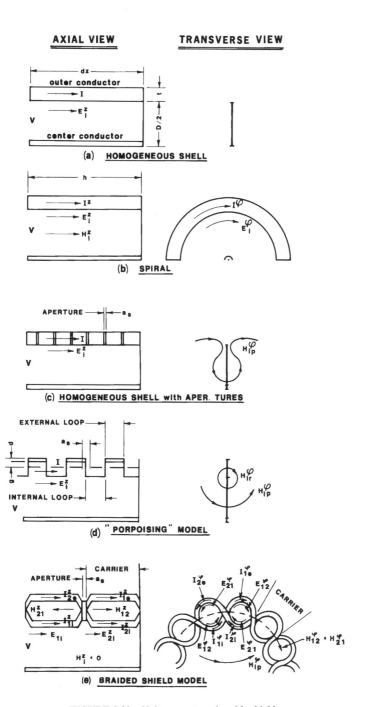

FIGURE 3.39 Voltage contours in cable shields

We have already been through similar considerations, and they lead us to familiar Eqs. (3.50) and (3.53).

The next mechanism is spirality. The contour is shown in Fig. 3.39b. We have been through the corresponding derivations in the previous section. However, there is no need to repeat them. Didn't we say that the two oppositely woven sets of strands produce compensating axial magnetic fields in a braid? So, perhaps, the solid homogeneous shield model for a braid may suffice. The only thing to watch, is that the spiral shield has larger dc resistance, $R_{\alpha 0}$, than a homogeneous tube R_{0h}, and at dc this resistance is equal to the transfer and loss impedances of the braid:

$$Z_{t\alpha 0} = Z_{s\alpha 0} = \frac{R_{0h}}{\cos^2 \alpha} \quad (3.107)$$

But, not so fast! A quick look at Fig. 3.9 proves that there is a large difference between the transfer impedance frequency characteristics of the braid and solid homogeneous tube. What's even worse, this is a difference in principle: whereas the frequency characteristic of a solid tube is declining, that characteristic of a braid is increasing, just as it was the case of spiral shields and shields with apertures.

Well, taking a second look at Fig. 3.38, we notice the diamond-shaped apertures formed at the crossovers of the carriers. We already know how to handle these apertures (see Section 3.4.5). Fig 3.39c shows the corresponding voltage contour. Expressing the number of holes, v, aperture size, and cable dimensions as a function of braid design characteristics, and introducing magnetic and electric polarizability parameters for ellipse approximation of a diamond-shaped holes, Vance [3.3] obtained the braided shield mutual inductance and mutual capacitance due to the aperture, joined all this with the homogeneous tube diffusion component, and proposed formulas:

$$Z_t = Z_{td} + Z_{tp} = \frac{R_0 \gamma t}{\sinh \gamma t} + j\omega \frac{\pi \mu_0}{6n}(1-K)^{\frac{3}{2}} \theta_m \quad (3.108)$$

$$Y_t = j\omega \frac{C_1 C_2 (1-K)^{\frac{3}{2}}}{6n\varepsilon_1 \, v_0} \theta_e \quad (3.109)$$

Here the item Z_{th} reflects diffusion processes in the shield and represents the transfer impedance of a homogeneous pipe having the same with the braided shield dc resistance and diameter, but with the wall thickness equal to shield strand diameter d. The items $Z_p = j\omega M_{12}$ and $Y_t = j\omega C_{12}$ are due to the fields penetrating through the apertures. The other braid design parameters were discussed above. Parameter θ is determined from electrolytic tank measurements and depends on braid weave angle α.

For engineering purposes, approximate formulas can be proposed (with α in radians):

$$\theta_m \approx 0.66\alpha^2 - 0.11\alpha + 1, \quad \theta_e = 2 - \theta_m \qquad (3.110)$$

Considering Eq. (3.108), one can notice that, whereas the first item decreases with frequency, the second item increases. Thus, the pattern of the transfer impedance frequency characteristic is decreasing (or almost constant) in the low frequency range and increasing at the high frequencies. The larger the value of K, the later the rise begins. If K = 1 (i.e., 100 percent—an ideal case), the rise would not occur at all, the characteristic being that of a homogeneous tube.

Now, at least we have the shape of the transfer impedance frequency characteristic corresponding to the experimental data (and to the logic). Even so, the divergence between calculated and measured values often reaches 300 to 1000 percent! A more detailed analytical study of the perforated braided shield was given by Latham [3.4], who took into account the exact hole shapes, presence of nearby conductors, shield surface curvature, and interaction between neighboring holes. Another work in this area was done by Cathey [3.5]. However, even more accurate formulas did not bring the calculated and measured values significantly closer. Obviously, something was still missing.

✦ To explain the unexpected braided shield behaviors that cannot be predicted by theory, Fowler [3.6] proposed another coupling mechanism, termed "porpoising." According to Fowler and to Madle [3.7], braid geometry leads to individual wires effectively forming small loops of height g external to the braid and connected to similar loops internal to the braid. For the "porpoising" coupling, contour L (Fig. 3.39 d) is assumed to be like one for the perforated homogeneous shell in which external and internal loops exist, connected through the holes (Madle has actually physically modeled these loops). Magnetic field H_e^ϕ (consider external return path) induces in external loops currents that reach the internal loop and produce an internal field H_{ir}^ϕ. As the field H_{ir}^ϕ is opposite in direction to field H_{ip}^ϕ penetrating through the apertures, this leads to the decrease of braided shield transfer impedance. When the external loops' size is large enough (in the experiment described in Ref. [3.8], this was found to be 1.6–2 mm), compensation is achieved between the two fields.

Yet, many electronic cables with braided shields exhibiting this unexpected behavior are too small to provide such large loops. In these cables, the lower and upper carriers keep quite close at the place of their intersection, bringing the air gap loop height to zero. Thus, the loop surface for the external magnetic flux is limited only to the aperture area and to the lower carriers' individual wire section area with height d, and that under the angle $\pi/2 - \alpha$. This flux originates eddy currents in the lower carrier wires. In this case, it is not so much the loop's external inductance that has to be considered when defining the current in the loop, but rather the frequency-dependent internal inductance and losses in wires, as well as the secondary magnetic field. Besides, as it will be shown, in a braided shield, there also exists an axial magnetic component flux though the braid which is due to the spirality of braid. This also should have been accounted for.

All of this means that, at least in small cables, the porpoising mechanism cannot explain the difference between the theory and practice. There have been and are still being suggested other models of braid performance, with different level of complexity and sophistication. You can find them in original literature (e.g., see Refs. [3.8–10]).✦

3.5.3 Engineering Model of Braided Shield

If, after reading the preceding section, you acquired an impression that braided shield is not simple, you are right again. (How many times in this book have you come to similar conclusions?) And we still have not even discussed the effects of production tolerances! For this reason, it appears that a combination of theoretical and empirical approaches will best serve the practical applications both in design and selection of braided shields. Such an engineering model of a braided shield was suggested in Ref. [3.11].

First of all, the "missing link" in the physical mechanisms of braid performance had to be found. Therefore, while incorporating the "usual" mechanisms of diffusion and aperture penetration, an important new mechanism was added, addressing the special kind of spirality effect (or a different approach to "porpoising," if you will). The corresponding geometry of the voltage contour is presented in Fig. 3.39e. The braid is treated as a specific two-layer shield with intermitting spiral layers. Each unidirectional spiral layer is being formed by all strands weaved in the same direction. As it follows from Fig. 3.39e, both the axial and transverse views are some periodic structures mimicking this two-layer shield. The interchangeable layer positions form some special kind of coils in reciprocally perpendicular planes. It should be noted that, in fact, the adjoining coils in Fig. 3.39 e are formed not by the same but by different carriers (you may return to this later, when reviewing Fig. 3.42). But the currents "don't know this," and since all the unidirectionally weaved carriers begin from the same cable end, have identical parameters, and run in the same interchangeable mode, they are fully equivalent electrically. The currents, voltages, and electromagnetic field intensities, which are related to different unidirectional carriers at the same distance from the cable end, are identical, independent on carrier position on the shield's circumference. Therefore, the adjoining coils on both the axial and transverse views are shown connected electrically and can be considered as one continuous circuit.

From the equivalency of both sets of unidirectional carriers and according to the first Kirchhoff's law, it can be written:

$$I_1 = I_2 = \frac{I}{2}, \quad I_{1i}^z = I_{1e}^z = I_{2i}^z = I_{2e}^z = \frac{I\cos\alpha}{2}$$

$$I_{1i}^\phi = I_{1e}^\phi = -I_{2i}^\phi = -I_{2e}^\phi = \frac{I\sin\alpha}{2} \tag{3.111}$$

According to the model in Fig. 3.39e, five areas have to be studied: the air space outside the shield, the section area of the first (that is, in conductor metal) layer, the air gap between the layers, the section area of the second layer, and air space inside the shield. There is also the sixth area, formed by the shielded conductors inside the shielded space, but their interaction with penetrating fields is neglected. Because the electric and magnetic field intensities at the layer boundaries are related by Eqs. (3.90) and (3.91), it is sufficient to define the magnetic field only in the three air spaces.

Again, Ampere's law can be used. The currents I^z act in the same manner as in a homogeneous shield. Because all the $I_{1i}^z = I_{2i}^z$, it is sufficient to study only one coil in axial view—the results could be extended to any cable length. For the air spaces outside and inside the shield, the total current $I = I_1 + I_2$ must be considered, which is generating magnetic fields according to Eqs. (3.36) and (3.39). The fields between the layers are equal for external and internal current I return paths:

$$H_{12}^\phi = H_{21}^\phi = \frac{I}{2\pi D} \qquad (3.112)$$

The field H_{ip}^ϕ penetrating the shield apertures can be determined from Eq. (3.95). However, here "real life" interferes with theory. Experience proves that, at high enough frequencies, the field penetration item in Eq. (3.108) sometimes yields a number that is several times larger than the whole transfer impedance value (see also Ref. [3.3]). Curves $|Z_t|$ of a sample of measured data, and the item ωM_{12} in Fig. 3.40, illustrate the situation for one of the tested cables.

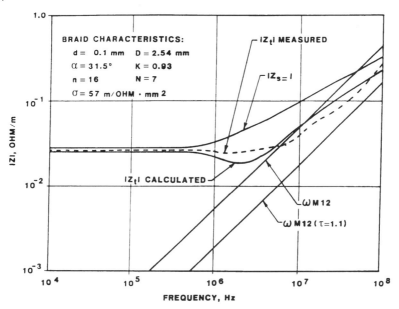

FIGURE 3.40 Transfer and transmission frequency characteristics of a braided shield

Because almost all, and quite different, theoretical models yield similar large values for the field penetration item, production tolerances could be suspected as a source of the inaccuracy. It is obvious from Eq. (3.95) that the most critical parameter is aperture diameter d_0, since it is in the cube power. The size of the aperture can be affected in several ways, but one of the most important seems to be that of the individual strands' spacing in the carrier. In Fig. 3.41, an aperture microscopy photo is presented of a braided shield (K = 86 percent). Due to the gaps between individual strands in the carrier, the aperture area is significantly decreased compared to its theoretical value.

FIGURE 3.41 Microscopic view of a diamond-shaped aperture in a braided shield

You may notice, of course, that instead of the main aperture with a larger area, multiple additional air gaps appear between the strands, so that the total optical coverage of the braid did not change. However, the gaps between the strands are much smaller than the main aperture, and it can be shown that their effect is comparatively small. Indeed, if the main aperture area is decreased by 50 percent, this will lead to around 30 percent (equivalent) diameter decrease, and to 65 percent of M_{12} decrease [see, for example, Eq. (3.95)]. Let this decrease of main aperture area be uniformly distributed between the four surrounding belts of wires, assume seven strands in each belt, and suppose that only the half of belt's strands are affected by this aperture decrease (the other half being affected by the opposite aperture). This will give 12 new secondary apertures, each with equivalent diameters of only 20 percent of the initial aperture. The total effect of all the 12 new apertures will constitute only 10 percent of the mutual inductance of the initial aperture. Thus, the total effect of the decreased main aperture, along with secondary apertures, will be only 45 percent of original aperture action. Such a decrease in aperture size can be viewed as an equivalent coverage increase (while the physical coverage remains the same).

The resulting equivalent coverage, K_{eq}, is defined from a comparison between the carrier widths—actual W and theoretical Nd:

$$\tau = \frac{W}{Nd} - \eta \qquad (3.113)$$

where η stands for the secondary aperture action.

The magnitude of τ depends on the shield design and production tolerances and, for not extremely dense braids (K < 95–97 percent), changes according to empirical data within the limits: $1 \geq \tau \geq 1.2$.

For more dense braids, the value of τ decreases and can even be less than 1; that is, the equivalent coverage becomes smaller that the initial values. This is the result of the geometry distortion because of "congestion" between the neighboring strands. As a result, the secondary holes can become even larger than the "main" aperture. Using all these relationships, it can be shown that

$$F_{eq} = \tau F, \quad K_{eq} = 2F_{eq} - F^2 = 2\tau F - \tau^2 F^2 \tag{3.114}$$

Now, the mutual inductance caused by apertures can be calculated according to the second term in Eq. (3.108), when substituting K_{eq} instead of K. The respective curve (assuming $\tau = 1.1$) for the previously discussed cable is presented in Fig. 3.35. (According to measurements, it is assumed that $\tau = 1.1$.)

Magnetic fields in transverse view do not depend on the current I return path because they are generated by the circular components of the current. Then, according to Ampere's law:

$$H_i^z = H_e^z = 0, \quad H_{12}^z = H_{21}^z = \frac{I}{2h} \tag{3.115}$$

✦ These relationships can be explained as follows. There is no solenoidal field outside the shield, and solenoidal fields from both layers are compensated inside the shield. The magnetic field H^z between the layers (and in the gap between them, if it is present—that's the "original porpoising") is generated only by the current in what is considered to be the coil. Because in the adjacent coils the external layers (1 or 2) are carrying opposite currents, $I_{1e}^\phi = -I_{2e}^\phi$, the directions of H_{12}^z and H_{21}^z are opposite, too. But while the sign of H^z between the layers changes simultaneously with the layer position (external or internal), the electric field intensities on the surface of each layer that faces inside the gap (e.g., $E_{12}^\phi = E_{21}^\phi$) retain their direction (and volume). This follows from Eqs. (3.34), (3.35), (3.90), and (3.91). In this way, it is evident that the intensities E_{12}^ϕ and E_{21}^ϕ "shuttle" from one to another surface of the respective carriers (according to their external or internal position in the shield) so as to face the opposite layer. Physically, these components are related to the five in the shield of axial magnetic field energy, which is concentrated between the layers. (As we know, the direction of energy flow in the diffusion process is determined by the Poynting vector.)

To make these considerations more graphic, imagine a braided shield with only two wide carriers at the lay length. In transverse view, they form two adjacent circular coils, schematically represented in Fig. 3.42. It is clear from the drawing that the voltage V_{spir} contour comprises both the upper and lower coils, while the middle loop corresponding to the shielded space is out of the contour. Though the directions of fluxes Φ_{12}^z ("from us") and Φ_{21}^z ("to us") are opposite, they induce the same direction currents in the respective layer (1 or 2) in each coil. This is because, with the change of the direction of the field, the position of the layer is also simultaneously changed. It is interesting to note that if one layer is eliminated (that is, in the case of a single spiral shield), the contour of the voltage V_{spir} will comprise the shielded space in which this time $\Phi_i^z \neq 0$ [see Eq. (3.86)]. Both unidirectional layers are switched in parallel. Therefore, to find voltage V, either of them may be chosen.

The electric field intensities E_{12}^ϕ or E_{21}^ϕ have to be taken on that surface of the chosen layer, that faces the second layer, and we must take into account the contribution of the magnetic fluxes Φ_{12}^z and Φ_{21}^z through the air gap between the layers (if present) and through the second layer. ✦

218 Electromagnetic Shielding Handbook

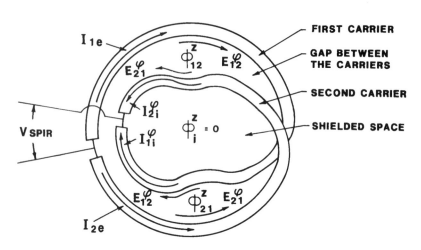

FIGURE 3.42 Spirality effects in braided shield

Now, the contour L for a braided shield can be built (Fig. 3.39 e): along the inner surface of the inner layer (axial view) in axial direction on the length of the lay h; then describes the revolution (transverse view) along the same layer on the shield circumference (while "shuttling" each time to its surface facing the opposite layers) with summary path length approximately πD (this neglects the change of length due to "porpoising"); then passes to the surface of the inner conductor and travels to the starting point (like in a spiral shield).

Thus, for and external return path of the current, and neglecting losses in the inner conductor, we obtain the voltage per cable unit length of

$$V_{it} = E_{1i}^z + \frac{\pi D}{2h}(E_{12}^\phi + E_{21}^\phi) + j\omega\left(\frac{1}{h}\Phi_{12}^z + \Phi_{12}^\phi + \Phi_{ip}^\phi\right)\upsilon \qquad (3.116)$$

Or, neglecting axial and circular magnetic fluxes in the gaps between the layers (that is, axial and circular "porpoising" effects) and assuming $E_{12}^\phi = E_{21}^\phi$, we obtain

$$V_{it} = E_{1i}^z + \frac{\pi D}{2h}E_{12}^\phi + j\omega\Phi_{ip}^\phi\upsilon \qquad (3.117)$$

If, in some specific case, the mentioned air gaps are present, the proper items can be included in Eq. (3.117). According to the developed procedure, the electric field intensities are found from Eqs. (3.90) and (3.91) via the magnetic components, while

Transfer Parameters of Electromagnetic Shields and Enclosures 219

these components are determined by the currents in the layers according to Ampere's law. Thus, substituting the braided shield carriers with equivalent spirally weaved tapes of thickness d, and expressing their characteristics (as well as apertures) via the braid design parameters, from Eq. (3.25) we obtain

$$Z_t = \frac{2}{\pi^2 \sigma FD}\left(\frac{\gamma}{\sinh(\gamma t)} + (\gamma \coth(\gamma t))\tan^2\alpha\right)$$

$$+j\omega\frac{\pi\mu_0}{6}\frac{1}{n}(1 - 2F_{eq} + F^2_{eq})^{3/2}\theta_m \quad (3.118)$$

Here, $D = D_i + 2d$, where D_i is the cable diameter under the shield; F is taken according to Eq. (3.104), and F_{eq} is from Eq. (3.114). Similar analysis for the loss impedance yields

$$Z_s = \frac{2\gamma}{\pi^2 \sigma FD \cos\alpha}\coth\gamma t + j\omega\frac{\pi\mu_0}{6}\frac{1}{n}(1 - 2F_{eq} + F^2_{eq})^{3/2}\theta_m \quad (3.119)$$

For capacitive coupling impedance it can be derived that

$$Z_F = j\omega Z_{c1} Z_{c2} C_{12} = j\omega \frac{\pi\sqrt{\mu_1\mu_2}}{6n}\sqrt{\frac{\varepsilon_1}{\varepsilon_2}}(1 - 2F_{eq} + F^2_{eq})^{3/2}\theta_e \quad (3.120)$$

3.5.4 Braided Shield Performance

To facilitate the analysis, it is convenient to deal separately with the items in the equations for the transfer and loss impedance, which correspond to different physical mechanisms. For this purpose, Eqs. (3.115) and (3.116) can be presented as follows:

$$Z_t = \frac{1}{2}Z_{the} + \frac{1}{2}Z_{she}\tan^2\alpha + Z_p \quad (3.121)$$

$$Z_s = \frac{1}{2}\frac{Z_{she}}{\cos^2\alpha} + Z_p \quad (3.122)$$

where Z_{the} and Z_{she} are transfer and loss impedances of a homogeneous solid tube having the same dc resistance as the braid's one unidirectional layer, and thickness equal to the strand diameter d. The first two items in Eq. (3.121) and (3.122) describe the diffusion, while Z_p accounts for the equivalent coverage increase for a magnetic

220 Electromagnetic Shielding Handbook

field penetrating the apertures. Each of the items in these equations, being a complex quantity, can be described by a vector diagram.

For the shield with parameters corresponding to Fig. 3.40, the transfer impedance ingredients vector diagram is presented in Fig. 3.43. The imaginary components are marked on the vertical scale, the real components are on the horizontal scale, and the numbers beside the curves denote the frequency in MHz. Of course, the Z_p item, being purely imaginary and inductive, is equal to 0 at dc (f = 0), and its magnitude is directly proportional to the frequency. (By the way, the same must be said about the capacitive coupling, except that it is located on the negative part of the imaginary axis.)

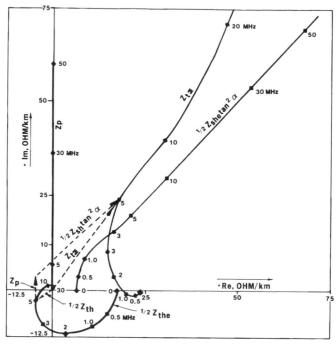

FIGURE 3.43 Vector diagram of braided shield

At dc, Z_{the} has its maximum value equal to half the equivalent tube dc resistance (real number); with the frequency rise, the real part is decreasing and the imaginary part appears, first increasing then also decreasing, while both parts periodically change their signs. Above a certain frequency (10 MHz for the cable under discussion), this item becomes so small that it can be neglected.

3.6 MULTILAYER SHIELDS

To obtain both reasonable flexibility and high electrical performance, multilayer shields are commonly used. These are usually combinations of copper or aluminum tapes, longitudinally and/or spirally applied (in some cases solid tubes also are used, but mainly in coaxial cables for microwave antennas, trunk telecommunications, and cable TV). As shown in Fig. 3.9, the performance of even a nonhomogeneous multilayer shield can approach the performance of solid tubes, albeit only in the lower part of the frequency range.

3.6.1 Multilayer Homogeneous Shields

Multilayer homogeneous shields are often used in shielding rooms and certain types of coaxial cables, when the application calls for a high degree of shielding effectiveness. The conducting layers can be physically distinct, may have an air, dielectric, or absorptive dielectric between them, or made of mechanically bonded different metals.

Similar to a single-layer solid shield, the transfer and loss impedances of multilayer homogeneous shields are determined by the ratio of the axial components of the electric field intensities at the respective shield surfaces, as related to the current in the shield. In general, a multilayer shield may consist of any number, n, of layers, with n = 2 to 4 the most common. Figure 3.44 shows a longitudinal cut of an n-layer shield, with the count beginning at the internal first layer. From Kirchhoff's law, the total current in the shield is

$$I = \sum_n I_i \qquad (3.123)$$

where each ith layer carries current I_i.

We will consider here two methods to determine the transfer parameters of a multilayer shield with homogeneous layers: one suggested by Schelkunoff [P.1], and the other suggested by this author [3.12].

Schelkunoff considered the external current return path. The calculations are performed using an iterative procedure. The derived expressions establish the link between the loss impedance (Schelkunoff called it surface impedance—remember?) of the shield, $Z_{sc(i+1)}$, consisting of $i + 1$ layers, and the loss impedance of the shield with the i layers, Z_{sci}, thus finding the effect of the additional layer on the shield impedance as follows:

$$Z_{sc(i+1)} = z_{sc(i+1)} - \frac{z_{t(i+1)}^2}{z_{sb(i+1)} + Z_{sci}} \qquad (3.124)$$

where $z_{sc(i+1)}$ and $z_{sb(i+1)}$ respectively, are the external (at the external return) and internal (at the internal return) impedance of the $(i + 1)^{th}$ layer.

Starting with the first layer (for which $Z_{sc1} = z_{sc1}$) and adding the rest of the layers one by one, we can obtain the impedance of the composite conductor as a continuous fraction.

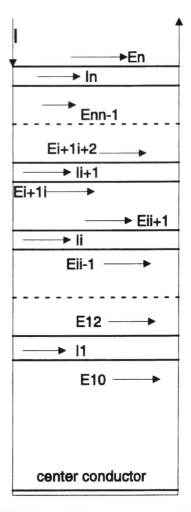

FIGURE 3.44 Multilayer homogeneous shield

A similar expression was obtained for transfer impedance $Z_{t(i+1)}$ of such a shield:

$$Z_{t(i+1)} = \frac{z_{t(i+1)} Z_{ti}}{Z_{sb(i+1)} + Z_{sci}} \qquad (3.125)$$

Proceeding from the fact that $Z_{121} = z_{121}$, one can determine consequently the transfer impedance of layers 2, 3, and so on.

The calculations in this case are even more cumbersome than for the loss impedance because of the presence of the items Z_{sci} in the denominator. For this reason, an alternative methodology was developed. While the derivation of Schelkunoff's method can be found in Ref. [P.1], the following is derived from alternative method-

Transfer Parameters of Electromagnetic Shields and Enclosures 223

ology. As usual, it is given in "fine print" so that those not interested in theory may skip directly to the practical formula, Eq. (3.138)

✦ Assume the external return current path. The transfer and loss impedances, respectively, of this shield can be determined as

$$Z_{tn} = \frac{E_{10}}{I}, \quad Z_{sn} = \frac{E_n}{I} \tag{3.126}$$

From Eqs. (3.61) and (3.62), the electric field intensities at the surfaces of separate layers can be determined as follows:

- The axial component of the electric field intensity at the external surface of the nth layer,

$$E_n = (I_1 + I_2 + \ldots + I_n)z_{sn} - (I_1 + I_2 + \ldots + I_{n-1})z_{tn} \tag{3.127}$$

- The field intensity at the internal surface of the nth layer,

$$E_{n(n-1)} = (I_1 + \ldots + I_n)z_{tn} - (I_1 + \ldots + I_{n-1})z_{sn} \tag{3.128}$$

- The field intensity at the external surface of the ith layer,

$$E_{i(i+1)} = (I_1 + \ldots + I_{i+1})z_{t(i+1)} - (I_1 + \ldots + I_i)z_{i+1} \tag{3.129}$$

- The field intensity at the internal surface of the ith layer,

$$E_{i(i-1)} = (I_1 + \ldots + I_i)z_i - (I_1 + \ldots + I_{i-1})z_{ti} \tag{3.130}$$

- The field intensity at the external surface of the 1st layer,

$$E_{12} = I_1 z_1 \tag{3.131}$$

- The field intensity at the internal surface of the 1st layer,

$$E_{10} = I_1 z_{t1} \tag{3.132}$$

In the above expressions, the *small* "z" items denote the respective parameters of single layers. From all these equations,

$$Z_{tn} = \frac{I_1}{I} z_{t1}, \quad Z_{sn} = z_n - z_{tn} + \frac{I_n}{I} z_{tn} \tag{3.133}$$

Let us determine the ratios of currents I_1 and I_n to current I. If the separate layers of the shield are in contact, or if the gap between them is small (which is the usual case), then at the border between two layers, the axial components of the electric field at the internal surface of the external layer and the external surface of the internal layer are equal. In an n-layer shield, one can write down n−1 such equalities. These equalities can be viewed as equations with regard to the currents $I_1, I_2, \ldots, I_n$ in each layer. These equations, along with Eq. (3.127), will produce a nonuniform system of n linear equations with regard to n unknown currents. In a matrix form this system can be written as

$$[y] = [\alpha][x] \tag{3.134}$$

where y is the matrix composed from the system's free members, x is the matrix of unknown currents, and α is the matrix of the coefficients by the currents:

$$y \equiv \begin{bmatrix} y_1 \\ y_2 \\ \ldots \\ y_n \end{bmatrix}, x \equiv \begin{bmatrix} I_1 \\ I_2 \\ \ldots \\ I_n \end{bmatrix}, \alpha \equiv \begin{bmatrix} \alpha_n & \cdots & \alpha_{1k} & \cdots & \alpha_{1n} \\ \cdots & \cdots & \cdots & \cdots & \cdots \\ \alpha_{i1} & \cdots & \alpha_{ik} & \cdots & \alpha_{in} \\ \cdots & \cdots & \cdots & \cdots & \cdots \\ \alpha_{n1} & \cdots & \alpha_{nk} & \cdots & \alpha_{nn} \end{bmatrix} \tag{3.135}$$

Finding from this system the ratios of currents I_1 and I_n to I, determine Z_{tn} and z_n. To determine the elements in expressions (3.135), write down the balance of longitudinal components of the electric field-intensity for the boundary between the ith and $(i+1)^{th}$ layers, accounting for Eqs. (3.129) and (3.130):

$$(I_1 + \ldots + I_i)z_i - (I_1 + \ldots + I_{i-1})z_{ti}$$
$$+ (I_1 + \ldots + I_i)z_{i+1} + (I_1 + \ldots + I_{i+1})z_{t(i+1)} = 0 \tag{3.136}$$

From Eq. (3.136), obtain the ith equation of the system Eq. (3.134):

$$(I_1 + \ldots + I_i)(z_i - z_{ti} + z_{i+1} - z_{t(i+1)})$$
$$+ I_i(z_i + z_{i+1} - z_{t(i+1)}) + I_{i+1}(-z_{t(i+1)}) = 0 \tag{3.137}$$

In Eq. (3.137), absent are the items that contain the currents in the layers with the numbers, larger than i +1. Therefore, the coefficients at the respective currents must be zeroes. From Eqs. (3.137) and (3.123), all the coefficients of the matrices can be determined.
For matrix y:

$$y_1, \ldots, y_{n-1} = 0, \; y_n = I$$

and therefore,

$$y \equiv \begin{bmatrix} 0 \\ 0 \\ \dots \\ I \end{bmatrix}$$

The elements of the matrix α are:

$$\alpha_{ii} = z_{si} + z_{s(i+1)} - z_{t(i+1)}, \quad \alpha_{i(i+1)} = -z_{t(i+1)}$$

$$\alpha_{ik} = 0 \text{ (at } k \geq i+2), \quad \alpha_{ik} = z_{si} - z_{ti} + z_{s(i+1)} - z_{t(i+1)} \text{ (at } k < i)$$

$$\alpha_{n1} = \alpha_{n2} = \dots = \alpha_{nn} = 1$$

The determinant of the system (3.134) is

$$\Delta = \begin{bmatrix} z_{s1}+z_{s2}-z_{t2} & -z_{t2} & 0 & \dots & 0 & 0 \\ z_{s2}-z_{t2}+z_{s3}+z_{t3} & z_{s2}+z_{s3}-z_{t3} & -z_{t3} & \dots & 0 & 0 \\ \dots & \dots & \dots & \dots & \dots & \dots \\ z_{s(n-1)}-z_{t(n-1)}+z_{sn}-z_{tn} & \dots & \dots & \dots & (z_{s(n-1)}+z_{sn}-z_{tn}) & -z_{tn} \\ 1 & 1 & 1 & 1 & 1 & 1 \end{bmatrix}$$

(3.138)

Currents I_1 and I_n can be found using the Kramer formulas:

$$I_1 = \frac{\Delta_1}{\Delta}, \quad I_2 = \frac{\Delta_2}{\Delta} \tag{3.139}$$

where Δ_1 and Δ_2 are obtained by substitution of the respective column of the determinant Eq. (3.138) with the elements of the matrix y. Substituting Eq. (3.139) into Eq. (3.133), we obtain the Z_{tn} and Z_{sn}.

As an example, consider an important case of a two-layer shield. Per Eq. (3.133),

$$Z_{s2} = z_{s2} - z_{t2} + \frac{I_2}{I} z_{t2}, \quad Z_{t2} = \frac{I_1}{I} z_{t1} \tag{3.140}$$

From (3.139) find:

$$I_1 = \frac{\begin{bmatrix} 0 & -z_{t2} \\ I & 1 \end{bmatrix}}{\begin{bmatrix} z_{s1} + z_{s2} - z_{t2} & -z_{t2} \\ 1 & 1 \end{bmatrix}} = I \frac{z_{t2}}{z_{s1} + z_{s2}}, \text{ and}$$

$$I_2 = \frac{\begin{bmatrix} z_{s1} + z_{s2} - z_{t2} & 0 \\ 1 & I \end{bmatrix}}{\begin{bmatrix} z_{s1} + z_{s2} - z_{t2} & -z_{t2} \\ 1 & 1 \end{bmatrix}} = I \frac{z_{s1} + z_{s2} - z_{t2}}{z_{s1} + z_{s2}} \quad (3.141)$$

Substituting the obtained values into (3.140), obtain

$$Z_{t2} = \frac{z_{t1} z_{t2}}{z_{s1} + z_{s2}}, \quad Z_{s2} = \frac{z_{s2}^2 + z_{s1} z_{s2} - z_{t2}^2}{z_{s1} + z_{s2}} \quad (3.142)$$

Similarly, the expressions for any number of layers can be obtained.

The transfer effectiveness of a two-layer shield is determined by substituting Eq. (3.142) into (3.7):

$$T_{12} = \frac{z_{t1} z_{t2}}{z_{s2}^2 + z_{s1} z_{s2} - z_{t2}^2}$$

$$= \frac{1}{\cosh \gamma_1 t_1 \cosh \gamma_2 t_2 + \frac{\gamma_1 \sigma_2}{\gamma_2 \sigma_1} \frac{\sinh \gamma_2 t_2}{\sinh \gamma_1 t_1}(\cosh \gamma_1 t_1 - 1)}$$

$$T_{12} = \frac{1}{\cosh \gamma_1 t_1 \cosh \gamma_2 t_2 + \frac{Z_{m2}}{Z_{m1}} \sinh \gamma_1 t_1 \sinh \gamma_2 t_2} \quad (3.143)$$

At low enough frequencies ($t < \delta$), from Eqs. (3.50) and (3.53) we obtain:

$$R_t = R_s = \frac{R_1 R_2}{R_1 + R_2}, \quad (3.144)$$

that is, the transfer and loss impedances of a two-layer shield are identical and equal to the dc resistances of the two layers connected in parallel.

With the frequency rise, the transfer impedance is fast decreasing, and the loss impedance is fast increasing, because the magnitudes of z_t are decreasing and the magnitudes of z_s are increasing.

What about reciprocity? Consider Eq. (3.142). As far as the transfer impedance is concerned, it is obvious that $Z_{t1} = Z_{t2}$, since the only difference will be in the items in the numerator to change places (which, as we remember from elementary mathematics, does not change the product). A more interesting result follows for the loss impedance: you can easy check that this impedance is determined mainly by the layer facing the return current. Now, it doesn't take too long to reach a very important conclusion:

To reduce impedance to the "useful" signal inside a shielding enclosure or a coaxial cable and simultaneously increase impedance to the interference current at the shield's external surface, the internal layer must have low loss impedance, and the external layer must have high loss impedance. The transfer impedance is not affected by the order of the layers. For example, if we have a combination of copper and steel layers, the copper layer should be internal, and the steel layer should be external.

◆ Between the shield *conducting* layers there can be placed *insulating* layers and/or exist *air* interstices. If the thickness of those is not large, the derived formulas can still be used by considering such separated by dielectric layers as conductors of a coaxial line and introducing a correction accounting for such circuit ac resistance R_{12}, capacitance C_{12}, and inductance L_{12}. Thus, for example, for a two-layer shield, we obtain

$$Z_{t2} = \frac{z_{t1} z_{t2}}{z_{s1} + z_{s2} + Z_{cor}} \tag{3.145}$$

where

$$Z_{cor} = \frac{1}{\dfrac{1}{R_{12}} + \dfrac{1}{j\omega L_{12}} + j\omega C_{12}}$$

$$R_{12} = \frac{\ln\dfrac{r_2}{r_1}}{2\pi\sigma l}, \quad L_{12} = \frac{\mu_r \mu_0}{2\pi} \ln\frac{r_2}{r_1}$$

$$C_{12} = \frac{2\pi\varepsilon l}{\ln\dfrac{r_2}{r_1}}$$

(3.146)

228 Electromagnetic Shielding Handbook

Here, r_1 and r_2 are the radii of the internal and external tubes respectively, and l is the shield length.✦

✦ As an exercise, prove that the transfer impedance of a *two-layer* shield with *identical* layers (1 and 2) and *no gap* between the layers, is identical to the transfer impedance of a *single-layer* shield with the thickness equal to the *sum* of the two layers and identical material and radius. Use Eqs. (3.145), (3.50), and (3.52). (For simplicity, you may omit the L_s, but that's not necessary.)✦

✦ It should be obvious that there is no difference between the performance of multilayer solid shields and equivalent composite metal shields (e.g., clad metals, multilayer coatings, etc.).✦

✦ This was theory. Using it, the calculations often yield microohms of transfer impedance, and corresponding hundreds of decibels of shield transfer effectiveness. But we have already seen in the examples in chapter 2 (see sections 2.2.3 and 2.3.3), that practical shield performance can be a far cry from theoretical derivations. Because of the inherent to the *physical* shielding seams, gaps, overlaps, apertures, spiralities - you name it- the performance of actual shields was much worse than the calculations would yield for ideal shields. The developed in the section 2.5.2 "Parallel Bypass" model of energy transfer through the shield, together with an experimental numerical example, show the more often than not, the shield effectiveness is determined by these "parallel bypasses".
We may finish this section with an old "adage":the shielding system performance is only as good, as the "weakest" element in the system. That's what palallel model is all about. ✦

3.6.2 Multilayer Nonhomogeneous Shield Performance

Performance and Problems of Multilayer Nonhomogeneous Shields

This is one of the most challenging (and also the most rewarding) subjects in electromagnetic shielding. Indeed, multilayer nonhomogeneous shields can provide very high levels of EMI protection (see Fig. 2.8 and 3.45). For example, Fig. 3.45 may create an impression that increasing the number of layers brings about consistent improvement in shield performance. But this is not necessarily so. As shown in Fig. 3.46, a four-layer shield (this happens to be a sample of the Ethernet data transmission trunk cable) is worse by far than a three-layer shield whose dc resistance is actually higher than that of the first one.

The "secret," of course, is in the specific parameters of individual layers, their relative positioning in the shield, and their interaction with each other. Thus, the foil layer in the three-layer shield of Fig. 3.46 has a shorting fold, while the two foil layers in the four-layer shield are not shorted.

Another factor is the magnetic permeability. By incorporating in the shield a high magnetic permeability layer, a further improvement in the shield performance is achieved [3.6,12,14]. The higher performance level is achieved for three reasons. First, it's "business as usual": due to higher magnetic permeability, penetration depth δ into the shield decreases. The second reason is related to the magnetic flux between the layers: when placed between two other layers, the magnetic layer results in a dramatic increase of the interlayer inductance and reduction of the transfer impedance [see Eqs. (3.145) and (3.146)]. The third reason is the most "hi-tech." Consider

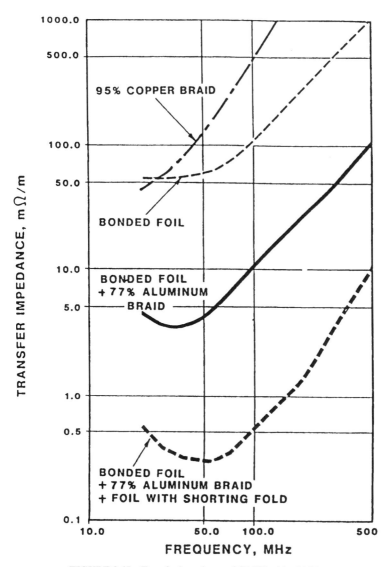

FIGURE 3.45 Transfer impedance of CATV cable shields

the vector diagram in Fig. 3.43. In the low-frequency band, the transfer impedance is determined mainly by the Z_{th} ingredient of $Z_{t\Sigma}$, which has a negative (that is, capacitive) imaginary part. By introducing a high magnetic permeability layer, the inductive (that is, positive) imaginary part of the transfer impedance is enhanced, which compensates the capacitive part. By properly designing the layers, the performance of such so-called superscreened cables is significantly improved, especially in the lower frequencies (see Fig. 3.47).

As the last example of multi-layer shield performance, we will compare the two double-layer shielded rooms: one with copper+steel solid confines, and the second - with brass mesh+ brass mesh confines. Fig. 3. 48 presents experimental data.

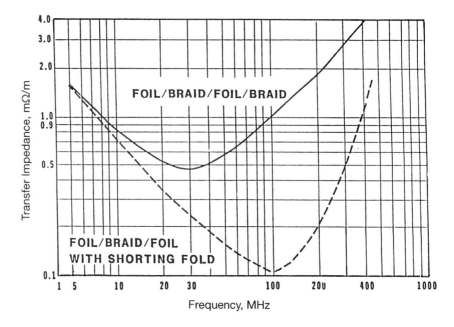

FIGURE 3.46 Multilayer shield efficiency

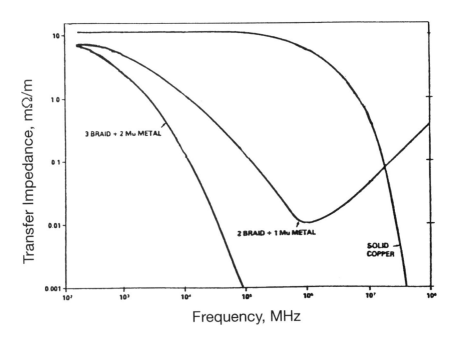

FIGURE 3.47 "Superscreened" shield performance

Transfer Parameters of Electromagnetic Shields and Enclosures 231

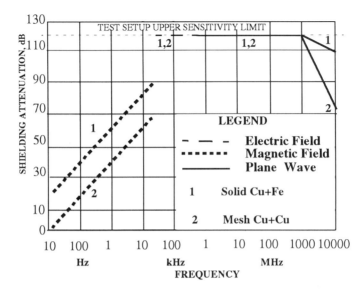

FIGURE 3.48 Double-layer solid copper/steel vs mesh copper/copper shielded room effectiveness

There are certainly important lessons to learn. First, even if, as we saw, a single-layer mesh shield transfer impedance was much worse than a solid shield (e.g., see Fig. 3.37 (because of the apertures, of course), a double-layer mesh provided for the shield room effectiveness at a comparative level as a solid coper/steel shield. How come? Two reasons can be thought of:

- The rooms' performance is much better than the test setup upper sensitivity limit (remember the theoretial "many hundreds of dB of shield effectiveness?" — so what you see, is the upper limit of the test setup dynamic range, while the shielding effectiveness is actually much higher, especially for solid shield)
- Remember the "parallel model" of the shield transfer parameters? The "bypass" EMI energy paths were certainly present in the tested rooms. The only question is how large the bypass energy is comparing to the "direct shield transparency". If the bypass energy is large, it may not matter whether the shield is solid or perforated,)

We do see significant difference in the magnetic shielding performance (of course, since the solid shield contains a high magnetic permeability layer - compare with Fig. 3.47) and in the centimeter wavlenght band (where the apertures in the mesh may approach a significant fraction of the signal wavelength).

Along with this physical insight, you may use the presented data for a very rough evaluation of you shielding room needs, depending on applications.

Now, how do we account for all the complex factors and predict the performance of multilayer nonhomogeneous shields or, even more important, make sure that the shield design is optimal? This can be done using two approaches: the "hard way," by actually manufacturing and testing a large number of samples, or the "intelligent" way, by applying theoretical analysis. Hopefully, you will select the right approach.

The Theory of Multilayer Nonhomogeneous Shields

But even if we do want to use the theory, why not to utilize something available and relatively simple? For example, we have already reviewed the theory of multilayer homogeneous shields. The "brute force" approach would be to use these formulas for nonhomogeneous shields. However, those who try this are quickly

There is an explanation why the "brute force" approach wouldn't work. Revisit the derivation of the formulas for multilayer homogeneous shields in Section 3.6.1. The critical element of the routine was to compare the longitudinal components of the electric field intensity at the boundaries between the adjacent layers [Eqs. (3.127) through (3.132)]. Because of the specific field configuration generated by currents in the homogeneous tubes, we were able to write down the balances [Eqs. (3.136) and (3.137)] of the electric field intensities. However, we saw that in nonhomogeneous shields, to deal only with longitudinal components of electric field intensity is insufficient because of the presence of the other field components. For example, a shield with apertures produced a circular component of magnetic field inside the shielded space, even at the external return current path. Also, a spiral shield generated the longitudinal component of magnetic field inside the shielded space, independent of the return current path.

As you remember, this was the main reason that we had to depart from the Schelkunoff's definition of the transfer and loss impedances via the longitudinal electric field intensities, and switch over to the voltage-related definitions. We may try the same approach in this case of multilayer shields. Indeed, if instead of writing the balances of electric field intensities at the boundaries between the layers, we write down the balances of the voltages at the same boundaries between the layers, we will automatically account for all the field components. Of course, to do that, we must correctly evaluate the voltages in terms of these field components, but by now we have good experience in doing that.

As an important practical example, consider [3.12,13] a coaxial cable with the outer conductor consisting of a copper solid wall tube as an internal layer with two steel spiral tapes as external layers. Although this example may seem too specific, it really is not. We will show that the derived formulas can be readily transformed into formulas for one-, two-, and three-layer solid wall shield, for one- and two-layer spiral shield, and for a composite solid tube + spiral two-layer shield. Moreover, the method we will use in the process of solving our problem appears to be general enough to be applicable to different types of nonhomogeneous shields. This method of evaluating the balance of the voltages at the boundaries between the layers was originally utilized by Kaden to solve a specific problem of a spiral tape enclosed in a solid wall tube. By generalizing this approach, we will be able to transform it into a universal tool, applicable to a wide class of multilayer nonhomogeneous shields including solid and perforated tubes, serve and spirally applied tapes, and braid.

Those not interested in the following derivation can skip directly to the formulas of Eq. (3.171), describing the current distribution in the layers of this shield.

♦ Fig. 3.49 presents the three-layer shield with external return current path. For simplicity, the thickness of the interstices between the layers is zero. Also assume that the currents in the spiral layers mimic the spiral shape (that is, we neglect the contact conductivity between the layers—we have seen that this corresponds to the experimental data). The directions of the currents are as shown in the drawing.

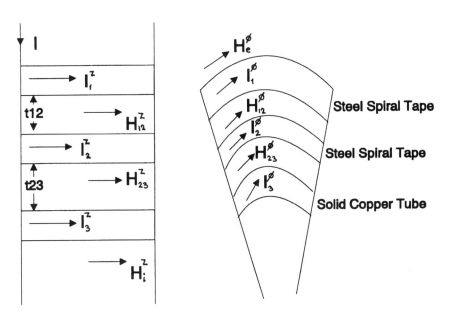

FIGURE 3.49 Three-layer composite shield

From Kirchhoff's current law, the total current in the shield is equal to the sum of the ingredient currents in the layers.

$$I = I_1 + I_2 + I_3 \tag{3.147}$$

Voltages U_e at the external surface of the outer tape, and U_i at the internal surface of the inner tube, will define the loss and transfer impedances.

$$Z_t = \frac{U_i}{I}, \quad Z_s = \frac{U_e}{I} \tag{3.148}$$

Due to the fact that the internal (third) layer is homogeneous solid tube in which $U_i = E_i l = E^z{}_{i3} l$, we can simplify the expression for the transfer impedance.

$$Z_t = \frac{E^z{}_i}{I} = \frac{I_3}{I} Z_{t3} \tag{3.149}$$

Unfortunately, this "trick" will not work for the loss impedance, because in our case (i.e., external return) it is determined by the voltage drop at the external surface of the external spiral tape. To find the voltages, we will use the "old" Eq. (3.10). From Ampere's law, we find the magnetic field intensities via the currents in the layers and their lay lengths (for H^z), similarly to what was done in the analysis of single and multilayer homogeneous shields and a spiral shield.

234 Electromagnetic Shielding Handbook

Consider first the voltage V_{1i} at the boundary between the internal surface of the first layer (outer tape) and the external surface of the second layer. At each spiral step, the surface S is a spiral surface bounded by the spiral outside and the center conductor, inside. The integration path follows the internal surface of the first layer in axial direction, then commences complete revolution (the revolution path length is $2\pi r$). Furthermore, the integration path switches to the surface of the cable center conductor, moves axially back, commences a revolution in the opposite direction at the center conductor surface (we will disregard both axial and circular components associated with the center conductor, because we neglect the losses in the center conductor), then closes the contour at the starting point.

Dividing by the external tape lay length h_1, we obtain:

$$V_{1i} = E^z{}_{12} + \frac{2\pi r}{h_1} E^\phi{}_{12} + \frac{1}{h_1} j\omega \Phi^z{}_{1i} + j\omega \Phi^\phi{}_{1i} \tag{3.150}$$

Here, the voltage in the contour, associated with the axial component of magnetic field at one spiral lay length, is

$$V_1 = j\omega \Phi^z_{1i} = U^a{}_1 + U^b{}_1 + U^c{}_1 \tag{3.151}$$

where the expressions

$$U^a{}_1 = j\omega \mu_0 \pi r^2 H^z_i \tag{3.152}$$

$$U^b{}_1 = j\omega \mu_0 \mu_2 2\pi r (H^z{}_{23} + H^z{}_{12}) \frac{1}{\gamma_2} \tanh \frac{\gamma_2 t_2}{2} \tag{3.153}$$

$$U^c{}_1 = j\omega \mu_0 \mu_3 2\pi r (H^z{}_{23} + H^z{}_i) \frac{1}{\gamma_3} \tanh \frac{\gamma_3 t_3}{2} \tag{3.154}$$

account for the *axial* magnetic fluxes, respectively, through the spaces bounded by the internal surface of the first layer, namely, "free space" enclosed by the third layer (i.e., cable insulation), through the body of the second layer, and through the body of the third layer. The effect of the axial magnetic flux through the third layer (copper tube) is neglected.

The voltage $j\omega \Phi^\phi_{1i} h_1$ accounts for the circular magnetic field in the contour for one spiral lay length.

Similar expressions can be developed for the external surface of the second layer, per the cable unit length.

$$V_{2e} = E^z{}_{21} + \frac{2\pi r}{h_2} E^\phi{}_{21} + \frac{1}{h_2} j\omega \Phi^z_{2e} + j\omega \Phi^\phi_{2e} \tag{3.155}$$

Because, at the cable ends, all the layers of the outer conductor are shorted with each other, we obtain

$$V_{1i} = V_{2e} \tag{3.156}$$

From this we obtain

$$E^z{}_{12} - E^z{}_{21} + \frac{2\pi r}{h_1}E^\phi{}_{12} - \frac{2\pi r}{h_2}E^\phi{}_{21}$$

$$+ j\omega\mu_0\pi r \left[rH^z{}_i + 2\mu_2(H^z{}_{23} + H^z{}_{12})\frac{1}{\gamma_2}\left(\frac{1}{h_1} - \frac{1}{h_2}\right)\tanh\frac{\gamma_2 t_2}{2}\right] = 0$$

(3.157)

Similarly, obtain for the boundary between the second and the third layers:

$$E^z{}_{23} - E^z{}_{32} + \frac{2\pi r}{h_2}(E^\phi{}_{23} - E^\phi{}_{32}) = 0 \tag{3.158}$$

The ferromagnetic spiral tapes have seams and gaps in the path of the magnetic flux, leading to the increase of magnetic resistance and reduction of the magnetic flux. The magnetic flux weakening can be estimated by the so-called de-magnetization factor, N. When the nonmagnetic gap length, l_b, is small compared to the full length of the magnetic induction lines, then

$$N = \frac{l_b}{l} \tag{3.159}$$

Hence, the actual magnetic field intensity H_a in the layer

$$H_a = \frac{H}{1 + \frac{l_b}{l}(\mu_r - 1)} \tag{3.160}$$

Now, let us express the electric and magnetic field intensities vie the currents in and parameters of the tapes. Thus we obtain

$$H^z{}_i = \frac{z_{123}\left(\frac{I_1}{h_1} + \frac{I_2}{h_2}\right)}{j\omega\frac{\mu_0}{4\pi} + z_3} \tag{3.161}$$

$$E^z{}_1 = I(z_1 - z_{t1}) + I_1 z_{t1}$$

$$E^z{}_{12} = Iz_{t1} - (I_2 + I_3)z_1$$

$$E^z{}_{21} = I_2 z_{s2} + I_3(z_2 - z_{t2}) \tag{3.162}$$

$$\frac{2\pi r}{h_1} E^\phi{}_{12} = I_1 z_1 \tan^2\alpha$$

$$\frac{2\pi r}{h_2} E^\phi{}_{21} = I_1 \tan\alpha_1 \tan\alpha_2 (z_{t2} - z_{s2}) + I_2 z_{t2} \tan^2\alpha_2 \tag{3.163}$$

$$E^z{}_{23} = I_3(z_{t2} - z_{s2}) + I_2 z_{t2}, \quad E^z{}_{32} = I_3 z_{s3} \tag{3.164}$$

$$\frac{2\pi r}{h_2} E^\phi{}_{23} = I_1 \tan\alpha_1 \tan\alpha_2 (z_{s2} - z_{t2}) + I_2 z_{s2} \tan^2\alpha_2 \tag{3.165}$$

$$\frac{2\pi r}{h_2} E^\phi{}_{32} = (I_1 \tan\alpha_1 + I_2 \tan\alpha_2) \left(\frac{z^2{}_{t3}}{j\omega\dfrac{\mu_0}{4\pi} + z_{s3}} - z_3 \right) \tan\alpha_2 \tag{3.166}$$

$$H^z{}_{12} = \frac{I_1}{h_1}, \quad H^z{}_1 = 0, \quad H^z{}_{23} = \frac{I_1}{h_1} + \frac{I_2}{h_2} \tag{3.167}$$

After substituting the field intensity expressions into the voltage balance equations, obtain a system of linear equations with regard to currents I_1, I_2, and I_3:

$$AI_1 + BI_2 + CI_3 = 0 \tag{3.168}$$

$$DI_1 + EI_2 + FI_3 = 0 \tag{3.169}$$

$$I_1 + I_2 + I_3 = I \tag{3.170}$$

where it is designated

$$A = z_{t1} + z_{s1} \tan^2\alpha_1 + (z_{s2} - z_{t2}) \tan\alpha_1 \tan\alpha_2$$

$$+ j\omega\frac{\mu_0}{4\pi} \left(\frac{z_{t3}}{j\omega\dfrac{\mu_0}{4\pi} + z_{s3}} + \frac{4\mu_2}{r\gamma_2} \tanh\frac{\gamma_2 t_2}{2} \right) (\tan\alpha_1 - \tan\alpha_2) \tan\alpha_1$$

$$B = z_{t1} - z_{s1} - z_{s2} - z_{t2}\tan^2\alpha_2$$

$$+ j\omega\frac{\mu_0}{4\pi}\left(\frac{z_{t3}}{j\omega\frac{\mu_0}{4\pi} + z_{s3}} + \frac{4\mu_2}{r\gamma_2}\tanh\frac{\gamma_2 t_2}{2}\right)(\tan\alpha_1 - \tan\alpha_2)\tan\alpha_2$$

$$C = z_{t1} - z_{s1} - z_{s2} + z_{t2}$$

$$D = \left(z_{s2} - z_{t2} + \frac{z_{t3}^2}{j\omega\frac{\mu_0}{4\pi} + z_{s3}} + z_3\right)\tan\alpha_1\tan\alpha_2$$

$$E = z_{t2} + \left(z_{s2} - z_{s3} + \frac{z_{t3}^2}{j\omega\frac{\mu_0}{4\pi} + z_{s3}}\right)\tan^2\alpha_2$$

$$F = z_{t2} - z_{s2} - z_{s3}$$

Using Kramer's formulas, we obtain

$$I_1 = \frac{\Delta_1}{\Delta}, \quad I_2 = \frac{\Delta_2}{\Delta}, \quad I_3 = \frac{\Delta_3}{\Delta} \tag{3.171}$$

The system determinant

$$\Delta = \begin{vmatrix} A & B & C \\ D & E & F \\ 1 & 1 & 1 \end{vmatrix} \tag{3.172}$$

and the $\Delta_1, \Delta_2, \Delta_3$ are obtained by substituting the respective columns of the determinant Δ with the free items in the equation system.

From here,

$$\zeta_1 = \frac{I_1}{I}, \quad \zeta_2 = \frac{I_2}{I}, \quad \zeta_3 = \frac{I_3}{I} \tag{3.173}$$

Also, from Kirchhoff's law

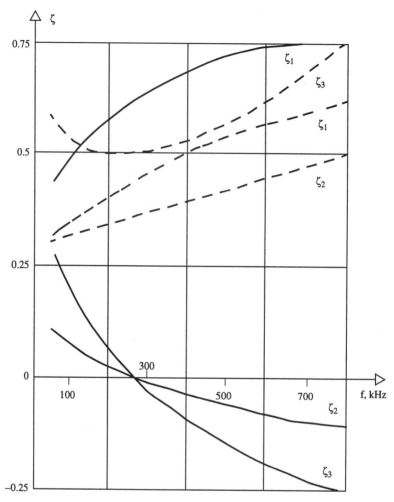

FIGURE 3.50 Current distribution in the layers of a composite shield (solid line, $\alpha_1 = -\alpha_2$; dashed line, $\alpha_1 = \alpha_2$)

$$\zeta_1 + \zeta_2 + \zeta_3 = 1 \tag{3.174}$$

Hence, the current distribution in the layers of a composite shield is expressed as a function of the electrical and construction parameters of the shield layers, as well as the signal frequency. As an example, Fig. 3.50 shows the frequency characteristics of the values of the parameters ζ for two cases of the spiral tape winding: $\alpha_1 = \alpha_2 = \pi/4$ and $\alpha_1 = -\alpha_2 = \pi/4$, 0.1 mm thickness of the steel tapes, and 0.15 mm thickness of the copper tube.

When the weave angles of the tapes are opposite, the currents in certain sections of the spiral layers can move in directions that are opposite to the direction of current I (this is the meaning of the negative ζ). This happens because mutual inductance

Transfer Parameters of Electromagnetic Shields and Enclosures 239

exists between the spiral layers, and the direction of the current associated with this inductance depends on the relationship between the lay angles of the tapes. When these angles are opposite, the bulk of the current flows in the external tape, and the current in the third layer (solid copper tube) is minimal. But per Eq. (3.149), this is what we want in order to reduce the transfer impedance of the shield! And when the angles are identical, the bulk of the current flows mainly in the internal layer, leading to the increase in the shield transfer impedance (compare this with the unfavorable performance of a single spiral).

But just "how much opposite" should the tapes be (i.e., what specific weave angles should be used) depends on the signal frequency, construction and electrical parameters of the tapes, and mechanical limitations. For example, in Fig. 3.50, at frequencies below 200 kHz, $\zeta_3 > \zeta_2$, which clearly indicates that the design is not optimal. We will discuss the shield design and optimization principles in more detail in Chapter 6.

Substituting the derived expressions into Eq. (3.149), for the transfer impedance of our three-layer shield we finally obtain:

$$Z_t = z_3 \frac{\begin{bmatrix} A & B \\ D & E \end{bmatrix}}{\begin{bmatrix} A & B & C \\ D & E & F \\ 1 & 1 & 1 \end{bmatrix}} \qquad (3.175)$$

We were able to "get away" with relatively "simple" Eq. (1.175) because we had to deal with fields at the internal surface of a homogeneous tube (with external return), so that all other field components except only the longitudinal electric field intensity at the internal surface of the inner homogeneous tube, are nonexistent. That's why, in the first place, we needed only the current I_3 and, as a result, we could use Eq. (3.149).

Unfortunately, to obtain the loss impedance of our composite shield, we have to evaluate voltage U_e at the external surface of the external spiral tape, which is associated with all three currents: I_1, I_2, and I_3. In this case we must deal with the full Eq. (3.148). (Didn't it cross your mind that we deliberately selected our example to give you the flavor of both outcomes?)

Substituting the derived above relationships into Eq. (3.148), we obtain the loss impedance for our three-layer shield (external return):

$$Z_3 = z_{s1} - z_{t1} + \zeta_1(z_{t1} + z_{s1}\tan^2\alpha_1) + j\omega\frac{\mu_0}{4\pi}[\frac{z_{t3}(\zeta_1\tan\alpha_1 + \zeta_2\tan\alpha_2)\tan\alpha_1}{2(j\omega\frac{\mu_0}{4\pi} + z_{t3})}$$

$$+ \frac{\mu_2}{r\gamma_2}(2\zeta_1\tan\alpha_1 + 2\zeta_2\tan\alpha_2)\tan\alpha_1\tanh\frac{\gamma_2 t_2}{2} + \frac{\mu_3}{r\gamma_3}(\zeta_1\tan\alpha_1$$

$$+ \zeta_2\tan\alpha_2)\tan\alpha_1\left(\frac{z_{t3}}{j\omega\frac{\mu_0}{4\pi} + z_{t3}} + 1\right)\tanh\frac{\gamma_3 t_3}{2}]$$

(3.176)

3.6.3 Calculating "Derivative" Shields

The techniques we used to obtain the expressions for our three-layer shield are universal in the sense that the same approach can be applied to an arbitrary shield consisting of different numbers and types of layers. Owing to this generality, as soon as the expressions for any composite shield design are obtained, they can be used to define the transfer impedance and loss impedance of derivative shields obtained by eliminating or "simplifying" the layers of the "progenitor" composite shield. To do this, note that when a layer is absent, its thickness t and current in it can be assumed to be zero, the values $z_t = z_s \to \infty$, and any fields generated by this layer can be disregarded (e.g., also assumed to be zero). If a solid tube is substituted instead of a spiral layer, the angle $\alpha \to 0$. And if a perforated shield is used, then the fields penetrating through the apertures must be accounted for at the corresponding boundaries.

For example, assuming $\alpha_1 = \alpha_2 \to 0$, determine from Eq. (3.175) the transfer impedance of a three-layer shield with solid homogeneous layers:

$$Z_t = \frac{z_{t1}z_{t2}z_{t3}}{z^2_{s2} - z^2_{t2} + z_{s1}z_{s2} + z_{s1}z_{s3} + z_{s2}z_{s3}}$$ (3.177)

From Eq. (3.178), familiar expressions can be deduced for one- and two-layer shields with homogeneous tubes. It is easy to see that with the frequency rise, the transfer impedance Z_t is rapidly decreasing because the numerator is decreasing in "cube" power order, while the denominator is increasing. In the lower frequencies ($t < \delta$), this reduces to

$$R_{0t} = R_{0s} = \frac{r_{01}r_{02}r_{0(3)}}{r_{01}r_{02} + r_{01}r_{03} + r_{02}r_{0(3)}}$$ (3.178)

When the internal layer (the third layer is copper tube) is absent, obtain formulas for a two-layer spiral shield:

$$Z_t = \zeta_1\left(z_{s2} - z_{t2} + j\omega\frac{\mu_0}{4\pi}\right)\tan\alpha_1\tan\alpha_2 + \zeta_2\left(z_{t2} - z_{s2}\tan^2\alpha_2 + j\omega\frac{\mu_0}{4\pi}\tan^2\alpha_2\right) \tag{3.179}$$

$$Z_s = \zeta_1\left(\frac{z_{s1}}{\cos^2\alpha_1} + \frac{j\omega}{\pi}\left(\frac{\mu_0}{4} + \frac{\mu_{w2}t_2}{r}\right)\tan^2\alpha_1\right)$$

$$+\zeta_2\left((z_{s1} - z_{t1}) + \frac{j\omega}{2\pi}\left(\frac{\mu_0}{2} + \frac{\mu_{w2}t_2}{r}\right)\tan\alpha_1\tan\alpha_2\right) \tag{3.180}$$

where

(3.181)

$$\zeta_1 = \frac{I_1}{I} = \frac{1}{\chi}\{z_{t1} + z_{s1}\tan^2\alpha_1 + (z_{s2} - z_{t2})\tan\alpha_1\tan\alpha_2$$

$$+ j\omega\left(\frac{\mu_0}{4\pi} + \frac{\mu_{w2}t_2}{\pi r}\right)(\tan\alpha_1 - \tan\alpha_2)\tan\alpha_1\}$$

$$\zeta_2 = \frac{I_2}{I} = \frac{1}{\chi}\{z_{s1} - z_{t1} + z_{s2} + z_{t2}\tan^2\alpha_2$$

$$+ -j\omega\left(\frac{\mu_0}{4\pi} + \frac{\mu_{w2}t_2}{2\pi r}\right)(\tan\alpha_1 - \tan\alpha_2)\tan\alpha_2\} \tag{3.182}$$

$$\chi = \frac{z_{s1}}{\cos^2\alpha_1} + z_{s2} + z_{t2}\tan^2\alpha_2 + (z_{s2} - z_{t2})\tan\alpha_1\tan\alpha_2$$

$$+\frac{j\omega}{\pi}(\tan\alpha_1 - \tan\alpha_2)(\frac{\mu_0}{4}(\tan\alpha_1 - \tan\alpha_2) + \frac{\mu_{w2}t_2}{r}\left(\tan\alpha_1 - \frac{1}{2}\tan\alpha_2\right)) \tag{3.183}$$

It is easy to verify that when the second and the third layers are absent, familiar formulas for a single-layer spiral shield are obtained. Another practical design of a copper tube with a one-layer spiral over it is obtained when the first (or the second) spiral is eliminated from the equations. For example, assuming for simplicity that

$$z_{s3} = z_{t3} = r_{03} \leq j\omega \frac{\mu_0}{4\pi}$$

we obtain

$$Z_t = \frac{z_{t2} + (z_{s2} + r_{03})\tan^2\alpha_2}{\dfrac{z_{s2}}{\cos^2\alpha_2} + (z_{s2} + r_{03})\tan^2\alpha_2} \tag{3.184}$$

Similarly, for the loss impedance of a tube + spiral shield:

$$Z_s = z_{s1} - z_{t1} + \zeta_1(z_{t1} + z_{s1}\tan^2\alpha_1) + j\omega\frac{\mu_0}{2\pi}\tan^2\alpha_1\left[\frac{z_{t3}}{2\left(j\omega\dfrac{\mu_0}{4\pi} + z_{t3}\right)}\right.$$

$$\left.+ \frac{\mu_2}{r\Upsilon_2}\left(\frac{z_{t3}}{2\left(j\omega\dfrac{\mu_0}{4\pi} + z_{t3}\right)} + 1\right)\tanh\frac{\gamma_3 t_3}{2}\right] \tag{3.185}$$

✦ The suggested method of solving the problems of a nonhomogeneous multilayer shield is not the only choice. As an alternative, the use of modern numerical techniques opens great opportunities for direct application of Maxwell's equations. Unfortunately, the numerical techniques are not always conducive to the analysis of the shield behavior trends. There are also different analytical methods available. One such method, utilizing the concepts of layer parameter anisotropy, was briefly mentioned in Chapter 2. We will conclude with a brief summary of this method.

The author [3.15] considered a composite three-layer shield consisting of a copper tube and two spirally applied outer steel tapes, a design similar to the one we evaluated previously. It was assumed that the contact resistance between the layers was zero, so that the currents in the spiral tapes could flow in the direction of both along and across the layer. (Actually, this was a fatal mistake because, as we saw from the experiments, the contact resistance between the layers in practical shields, especially those manufactured of steel and/or aluminum, is much larger than the resistance along the layer direction (say, spiral, so that the bypass currents can be neglected altogether. But this does not at all discredit the method itself.)

Considering a Cartesian coordinate system, the layers are characterized by the tensors of conductivity and magnetic permeability σ_l, μ_l in the direction of the tape length, and σ_g, μ_g in the direction of the tape width. Adopting the Cartesian coordinate system, Maxwell's equations can be written as follows:

$$\begin{bmatrix} I^y \\ I^z \end{bmatrix} = \begin{bmatrix} \sigma_{yy} & \sigma_{yz} \\ \sigma_{zy} & \sigma_{zz} \end{bmatrix} \begin{bmatrix} E^y \\ E^z \end{bmatrix} \qquad (3.186)$$

$$\begin{bmatrix} I^1 \\ I^g \end{bmatrix} = \begin{bmatrix} \sigma_1 & 0 \\ 0 & \sigma_g \end{bmatrix} \begin{bmatrix} E^1 \\ E^g \end{bmatrix} \qquad (3.187)$$

Transforming the coordinates of electrical fields and currents, we obtain

$$\begin{bmatrix} E^y \\ E^z \end{bmatrix} = \begin{bmatrix} \sin\varphi & -\cos\varphi \\ \cos\varphi & \sin\varphi \end{bmatrix} \begin{bmatrix} E^1 \\ E^g \end{bmatrix} \qquad (3.188)$$

$$\begin{bmatrix} I^y \\ I^z \end{bmatrix} = \begin{bmatrix} \sin\varphi & -\cos\varphi \\ \cos\varphi & \sin\varphi \end{bmatrix} \begin{bmatrix} I^1 \\ I^g \end{bmatrix} \qquad (3.189)$$

where φ is the angle formed by the tape and direction "Z."
From these equations, the components of the *conductivity tensor* can be obtained.

$$\sigma_{yy} = \sigma_1 \sin^2\varphi + \sigma_g \cos^2\varphi$$

$$\sigma_{zz} = \sigma_1 \cos^2\varphi + \sigma_g \sin^2\varphi$$

$$\sigma_{yz} = \sigma_{zy} = (\sigma_1 - \sigma_g)\sin\varphi\cos\varphi$$

If in the previous expressions, the E, I, and σ are substituted with B, H, and μ, respectively, then we obtain the equations linking the electric and magnetic field intensities, magnetic induction, and magnetic permeability. For instance, with regard to E^z, we get

$$\left\{\frac{d^2}{dx^2} - \gamma_1^2\right\}\left\{\frac{d^2}{dx^2} - \gamma_g^2\right\} E^z = 0 \qquad (3.190)$$

where Υ_1 and Υ_g are intrinsic propagation constants for the currents flowing along and across the tapes, respectively.
As a result, from Eq. (3.190) we obtain

$$E^z = A\sinh\Upsilon_1 x + B\cosh\Upsilon_g x + D\cosh\Upsilon_g x \qquad (3.191)$$

Similarly, the solutions for other field components E^y, H^y, and H^z can be obtained.

The integration constants are found from the corresponding four boundary conditions, originating from Ampere's law for the external surface of the external conductor, Faraday's law for the internal surface of the external conductor, and from the equality of the axial ("z") and tangential ("y") components of the electric field at the boundaries between the adjacent layers.

Unfortunately, the estimation of the values of the μ_g and σ_g, itself presents a not-so-trivial problem.✦

4

Emi Environment and Electromagnetic Coupling in Shielding

4.1 DEFINING SHIELDING COUPLING

4.1.1 Shielding To Decouple And Coupling To Shield

What is electromagnetic coupling in shielding problems' context? "What" couples to "what", how, and why?

The role, effects, and function of electromagnetic coupling are among the most important topics in electromagnetic shielding. They are also among the most confusing and challenging issues. The existing approaches span two extremes: (1) reducing all the shielding mechanisms to different forms of coupling, and (2) ignoring completely the role of coupling in shielding. Although, as usual, the "truth" resides between the extremes, it is worthwhile to briefly investigate what they amount to.

The first "school of thought" often treats coupling as a "black magic" art, which explains "everything" about shielding. Notwithstanding the "mystique," there is nothing wrong in treating shielding-related phenomena as manifestations and transformations of various coupling mechanisms. Indeed, the objective of shielding is to serve as a barrier to unwanted coupling to or between the circuits, or the circuits and the electromagnetic environment, that is to be a *"decoupling device"*. On the other hand, being a system element itself, the shield couples into both the source and the victim of EMI, whether they are circuits or ambient environment. Even the shield transfer function may be thought of as representing signal coupling between the shield's two opposite surfaces and through its "body". But, of course, to make sure of their "legitimacy," these concepts must be implemented in conjunction with the proper analysis. Because this is not always done, many misunderstood and/or irrelevant effects are too often "blamed" on coupling phenomena.

The opposite extreme is to completely neglect the contribution of coupling to shielding process. One often-used such "technique" is to reduce the scope of shielding effectiveness to the shield transfer function. Really, isn't the knowledge of a shield transfer function sufficient for our shielding needs? Unfortunately, it isn't (at least, not always). However useful and convenient are the transfer parameters, they represent only one, albeit important, dimension of the electromagnetic shielding performance. In Chapter 2, by comparing theoretical and experimental data in practical examples, we had ample opportunities to observe the potential catastrophic consequences of errors originated by such a "rationale" that neglects the role of EMI environment and coupling. But at that time, we did not have enough background to trace down the causes of the discrepancies.

A common denominator to these and many other misconceptions is heavy reliance on "common sense," combined with disregard to system EMC analysis. Previously, we proved that the ultimate worth of a shield is determined by its system shielding effectiveness, which is a function not just of the shield transfer function but also of the ambient environment and electronic system specifics. Thus, the shielding effectiveness may be affected by factors that are indirectly related to, or even independent of, the particular shield design and transfer characteristics (e.g., think of a grounding system). To account for the most important such factors, we have identified five basic ingredients/steps involved in electromagnetic shielding and used them to build a system model of electromagnetic shielding performance (see Chapter 2). And even if the shield transfer function is conceptually and "graphically" placed at the center of our model, these other factors are also crucial to the shielded circuit EMC performance.

EMI environment and electromagnetic coupling to the shield are among the most decisive such other factors.

4.1.2 Electromagnetic Environment and Coupling in Shielding Model

A shield separates the "space" in three areas characterized by different electromagnetic environment (e.g., see Figs. 2.17 and 2.23): within the shield's "body" and adjacent to the shield's surfaces. Surprisingly, the two last areas are not simple to define. True, for a shielding enclosure of a limited size, these areas are readily visualized as "internal" and "external". However such visualization may not be as easy for "open" shields (say, a shielding plane, or parabolic shape of finite or infinite size), or even "closed" shields of large size. Also as we have seen (see section 3.4.2), even in such traditionally "closed" shape enclosure as a spiral shield, there may be difficulties just to relate the environment to the respective shield surfaces! Nevertheless, since we do need some terminology to refer to the two parts of space separated by shield, we will apply the terms "internal" and " external" space with regard to the shield to all shield configurations, unless specifically mentioned otherwise, where appropriate.

Given this terminology and keeping in mind the system electromagnetic shielding model introduced in chapter 2, the EMI coupling to the shield represents the electromagnetic link between the shielding enclosure and the electromagnetic environments in internal and external space. In chapter 2, respective components were introduced in two models - for the emission and immunity shielding - to account for this link. We referred to these components as *electromagnetic coupling to the shield and* specified them in terms of EMI energy exchange between the environments adjacent to the "inside" and "outside" shield surfaces:

- *Immunity Shielding Model:* transmitter and / or external EMI environment ⇒ external coupling to the shield ⇒ EMI transfer through the shield ⇒ internal coupling to the shield ⇒ internal EMI environment and / or receiver
- *Emissions Shielding Model:* transmitter and / or internal EMI environment ⇒ internal coupling to the shield ⇒ EMI transfer through the shield ⇒ external coupling to the shield ⇒ external EMI environment and / or receiver

In these models, the processes of EMI coupling to the shield are placed "in series" at both sides of the shield transfer function, providing the input to and receiving the output from it. Thus, any inconsistencies in identifying EMI environment, formulating and applying the coupling functions will result in the failure of the whole system shielding effectiveness model, even if the transfer function is adequate ("garbage in, garbage out").

To simplify the task of analyzing the shield transfer function, in the previous chapter we "fixed" the EMI enfironment at both sides of the shield and sort of "bypassed" the shield interaction with the environment by considering a given current in the shield when evaluating the transfer impedance, and a given voltage between the shield and other conductors (e.g., external conductor of a triaxial line) when evaluating the transfer admittance (see Fig. 3.2). We fixed the current and voltage independently of the shield's ability to generate them in a specified environment. We then considered the effects associated with the respective shield surfaces, which we used to determine the shield transfer parameters. On this track, the shield transfer effectiveness parameter was as close as we came to the shielding effectiveness. However the *shield transfer effectiveness* is a far cry from the *system shielding effectiveness*, which is affected by the shield ability to generate the EMI currents and voltages. To see how important is this attribute, one can imagine "good" shields (that is, possessing favorable transfer parameters) that are strongly coupled to the ambient EMI and develop large EMI currents in the shield, thus resulting in poor system shielding effectiveness. And conversely, "bad" shields (that is, possessing unfavorable transfer parameters) that are weakly coupled to the same ambient and develop small EMI currents in the shield, manifest excellent system shielding effectiveness.

Our system shielding model identified two major coupling-related ingredients/ steps: external coupling between the shield and the ambient environment outside the shielding enclosure, and internal coupling between the shield and the enclosed environment inside the shielding enclosure. We also saw that, in certain shield topologies, it is not a trivial matter to tell the "outside" from the "inside" environments. E.,g., remember the longitudinal magnetic field in a spiral shield with external and internal return path? Another example is a shielding plane, where the "inside" cannot be told from "outside". But then, how do the internal and external, whatever they are, environments are "connected" in the shield's presence?

Perhaps, the problem is better clarified when you remember, that even if in our model we have separated the shielding problem at large into ingredient processes / steps, in reality all these processes are very "intimately" related. For this reason, neither of these processes can be considered in complete isolation. Thus, while looking at allegedly "pure" transfer parameters (e.g., transfer impedance in immunity case), we still had to introduce somehow the current into the shield "from the outside", and on the other hand, reference the shielded conductors to the shield - "inside" - that is, exersize typical "coupling" actions.

When defining the shield transfer parameters previously, we have "succeeded" in completely ignoring the external coupling mechanism. Even if later we had to present some practical means of introducing the current in the shield, e.g., by conduction coupling per Fig. 3.2, the current really "did not care" about its origin, and thus the generality of the derived transfer impedance parameter was never affected. Unfortunately, there was no such "easy way out" when referring the potential generated in the shielded conductors with regard to the shield, that is, when measuring the voltage between them. There, we had to rigidly specify the characteristics of the internal circuit formed between the shield and the shielded conductors. Although we did it identically for all cases - shorted at one end and open at the other end (e.g., see Fig. 3.2) - such a configuration of the internal circuit provided a useful common base for transfer parameter comparison of different shields, but it also did not necessarily reflect many real life situations. For example, the internal circuit in a coaxial line (which also can be viewed as a shielded, single-ended circuit) is usually matched to its characteristic impedance. But, in general, the characteristic impedances and loads at the ends of the "real life" internal circuits are arbitrary and not necessarily coaxial. Also with regard to the external circuits formed by the line shield, the shield termination and grounding situation can be very diverse and often unpredictable, but nevertheless having drastic effect on shielding performance (e.g., see Fig. 2.10). Later, we will give a detailed theoretical analysis of these effects with respect to crosstalk between coaxial cables.

How does the inclusion of the coupling function affect the system shielding model generality and relevancy? In Chapter 2, we showed that the shield transfer function relates to the most general model: "just" the shield itself. (Well, as you see, it is "almost the most general"; we did specify an internal coupling circuit, although of a special type, as we have just discussed). In contrast, the coupling function presents the first step from the transfer function toward the evaluation of the system shielding performance. To this extent, the shield external and internal coupling must be studied to obtain an evaluation of the system shielding effectiveness at all system levels up to the most specific (i.e., the actual electronic system as a whole). This is a direct consequence of the series character (and the respective model) of electromagnetic energy transfer through the shield.

From the physical point of view, "the field does not know what it is coupled to," so the shield coupling is just a specific case of electromagnetic coupling to line circuits "at large" (e.g., see Refs. [P.2, 5, and 4.1–3]). Whenever there exist electrical charges, currents, and/or electric and magnetic fields, they will interact with (or we may say couple to) the surrounding dielectrics and conductors and induce in them corresponding charges, currents, and electromotive forces. This is how the electrical charges in different physical objects "communicate."

Why and how does this happen? Although the answer may be "buried in the depths" of the relativity theory and quantum mechanics, the *macroscopic* laws of electromagnetism yield phenomenal models and mathematical tools for numerical evaluation of coupling problems, at least at the level these problems are treated in this book. In this respect, the concepts of electromagnetic coupling can be traced down to the laws discovered by Coulomb, Ampere, and Faraday, and then generalized by Maxwell. According to these laws, the signal coupling between the circuits and interconnections in the same or different systems occurs via common resistances, capacitances, and/or inductances, or by the "energy exchange" of radiated fields. The signal is generated by a source, propagates via propagation medium, and reaches the receiver. The character of the signal, the parameters of the propagation environment (including the source and receiver), and the boundary conditions at the interface of different signal propagation media determine the nature and numerical value of coupling.

4.1.3 Definitions: What Is "Shielding" Coupling?

So, how can we define the electromagnetic coupling within the context of shielding processes? Again, as is "typical" for electromagnetic shielding, the problem is not in the lack of the definitions but in their abundance. For example, our "old acquaintance," the IEEE Standard Dictionary of Electrical and Electronic Terms (ANSI/IEEE Std. 100), dedicates two pages to the term coupling, for a total about three dozen definitions! And, in principle, many of the suggested definitions are applicable to shielded circuit or line interference. Do we need all of them? Judge for yourself:

- coupling (1) (electric circuits)
  The circuit element or elements, or the network, that may be considered common to the input mesh and the output mesh and through which energy may be transferred from one to the other.
- (3) (interference terminology) (electric circuits)
  The effect of one system or subsystem upon another.
  (A) The effect of an interfering source on a signal transmission system.
  (B) The mechanism by which an interference source produces interference in a signal circuit.
- coupling capacitance (1) (general)
  The association of two or more circuits with one another by means of capacitance mutual to the circuits.

- (2) (interference terminology)
  The type of coupling in which the mechanism is capacitance between the interference source and the signal system, that is, the interference is induced in the signal system by an electric field produced by the interference source.
- coupling, conductance (interference terminology)
  The type of coupling in which the mechanism is conductance between the interference source and the signal system.
- coupling, inductance (interference terminology)
  The type of coupling in which the mechanism is mutual inductance between the interference source and the signal system, that is, the interference is induced in the signal system by a magnetic field produced by the interference source.
- coupling, radiation (interference terminology)
  The type of coupling in which the interference is induced in the signal system by electromagnetic radiation produced by the interference source.

Similarly, the CCITT [4.1] defines three coupling mechanisms: capacitive, inductive and conductive coupling. Based on the approaches developed in the previous chapters, all these definitions seem quite relevant, dealing with different levels of the shielding model generality and addressing different physical EMI processes. They are also associated with specific domains of practical applications. For example, the CCITT definitions appear in the "Directives concerning the protection of telecommunication lines against harmful effects from electric power and electrified railway lines." Still, before using these concepts and definitions, very important questions should be answered:

1. Is the above set of definitions complete?
2. If not, then why were these particular concepts and definitions selected, and what's missing?
3. When and how do we use these and other definitions? What are the criteria for their selection?
4. How do these concepts match the adopted system shielding effectiveness models?

Consider first the issue of the completeness of the definition set. Really, do these definitions cover all the aspects of coupling to electromagnetic shields? The answer is not trivial, depending on the involved physical processes and also on the shielding effectiveness models that reflect the specific applications. For example, the majority of the definitions above focus on interference between particular systems. True, in Chapter 1, we also considered interference between the identifiable systems (or their elements) and named this phenomenon crosstalk. But we emphasized that crosstalk is just a specific case of four, more general, EMC parameters which were introduced as "primary" ones: radiated emissions, conducted emissions, radiated susceptibility (or immunity), and conducted susceptibility (or immunity). These EMC parameters are defined in terms of coupling to the EMI environment; that is, they are characterized by the fields and/or voltages, without regard to their sources (in immunity problems) or receivers (in emission problems). How, then, do we apply the crosstalk-based definitions to the coupling of a circuit (shielded or unshielded, for that matter) into the EMI environment when one (or more) of the interfering systems cannot be identified? By now, at least we know where to look for and how to proceed with the answers to these questions: we look to the model of system shielding effectiveness and use the laws of electromagnetism.

Clearly, all these definitions won't do much good unless we define the criteria for their selection and use. To arrive at such criteria, we will use the following strategy: we start with a global and general definition of electromagnetic coupling to the shields, which is invariant to the particulars of specific tasks and thus is applicable to a wide class of problems. Using this global shielding coupling definition, we could derive and evaluate the local shielding coupling definitions by considering and accounting for the particulars of the respective specific problems. Then, the local definitions will be used to segregate the problems of electromagnetic coupling to the shield by introducing different coupling mechanisms that reflect specific physical processes.

We define the shield coupling as

the EMI signal generation in the shield by internal or external source, or respective EMI environments, and/or EMI signal generation in the internal or external sensor, or respective EMI environments, by the electromagnetic processes in the shield

We will use the term coupling to describe and quantify both the process and effect of emission- and immunity-related coupling mechanisms, identifying the particular application of the term when necessary. We will understand the source to be any concentration of electromagnetic energy that is or is not associated with a particular generator and/or radiator (see Chapter 1); electric currents, voltages, and electromagnetic fields. Also, we have introduced two *locators* of coupling fields and sources or sensors: internal, in the space enclosed by the shield, and external, outside the shield. In the process of *internal* coupling, the shielded circuits carrying currents and voltages and producing the electromagnetic fields inside the shielding enclosure induce the EMI signal in the shield, and vice versa. In the process of *external* coupling, the circuits carrying currents and voltages and producing the electromagnetic fields outside of the shield induce the EMI signal in the shield, and vice versa.

The applicability of our global definition to both external and internal coupling processes should not obscure the fact that, qualitatively and quantitatively, these processes can be different; that is, the internal and external coupling processes may be not symmetrical with regard to the shield. The differences are affected by three main factors: the different propagation parameters of the media inside and outside the shield, the different boundary conditions of the media inside and outside the shield, and the shield material linearity, which affects the reciprocity of the physical processes. Even when the external and internal processes are qualitatively identical, their figures of merit may differ. These differences contribute to the asymmetry of our shielding effectiveness model with regard to the external and internal coupling mechanisms, which for this reason have to be considered separately.

252 Electromagnetic Shielding Handbook

4.2 A ROADMAP TO SHIELDING COUPLING MECHANISMS: EMI SIGNAL PROPAGATION REGIMES

4.2.1 A Roadmap to Roadmap: Maxwell's Equations and Coupling Mechanisms

As we have discovered, the shielding coupling problems are far from simple. For this reason, not only that we need a roadmap to shielding coupling mechanisms, but we also had to include a guide on how to use this roadmap - a kind of a "roadmap to roadmap". Hopefully such thorough approach will facilitate the clarity of hte discussion.

As always in electromagnetics, the vehicle to develop the shielding coupling understanding and quantification is provided by Maxwell's equations, which relate the spatial field distributions to the signal time variations. The time variations of the EMI signal are described by the $\partial/\partial t$ (partial derivative by time) terms in the Maxwell's equations, which transforms into the $j\omega$ product items for monochromatic fields in phasor notation. It appears that in general case, a closed form analytical solutions of Maxwell's equations are unavailable, and often even numerical solutions are extremely complicated. Thus as a rule, only when certain assumptions and approximations are made, can important practical problems be resolved and analyzed. Hence, the respective subdivision and classification of the problems become attractive alternatives, which in our case result in identification and characteristics of a set of different electromagnetic coupling mechanisms.

Although different approaches are possible, the most obvious shielding problem subdivision is that by frequency regimes of the EMI signal propagation. From there, using frequency-based process classification, we can identify and physically interprete the coupling mechanisms in such a way that they mathematically represent specific solutions of Maxwell's equations. Then in the next section, we will apply the identified frequency regimes of signal propagation to reduce the global shielding coupling process to a respective set of local shielding coupling processes, corresponding to the respective coupling mechanisms. We will also compare the results to the IEEE and CCITT definitions quoted above.

When identifying the coupling mechanisms in accordance with our strategy, we will be guided by the following fundamental principles:
- *The formulated coupling mechanisms should make sense from the physical point of view in a given EMI environment.*
- *As such, we will model the coupling mechanisms after the EMI signal propagation regimes, which in turn, are dependent on the signal bandwidth and propagation media parameters.*

Consider now the set of Maxwell's equations in Appendix (Eq. [A.16]). In general, the electric and magnetic fields are related. The top equation [Eq. (A.16a)] or the three top equations in (A.16b) or (A.16c), which are interchangeable with (A.16a), establish the dependence between the magnetic field frequency (that is, time variations) and electric field distribution in space. This dependence is also affected by the magnetic permeability of the propagation media. Similarly, the second equation of Eq. (A.16a), and the bottom three equations in (A.16b) and (A.16c), establish the dependence between the electric field frequency (time) variations and magnetic field distribution in space, which is also affected by the propagation media characteristics: dielectric constant and conductivity.

The difficulties in solving the Eq. (A.16) system originate from the intimate and intricate mutual effects between the space distributions and time variations of the electric and magnetic field components and media characteristics. In general, trying to solve the four Maxwell's equations simultaneously leads to inhomogeneous differential equations of the second order, with the space coordinate-dependent and time-dependent variables. Because the analytical solutions for the majority of such formulations are not available, this problem is approached in two possible ways: the use of numerical modeling techniques and the simplification of analytical tasks. Modern computer technology has eliminated the difficulties with long, cumbersome calculations, so quite accurate numerical solutions can be obtained for almost any specific problem. But then again, the numerical solutions are not all that convenient for general analysis of the phenomena and the expansion of the obtained results over a wide class of problems.

Fortunately, for most practical problems, many physical dependencies can be either simplified or neglected altogether, thus facilitating the solutions of Maxwell's equations in theoretical analysis of separate coupling mechanisms. In this respect, it is important to realize that the use of mathematical simplifications must be based on physical models and checked by experimental testing. As a rule, the separation of the general solution on specific coupling mechanisms is being accomplished along the EMI signal propagation regimes, which depend on media parameters and signal bandwidth. In this section, we will try to develop a classification of the shield coupling phenomena based on these characteristics. In the following sections of this chapter, the developed classification will be applied to important practical problems of electromagnetic shielding.

4.2.2 Signal Propagation Media Parameters

Resolving Maxwell's equations for specific media parameters is a "usual trick" to simplify the task. Maxwell's equations operate with the following propagation media characteristics: conductivity σ, dielectric constant ε, and magnetic permeability μ. The medium may be isotropic or anisotropic, linear or nonlinear, and of high or low magnetic permeability. In principle, the conductivity of the media may vary from zero (ideal dielectric, $\sigma = 0$) to infinity (superconductors, $\sigma = \infty$). In the following analysis, we will distinguish between two major types of materials: dielectrics ($\sigma = 0$) and conductors $\sigma \gg$. Dielectrics and conductors may have different magnetic permeability characteristics, from $\mu = 1$ to $\mu \gg$ (ferromagnetics). The relative dielectric constant of the dielectrics may change from $\varepsilon = 1$ (e.g., air) to $\varepsilon \gg$ (segnetoelectrics), while the dielectric constant of the metals is usually neglected.

Although a uniform propagation medium that is infinite in space is a convenient and useful theoretical assumption, practical signals propagate in a nonuniform envi-

254 Electromagnetic Shielding Handbook

ronment consisting of areas with different propagation properties. As a matter of fact, the whole idea of electromagnetic shielding is drawn from the notion of the incident signal reduction at the interface with a barrier (i.e., shield) possessing certain characteristics. Thus, for the solution of the shielding problems, especially important are the boundary conditions at the interface between the areas with different properties. Because the propagating signals behave in a known way at such interfaces, the boundary conditions are used to determine the EMI signal in the system. This is true in the general case of applying Maxwell's equations when the boundary conditions are used to figure out the integration constants, and in the more limited case of circuit theory when relations between the transmission line impedances affect the voltages and currents in the circuits. In Chapters 2 and 3, we have already applied these concepts to the solution of practical problems of electromagnetic shielding. As we saw, especially important are conditions at the border between the conductors and dielectrics in general, and specifics of shield termination and grounding.

4.2.3 Role of Time Variations: Solid Homogeneous Shield in a Wide Frequency Band

✦Previously, we have devoted considerable attention to the limitations of the transmission theory of shielding (e.g., see Sections 2.1.3 and 2.3.1). Indeed, we deemed the associated constraints and problems of this theory so severe as to warrant the "alert" of its would-be users. So why would we want to revisit this theory? To be sure, we have not changed our mind about this, and later in this chapter, we will present a shielding coupling theory based on our adopted previously model. We keep returning to the transmission theory because, with all its problems and limitations, it presents an excellent opportunity to *illustrate* (as opposed to *study* which, as we saw, in general case cannot be done on the basis of this theory) important effects of the frequency regimes on electromagnetic shielding. Moreover, we will even improve on the previously presented results by introducing a *cylindrical* shield of final diameter D_s instead of the former infinite-size plane, as well as three different source field configurations (TE, TM, and TEM) instead of former plane wave only (i.e., TEM) case. Otherwise, the shield remains solid homogeneous, and the "back" effect of the shield on the source is disregarded. We will follow the unified "broadband shielding" approach to shielding in different field configurations, developed in Ref. [P.3], which is still based on the already familiar transmission theory of shielding.

Consider an electromagnetic field generated inside a cylindrical enclosure. According to the transmission theory of shielding, the shielding effectiveness, A (understood as the logarithmic measure of the ratio of the field intensities in a point of space separated from the source by the shield, with the shield present and absent respectively), can be found from an expression similar to Eq. (2.33):

$$A = 20\log|\cosh\gamma_c t| + 20\log\left|1 + \frac{1}{2}\frac{Z_d}{Z_m}\tanh\gamma_c t\right| \tag{4.1}$$

where the first item in the right-hand part denotes the absorption losses in the shield, and the second item denotes the reflection losses at the interface between the shield and the adjacent dielectric. Similar expressions were analyzed in Chapters 2 and 3; therefore, here we will concentrate only on the reflection losses, which can be related to the EMI coupling to the shield.

The wave impedance for different field configurations can be found from the following expressions [P.3]:

- for a TM wave,

$$Z_d^H = Z_0 j\pi\gamma_d I_1\left(\gamma_d \frac{D}{2}\right) H_1\left(\gamma_d \frac{D}{2}\right) \quad (4.2)$$

- for a TE wave,

$$Z_d^E = Z_0 j\pi\gamma_d I_1^1\left(\gamma_d \frac{D}{2}\right) H_1^1\left(\gamma_d \frac{D}{2}\right) \quad (4.3)$$

where I_1 and H_1 are cylindrical functions of the complex argument of the first order, modified first kind (Bessel) and third kind (Hankel), respectively. The I_1^1 and H_1^1 are the derivatives of the same functions. The intrinsic propagation constant of the dielectric $\gamma_d = \omega\sqrt{\mu\varepsilon}$.

And of course, for the plane wave,

$$Z_0 = \sqrt{\frac{\mu}{\varepsilon}} \quad (4.4)$$

which equals 377 Ω for free space.

Typical frequency characteristics of the impedances per Eqs. (4.2) through (4.3) are presented in Fig. 4.1, and the corresponding reflection shielding effectiveness characteristics are presented in Fig. 4.2.

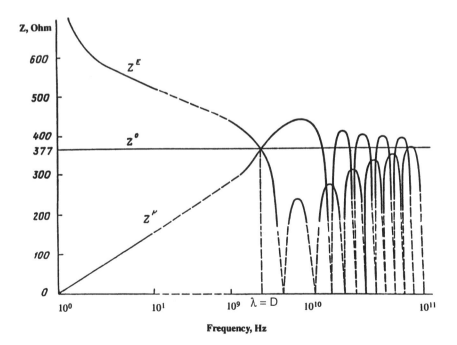

FIGURE 4.1 Free-space impedance to propagating fields

256 Electromagnetic Shielding Handbook

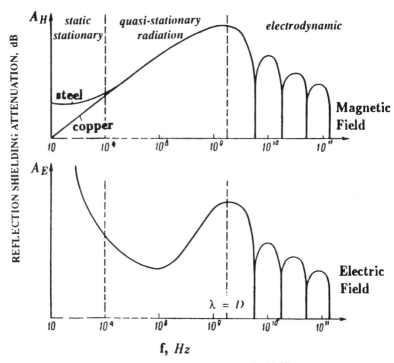

FIGURE 4.2 Frequency regimes in shielding

As shown, in the stationary regime, the electric field is completely reflected (electrostatic shielding), while the shield is almost transparent to the magnetic field. One exception is ferromagnetic material, which exhibits a "modest" degree of protection throughout this frequency range (magnetostatic shielding). With the rise of frequency, the shield performance is determined by the quasi-stationary regime, which is characterized by the absorption losses in the shield (not shown here—but see Chapters 2 and 3) and the reflection losses due to the eddy currents in the shield. At the higher part of this range, the shielding effectiveness is determined by the electromagnetic coupling mechanisms resulting in a similar order of reflection losses for electric and magnetic fields. And with the further rise in frequency, when the wavelength becomes of the same order or smaller than the shield diameter, the electrodynamic frequency regime results in an oscillating frequency characteristic of the reflection losses.✦

4.2.4 Coupling Regimes in Conductors and Dielectrics

Reactive Fields in TEM Regime

By considering the importance of contributions and relative "weight" of the conduction and displacement currents in particular propagation media - conductors and dielectrics, - we can identify specific physical phenomena that lead to further detailed analysis of the respective coupling mechanisms. But the subdivision on these mechanisms may be quite arbitrary, with the boundaries between them often "blurred."

In dielectrics, the right-hand part of the second Eq. (A.16a) is equal J_d, while J_c is negligibly small; in conductors, the right-hand part of the second Eq. (A.16a) is equal J_c, while J_d is negligibly small. If $J_d \ll J_c$, then the second Eq. (A.16a) is formally no different from the one in stationary case. [We emphasize that the similarity is formal because, in the monochromatically varying case, we deal with phasors. Also, don't forget the fundamental difference due to the first equation in (A.16a)]. Although occurrence of this quasi-stationary regime is more probable when the signal frequency is "low enough," as we see, it is not the frequency per se that is important, but the variation of the field within the area of interest, which is affected by the difference between the conduction and displacement currents. Thus, depending on the conductivity of the conductors and dielectric constant of the dielectrics, the same frequency may be considered "low" in one case and "high" in another.

If the wave propagates in a highly conductive medium (e.g., metal), then $J_d \ll J_c$ up to extremely high frequencies, and the quasi-stationary propagation regime is sustained at bandwidths of many gigahertz! In the dielectric around the conductors, the current J_c generates two Poynting vector components meeting such quasi-stationary conditions, but not necessarily in the same frequency bands: P_z directed along the shielded line axis, and P_r directed perpendicular the shielded line axis into the conductors. For example, see Fig. 2.18, where P_z is generated by the field components E^r and H^ϕ, and P_r is generated by the field components E^z and H^ϕ. (However, unlike Fig. 2.18, in the process of energy propagation along the line, the field E^r is generated by the voltage applied between the circuit conductors.) Note that both the longitudinal and transverse energy fluxes are generated by a TEM E-H field configurations. Both Poynting vector components, P_z and P_r, together play important role in the signal propagation in the transmission lines.

With regard to practical lines, two important comments must be made:

1. The cables are designed to transmit the energy along the line while minimizing the losses in the conductors. The electric fields $E^z \ll E^r$, and the signal energy is, as a rule, contained in the dielectric between the direct and return current paths and propagates in the dielectric along the conductors. The role of circuit conductors is only to channelize the signal energy, but not transmit it: the conductors actually "siphon" the energy off the signal; this is the way the energy is "tied" to the line.
2. In generating the field H^ϕ, the role of displacement current in the dielectric is insignificant compared to the conduction current in the conductors which channelize the circuit energy. The component E^r in a trsansmission line in TEM regime is created by the applied voltage between the conductors and is "terminated" at the corresponding positive and negative charges at the direct and return path conductors. So, for the TEM wave to propagate along the transmission line, there is no need for the displacement currents. In that case, the displacement current between the line conductors, if present, only tends to reduce (by radiation) the energy propagating along the line and reaching the load at the opposite end of the line.

258 Electromagnetic Shielding Handbook

When the transversal dimensions of the circuit are "small" and/or the frequencies are "low enough," the transverse electric field phase in the dielectric across the line is almost constant, and the electric lines of force are terminated only at the line conductors. (The reasons why the "dielectric" is emphasized will become clear a little bit later, when we will discuss wave propagation in conductors.) Variations of such electric fields do not generate any significant magnetic fields (which lines would encircle the electric field lines and thus have some components along the line axis), and the radiation from the line can be neglected. If, additionally, the losses in the conductors are relatively small (i.e., $E^z \ll E^r$), then the fields are approximately transverse: the E and H vectors are in the plane, which is perpendicular to the line axis. Then, the Maxwell's equations for a quasi-stationary regime can be written as follows:

- for magnetic field time variations,

$$\nabla \times E = -j\omega\mu H \qquad (4.5)$$

- in conductors carrying currents,

$$\nabla \times H = \sigma E \qquad (4.6a)$$

- in dielectrics,

$$\nabla \times H = 0 \qquad (4.6b)$$

$$\nabla \cdot D = \nabla \cdot \varepsilon E = \rho \qquad (4.7a)$$

$$\nabla \cdot B = \nabla \cdot H = 0 \qquad (4.7b)$$

Wave Regime in Dielectrics

Consider now the energy flux out of the transmission line in the dielectric across the line. This energy flux is described by the Poynting vector ($-P_r$), which can be generated by a combination of E^ϕ and H^z. In straight conductors, such a combination of fields is generated by the displacement currents in the dielectric between the conductors. In this case, the media conductivity can be neglected, and there is only displacement current present in the right-hand part of the second equation in Eq. (A.16a). Due to the "coupling" and mutual "transformations" between the electric and magnetic fields, the wave can propagate in a dielectric without the conduction currents which, even if present, results in the loss of the wave energy. Quite logically, we will name this mode of propagation the wave regime.

This displacement current "connection" between the electric and magnetic fields is always present in the line dielectric, but at lower frequencies its effect is usually is small compared to the effect of longitudinal current in the line conductors. When the transverse dimensions of a circuit are large and can accommodate a significant part of the wavelength, the transverse electric lines of force can be terminated "at themselves" (as well as at the line conductor charges, the difference now depending on the distance between the line conductors). But this is the "classical" description of the wave radiation mechanism, and it means that our "large circuit" radiates! Just how "large" should the distance be between the conductors for the radiation to become significant? The answer depends on the signal wavelength, the media parameters, and the media impedances at the boundaries of the "inter-conductor" space. Consider first the case when the signal propagates (radiates) in a boundless dielectric without being channelized by conductors. The conduction current $J_c = 0$, the quasi-stationarity conditions are no longer met, and the first two Maxwell's equations become:

$$\nabla \times E = -j\omega\mu H \qquad (4.8)$$

$$\nabla \times H = j\omega\varepsilon E \qquad (4.9)$$

These well known equations describe the radiated waves in dielectrics, including free space. From Eqs. (4.8) and (4.9), we can easily obtain the familiar wave equations for electric and magnetic components of the electromagnetic wave propagating in a dielectric:

$$\nabla^2 E + \gamma_d^2 E = 0 \qquad (4.10a)$$

$$\nabla^2 H + \gamma_d^2 H = 0 \qquad (4.10b)$$

and $\gamma_d = \omega\sqrt{\varepsilon\mu}$. Note that in Eqs. (4.5) through (4.10), the vectors E, H, and J are phasors which vary with frequency ω.

Wave Regime in Conductors

Consider now electromagnetic wave propagation in metals (displacement current disregarded). Mutual substitutions between Eqs. (4.5) and (4.6a) lead to two equations, respectively:

$$\nabla^2 E = j\omega\mu\sigma E \qquad (4.11a)$$

$$\nabla^2 H = j\omega\mu\sigma H \qquad (4.11b)$$

We have already discussed these equations when addressing the transfer parameters of homogeneous solid shields, and we know their solutions [e.g., see Eqs. (3.20) and (3.21)]. In the case of a homogeneous solid tube, they were used to determine the electromagnetic wave propagation in a conducting medium as a function of its intrinsic propagation constant, γ_c (or skin-depth, δ). However, here we are looking at these equations from the electromagnetic wave propagation and coupling perspective. In this regard, we now will introduce the concept of the wavelength of a wave propagating in conducting medium as the distance in the medium along the direction of the wave propagation, on which the wave's phase varies by 2π.

Because

$$e^{\gamma_c r} = e^{\frac{1+j}{\delta}r} = e^{\frac{r}{\delta}} e^{j\frac{r}{\delta}}$$

then, obviously, for $\lambda/\delta = 2\pi$, it should be

$$\lambda_c = 2\pi\delta \qquad (4.12)$$

To see the implications of this result, just try to substitute into Eq. (4.12) the values of δ for several common metals at useful frequencies (you can use the data from Table 2.1). Some readers may be in for a big surprise. For example, for signals with frequencies of 1000 Hz, 10 kHz, 1 MHz, and 100 MHz, propagating in an aluminum shield, the wavelengths will be around 1.8 cm, 0.54 cm, 0.54 mm, and 0.054 mm, respectively. Compare these to the wavelengths in dielectric! What this means is that, in metals, the propagation regime may be considered quasi-stationary only at relatively low frequencies — much lower than in dielectrics.

But at higher frequencies (albeit, they can be quite "low" in absolute terms), the process looks like a wave, behaves like a wave, and calculates like a wave, so we will call it the wave propagation regime in conductors (remember the proverbial duck?). However, unlike the propagation in a dielectric, in this case, we deal only with conduction current.

4.2.5 "Electrical Geometry" in Near-Zone / Far-Zone Radiating Fields

> ✦ It may be a good time to address the *relative* nature and context dependence of such terms as *high / low* frequencies, *short / long* wavelength, *short / long* line. Indeed, a 30 kHz *audio* signal is so "high" that a human ear cannot hear it, while a 30 kHz electromagnetic signal is placed between the "very low" (VLF) and "low" (LF) frequencies in the spectrum scale (see Fig. 1.2). Or consider clock distribution lines in the ENIAC computer ("circa" 1946). With clock speed 400 kHz ($\lambda \approx 500$ m in a good cable with low dielectric constant insulation), a 10 m long I/O line is only (1/50) λ - an electrically "infinitesimal" length by the most restrictive standards. On the other hand, this book is being written on a 166 MHz desktop ($\lambda \approx 0.1$ m). At such speed to be smaller than a 1/50 λ ratio, the line length should not exceed 2 mm! Thus, even a modest 5 cm long interconnection is comparable to the wavelength, can be considered an electrically "long line", and may sometimes act as a "tuned" half-wavelength radiator. Looks, like the problem of "electrical scaling" certainly deserves our attention.✦

It is well known that if a radiator (transmitting antenna) generates electromagnetic waves, the field distribution in space around it depends on the distance (in terms of wavelength) from the observation point to the source. While we will address this issue in more detail a little later, we now want to introduce the "electrical" measures of geometrical distance and length. As we have just seen, the propagation regime in a transmission line depends on the relationship between the line size and the wavelength. Thus, although the Maxwell's equations are formulated in terms of signal frequency (that is, time variations), the physical meaning of different frequency bands becomes more obvious if related to the signal wavelength, $\lambda = v/f$, where v is the wave propagation speed.

We will introduce two "electrical geometrical" parameters: electrical distance and electrical size. First, the electrical distance between the EMI field source or sensor and the shield is the ratio of geometrical distance r to signal wavelength λ. As boundary cases, we will distinguish between the near distance,

$$(r/\lambda \ll 1) \tag{4.13}$$

and far distance,

$$r/\lambda \gg 1 \tag{4.14}$$

Second, the electromagnetic properties of physical objects also depend on the ratio of the object size to the wavelength. Again, as with boundary cases, we will distinguish between the electrically small objects and large objects. However, as we saw on the example of longitudinal and transverse direction of energy propagation in a transmission line, in each specific case, the most important object dimensions Λ are those related to the direction of energy propagation. By the critical electrical size we will understand the ratio Λ/λ. Thus, we will distinguish between electrically small objects ($\Lambda/\lambda \ll 1$) and electrically large objects ($\Lambda/\lambda \gg 1$).

Returning now to the quasi-stationary transmission regime in a circuit, we see that it takes place when for the signal TEM fields within it the circuit acts as electrically small:

$$\lambda \gg \Lambda \tag{4.15}$$

The near fields which meet the conditions of Eqs. (4.15) and (4.13) constitute the so-called induction fields. The induction fields that do not exhibit any significant radiation are reactive induction fields. The reactive fields can be analyzed on the basis of the circuit theory, which is operating with lumped coupling components: capacitors, inductors, and resistors. Correspondingly, the effects of these fields are reduced to such coupling mechanisms as capacitive, inductive, and galvanic coupling. In particular, the static fields can be considered as a boundary case of reactive fields.

In reactive fields, the dominant propagation mode is transverse electromagnetic (TEM), where the field components are located in the plane perpendicular to the direction of the electromagnetic energy propagation. The relationship between the electric and magnetic fields (i.e., media impedance) is variable, depending on the system configuration: the closer are the direct and return current paths, the smaller is the circuit characteristic impedance, and vice versa. The low-impedance circuits emphasize the magnetic field, the high-impedance circuits emphasize the electric field.

262 Electromagnetic Shielding Handbook

The static, induction, and, (partially) near radiation zone fields are characterized by the inverse square law of the amplitude change with the distance.

On the other hand, when Eq. (4.14) is true, we are dealing with the radiated far-field zone. In a homogeneous propagation medium, the far-zone radiating fields are reduced to plane waves. We know that the field configuration in a plane wave is also TEM. However, there are two importat distinctions:
- while the induction field is "purely" reactive resulting in the *energy exchange* between the radiator and the environment with no active energy losses for the radiator, the radiation field is "purely" resistive, i.e., lossy, with regard to the radiator
- unlike the induction zone characterized by the inverse *square* (or *cubic*) law of the amplitude change with the distance, the far radiating field is characterized by the inverse *linear* law of the amplitude change with the distance.

Now, it looks like the induction and radiating zones are two "extremes". But what about the region in the vicinity of

$$r/\lambda \cong 1 \qquad (4.16)$$

which is between the induction reactive and far field radiation zones? The fields that meet Eq. (4.16) conditions and propagate in the dielectric are considered near-zone radiated fields. Such fields exhibit a combination of both induction and radiation properties. In particular, at close distances to the source, the propagation impedance is determined mainly by the source impedance, while at larger distances it is by the impedance of the propagation media.

Since both the electrical distance between the radiators and the electrical size of the radiators are related to wavelength and affect the field character, a valid question is whether there is a meaningful relationship between all three of these characteristics. Indeed, such relationship does exist and is often used as an approximate criterion for the border distance, R, between the near and far field radiation zones:

$$R = \frac{2\Lambda^2}{\lambda} \qquad (4.17)$$

Equation (4.17) is based [4.5] on the analysis of Fresnel zones generated by a radiator of size Λ at a distance R. It is often used to define the far-field conditions for antenna measurements.

✦ Perhaps, you wonder if the constraints (4.15) and (4.17) do not contradict each other. The answer is: no. While (4.15) specifies TEM induction field conditions for a signal propagating *within* the circuit, (4.17) deals with the processes *outside* the circuit ✦

✦ In general, there is no strict " border" between the near and far field zones. Thus, any definition of such a border is *necessarily* an approximation, which in certain cases may be vague or even ambiguous. For this reason, when discussing the near / far field zone differences, it is *always* worthwhile to start with an appropriate definition. The definition we gave, is not the only one possible. For example, often a convenient *practical* alternative to the combination of constraints based on

(4.13) through (4.17), is to define a source far field zone as such, where the radiation pattern does not vary with the change of the distance from the source to receiver. ✦

In discussing the radiated fields, we assumed that the propagation medium is unbounded (or at least, semi-infinite). However, when the dielectric propagation medium is bounded by conducting surfaces, the wave reflections from them may result in higher order mode generation, i.e., in the waves of transverse electric (TE) and transverse magnetic (TM) types. Thus inside a shield of large enough diameter

$$D/\lambda > \frac{1}{4} \tag{4.18}$$

the higher modes can propagate in the dielectric enclosed in the shield, and the circuit acts as a waveguide. Then, in general case, both items, $\partial D/\partial t$ and J, must be accounted for in the right-hand part of the second of Maxwell's equations, and a "mixture" of different waves propagates in the line. The specific feature of this regime is that along with TEM mode, the TE and TM transmission modes of different order also must be considered where either magnetic or electric field component, respectively, is pointing in the direction of the wave propagation. For instance, such a regime is possible in a coaxial line. However, for these higher-order propagation modes, there is no need for the current-carrying center conductor. Corresponding propagation regime is sometimes called electrodynamic.

4.2.6 Shielding Coupling / Propagation Regimes: Summary

✦ We can now generalize the fundamental theoretical and even philosophical implications of the difference between the static, induction reactive, and, to a degree, near radiated fields on one hand, and the far radiated fields, on the other hand. When the interacting elements of the interfering circuits are electrically close, an instant action between the elements is assumed. The phase of the EMI signal at the receiving circuit is exactly that of the phase at the source circuit. This "exactness" is, of course, the result of mathematical approximations which were used to simplify the Maxwell's equations. Actually, the signal "takes time" to propagate from the source to the receiver.
This retardation of the effect with regard to the cause is accounted for in the radiation regimes. The larger the electrical distance between the radiation-coupled circuits, the greater will be the phase difference between the signals at the source and the receiver. For example, we can imagine a situation wherein the phase difference between the source and receiver is 180°. This is not just a "curious theoretical" fact! To see the practical implications of this phenomenon, consider a shield illuminated by fields from two identical EMI sources which are placed at different distances from the shield. If the field action is instantaneous, then the phases of the fields arriving from both sources are identical, and the total received field is determined by the sum of both incident fields. On the other hand, in radiated fields, the sources may appear at such distances from the shield that the fields arriving from each of them are shifted 180°, and the total received field will be determined by the difference of both incident fields! Some food for thought. ✦

✦ Consider a coaxial line with center conductor radius r and the internal radius of the outer conductor R. The line is directed along axis z in cylindrical coordinate system. A signal with voltage V and frequency ω is applied between the conductors, driving current I in the line. This signal results in the radial electric field intensity E^r across the line and tangential magnetic field intensity H^ϕ

264 Electromagnetic Shielding Handbook

around the center conductor, in space enclosed by the outer conductor. When the radius $R \ll \lambda$, this a "typical" coaxial line supporting the TEM transmission mode. However, we know from the theory of wave propagation in a waveguide, that when $R > \lambda/4$, over cutoff frequency, then the higher-order modes, TE and/or TM, appear with magnetic or electric components, respectively, directed along axis z. Moreover, these modes are supported by the displacement currents and do not need the center conductor to propagate along the line. ✦

Fig. 4.3 presents a "graphic" summary of the EMI signal propagation / shield coupling concepts and definitions which will be developed in the following sections. Again, we have to emphasize that while such classification is quite useful and widely applied, it is just a useful tool which facilitates the physical understanding and numerical evaluation of the underlying processes. Also, the limits between the different zones and mechanisms are not "cut in stone", and alternative treatments of the same problems are usually possible.

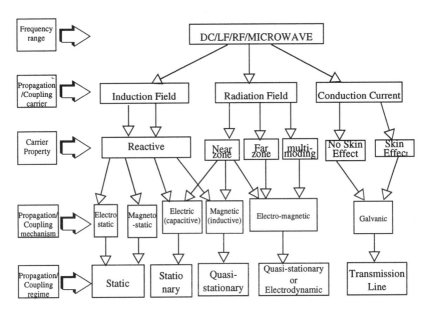

FIGURE 4.3 EMI Propagation / Shield Coupling Mechanisms

4.3 INDUCTION FIELD COUPLING

4.3.1 Static and Stationary Field EMI Environment and Coupling

The simplest regime is that of static signals ($\omega = 0$), as opposed to time-varying signals ($\omega \neq 0$). In such environment, the electric and/or magnetic fields do not vary with time, or their time variations are slow, the time derivatives of the field components are equal to zero. These fields are called, respectively, electrostatic and magnetostatic. Not surprisingly, the respective coupling and shielding mechanisms are also named electrostatic and magnetostatic. Analysis of Maxwell's equations in static regime (i.e., assuming ω = 0), reveals that electric and magnetic fields are mutually independent! This means that coupling (and shielding, for that matter) can be considered separately in electric and magnetic fields.

✦ In static conditions (ω = 0), the set (A.16a) of Maxwell's equations reduces to:
In conductors and dielectrics,

$$\nabla \times E = 0 \quad (4.19)$$

In conductors, carrying DC "conductivity" currents with density J_c,

$$\nabla \times H = J_c \quad (4.20)$$

In dielectrics and conductors carrying no current,

$$\nabla \times H = 0 \quad (4.21)$$

Also,

$$\nabla \bullet D = \nabla \bullet \varepsilon E = \rho \quad (4.22)$$

$$\nabla \bullet B = \nabla \bullet H = 0 \quad (4.23)$$

Note, that in static conditions you don't need to operate with phasors, since the currents and voltages do not vary.

Equations (4.19) and (4.22) describe the electric field at a point of space as a function of the source electric charge density, ρ, distribution, and Eqs. (4.20), (4.21), and (4.23) describe the magnetic field at a point of space as a function of the source current density, J, distribution (if present). Note, that E, H, and J in this case (static signals) are time-independent vectors. But, as we will see later, practically, these equations are applicable even to some fast varying fields that meet certain conditions.

Comparing the system of Eqs. (4.19) through (4.23) with Eq. (A.16), we see that the most striking difference between them is independence of electric and magnetic fields! This means, that the coupling (and shielding, for that matter) can be considered separately in electric and magnetic fields, as we mentioned above.✦

✦ Actually, only the electrostatic fields, which are generated by electrical charges that are fixed in space, can be called "truly" static. No "active" energy is associated with the electrostatic field—only the "potential" of it. In contrast to the electric field, the magnetic field (even when "static") is by its nature related to the movement of charges: just think of the definition of an electric current as

$$I = \frac{dQ}{dt} \quad (4.24)$$

where Q is the electric charge. What makes the difference is that the magnetostatic fields are created by the direct current (dc), which is a steady flow of electrical charges; i.e., I = const and

$$\frac{dI}{dt} = 0 \quad (4.25)$$

The magnetic fields, generated by constant in time currents are also called *stationary*.

Of course, along with magnetic field, the moving charges also generate electric field, which in case of steady currents is called stationary electric field. A reference to the top equation of Eq. (A.16a) proves that there is no difference between the electric effects of static and stationary electric fields [in both cases, $\omega = 0$, which leads to the same Eq. (4.5)]. For this reason, no distinction is usually made between the static and stationary fields.✦

4.3.2 Quasi-Stationary Field Coupling

In time-varying signals, the electric and magnetic fields are "coupled" to each other, and mathematically the electric and magnetic field components cannot be separated as easily. It is no longer a case of "two fields" (electric and magnetic) but two different manifestations of a single unified electromagnetic field, leading to electromagnetic coupling and shielding mechanisms. This follows from the top two equations of Eq. (A.16a). These equations unconditionally link the magnetic field variations with the generation of an electric field of a certain configuration, and the electric field variations with generation of a magnetic field of a certain configuration. But there remains room for useful simplifications that lead to respective coupling and shielding mechanisms.

To start with, there is a wide class of practical problems that exhibit field space distributions similar to those in static and stationary fields, although they may vary quickly in time. Because of this similarity, such fields are dubbed quasi-stationary.

But what is it in the field space distribution that makes the static fields so "special?" Mathematically, the static electric and magnetic field distributions in space can be found by applying Laplace's equations to the scalar electric (in a charge-free area) and magnetic potentials. We don't have to dwell on mathematical details since they can be found in the available abundant literature on the subject. What we are interested in here is a qualitative understanding, based on physics. Consider a number of constant (static) electrical point charges distributed in space at different distances from the point of interest (say, an electric field intensity meter). The field produced by each charge in the direction of the meter monotonously reduces with the distance from the meter. At any moment in time, the total field registered by the meter will be the same: a geometric sum of the fields produced by each of the charges. Time is really not a factor in this case.

Imagine then that the charge magnitudes sinusoidally vary in time with the same frequency (since we are discussing the monochromatic fields; for transient fields, we can expand the signal into Fourier series or integral and again deal with each harmonic separately). Now, at any moment in time, the field intensity produced by each of the charges depends not only on the amplitude of the variations (reduced by the respective distance change law), but also on the phase of the field reaching the meter. If the signal propagation speed were infinitely large, the signals from all the charges in the system would have reached our meter instantaneously, that is, with identical phase, and there wouldn't have been any fundamental difference between such a time-varying field and static field: just substitute the field vector with a phasor. That's what the "pure" quasi-static field is all about.

But in nature, all signals propagate with finite speeds. Because the signals from each of the charges in our system travel a different time span (to cover the different distances), their phases at the meter location are also different. For example, we know how the phase of a traveling electromagnetic wave is distributed in space: at the same moment of time it varies by 2π at the distance equal to the wavelength, λ. For this reason, if the distances from any two charges to the meter differ by $\lambda/2$, their effects on the meter will be compensatory, even if the amplitude of the signal reaching the meter from either of these charges alone is large, and the charges are *in the same phase*! Compare this to the case of statics, wherein two such signals will always geometrically add. To summarize, in time-varying fields, three signal factors must be accounted for: phase, magnitude, and direction. In statics, however, only the latter two of these are factors.

So, are quasi-stationary fields possible at all? It follows from the above considerations that time-varying fields distributed in space similarly (but not exactly!) to static fields are possible only at distances much smaller than the signal wavelength:

$$r \ll \lambda = \frac{v}{f} \qquad (4.26)$$

where v is the signal propagation speed in a medium. Obviously, to meet this condition at a given distance, the signal frequency must be low enough. And vice versa, to meet this condition at a given frequency, the distance must be small enough.

✦ We will define the quasi-static regime in dielectrics for slowly varying fields following the approach presented in Ref. [4.4]. As in the "true" statics, the quasi-static potential (or scalar electric potential) depends directly on the charge distribution. However, in statics, the negative gradient of the potential [see Eqs. (A.6) and (A.12)] yields the total electric field intensity, whereas in quasi-statics, it accounts for only a part of the field intensity. This difference is caused by the necessity in quasi-statics to account for the phase retardation factor, βr, which depends on the distance r between any element charge and the observation point. For example, as we saw in dielectric, $\beta = \gamma_d = \omega\sqrt{\mu\varepsilon}$. In statics, there is no variation of the charge, and the phase parameter is inapplicable (or the phase may be assumed zero in all the cases). The potential V_0 at distance r from the charge q is $V_0 = q/(4\pi\varepsilon r)$. On the other hand, the time-varying field potential, V_t is as follows:

$$V_t = (qe^{-j\beta r})/(4\pi\varepsilon r) \qquad (4.27)$$

268 Electromagnetic Shielding Handbook

Thus, the time-varying field potential may be considered as consisting of a sum of two ingredients: scalar ("static") electric potential and a dynamic component of electric field intensity F, which diminishes as the frequency approaches zero:

$$F = -j\omega\mu[(I_s e^{-j\beta r})/(4\pi r)] \quad (4.28)$$

where Is = the moment of the current element.✦

✦ In general, we can arrive at the quasi-stationarity conditions by analyzing Maxwell's equations. Consider again the system of Eq. (A.16a). We can arrive at the quasi-stationarity conditions by analyzing Maxwell's equations. In Eq. (A.16a), the current in the right-hand part of the second equation is a sum of displacement current I_d and conduction current I_c. Assuming, for simplicity, and isotropic and linear environment (that is, vector D is parallel and proportional to vector E), the current densities can be expressed via the electric field intensity:

$$J_c = \sigma E, \quad J_d = j\omega\varepsilon E \quad (4.29)$$

Also, in this isotropic and linear environment, B = μH. Then, expressing H as a function of E in the first equation of Eq. (A.16a) and substituting it into the second equation, and on the other hand, expressing E as a function of H in the second equation of Eq. (A.16a) and substituting it into the first equation, applying the rules of vector operator multiplication, and assuming the absence of free charges in the investigated area (i.e., $\mathrm{div}\rho = 0$), we obtain:

$$\nabla^2 E - \gamma^2 E = 0 \quad (4.30a)$$

$$\nabla^2 H - \gamma^2 H = 0, \quad (4.30b)$$

where [see also Eqs. (A.55) through (A.57)]

$$\gamma^2 = \gamma_c^2 - \gamma_d^2 = j\omega\mu\sigma - \omega^2\mu\varepsilon \quad (4.31)$$

Equations (4.30) are familiar Helmholtz equations. It is easy to see that when the frequency is zero, Eqs. (4.30) transform into the Laplace's equations which are similar to those obtainable for static fields from Eqs. (4.19) through (4.21). That's expected. But we are interested in the conditions that will reduce Eq. (4.30) "almost" to Laplace's equations when $\omega \neq 0$. To do this, let us review Eq. (4.30 a) or (4.30 b) (mathematically, they are identical). For example, expanding the operator ∇^2 in Eq. (4.30a) in Cartesian coordinates (see Chapter 2), and expressing the frequency via the wavelength, we obtain:

$$\frac{\partial^2 \vec{E}}{\partial x^2} + \frac{\partial^2 \vec{E}}{\partial y^2} + \frac{\partial^2 \vec{E}}{\partial z^2} = \left[\frac{2\pi\upsilon^2\mu\varepsilon}{\lambda^2} - j\frac{2\pi\upsilon\mu\sigma}{\lambda}\right]\vec{E} \quad (4.32)$$

Now, to be able to approximate Eq. (4.32) by a Laplace equation (which describes the static fields), it is necessary that the right-hand part in Eq. (4.32) be much smaller than the left-hand part. Substituting the derivatives by differentials, we see that this can happen when the "denominators" $\Delta x, \Delta y, \Delta z$ corresponding to $\partial x^2, \partial y^2$, and ∂z^2 are much smaller than λ. Just how much smaller depends on the propagation medium parameters ε, μ, and σ. We already know this condition for the quasi-stationary regime.✦

4.4 CROSSTALK EMI ENVIRONMENT AND COUPLING

4.4.1 Crosstalk Interactions

Generally, the crosstalk phenomenon can be caused by electromagnetic (i.e., field-related) and galvanic (i.e., current- and voltage-related) interaction between the disturbing and disturbed circuits. Two considerations determine the approaches to crosstalk coupling. First, both the disturbing and disturbed circuits are considered to be known. Second, it is usually assumed (although it is not necessary in general) that the distances between the conductors in each circuit, as well as the distance between the interfering circuits, will be small enough in comparison to the signal wavelength that the fields in the dielectric between the interfering circuits can be treated as induction fields [see Eqs. (4.13) and (4.15)]*. Also, weak interaction is assumed when the effect of the victim on the source can be neglected. In the quasi-stationary regime the electric and magnetic fields are mutually independent, so the electric (capacitive) and magnetic (inductive) coupling, along with the common impedance (conductive) coupling, can be considered separately. The crosstalk interaction is represented by the respective coupling mechanisms, which can be present separately or simultaneously within and/or between the lines in the same or different systems. And in addition to the immediate interaction of both disturbing and disturbed circuits, we must also account for the interaction between them via other conductors in the line (including shields and ground wires). In crosstalk theory, this is usually referred to as crosstalk via third circuits, or tertiaries. Both interfering circuits interact with the tertiaries via the same electric, magnetic, and galvanic coupling mechanisms. Several important examples of these effects for balanced and unbalanced lines are illustrated in Fig. 4.4.

In the quasi-stationary regime, the electric and magnetic field configurations are independent and determined by the electromagnetic properties and geometry of the EMI source, victim, and propagation medium. The electric field intensity is proportional to the potential difference (voltage) between the line conductors carrying the signal and its return. Magnetic field intensity is proportional to the current in the line. Given the same signal energy, the higher the line characteristic impedance, the larger will be the voltage and the smaller will be the current, leading to larger effect of electric coupling and smaller magnetic coupling. For low-impedance circuits, it is just opposite. Other factors affecting coupling are the circuit configuration and transmis-

---

*In chapter 1, we have introduced the most general concept of crosstalk between two or more *identifyable* electronic systems. These included not only induction, but also radiating fields. But *in this section*, we limit the discussion *to induction only* fields.
On a "philosophical note, let's seize this opportunity to once more emphasize the importance of properly formulating and adhering to respective definitions, *before* any discussion takes place."

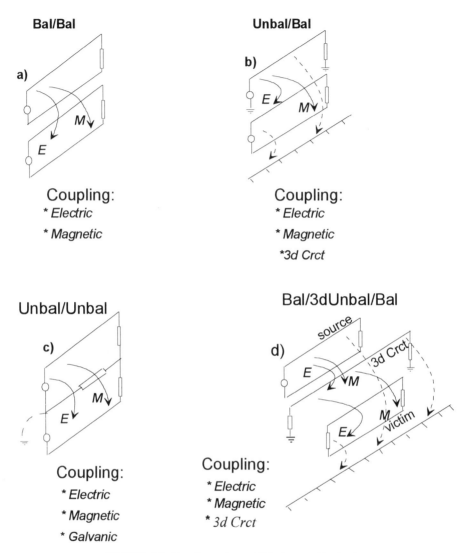

FIGURE 4.4 Coupling and crosstalk between cable circuits

sion mode. In Chapter 1 (see Section 1.2.1) we dedicated significant attention to these subjects while discussing the EMI environment as related to the cabling and interconnections. In that text we also indicated the fundamental difference between the open (balanced or unbalanced) and closed (mainly, coaxial) circuits.

4.4.2 Transverse Reactive Field Environment and Coupling

The distribution of electric and magnetic fields created by lossless non-twisted balanced and coaxial pairs is shown in Fig. 4.5, a and b.

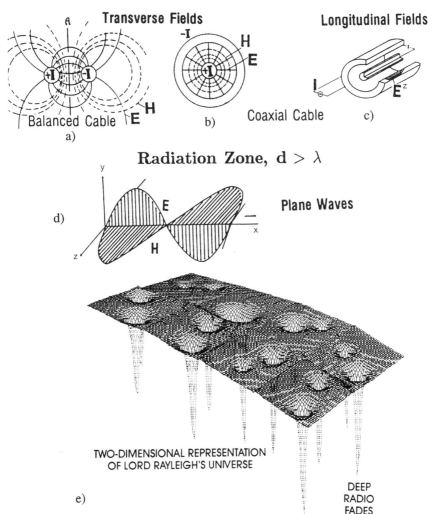

FIGURE 4.5 Propagating electric and magnetic field distributions

The lines of electric and magnetic fields are orthogonal to each other and run in the plane perpendicular (transverse) to the line axis. The fields are shown in a moment of time when the electric charge on wire 1 (the left wire in a balanced pair, and the center conductor in coaxial line) is positive and the current flows "into the page" along the wire. (Correspondingly, the charge is negative and the current flow is reversed, or "out of the page," for the wire 2, the right wire of the balanced pair and the outer conductor of the coaxial pair). In each point of space, the E- and H-field intensities are defined by the combined action of wires 1 and 2. The field configuration is symmetrical with regard to both horizontal and vertical axes.

In a balanced pair, the plane running along the line through its symmetry axis, and perpendicularly to the plane formed by the pair conductors, is an equipotential surface with zero potential. Therefore, applying the method of images, it can be proved that if our balanced pair is transformed into a single-ended circuit by using, instead of a return conductor (with the negative sign, at the right), this plane (in Fig. 4.5 a, we see its trace A), the field in the left "semi-plane" will have the same configuration. Of course, the same is true for the right semi-plane. In our case, this well known fact of elementary electromagnetics means that there is no fundamental qualitative difference between the electric and magnetic fields generated by a balanced line and a single-ended line in which the "negative" current returns in an infinite ground plane instead of a return conductor. As a matter of fact, we can view a balanced line as two single-ended lines, symmetrically (mirror image) connected at their grounding points.

Consider another circuit which is brought into the "zone of action" of the electric field, E, generated by the source circuit. As a result of electrical induction in the E field, electrical charges are formed on the conductors of the disturbed circuit. This process is referred to as electrical coupling. Because the radiation is neglected, the use of circuit theory can significantly simplify the analysis. Thus, since the ratio of a conductor's electric charge to the potential difference between conductors (required to maintain that charge) is the general definition of capacitance, electrical coupling is by nature capacitive coupling. Previously, we have seen that the electric quasi-stationary field behaves similarly to the stationary field. Therefore, the considerations and expressions describing this field are applicable here, while the effect of signal quasi-stationary time variations can be accounted for easily by using phasors instead of time-independent vectors.

It is common to define the electrical coupling coefficient, K, as the ratio of the current induced in the disturbed circuit to the voltage (potential difference) in the source circuit. It depends on the line geometry, as well as line insulation properties (dielectric constant and dissipative losses):

$$K = \frac{dI_r}{dV_s} = g + j\omega c \qquad (4.33)$$

In the same way, the magnetic coupling is introduced, and it is attributed to the electromotive force in the disturbed circuit due to magnetic induction in field H generated by the disturbed circuit. This electromotive force e is generated due to the time variations of the disturbing magnetic flux (originated by the current in the disturbing circuit) and is directed to oppose the change in the current that generated it (sounds familiar, doesn't it?):

$$e = -\frac{d\Phi}{dt} = -j\omega \int_S H_n ds \qquad (4.34)$$

The surface, S, is limited by the contour formed by the conductors of the disturbed circuit, and H_n is the interference magnetic field component, normal to this surface. Since this process presents an inductive effect, this coupling is called inductive coupling. The magnetic (inductive) coupling coefficient, M, is defined as the ratio of the electromotive force induced in the disturbed circuit to the current in the source circuit. It depends on the line geometry, as well as conductor properties: conductivity and magnetic permeability.

$$M = \frac{dV_r}{dI_s} = r + j\omega m, \; \Omega/m \qquad (4.35)$$

The coefficients c, m, g, and r in definitions (4.33) and (4.35) are measures of the *electrical and magnetic coupling asymmetry*, expressed, respectively, in terms of the differences in *partial* capacitances, mutual inductances, dielectric losses, and galvanic (eddy current) losses between the conductors of the disturbing and disturbed lines.

Why the derivatives in Eq. (4.33)? Because transmission lines are circuits with distributed parameters. In the quasi-stationary regime, the phase of the currents and voltages along the line vary with the line length l (it is proportional to βl). For this reason, the coupling coefficients are usually determined for an infinitely small line length and referred to the unit length. To calculate the coupling at an arbitrary line length, we will have to integrate the contributions from all infinitely small ingredients of the line.

The simultaneous action of electric and magnetic coupling between two circuits is illustrated in Fig. 4.6.

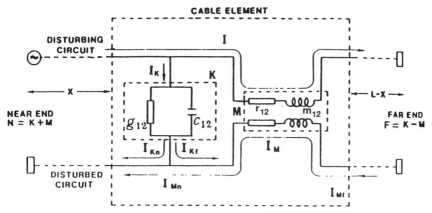

FIGURE 4.6 Electric and magnetic coupling between crosstalking circuits

As shown, the electric coupling interference currents in the disturbed circuit have opposite directions at the near and far ends, I_{kn} and I_{kf}, respectively, while the magnetic coupling interference currents, I_{mn} and I_{mf}, have the same direction at both ends. At the near end, the electric and magnetic coupling add, while at the far end

274 Electromagnetic Shielding Handbook

they deduct. Thus, we can expect the near-end coupling, N, and crosstalk, A_n, to be larger than the far-end coupling, F, and crosstalk, A_f!

$$N = \left(\frac{g}{j\omega} + c\right) + \left(\frac{r}{j\omega} + m\right) \cdot \frac{1}{Z_c^2} \qquad (4.36a)$$

$$F = \left(\frac{g}{j\omega} + c\right) - \left(\frac{r}{j\omega} + m\right) \frac{1}{Z_c^2} \qquad (4.36b)$$

The circuit characteristic impedance Z_c (which was assumed here to be identical for both interfering circuits) is introduced into the definitions (4.36) to bring the electric and magnetic coupling to a "common denominator", i.e., express them in the same dimensions, 1/Ohm*.

At an electrically small length, l, of line, the near-end (A_o) and far-end (A_l) crosstalk attenuation can be determined easily from the coupling coefficients:

$$A_o = 20\log\left|\frac{2}{\omega Z_c Nl}\right| \qquad (4.37a)$$

$$A_l = 20\log\left|\frac{2}{\omega Z_c Fl}\right| \qquad (4.37b)$$

Fig. 4.7 presents experimental data illustrating the relationship between the near- and far-end crosstalk in a flat-cable line.

Per this graph (of course, the numbers are true only in this particular case - and may differ for other line types and system configurations), equations (4.37) appear valid up to 40 MHz, above which the line becomes electrically long. Now the crosstalk is determined by the contributions of each elementary section of the line, which propagate to the near or far end, respectively. Note that around 75 MHz, the trend reverses, and the near-end crosstalk attenuation becomes larger than the far-end crosstalk attenuation. The reason is that the farther from the near end any given elementary line section is, the larger loss (and also, the larger phase change) it incurs while propagating to the beginning of the line. Thus, the contributions from the elements farther down the line can be severely attenuated and do not affect the near-end crosstalk. On the other hand, the contributions of all the line elements reaching the far end of the line incur an equal amplitude loss and phase change, and with the increase of the line length their amplitudes add (although strictly speaking, this is true only for crosstalk between identical circuits).

---

*This is not the only way to express the coupling parameter units. As a matter of practical convenience, these parameters can be expressed in terms of impedances, conductivities, or as unitless ratios.

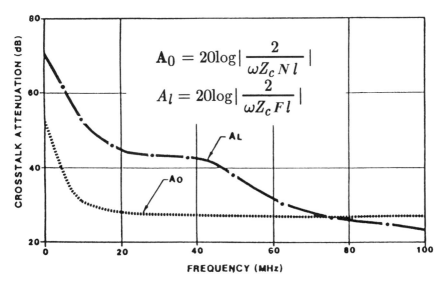

FIGURE 4.7 NEXT and FEXT frequency characteristics (laminated flat cable with twsted pairs, 28 AWG, l = 10 m)

Comparing transverse reactive field coupling in a coaxial line and an open line, the coaxial structure has two advantages: it is shielded by virtue of its design, and its concentric circular configuration yields to a relatively simple analysis. The transverse electric and magnetic fields generated by a lossless coaxial line with solid outer conductor in quasi-stationary regime are limited to the area enclosed by the shield (Fig.4.5 b): there are no transverse electric and magnetic fields outside of coaxial line! The electric field lines start at the external surface of the center conductor and end at the internal surface of the outer conductor. The direction of the lines depends on the conductor charges, from + to −. In any case, no electric field "leaks" outside. The magnetic fields generated by the currents in the center and outer conductors compensate, again resulting in the absence of external field. This is always true for isolated in space *concentrically* shielded objects (see Fig. 4.8), with or without the skin effect taken into account.

✦ Indeed, according to Ampere's law, a line integral of static magnetic field H taken over any closed path L enclosing a current I, is equal to that current:

$$\oint_L \vec{H} \vec{dl} = I$$

From this, the transverse magnetic field intensities generated by currents +I and −I in the center and external conductors, respectively (Fig. 4.8), at some point of space situated at distance r from the center of the current-carrying conductor, can be determined:

$$H^\phi_c = I/(2\pi r), \quad H^\phi_o = -I/(2\pi r)$$

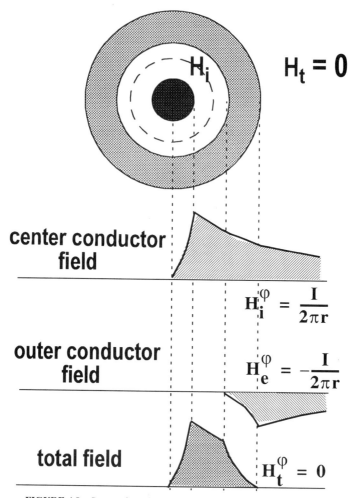

FIGURE 4.8 Summation of magnetic fields generated in coaxial cable conductors

where indices c and o relate to the center and outer conductors. In the space bounded by the inner surface of the outer conductor, only the center conductor current can be enclosed in any path. Therefore, the magnetic field in this space is determined only by the current in the center conductor. In the space outside the external surface of the outer conductor, the field is determined by the currents in both conductors and thus is compensated. The above considerations and derivations were done for the line considered as a source–emission problem. Applying the principle of reciprocity, it easy to see that the immunity (susceptibility) case brings the same results. ✦

But "real life" cables are coupled into other cables, conductors, and ground planes, which may detract part of the current from the given shield (e.g., see Fig. 4.4 b -d). Won't this violate the compensation of the fields generated by the currents in the shielded conductor and the shield? Fortunately (for electromagnetic shielding) we can make a very important statement:

With the exception of relatively low frequencies of up to several kilohertz or several tens of kilohertz (depending on the line shield impedance), it can be assumed that all the current in the shielded line returns through the shield.

✦ Really, consider first an ideal (perfect symmetry, homogeneous uniform solid and zero resistance external conductor) shielding structure (coaxial cable), just like in the previous author's note. As you saw, *all* the current in such shielding structure returns *internally* and produces neither high nor low frequency external electric or magnetic transverse or longitudinal fields. Bringing such a shield into the vicinity or even connecting it to, say, a ground plane, will change nothing in the current and field distribution! No current will flow in the shield with an external return path.

Now, let the external conductor (shield) has a finite conductivity. Due to the shield impedance, there will be a voltage drop at the external conductor. Now, if a ground plane or other conductors are galvanically (or capacitively and inductively) connected to ourshield, this voltage drop becomes *coupled* into these *external* metallic elements. In this case, a current will flow in the shield, with an *external* return path. In chapter 3, we considered this situation in many details, and even came up with a set of three theorems, dealing with current distribution in a shield with external and internal return path. As we saw, in this respect, the shielding circuit behavior is determined by the shield transfer impedance (admittance).

The simplest situation is at DC, or low enough frequencies, where the shield's transfer impedance can be approximated with its DC resistance. There, the system can be analyzed using a system of simple mesh equations for the current contours formed by the shield with the involved other conductors and ground planes (e.g., see Refs. [4.9, 1.20]). The solution yields the relationship between the total returning current, I_{tot} (which of course is equal in magnitude and opposite in sign to the direct path current in the shielded conductor), and the current returning in the shield, I_s, is

$$I_s = I_{tot} \frac{j\omega}{j\omega + \omega_c} \qquad (4.38)$$

In this case, the cutoff frequency, represented by

$$\omega_c = Z_s/L_s \qquad (4.39)$$

depends on the shield loss impedance and the impedance of the conductors in parallel with the shield, including the ground plane (the original derivations are modified to include these factors; see Refs. [4.9, 1.20]. The larger the shield's loss impedance (including the impedance of the conductors terminating the shield), the greater the portion of the current that returns via ground instead of the shield. In practical terms, if the shield-to-line termination impedances are small, the return via the other conductors and the ground can be disregarded, even if present, at frequencies much larger than ω_c. However, in the extreme case when the shield is disconnected and "decoupled" from the line termination at one or both ends, all the current will return in the ground plane or whatever conductors the line is connected to, until the capacitance between the shield and the shielded conductor will not provide sufficient shield termination impedance at higher frequencies.

Of course, at higher frequencies the amount of current in the shield external circuit is mainly determined by the shield's transfer impedance. While conducting the derivations of the transfer and loss impedances of the shields in chapter 3, we did touch upon this subject. But then, we were looking at the involved processes mainly from the point of view of the shield transfer function. In this chapter, we will return to this subject from the point of view of electromagnetic coupling, while discussing the crosstalk and far field coupling. ✦

◆ The above considerations are pretty much simplified: balanced pairs are often twisted, coaxial cables may posses certain eccentricity, both balanced and coaxial lines have losses. A large body of research has been conducted on the subject (e.g., see Refs. [4.8, 4.11, 4.14, P.2]).◆

◆ How large can the signal frequency be before the quasi-stationary regime in coaxial line is sustained? Consider a coaxial line with a homogeneous and electrically linear center and outer conductors, between which the voltage V_0 is applied. Current I in the center conductor returns –(–I) in the outer conductor, and the line dielectric within and outside of the outer conductor is "free space" ($\varepsilon = \varepsilon_0$, $\mu = \mu_0$, $\sigma = 0$). Select a cylindrical coordinate system with the axis z coinciding with the line center conductor axis.

For coaxial line, the transverse electric and magnetic fields obey the quasistationary laws up to extremely high frequencies, i.e., until the signal wavelength becomes comparable with the cross-sectional dimensions of the line. Thus, for a line with 5 mm diameter (0.005 m), even a 3 GHz signal corresponds to only $\lambda/20$. There, electric field intensity E_{irs} in the dielectric surrounding the center conductor inside the line has only a radial component whose value follows directly from Gauss's law, leading to a simple expression:

$$E_{irs} = q_1/(2\pi\varepsilon r)$$

where E is in V/m; the index $_{irs}$ stands for internal-radial-static (nothing to do with taxes!); q_1 is the linear charge density in the line in coulombs/m; $\varepsilon = \varepsilon_0\varepsilon_r$ is the line insulation dielectric constant consisting of the insulation relative dielectric constant and the dielectric constant $\varepsilon_0 = 1/36\pi 10^{-9}$ F/m for vacuum; and r is the distance from the center conductor center to the observation point inside the line.◆

4.4.3 Common Impedance (Galvanic) Coupling

When two lines have a common impedance, e.g., a common conductor (Fig. 4.4 c), any voltage drop developed at this impedance by one of the circuits also is applied to the other one. This may, or may not be the signal line. For example, conduction coupling often develops in lines and interconnections when the lines use a common return (e.g., PCB traces with a common return via the same conductor or a ground plane, or the SG circuit formation in a flat cable).

While in the circuit "originator" this voltage drop acts as a voltage loss, for the other circuit it is a source of galvanic or conduction coupling. Thus in Fig. 4.9, the common impedance carries the sum of the currents from both circuits .

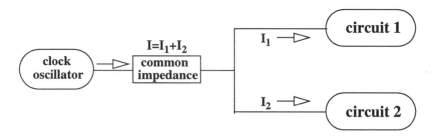

FIGURE 4.9 Common impedance coupling in clock signal supply

If the impedance is R, then the coupling voltage I_1R, generated by the current in line 1, is applied to the circuit 2, and the coupling voltage I_2R is applied to the circuit 1. Obviously, the smaller is the common impedance, the smaller is the voltage drop caused by the currents in it, and the smaller is the coupling. Thus, the magnitude of the common impedance can serve as a measure of galvanic coupling. The larger the current and the larger this common impedance, the stronger will be the coupling between the circuits.

(So, how is galvanic coupling related to the skin effect? - "food" for thought).

Other examples of common impedance coupling are the propagation of electrostatic discharge (ESD) currents in a common ground or other common elements of the circuits. The only requirement for the coupling mechanism to "qualify" as galvanic is that the coupled circuits be in electrical contact with the common conductor.

4.4.4 Coupling via Third Circuit

This coupling mechanism should be clear from Fig. 4.4 d. A source line affects not just a given victim line but also other circuits, if present. These other circuits serve as a transport which "transfers" an EMI signal to the victim circuit. This may occur in addition to the direct coupling, or it can be the only coupling mechanism present in a given circuit arrangement. For example, the latter situation may occur when the source and the victim do not have any parallel run, but the third circuit has partial parallel runs with the both of them.

✦ The principle of crosstalk via the third circuit seems simple enough (at least, in Fig. 4.4 d), but there is actually much more to it. The problems start piling up when a multiconductor line system is considered, with conductors in near proximity. Generalizing the coupling coefficients in Eqs. (4.33) and (4.35), we substitute, respectively, the ΣI and ΣV for a single source current and voltage. What this means physically, is that we look at the sum effect of all the currents and voltages in the combined source/victim system as affecting the disturbed line. But each of the conductors in the set can be viewed as a part of a circuit, providing either a direct or return path. Now, the question is which circuit?! Since this is a pretty old subject, it has been investigated down to the smallest detail (e.g., see Refs. [4.1,4.2,4.8-11,4.13-14]). However, even if the available analysis is not really complex, it is certainly cumbersome, requiring us to build large matrices of mutual coupling coefficients between the conductors in the system.

In the context of this book, it is important to understand the multiple ways of coupling to the shield. Therefore, while the readers are referred to the original literature for an extended discussion, here we will merely illustrate it. The main challenge with figuring out third circuits is of a "database processing" type: there may be multiple ways to form the third circuits within a line with a given number of conductors. In the literature, different ways of handling this problem have been proposed. For instance, in Ref. [4.14], the following order is suggested for forming the third circuits from z conductors in a line:

From z conductors, first all possible two-conductor balanced or single-ended unbalanced circuits are formed. Then two conductors of one balanced pair (direct path) and two conductors of another balanced pair (return) form a new four-conductor balanced pair, or two conductors of each balance pair form a direct path of a "single-ended" circuit with ground return. This process continues, until z circuits are not formed.

Let us apply this rule to a simple system consisting of two conductors and the ground ($z = 2$), with the currents +2 mA, +5 mA, and −7 mA, respectively. The following two combinations can be obtained:
1. Two single-ended circuits with the + 2 mA and +5 mA currents in the conductors (direct paths) and respective "negative" current returns, −2 mA and −5 mA, in the ground (return paths). This is equivalent to two common-mode currents of 2 mA and 5 mA.
2. One differential- and one common-mode circuit. A differential (balanced) mode circuit transmits +2 mA and −2 mA currents (opposite direction) in each conductor. The common-mode circuit transmits a +3.5 mA current in each of the conductors, with the −7 mA returning through the ground.

Then the coupling and crosstalk are considered superimposing the effects for each of these z circuits. The point is, that the end result will be identical for any of the z-circuit set combinations obtained in this way.◆

4.4.5 Coupling Between Shielded or Coaxial Lines

As a model, we will use crosstalk between coaxial lines. Although this problem itself presents an important practical case, its utility goes far beyond coaxial cables. As we stated before, a multiconductor shielded system or line can be viewed as a set of balanced and unbalanced circuits with their common-mode returns in the shield (of course, here we mean, the *crosstalk* common mode). Then, accounting for the system specifics (e.g., eccentricity with regard to the shield), a similar to coaxial line approach can be extended to the crosstalk between any shielded circuits carrying common-mode currents: closed or open design, single-ended, nonbalanced, or balanced.

The specific feature of crosstalk study in a wide frequency band is that, while the propagation regime in the dielectric remains quasi-static, within the shield body itself the transmission regime is that of wave propagation in conductors. But why would coaxial lines have crosstalk, anyway? Haven't we seen (Figs. 4.5 b and 4.8) that electric and magnetic fields are contained within the area enclosed by the shield (at least, an ideal coaxial outer conductor—solid, uniform tube)? So, allegedly there is "nothing to electromagnetically couple" to any "other" conductors, and no crosstalk should take place at all!

However, this is only correct with regard to the transverse reactive fields shown in Fig. 4.8. Indeed, we have never even mentioned the existance of the electric field intensity, E^z, *along* the shield (this designation implies that the line is placed along axis Z) - as shown in Fig. 4.5 c. By the meantime, unless a solid homogeneous tube is superconductive, it possesses finite impedance, which results in a non-zero external longitudinal electric field intensity produced by the current in the tube (you cannot see this current in Fig. 4.8, which shows the transverse section of the line). The same is true for the internal longitudinal electric field intensity produced by the current in the tube with an external return path. In Chapter 3, we dedicated enough time and effort to this subject. We found that the electric field intensity at the side of the shield facing the current return determines the shield loss impedance, while the electric field intensity at the side of the shield opposite the current return determines the shield transfer impedance. Isn't it curious that we have started with the notion of coupling but ended with the shield transfer parameters? But we already know that transfer parameters can be viewed as a specific case of coupling between the surfaces of the shield (see Section 4.1.2).

Now, if in Fig. 4.8 we imagine another coaxial line ("victim") with the external surface of its outer conductor in galvanic contact with the external surface of the shown coaxial line ("source"), the electric field, E^z, also will be applied to the victim, producing in it EMI and crosstalk. This is a typical case of galvanic coupling, which we discussed above. But the crosstalk between shielded (or coaxial) circuits carrying common-mode currents occurs even if the shields are not touching. In this case, the coupling can be considered as occurring via a third (tertiary) circuit formed by the outer conductors or shields of the coupled cables. The corresponding model is illustrated in Fig. 4.10 for two coaxial cables with solid tube outer conductors.

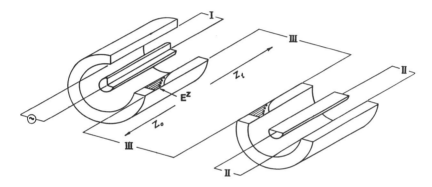

FIGURE 4.10 Crosstalk coupling between coaxial lines (coupling via 'third circuit)

It "works" as follows. When a signal propagates in a source coaxial line (or a shielded circuit carrying common-mode currents), a current with an internal return path flows in its outer conductor (or shield). This current generates a voltage drop at the external surface of the outer conductor, and it is proportional to the current in the outer conductor and its transfer impedance. With regard to the external environment, this voltage drop can be viewed as an EMI electromotive force (emf) which is applied longitudinally to the tertiary formed by the outer conductor and any external conductors, including the victim cables and lines, whether shielded or nonshielded. In this tertiary, the signal from any given element can propagate toward the near end of the line (encountering the input impedance Z_0) and to the far end of the line (encountering the input impedance Z_1).

The model in Fig. 4.10 can be also generalized on any kind of outer conductors and shields and the parameters of the third circuits. Thus, even the shields in continuous contact can be viewed as a specific kind of a tertiary. Moreover, a line or a line may consist of a number of shielded groups of conductors or coaxial pairs with grounded or nongrounded shields, in galvanic contact or isolated from each other, and may incorporate overall line shields as well as nonshielded isolated conductors forming balanced and nonbalanced circuits, and so on. The crosstalk in the victim circuits also depends on their design and electrical symmetry. For instance, if the victim circuit is balanced and its conductors are symmetrically situated with regard to

282 Electromagnetic Shielding Handbook

the source coaxial line, the common-mode currents induced in the conductors of the balanced circuit will be identical and compensate at the victim's load. This does not mean, however, that these common-mode currents will not propagate along the balanced circuit, which in this case may act as a third circuit distributing the EMI throughout the entire electronic system, while "crosstalking" with other circuits and/or radiating the EMI energy into the ambient environment ("electromagnetic pollution"—see Chapter 1). Of course, the situation is even worse if the victim circuit is single-ended; obviously all the induced common-mode will be applied to the victim's load. Other practical situations can be analyzed similarly.

4.4.6 Crosstalk Coupling Between Balanced vs Unbalanced Lines: Common and Differential Mode Regimes

We have already stated in section 1.2.3 that distinction should be made between the crosstalk (induction field) and radiation (propagation field) common-mode effects. In this section, we will deal with the crosstalk common-mode effects, while the radiation common mode is addressed in the next section.

Consider kind of multiple choice test - 'agree' or 'disagree':
1. The differential transmission mode results in less crosstalk than the common mode transmission.
2. A balanced line is more immune to crosstalk, than non-balanced.
3. Getting the statements 1. and 2. together, for differential-mode transmission in a balanced circuit, the superposition of the effects of the direct and return signal paths leads to a compensation of the fields generated in a victim line by the source conductors with the opposite charges and/or currents (compare Fig. 4.4 parts a and b).

Although the answer to the above statements is often"right", this is not always true. Fig. 4.11 presents several relevant cases.

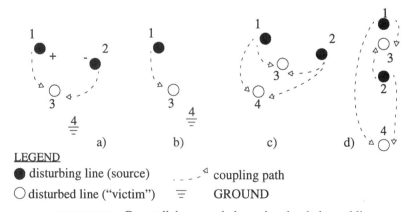

FIGURE 4.11 Crosstalk between balanced and unbalanced lines

In Fig. 4.11 a, the disturbing balanced line "12" is coupled to a single-ended line "3". Line "3" may, or may not have a current return path. When the line conductor is not terminated "on anything", the currents induced in the line are common mode currents which propagate toward its ends. In the crosstalk process such "unattached" conductor "3" is the cause of the crosstalk via third circuit, and in the radiation process - it acts as an antenna. Both ways - it is a source of strong EMI. Of course, when the line "3" does not have a return, it is practically unusable as a transmission line. Nevertheless, such "unattached" conductors do exist in electronic systems: unterminated traces on the PCB, disconnected conductors in the cables, unterminated shields and conducting elements of supporting structures, cabinets, etc. Practically, we should be careful with the idealization of such conditions, because even "unattached" conductors are really connected to the system other conducting elements, via the "stray" mutual capacitances and inductances. No matter how small are these mutual capacitances and inductances, at high enough frequencies they may provide effective terminations.

In functional single-ended circuits, the return path is usually provided via the "GROUND"). The "GROUND" can be realized as a line conductor, or a ground plane, or some kind of "unidentified ground object".

Whatever is the case in Fig. 4.11 a (with or without the return), when the couplings "13" and "23" are equal and since the currents in "1" and "2" are equal in volume and opposite in direction "+ and -"), we can expect the interferences produced by them in line "3" to cancel each other. However, this happens only when *electrical symmetry* takes place. If no other conductors are present, the electrical symmetry will be observed *only* when "12" is *geometrically* symmetrical with regard to 3 *and* the media parameters between each of these conductors (i.e., conductivities, dielectric constants, and magnetic permeabilities of respective conductive and insulating materials) do not violate the electrical symmetry. As a matter of fact, the coefficients c, m, g, and r in definitions (4.33) and (4.35) are just the measure of the electrical asymmetry, expressed, respectively, in terms of the *partial* capacitances, mutual inductances, dielectric losses, and galvanic (eddy current) losses between the conductors of the disturbing and disturbed lines.

Suppose, that conductor "2" is absent, resulting in the schematic of Fig. 4.11 b. Then, we have a case of interference between two single- ended lines (both lines will have returns via the GND). Now, we can expect an increase in interference, because the currents due to the coupling "13" are not compensated by the currents due to the coupling "23", as in Fig. 4.11 a. And vice versa, having both interfering lines balanced (as in Fig. 4.11 c) may result in even less interference than in the case of Fig. 4.11 a. Again, while this is often the case, there may be situations when the "gain" from the introduction of a second conductor as the return path in the line, may be "negated" due to a specific nature of the placement of the interfering line. One extreme such example is shown in Fig. 4.11 d.

We have also seen that crosstalk common mode can be considered as such only under certain circumstances. When the crosstalking lines have a return path via other conductors or ground planes, the common mode crosstalk is usually due not so much to the common mode nature of the transmission, but due to the electrical assymmetry of the interfering lines, including their return paths (even if not explicitly designed). Another thing, that in unbalanced circuits, the common-mode current may return in many possible and usually unidentified ways, without any "guarantee" that it will not create a situation, say, similar to the one in Fig. 4.11 d.

This also makes crosstalk to a large degree unpredictable. For an illustration of of such sudden effects, just revisit the twisted pair crosstalk data in Fig. 1.39 and the respective author's note. There, just by changing the termination conditions, the balanced lines were rendered unbalanced with "unidentified" return path and transformation of the differential mode crosstalk into a common mode crosstalk. As you saw, this produced a "catastrophic" effect. Such data is not an exception but rather a general rule.

4.4.7 Crosstalk Coupling Reduction Techniques

Line Separation Decoupling

The most obvious way to reduce crosstalk between the lines, is to separate them apart. Since the electric and magnetic field intensities reduce with the distance, so are mutual inductances and capacitances between coupled conductors. For example, the inductance L and capacitance C between two parallel conductors of radius r_0 and placed in *free space* at a distance a from each other can be *approximately* determined as:

$$L \approx \log\left|\frac{a}{r_0}\right|, \mu H/m; \quad C \approx \frac{1}{0.08 \cdot \log\left|\frac{a}{r_0}\right|}, pF/m \quad (4.40)$$

In general, the circuit types, system configuration, media parameters (ε and μ), and proximity to the metallic system components (including shields) significantly affect the quantitative and even qualitative relations in (4.40). However, the general proportionality to $\log|a/r_0|$ as a rule holds. Therefore, in certain circumstances this technique may be quite useful, e.g., determining the limiting geometry of PCB layout (see Fig. 4.12). Practically, the economical considerations usually prevent wide applications of this approach, since "real estate" in modern electronics is at premium.

Fortunately, the electromagnetic coupling theory suggests much more economical and efficient ways to reduce electromagnetic coupling between electronic circuits. Respective coupling reduction techniques are based on the optimal circuit layout in the system, and on electromagnetic shielding. In the following discussion, we will briefly consider the most important of them.

Decoupling vs Economics

First, we will address the conflict between the economical and coupling considerations. When laying out a PCB, or assigning conductors in a flat I/O cable, the most *economical* configuration is such, where only one conductor, or a limited amount of conductors serve as the current return, while the rest of the available conductors is used for the signal transmission direct path. Such circuit configuration is usually

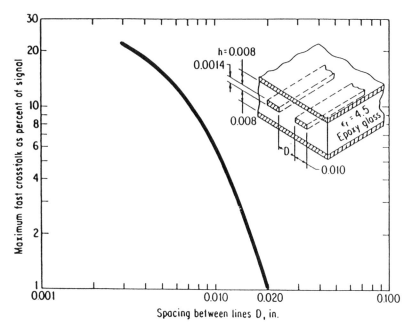

FIGURE 4.12 Crosstalk between strip lines

called *signal-single ground* (SG). To illustrate the effect of coupling between such lines, consider crosstalk between two SG lines 1G and 4G in a flat line (see Fig. 4.13, top configuration).

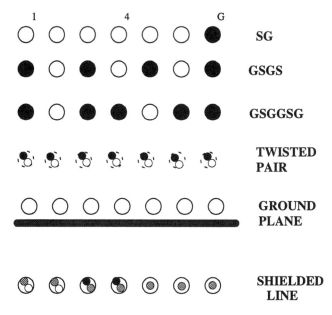

FIGURE 4.13 Circuits configured in a PCB or a multiconductor line

286 Electromagnetic Shielding Handbook

The common ground return provides an obvious common impedance for galvanic coupling: the larger is the ground conductor loss impedance, the larger is the coupling.

But common impedance coupling in the SG circuit configuration is only "the tip of the iceberg." It is easy to imagine all the conductors within the line linked with capacitors that model the electric coupling. Since the line 2G is completely enclosed in the area formed by the circuit 1G, any magnetic flux generated by the first (disturbing circuit), also "belongs" to the latter ("minus" the flux in the area 12). When the area $12 \ll 2G$, the EMI voltage in the line 1G will be almost equal to the source voltage 2G!

There are also "ample opportunities" for the crosstalk via the third circuit. Consider, for instance, crosstalk between the lines 1G and 4G. On the first sight, these circuits are relatively "far" from each other and, moreover, separated ("shielded") by conductors 2 and 3. We may expect that, at the very least, the electric coupling between 1G and 4G is somewhat mitigated. And that's really is the case. However, since conductors 2 and 3 are close to conductor 1, the EMI induced in the corresponding lines due to the electric coupling will propagate along these circuits and couple to line 4G. The closer to the interfering circuits and the more unbalanced the third circuit is, the larger this crosstalk will be. Experimental data (e.g., Fig. 1.39) confirms the problems with the SG circuit formation in a flat line.

As we see, the SG circuit configuration is really fraught with problems. As we have mentioned above, the only advantage of this configuration is its economy, which nevertheless may be crucial in many applications.

Layout for Reduced Coupling

Can we "fix" these problem? Fig. 4.13 illustrates several EMI coupling reduction techniques that are practically applied: better (from the EMI viewpoint) circuit layout, twisted pairs, ground plane, shielding. The GSGS and GSGGSG layout configurations are two such crosstalk coupling reduction techniques shown in Fig. 4.13. Indeed, by distributing the ground returns across the layout area and by bringing them closer to the respective direct paths, the common impedance coupling, the electric coupling between the lines, and the magnetic flux linkage may be reduced. In such configuration, the ground conductors (returns) can be viewed as partial shields, protecting the lines from crosstalk. The effectiveness of such *compensation* shielding is illusrated in Fig. 1.39. Such shielding may be also effective agains emissions and immunity problems, e.g., in PCB layout and flat cable emissions. In the next chapter we will address this subject in more details.

Another example of layout for reduced coupling is presented in Fig. 4.14. Here, by offsetting the pairs in connector matrix, the crosstalk was reduced more than by 8 dB.

Balanced and Twisted Pairs for Decoupling

Another way to reduce crosstalk coupling in the parallel line layout, is by balancing and, often, in conjunction with twisting the lines. Indeed, as illustrated in Fig. 4.15, due to the twisting, the distances from the disturbing line direct and return path

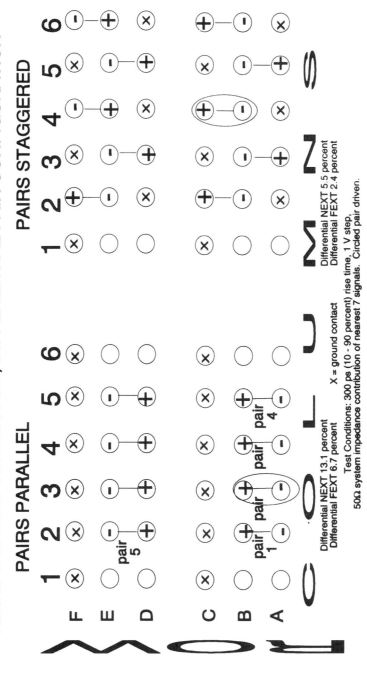

FIGURE 4.14 Pair offset reduces coupling

conductors (i.e., "+" and "-") to the "victim", alternate, and so do the signs of the electric and magnetic coupling coefficients.

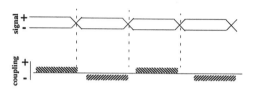

FIGURE 4.15   Coupling between twisted pair and a single-wire line

Generally, several important comments must be made with regard to the line balancing and twisting:
- While this technique was and still is the "staple" for round ("3-dimensional") cable EMI mitigation, it is not quite convenient for "flat" transmission lines: PCB traces, flat cables, etc. Although it is not impossible to realize such designs (for example, flat cables with continuously or periodically twisted pairs do exist), usually it is necessary to "pay a premium" in terms of the geometrical thickness and certain mechanical characteristics.
- The twisting pitch selection and matching in the crosstalking lines is practically not a "straight-forward", easy operation, but involves multiple considerations: frequency band, circuit geometry, physical properties of the insulation and conductors (especially, tensile strength and elongation properties)
- All the troubles with design and application of balanced and twisted pairs may be "for nothing", if the line balance is violated. A good example is the data in the Fig. 4.16 a, where the *same balanced* twisted pair performance deteriorated by a hundred times, when line ends were terminated at unbalanced loads (when connecting the unbalanced source and receiver to the lines).
- Of course, the *non-balanced single - ended* lines (that is, lines that were designed non-balanced, to start with) perform no better (see Fig. 4.16 b). Comparison with Fig. 1.40 proves that because of single-ended line terminations, the balanced twisted pairs terminated on unbalanced loads performed at the same level, as SG.

Coupling To and Coupling Through the Shield: Parallel and Separation Shielding

Functionally, in the presence of the shield (or any metallic object, for that matter), the mutual coupling between lines can fit two paradigms: both coupled circuits or lines are located at the same side of the shield - we called this *parallel shielding*, or both coupled circuits or lines are at the opposite side of the shield - *separation shielding* (see section 2.2.9). The two techniques at the bottom of Fig. 4.13 - ground plane decoupling and shielding decoupling - are based on these paradigms, respectively. How do they "work"? Calling upon our shielding model, as expressed graphically by Figs. 2.23 and 2.24, and keeping in mind the analysis of the model elements

(a) Fast crosstalk dependence on circuit balance

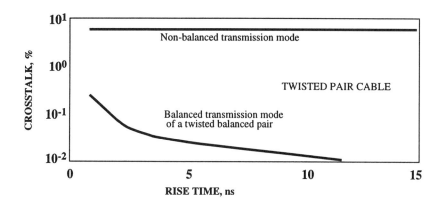

(b) Line balancing effect on crosstalk between the same lines

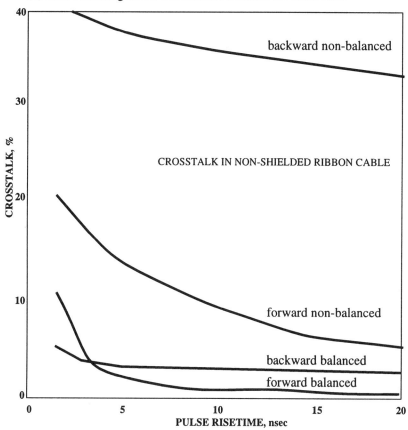

FIGURE 4.16 Circuit balancing and crosstalk

in chapter 3 and this chapter, we can now identify the similarities and fundamental differences between the two - something we weren't able to do before.

The similarity between these two shielding modes is related to coupling-to-shield regimes: in both cases shield works as decoupling means for both the EMI source and the victim. However, while in the parallel shielding the *direct* coupling between the interfering circuits is exhibited *along* with coupling to the shield, in the separation shielding such direct coupling is modified, hopefully, reduced or even eliminated by the shield. Thus, we have observed (see Fig. 4.8) that when the ideal separation shield encloses a current-carrying conductor and provides all the current return, such a structure acts as a coaxial line and does not produce an external transverse reactive field.

Obviously, the main distinction between the two shielding modes is related to the shield transfer function: it is present in the EMI propagation path in the separation shielding, and "doesn't participate" in the EMI propagation in the parallel shielding. But then, why would the parallel shielding work, anyhow? The answer is, that even if in the parallel shielding the shield does not carry any separation function, it is usually designed such, as to enhance the EMI protection by reducing the direct inductive and capacitive coupling as well as common impedance coupling between the interfering circuits.

Let's review the differences in shielding coupling mechanisms between these two paradigms. With regard to crosstalk, a separation shield can be viewed in two capacities: as a line conductor and as an energy barrier. (As we will see in Section 4.5.1, in the radiating regime a third function is added—shield as a radiator). On the other hand, the parallel shielding configuration exhibits only the line conductor (and radiator) properties, while *not* acting as an energy barrier between crosstalking lines. Of course, it still may act as an energy barrier against *other* fields impinging at the shield from the opposite to the crosstalking line location direction, but this is not the case that we consider here.

As a conductor, the shield is capacitively and inductively coupled to interfering circuits and often acts as common impedance (providing a return for common-mode currents, including single-ended circuits) and/or a third circuit element between lines. Here two important factors should be considered:

1. In separation shielding, the electric and magnetic mutual coupling are affected by the *partial* capacitances and inductances of the interfering lines to the shield, *and* are also modified according to both the shield transfer function (although this not necessarily may lead to the reduction of the near or far field coupling). In the parallel shielding, it is *only* the *partial* capacitances and inductances of the interfering lines to the shield, that affect the mutual coupling.
2. In the separation shielding the current return paths *in the shield* for the source and victim lines are located at the opposite sides of the shield respectively, while in parallel shielding - at the *same* side of the shield. Applying the "current return" theorems in chapter 3, it is easy to see that the common impedance in the first case (i.e., separation shielding) is the *shield's transfer impedance*, while in the second case (parallel shielding) - the *shield's loss impedance*.

Ground Plane Decoupling

A kind of "automatic" balancing of the unbalanced lines is achieved by proper use of ground planes. In subsection 3.3.3 of chapter 3 we have discussed the shielded conductor-to-shield coupling in 1-, 2-, and 3-D shields. Then in subsection 3.3.4 we illustrated how the current returns in the ground plane parallel to a signal line (direct path). We saw, that when the signal frequency is high enough, the ground plane return mimics the direct path, even if a *geometrically* shorter return path is available. Now, isn't this a "dream come true"?

And it is! No need to "bother" with the line return conductor, no need to observe the geometry (as long as the signal conductor is parallel to the ground plane), no need to strive for circuit balance or do "black" magic on the line twisting: all this is being done "automatically", by "the forces of nature". That's why one or more ground planes are usually included in many modern PCB layouts, and even flat cables are manufactured with integrated ground plane. Of course, an extra layer in a PCB adds cost to price-sensitive modern designs, and a cable with ground plane is heavier and less flexible, than without it. But this is the price many designers are willing to pay for reduced electromagnetic coupling.

To illustrate the ground plane (and also enclosing shield - see next sub-section) effectiveness, consider near-end pulse crosstalk measurement results between the lines in a flat line with a ground plane (see Fig. 4.17). This is a typical case of crosstalk between the lines with parallel shielding (that is, the coupled lines run at the same side of the shield).

As we have already indicated, the field configuration of a single-ended line over a ground plane is equivalent to the field of a balanced pair. We came to this conclusion qualitatively, on the basis of general theoretical considerations. (If you recall, we applied the method of images.) It was also shown both numerically and experimentally (e.g., see Refs. [3.1, 4.15]) that when a current-carrying conductor is placed over a ground plane which serves as a return current path, the return current in this ground plane tends to concentrate directly under the direct current, as closely as possible to it. The higher the frequency, the larger will be the share of the return current is flowing this way, but this phenomenon becomes dominant at frequencies as low as 1 MHz. In contrast, the dc and the low-frequency currents tend to spread over the entire ground plane. For this reason, we can expect both single-ended and balanced circuits to exhibit less high-frequency crosstalk when they are situated in a line with a parallel shield than without it. Indeed, the curves in Fig. 4.17 confirm this. By the way, when a line contains any grounded conductors (not just a ground plane), they also may act as a parallel shield, if properly situated.

Another problem with the ground plane may be its quality. For instance, a ground plane in modern multi-layer PCB may be pierced by hundred of vias - spetial apertures in the PCB, interconnecting the circuits placed in different PCB layers (see Fig. 4.18). When the signal line (say, a microstrip) is running *across* the vias, gaps and other conductivity interruptions in the ground plane, as well as close to the ground plane edges, the return currents in the ground plane must go around these obstacles. From the circuit theory point of view, this results in additional inductance of the line. Actually, these are old problems which were addressed in many details by Kaden [P.2]. In modern times such problems are again timely, due to the "electrical geome-

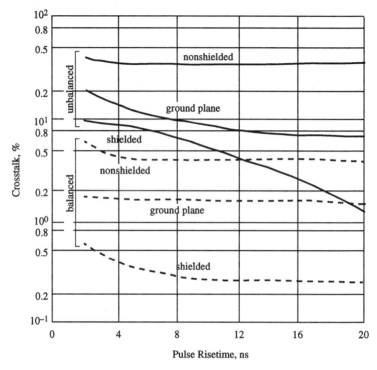

FIGURE 4.17 Near-end crosstalk in ribbon cables

try of scale". Fig. 4.19 illustrates the line inductance increase for several typical ground plane non-uniformities, based on the reference [4.16].

Separation Shielding for Crosstalk Decoupling

The most dramatic reduction of the crosstalk (and radiation) EMI coupling can be achieved by using separation shielding. Fig. 4.13 illustrates several applications of shielding to the coupling reduction: balanced lines, single-ended lines, coaxial cables. These applications are typical for PCB layout and cables. Another class of important applications is presented by product enclosures and shielding rooms (e.g., see Fig. 3.14). The following discussion fits all these applications, as long as the coupling regime is the same.

The separation shield function is to provide crosstalk protection to circuits situated at the opposite surface of the shield with regard to the EMI source. In this case, the working mechanism and the effectiveness of the shield depend on the signal propagation regime through the shield. At the relatively low frequencies, the EMI signal propagation regime through the shield can be assumed quasi-static in both the shield and the dielectric adjacent to it. In this regime, the shield transfer function will be different for electric and magnetic fields, while the shield transfer impedance is equal toits the DC resistance (see Chapter 3).

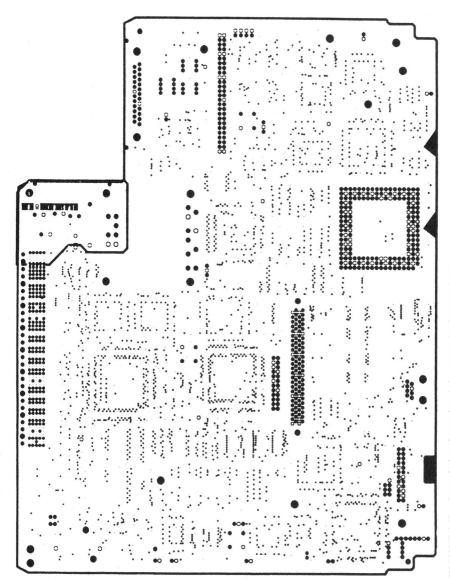

FIGURE 4.18 Ground plane layer view (telecommunications product)

294 Electromagnetic Shielding Handbook

GROUND PLANE DESIGN/ARRANGEMENT	INDUCTANCE CALCULATION FORMULA	EXAMPLE
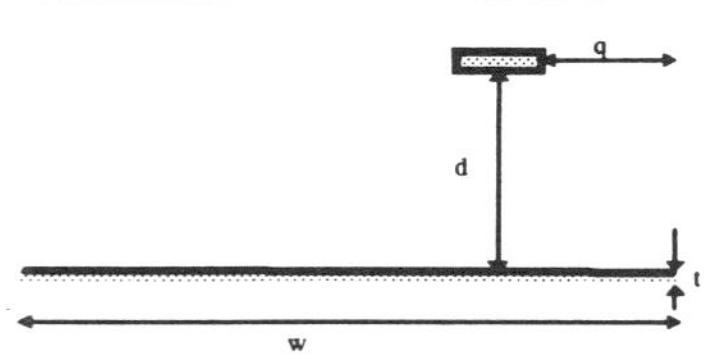	$$L = \frac{\mu l}{2\pi w} \int_t^d (a\tan\frac{q}{r} + a\tan\frac{w-q}{r})dr$$	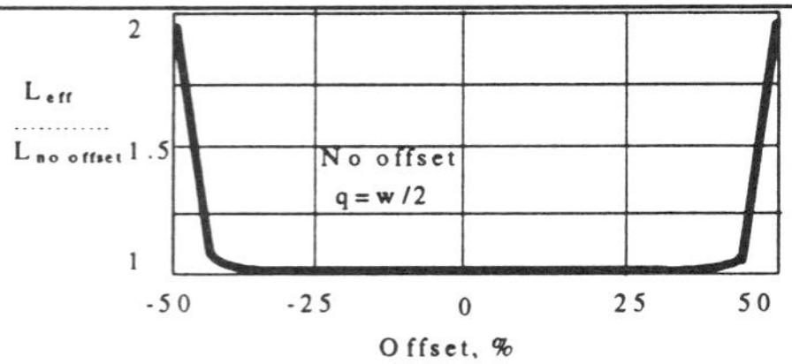 No offset; q = w/2
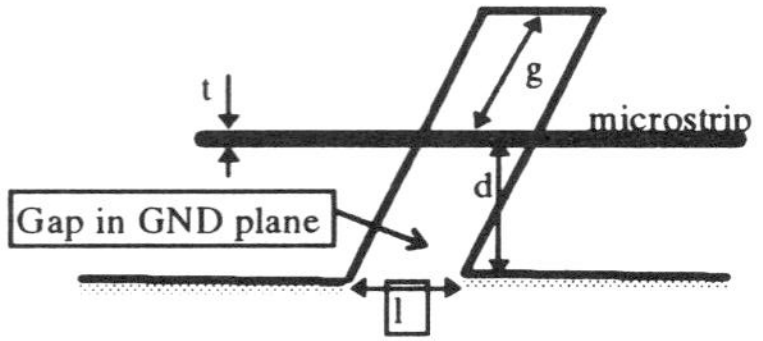	$$L = L_{plane} + \frac{\mu l}{2\pi} \ln\frac{g}{t}$$ if $g > d$, otherwise use d instead of g	$l=10\text{mm}; g=50\text{mm};$ $t=0.35\text{mm}$ $L_{gap}=14.5\text{ nH}$
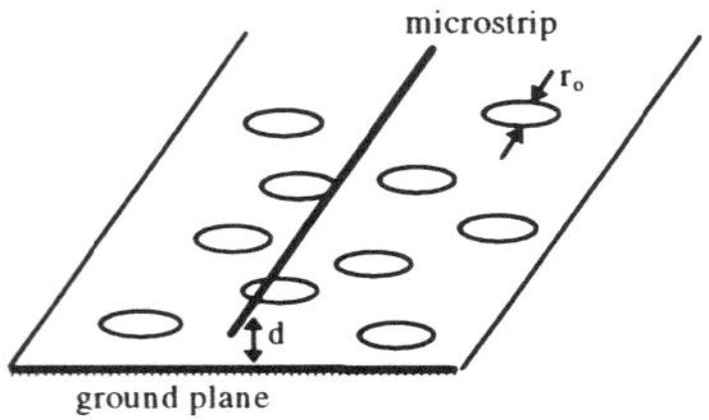	$$L = L_{plane} + \frac{\mu r_o^3}{3\pi^2 d^2} e^{-\frac{l}{r_o}}$$ $r_o \ll \lambda, \nu = $ number of holes per unit length	$r_o=1\text{mm}; d=1.6\text{mm}$ $L_{hole} = 17\text{ pH/hole}$ Leferink, EMC-95

FIGURE 4.19 Ground plane design effects on microstrip inductance

As we will see in the next chapter, under certain conditions ("grounding") any conductive shield *completely* "shunts" the electrostatic field, and any high magnetic permeability shield (conductive or non-conductive) *partially* shunts the magnetostatic fdield. In the quasi-stationary regime, shielding is defined mainly by the compensation of the currents developed by the direct magnetic field coupling of the source to the "victim" circuit and to a conductive shield.

Consider an ideal shield (solid uniform metallic enclosure of zero resistance) placed between the interfering lines at low enough frequencies that the propagation regime in conductors is quasi-stationary. The current distribution in conductors is assumed to be at dc. In the quasi-stationary regime, the skin depth in conductors is much larger than the conductor (signal-carrying and shields) thickness.

In the quasi-stationary regime, the shield transfer and loss impedances are equal to the shield's dc resistance. The magnitudes of common impedance and inductive couplings are close in parallel and separation shielding modes.This is a direct result of the shield thickness being smaller than the skin depth.

Since the quasi-stationary regime in conductors is valid only to much lower frequencies than in dielectric, in this case the regime is also quasi-stationary in dielectrics. The electric and magnetic fields in dielectric are reactive transversal.

With a frequency rise, the crosstalk EMI propagation regime in the line dielectric remains quasi-static up to extremely high frequencies (we have dedicated enough time to this analysis), while the propagation regime in a conducting shield may have to be treated as a wave regime, depending on the ratio of skin depth to shield thickness. In this regime, both the shield transfer function and the coupling mechanisms must be accounted for.

Figure 4.17 also illustrates the effect of separation shielding on crosstalk with the variation of the rise time (which is equivalent to the frequency variations—see Chapter 1). You may notice, that the separation shielding presents significant advantages in the performance, as compared to the parallel shielding. More illustrations will be given in chapter 5.

4.5 EMI ENVIRONMENT AND SHIELDING COUPLING IN RADIATING FIELDS

4.5.1 Problems and Strategies

Numerous questions and uncertainties originate from urgent practical needs, like:

- What kind of EMI environment are we dealing with and what are the respective effects on shielding?
- How does the coupling occur in radiating fields?
- Is there a difference between coupling in the emissions and susceptibility modes?
- How do the specifics of near and far radiating field coupling affect shielding performance?
- What is the effect of the wave incidence angle at the shield?

- Should we look at electric or magnetic components, or both?

To answer these (and many others) questions, we have to develop physical and mathematical models of the related coupling mechanisms. Often, the goals, formulations, and answers of such an investigation depend on the functions the shield performs in the circuits (whether we are aware of them or not) and on the shielding objectives. For example, reducing EMI in the line circuit load is only one EMC objective of electromagnetic shielding. It is very often even more important to prevent the shield radiation or re-radiation of the transmitted or picked up signal. Thus, although the use of a balanced line may provide good enough common-mode rejection for an amplifier or a gate at its end, the common-mode signal propagating in the line can produce extra crosstalk (crosstalk common mode) or radiation (radiation common mode). In both cases, the electronic system as a whole may "fail" the EMC test, while no EMI will be detected at this particular line load. On the other hand, common mode crosstalk is not necessarily an indicator of strong radiated EMI, because at far enough distance from such a line, the crosstalk common mode return may (but again, not necessarily will) compensate the direct path signal current, and thus reduce the radiation.

To study the radiating field "shielding coupling" we will adopt our "favorite" strategy: we will try to capitalize on widely known and well developed antenna theory. We developed the rationale for this in Chapter 1 when we formulated important EMC requirements in terms of electromagnetic fields and showed that a circuit or line in the EMI environment can be treated as an antenna. As a matter of methodology, we will attempt to draw a parallel with the theory of crosstalk between shielded lines, as presented in the previous sections. The comparison comes from the recognition of the similarities of, and the differences between, the field and circuit theories (see Chapter 2 and Appendix). Hopefully, the reliance on already developed antenna and crosstalk models will facilitate the problem treatment and help to avoid "reinventing the wheel."

We have already mentioned (see section 4.4.7 - "Coupling to and Through ..."), that a line shield generally should be viewed in three capacities: as a *line conductor*, an *energy barrier*, and a *radiator / sensor*. Comparing crosstalk between lines vs radiation, in principle we see no difference between the shield function as a line conductor. The shield coupling mechanisms within the line (internal returns) are determined by the near field effects and thus are no different from the behavior of other conductors. Except for electrically large enclosures (e.g., shielded rooms where internal resonances and multi-moding can occur), the shield is capacitively and inductively coupled to shielded circuits and often acts as common impedance. But, above all, it provides the return for crosstalk common-mode currents propagating in the lines inside the shield; without this there would be no compensation between the transmission line and shield currents and the line will generate external field leading to both crosstalk and radiation. In addition, once the EMI is induced at one surface of the shield, there is also a similarity between energy propagation through the shield (that is, the shield transfer function) in the crosstalk and radiation regimes.

The fundamental differences between shielded line performance in the crosstalk and the radiation modes originate from the properties of and the ways that the shield is coupled to the external environment. In the crosstalk mode, the interfering circuits were at an electrically small distance, which essentially resulted in induction fields and formation of transmission lines from the direct and return path conductors. Such an arrangement was conducive to the development of the mutual coupling concepts and the application of the circuit theory to identify and study the third circuits. For example, we treated the voltage drop at the external surface of the source line shield as an emf in the respective tertiary (the third circuit, formed by the shield with an external return). In the radiation mode (far-field zone), there are no conductors that would provide the return. Nevertheless, because of the radiation, the EMI energy does leave the source line, generating an EMI field in its vicinity and far beyond. As a result, an important question is how the electromagnetic waves couple to the shield in the radiation mode and how we can account for the radiation effects on the circuit characteristics.

The use of basic concepts of antenna theory permits us to address many of the posed questions even without a detailed analysis. Indeed, let us return to the questions at the section start. One logical way to evaluate shielding effectiveness in the *emissions* regime is to determine the ratio of the *field intensities* generated by the shielded and nonshielded circuits (as we saw, there may be alternative definitions, and we must also think about how to correlate them). On the other hand, another logical way to evaluate shielding effectiveness in the *susceptibility* regime is to determine the ratio of the *voltages or currents* generated by a specified field in the shielded and nonshielded circuits. The first impetus is to use the reciprocity theorem and declare that both approaches are equivalent. But aren't we comparing the "apples to oranges"? We saw in Chapter 2 that reciprocity is really applicable to the relationship between the voltage in one circuit and the current in another circuit (or equivalently, between the electric and magnetic field intensities related to these circuits), so it is not clear without proper analysis whether it can be applied here. In any case, it sure looks suspicious that we preferred to use different shielding effectiveness (coefficient) definitions for emissions and susceptibility regimes: the ratio of the fields in emissions case, and the ratio of the voltages in susceptibility case.

Probably, the reason that we tend to base the emission and susceptibility shielding effectiveness definitions on the different characteristics is that the EMI signal propagation conditions are different at both sides of the shield. And it does look more reasonable and practical to evaluate the signal voltages and currents (typical circuit theory entities) inside the shield, and electric and magnetic fields (typical field theory entities) outside the shield. Checking with Section 2.5.4 quickly proves that, by selecting these particular sets of characteristics, we violated the fundamental requirements of the reciprocity principle. Therefore, we cannot expect that the above emission and susceptibility definitions will yield identical figures of merit. This does not mean that reciprocal emission and susceptibility shielding effectiveness definitions are impossible, but they must be based on the valid principles.

Putting aside the question of whether it is the emission or susceptibility regime, what characteristics should we specify and study? Even if we limit ourselves only to the fields, there are crucial distinctions to consider between the electric and magnetic fields and between near and far radiating field zones. In the far radiating field, there is a well defined relation between electric and magnetic fields, and the energy associated with each of them is identical. For this reason, it is usually assumed that the shielding coefficients determined on the basis of the electric or magnetic components (that is E_1/E_2 or H_1/H_2) are identical. However, consider, for example, a cable shield. Even if outside of the cable the electromagnetic wave can be considered as a plane wave, inside the shield the field is inherently near field (unless we are dealing with some large line, like a TEM cell at high frequencies); the ratio of the electric and magnetic field components associated with the same transmitted energy is determined by the line characteristic impedance and thus varies. Only when the propagation media characteristics (including the boundary conditions) at both sides of the shield are identical are the shielding effectiveness(es) related to the electric or magnetic components of the plane wave equivalent. Thus, the widespread and "nice" Eqs. (2.1c) are almost never true for cable shielding because they do not include the system aspects of the line EMI environment.

Case in point: In Chapter 1, we have modeled the line radiators/sensors as antennas and used these models to define the line EMC performance. Since the EMC limits are as a rule specified at the inter-system level and deal predominantly with the far radiated field, at that time it was sufficient to formulate the models in terms of the far field. However, within the electronic system, the existing EMI mechanisms are not limited to the far field only, which requires a certain modification of those models. We have already reviewed such essentially near-field coupling mechanisms as line EMC performance in the static and induction fields, including crosstalk and shielding specifics. In this case, the model we used consisted of a source line which was capacitively, inductively, or common impedance-coupled to the victim line.

4.5.2 Radiating EMI Environment: Time and Space Domain Complications

There is EMI environment and EMI environment! Meaning that in "real life" systems - natural, electronics, and biological - there are "infinite" kinds of the field configuration and time variations. This "philosophical" statement is true for static, induction, and radiating fields. Just compare an "orderly" field configuration within a coaxial cable and balanced line in TEM mode or in a plane electromagnetic wave in free space (e.g., see Fig. 4.5, a - d) with fields generated in the same media in multi-moding regimes and in the presence of multiple reflections.

Multi-Path Propagation and Fading

A case in point with the latter is a multi-path propagation of radio signals, leading to slow and fast signal variations both in time at the same point of space, and in space - at the same point of time. To these you may want to add such "nasty" phenomena as the delay spread of the received signal, random modulation due to Doppler effect, variations in the reflecting objects location and properties*.

Take the so called Rayleigh fading of the radio signal amplitude. In principle, it's a 3D phenomenon. It results in *spatial* signal strength distribution interrupted by

---

*While the detailed discussion of these phenomena is beyond the topic of this book, we have at least to mention them to make the point, since the EMI environment is an ingredient of the shielding model.

many fades of varying depth. The fades repeat each half-wavelength, that is, every 20 cm for 800 MHz range, or 6 cm for 2.4 GHz range. An excellent metaphor for this complex process was suggested by George Calhoun (G. Calhoun, Digital Cellular Radio, Artech House, Inc, 1988, p. 2.17). He called the Rayleigh environment a "Swiss-cheese radio world". Fig. 4.5 e presents a two-dimensional representation of what Calhoun called "Lord Rayleigh Universe".

Metaphor or not metaphor, this process creates serious difficulties for radio engineers. But why *in this book* are we looking into all these complexities of mobile radio? Because the PCS and cellular industry are important users of electromagnetic shielding, and because it is not simple to deal with such spatial signal variations in evaluating the respective shielding performance. Indeed, even if a small enough device is illuminated by a 2 GHz field, its shield (or any long enough circuit within the housing) will "accomodate" simultaneously several fades of different depth, along with the signal "between" the fades. Now, how do we evaluate the results of such a filed coupling to the device's shield and circuits? Moreover, this environment keeps fast changing differently for each point of space, so that signal time variations also must be somehow accounted fo, along with spatial variations.

Unless you can come up with better suggestions, we have to evaluate the shielding performance of such devices either by somehow "integrating" these spatial and time variations, or on some statistical basis - both theoretically and experimentally! We can also simplify the task to some "manageable" conditions: average the signal (in time and/or space), or look for absolute maximums, or use certain signal distribution functions. Of course, the more we simplify, the less accurate we may be. Which emphasizes importance of proper planning of the shielding effectiveness evaluations and measurement.

Near Zone Radiating Field

We have mentioned previously and will continuously return later to the difficulties inherent to the treatment of the near zone radiating fields. Here, we will briefly mention one technique to address this issue. This approach utilizes the plane wave spectral representation of an *arbitrary* radiation field [4.28]. For instance, in a rectangular coordinate system,

$$E(x, y, z) = \iint \tilde{E}(k_x, k_y) \cdot e^{j(k_x x + k_y y + k_z z)} dk_x dk_y$$

where $k_z = \sqrt{k_0^2 - k_x^2 - k_y^2}$, $k_0 = \omega\sqrt{\mu_0 \varepsilon_0}$, and the tilde "~" is associated with the plane wave spectral components.

By using a plane wave expansion of an arbitrary located and oriented radiating source, the near field problem can be "reduced" to a plane wave integration problem. The word "reduced" is quoted because even if such approach is theoretically valid *in principle*, its practical realization may be associated with serious mathematical difficulties. Thus, almost all reported applications of this method to shielding tasks [4.29-30, 8.17] have required the use of numerical techniques.

Kind of an "inverse" problem arises when the EMI source far field behavior must be deduced from its near field configuration. This problem often arises when it is more convenient, or even only possible, to *measure* the field in the near zone. To ac-

300 Electromagnetic Shielding Handbook

complish this task, several methods were suggested, e.g., see [4.31-32] and can be used in shielding effectiveness applications.

4.5.3 How Do Radiating Fields Couple?

Coupling Energy Flux

The generation of the different field components in space by a current-carrying conductor (line or shield), or the effect of ambient field at the conductor, depend on the position and orientation of the line in the field. The energy flux of the field incident at and coupled to the line is described by the Poynting vector. Consider a susceptibility problem of a single-ended line horizontally stretched over the ground plane at height h (see Fig. 4.20). The line length is l, and the loads at the ends of the line are Z_{c1} and Z_{c2} (for simplicity, the line is assumed to be matched).

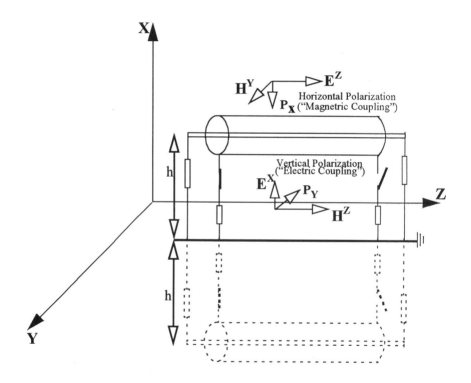

FIGURE 4.20 Shielded cable in radiating field over the ground plane

Suppose, the line is shielded and the loads at the ends of the shield are Z_{s1} and Z_{s2}. However, the conditions at the ends of the shield are less predictable: the loads may be considered matched (if the shield is long enough, or special effort is made to match it), open (shield "floating"), shorted to ground (shield "grounded"), or "anything else." In practice, the last is the most frequent case because, in real life, the shield can be imagined as capacitively and / or inductively coupled to the ground, or galvanically connected with conductors which have impedance of their own (e.g., see Refs. [4.17, 18]). Let the line be oriented along the z-axis in a three-dimensional space, and the incident on the line field have x, y, and z components. Depending on the field configuration, we can build three Poynting vectors carrying the energy toward the line from each of the three orthogonal directions along the coordinate axes: P_x generated by the incident field components (E^z, H^y) or (E^y, H^z); P_y generated by the components (E^x, H^z) or (E^z, H^x), and P_z generated by the components (E^x, H^y) or (E^y, H^x). The combinations containing the E^y and E^z component, correspond to "horizontal" polarization of the E-field, the combinations containing the E^x component to the "vertical" polarization. Since the line behaves as an antenna with a certain directivity, gain, efficiency, and so forth, its sensitivity to the energy fluxes generated by different combinations of electric and magnetic field components is different. Similar considerations of course are true for the emission regime.

How do these fields couple to the shield? In principle, we can answer the question using the basic laws of electromagnetism - in integral or differential form - see Appendix. Thus, applying the integral laws, it is possible to determine the effect of the magnetic flux generated, say, by the magnetic field component H^y and crossing the contour of the line (now you see why this is dubbed magnetic coupling), or the effect of electric field components E^x coupling across (or components E^z coupling along) the line. Not to say that it is easy to do, but there are no fundamental obstacles to this process. Indeed, consider the line circuit in Fig. 4.20. Applying Faraday's and Ampere's laws to the line, we can obtain the induced voltages and the currents, first at a small element of the line or shield, and then, integrating the elementary voltages and currents, along the line. Also, since the input impedance of the lines at each point can be determined with account of the terminations at the ends of the line, the obtained results can incorporate the effects of these terminations. Similarly, knowing the electric charge and current distributions along the conductor, the radiating fields can be determined (emission problem).

But, best of all, we don't need to do this, because this work has already been done in antenna and crosstalk theory!

Applying Antenna Analogy

We start with revisiting radiating emissions from simple elementary sources (see Chapter 1) that model the radiating fields of currents at the line conductor or shield. First, the current amplitude along this element is considered constant. The fields emitted by such sources are well known from the antenna theory. In particular, if a straight infinitesimal current element (a Hertzian dipole) of length l and carrying current I_o is placed symmetrically at the origin of a spherical coordinate system in a dielectric media with parameters $\gamma_d = \omega\sqrt{\mu\varepsilon}$, $Z_d = \sqrt{\mu/\varepsilon}$ and oriented along the z axis, the emitted fields at a distance r from the dipole can be described by well known equations:

$$H^\phi = j\frac{\gamma_d I_o l \sin\theta}{4\pi r}\left[1 + \frac{1}{j\gamma_d r}\right]e^{-\gamma_d r} \tag{4.41}$$

$$H^r = H^\theta = 0 \tag{4.42}$$

$$E^r = Z_d \frac{\gamma_d I_0 l \cos\theta}{2\pi r^2}\left[1 + \frac{1}{j\gamma_d r}\right]e^{-\gamma_d r} \tag{4.43}$$

$$E^\theta = j\frac{\gamma_d I_0 l \sin\theta}{4\pi r}\left[1 + \frac{1}{j\gamma_d r} - \frac{1}{(\gamma_d r)^2}\right]e^{-\gamma_d r} \tag{4.44}$$

$$E^\phi = 0 \tag{4.45}$$

Similarly, a small current loop of radius a and current I_o with constant amplitude (coordinate "z" perpendicular to the loop plane and current directed along the coordinate ϕ), can be represented as a magnetic dipole $m = I\pi a^2$, which results in

$$E^\phi = \frac{\omega\mu\gamma_d I_o a^2 \sin\theta}{4r}\left[1 + \frac{1}{j\gamma_d r}\right]e^{-\gamma_d r} \tag{4.46}$$

$$E^r = E^\theta = H^\phi = 0 \tag{4.47}$$

$$H^r = j\frac{\gamma_d I_o \cos\theta}{2Z_d r^2}\left[1 + \frac{1}{j\gamma_d r}\right]e^{-\gamma_d r} \tag{4.48}$$

$$H^\theta = -\frac{\omega\mu\gamma_d I_o a^2 \sin\theta}{2Z_d r}\left[1 + \frac{1}{j\gamma_d r} - \frac{1}{(\gamma_d r)^2}\right]e^{-\gamma_d r} \tag{4.49}$$

Reviewing these equations, we notice that they contain items that successively reduce with linear, square, and cubic distance from the source: $1/r, 1/r^2, 1/r^3$. Indeed, the significance of each of these items depends on the value of r, that is, on the distance from the radiating source. For example, at r = 0.1, these items are equal, respectively, to 10, 100, and 1000. But at r = 10, these items equal, again respectively, 0.1, 0.01, and 0.001!

What this tells us is that, at close distances (e.g. r = 0.1), the most significant item is $1/r^3$, which is 100 times larger than $1/r$. But at a greater distance (e.g., r = 10) it is just opposite: the item $1/r$ is 100 times larger than the $1/r^3$. Also, at a small distance from the source, the reactive component of the emitted power is large, signifying the lossless mutual exchange of energy stored in the electric and magnetic fields; at a great distance, the power is predominantly real, indicating the loss of energy with radiation. Then, unless we need very high accuracy, we don't have to bother with the linear item (i.e., 1/r) at a near distance from the source, or with the higher power items (i.e., $1/r^2$ or $1/r^3$) at far distances. For example, for a Hertzian dipole,

- field close to the source:

$$(E^r, E^\theta)_{max} \cong \frac{\kappa_1}{r^3}, \quad H^\phi_{max} \cong \frac{\kappa_2}{r^2} \qquad (4.50)$$

- and the far field:

$$E^\theta_{max} \cong \frac{\kappa_3}{r}, \quad H^\phi_{max} \cong \frac{\kappa_4}{r}, \quad E^r \ll E^\theta \qquad (4.51)$$

In previous sections, we have already dealt with the fields varying with the $1/r^3$ or $1/r^2$ laws. It can be shown that

the first law (that is, $1/r^3$) corresponds to the electrostatic field, while the second (that is, $1/r^2$) corresponds to stationary magnetic and quasi-static fields.

To obtain the solution for longer line runs, the contributions from all the elements of the line to the field at a given point must be integrated, just as we did for crosstalk. This leads to integrals, involving variable input impedances, which for themselves are quite intricate functions of line length and diameter, frequency, shield termination, and grounding, and even coiling of the line, which slows down the external waves propagating along the shield. As a rule, theoretical analysis is available only for some relatively simple designs. Considering the fields radiated by a long wire as composed of the fields generated by elementary line lengths, we can calculate the fields radiated by an arbitrary line length. Again, the results of antenna theory come in handy. For example, as far as electromagnetic radiation is concerned, a long line with losses (or a short matched for the common-mode line) presents a typical case of a traveling wave, long wire antenna. The summation of the fields from elementary lengths of such antenna in free space yields the far fields at distance r from the zero coordinate point of such antenna (e.g., see Ref. [1.14]):

$$E^r = E^\phi = H^r = H^\theta = 0 \qquad (4.52)$$

$$E^\theta \cong j\eta \frac{klI_0 e^{-jkr}}{4\pi r} e^{-j(kl/2)(K - \cos\theta)} \sin\theta \frac{\sin[(kl/2)(\cos\theta - K)]}{(kl/2)(\cos\theta - K)} \qquad (4.53)$$

$$H^\phi = \frac{E^\theta}{\eta} \qquad (4.54)$$

where η represents the impedance of the propagation medium to the wave, and $K = \lambda/\lambda_g$ is equal to the ratio of the wave length in free space to that along the transmission line.

Ground Plane Effects

The effects of the ground plane can be introduced as before, using the images.

If a ground plane is present then, applying the method of images, we can substitute for the ground plane with a mirror image of our circuit (see the bottom part of Fig. 4.20). Of course, we will have to generate the mirror image of "the whole picture," including the loads at the ends of the line and at the ends of the shield, and the incident fields (the respective components are "primed"). Then, the Faraday's and Ampere's laws must be applied to the line and its image. Depending on the LINE position with regard to the ground plane, two field configurations should be considered: horizontal and vertical. Suppose, the *nonshielded* radiating line element is placed horizontally at a height h over the ground plane (h « λ) and the observation point is at a distance r > λ (far field). Assume that all the current in the line returns via the ground plane. The substitution of the ground plane with the mirror image of the line (Fig.4.20 - but for now disregard the presence of the shield) results in a loop configuration that generates the radiating fields according to Eqs. (4.46) through (4.49).

$$E_{i1}^{\phi} \approx j\frac{\omega\mu\gamma_d I_0 h dm \sin\theta}{4\pi r}e^{-\gamma_d r} \qquad (4.55)$$

For LINE in vertical polarization, the ground plane results in a two-antenna end-fire array with the currents in the same direction. The far electromagnetic field generated by the current I_0 can be determined using Eqs. (4.41) through (4.45). Thus, we obtain

$$E_{i1}^{\theta} \approx j\frac{\gamma_d I_0 dm \sin\theta}{2\pi r}e^{-\gamma_d r} \qquad (4.56)$$

Another radiator useful for modeling the radiating properties of a set of parallel line shields at small distance from each other and/or from a ground plane is that of a folded dipole, whose analysis can be reduced to a straight wire or loop antenna.

✦ We doubled the field of the vertical line element to account for the effect of the "image" (even if in general case the direct and image signal amplitudes and phases at the observation point at distance r from the line are not equal). This has to do with practices adopted in EMC for measurement of radiating emissions (see Chapter 1): the receive antenna is height-scanned until the measured field is maximized. Obviously, the maximum is achieved when the phases of the direct and reflected rays produced by the source of radiation and its image in the ground plane are almost identical. On the other hand, you can easily calculate that typically in EMC measurements the effect of the different lengths of the direct and reflected rays is not significant. Thus, the fields produced by these two rays when the emissions are maximized, are almost equal and in the same phase. This comment gives us an extra opportunity to emphasize the dependence of the shielding effectiveness definition on the shielding objectives.✦

Accounting for System Configuration and Environment

In antenna theory, it is normal to represent the fields radiated by arbitrary antennas in arbitrary environments as functions of elementary fields, modified by special factors [1.14]: element factor A_e (which describes the field of a single radiating element in a free space environment), floor and housing ground plane array factors A_g and A_t, space factor A_s (which accounts for variable distances of small increment antenna elements from the observation point and ground planes), and so forth. For instance, the field at a given point of space, with regard to the specified radiator with standing or traveling wave distribution over its length L can be described as

$$E_\Sigma = \int_{-L/2}^{+L/2} A_s A_g A_t dE_e = E_e \int_{-L/2}^{+L/2} A_s A_g A_t dm \tag{4.57}$$

In the far radiating field, the energy is equally distributed between the mutually coupled electric and magnetic components of a "one" electromagnetic field. The impedance of the medium is constant (e.g., in free space it is 377 Ohm). At a large enough distance from the source, the electric and magnetic components form a plane wave, which propagates in the direction identified by Poynting vector. In the near radiating field, the energy associated with and the coupling between the electric and magnetic components is affected by the source and/or receiver properties: the electric field prevails in the high impedance circuits, and the magnetic field in the low impedance circuits. The ratio between the field electric and magnetic components is no longer 377 Ohm, and their interaction is further complicated by the field's nonuniform distribution. For this reason, the near-field models of EMC performance cannot be as general as those in the far field, requiring more system-related specifics.

4.5.4 Antenna Currents and Radiation Resistance of Lines and Shields

Perhaps you keep returning to that "nagging question": What is this I_o current and how can we determine it for real-life line conductors and shields? Indeed, this is a very valid and important question. As a matter of fact, this is *not* the differential-mode current in the line which is transmitted to its load and which has a return usually quite close to the direct path. Because of the compensating effect of the direct and return currents, mainly an induction field is created in the vicinity of the line transmitting the differential-mode, and radiating fields are extremely small. In contrast, the radiation common-mode current does not have an identifiable return. It is sometimes called the antenna current, emphasizing the "utility" of its far distance effects. But how do we determine this current without engaging in tedious and cumbersome mathematical derivations, often lacking physical "face"?

Fortunately, antenna theory helps with an effective and convenient way to do this.

✦ It is only logical, if we will look for the origins of the *antenna current* in the *antenna theory*. Indeed, electromagnetic radiation is the result of several factors:

- Flow of electric charges (current) from their larger concentration (the "source") to smaller concentration — universal law of Nature
- Finite propagation speed of electromagnetic signals
- Generation of magnetic field by an electric charge flow (i.e., by current)
- Inherent link and mutual transformations of alternating electrical and magnetic fields.

When a potential difference is applied to the (resonant) antenna terminals, a common mode current *starts* flowing, forming a standing wave pattern in antenna (see Fig. 4.21). But shouldn't this current "stop dead" right at the line start? Why would this current even "think" of flowing, when there is no load at the antenna ends and no return path? The answer is that "at the start" the "current does not know this nasty truth"! According to the relativity theory (remember Einstein?), the propagation speed of *any* signal is *limited* (to the speed of light - but that's immaterial in our case). So, when the potential difference is applied to the antenna terminals, the energy source has "no way to know" that there is "no return" and keep pumping the current into the antenna radiators. Of course, sooner or later, the current reaches the "end of the line", reflects from the "open" and comes back to the source to "report the trouble". But then, it's " too late" - the common mode standing wave pattern is already established in the conductors - and the radiation process endures, as predicted by Maxwell's equations!*

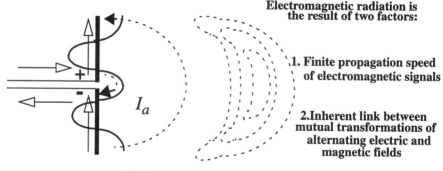

FIGURE 4.21 Einstein and radiated EMI

Of course, nothing "of this kind" would have happend in the *induction* fields, where signal propagation and interaction is assumed instantaneous. However, we know that instantaneous interaction is just a convenient model, but not really physical reality. Therefore, in principle, even the "slowest" varying induction fileds *must* be also *accompanied* by far field radiation, although it may be quite insignificant. Now, you may also want to revisit section 4.3.2 where we have already discussed some implications of the signal propagation finite speed.✦

---

*The described physical mechanisms are based on a model which considers the antenna current as the source of radiation. An alternative approach, much older and somewhat "abstract" makes use of Huygen's principle (developed originally for optics). It considers the electromagnetic fields as source of radiation. Such an approach can be instrumental, for example, to describe the radiation from apertures in the shield, as modeled by slot antennas. More on the subject can be found in [A.2]

"Borrowing" the ideas from the antenna theory, we can describe *in circuit terms* the *energy radiating* from a wire by introducing the shield's *radiation resistance* parameter. The three italicized expressions tell you "the whole story".

By definition, the radiation resistance of an antenna is equal to the ratio of the voltage at its terminals to the current, when the antenna is not loaded. Actually, the antenna radiation resistance, R_r, constitutes the real part of the radiation impedance, Z_r, which is modeled by a series element in a Thevenin equivalent circuit of an antenna (Fig. 4.22). Another series element of this equivalent antenna circuit is its loss impedance Z_l, which in the line radiator or shield radiator corresponds to their loss impedances.

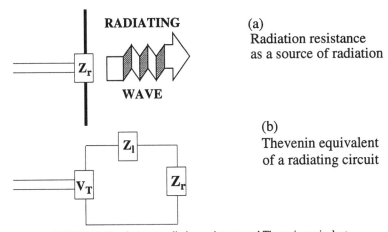

(a) Radiation resistance as a source of radiation

(b) Thevenin equivalent of a radiating circuit

FIGURE 4.22 Antenna radiation resistance and Thevenin equivalent

The voltage at a conductor element can be viewed as an emf at the conductor external surface that determines the antenna current. In this case, the conductor's loss impedance plays the role of the emf source impedance.

The antenna radiation impedance and resistance are mainly functions of the antenna geometry (electrical length, diameter, shape) and current distribution at the antenna surface. Even if the calculations of the radiation resistance generally are not trivial, through the years emerging antenna theory has produced readily available analytical and numerical solutions. The expressions and methodology of the radiation impedance and radiation resistance calculations can be found elsewhere (e.g., see the referenced literature on antennas). Also, to evaluate the general trends of shielding effectiveness, it is often sufficient to consider relatively simple geometries.

Such trend analysis could be based on several major antenna geometries leading to typical current distributions: a straight wire or a loop, along which are excited traveling or standing waves. Simple expressions exist for the radiation impedance and the radiation resistance of small antennas (see any reference book on antennas). For example, for linear dipole dm,

$$R_{re} \approx C_e \left(\frac{dm}{\lambda}\right)^2 \tag{4.58}$$

where

$C_e = 8 \times 10^2$, for $dm \leq \lambda/50$ (uniform current distribution along the conductor)

$C_e = 2 \times 10^2$, for $\lambda/50 < dm \leq \lambda/10$ (linear current distribution along the conductor)

Similarly, for a small loop,

$$R_{rm} = C_m \left(2\pi \frac{a}{\lambda}\right)^4 \tag{4.59}$$

where

$C_m = 6 \times 10$

a = the loop radius

For a large loop ($a \geq \frac{\lambda}{2}$ and current constant along the loop),

$$R_{rm} = 600 \left(2\pi \frac{a}{\lambda}\right) \tag{4.60}$$

The expressions are more complex for long radiators. For example, a lossless long wire traveling wave antenna of finite length m and current distribution

$$I_l = I_o e^{-j\gamma_d l}$$

along the line length l has the radiation resistance of [1.14]:

$$R_r = \frac{\eta}{2\pi}\left[1.415 + \ln\left|\frac{\gamma_d m}{\pi}\right| - C_i(2\gamma_d m) + \frac{\sin(2\gamma_d m)}{2\gamma_d m}\right] \tag{4.61}$$

where

$yd = \omega\sqrt{\mu\varepsilon}$

$C_i(x)$ = cosine integral function of x

From Eq. (4.53), the electric far field intensity generated by such current distribution is:

$$E^\theta = j30\gamma_d mI_o \sin\theta e^{jX} \frac{e^{-j\gamma_d r}}{r} \frac{\sin X}{X} \qquad (4.62)$$

where

$$X = \frac{\gamma_d m}{2}(1 - \cos\theta) \qquad (4.63)$$

Similar (although even more complex) formulas exist for standing wave long antennas:

$$R_r = \left\{ \frac{\eta}{2\pi} 0.5772 + \ln(\gamma_d m) - C_i(\gamma_d m) + \frac{\sin(\gamma_d m)}{2} \right\}$$

$$\times [S_i(2\gamma_d m) - 2S_i(\gamma_d m)] + \frac{\sin(\gamma_d m)}{2}$$

$$\times \left[0.5772 + \ln\left(\frac{\gamma_d m}{2}\right) + C_i(2\gamma_d m) - 2C_i(\gamma_d m) \right] \qquad (4.64)$$

Expressions (4.61) and (4.64) can be significantly simplified for specific conditions. For instance, if the length of a center-fed radiator is a large odd multiple n of the signal half-wavelength, than an approximate expression can be used [1.13]:

$$R_r \approx 30[0.57 + \ln(2\pi n)] \qquad (4.65)$$

Also important for electromagnetic shielding applications is another group of "antenna-derived" ideas and respective formulas that describe the radiation resistance of antenna arrays. These range from parasitic elements modeling the presence of other lines and conductors in the line to parallel (side-by-side) or collinear (end-fire) active elements modeling the ground plane effects. The respective expressions can be found using the general principle of pattern multiplication.

Thus, the electric field at a distance r and the radiation resistance of an infinitesimal horizontal radiator at height h over an infinite and perfectly conducting ground plane are equal [1.14]

$$E_{\psi h} = j\eta \frac{\gamma_d I_o l e^{-j\gamma_d r}}{4\pi r} \sqrt{1 - \sin^2\theta \sin^2\phi} [2j\sin(\gamma_d h\cos\theta)] \qquad (4.66)$$

$$R_{\psi h} = \begin{cases} \eta\pi\left(\dfrac{1}{\lambda}\right)^2\left[\dfrac{2}{3} - \dfrac{\sin(2\gamma_d h)}{2\gamma_d h} - \dfrac{\cos(2\gamma_d h)}{(2\gamma_d h)^2} + \dfrac{\sin 2\gamma_d h}{(2\gamma_d h)^3}\right] & \text{when } (\gamma_d h > 0.4) \\ \eta\dfrac{32\pi^2}{15}\left(\dfrac{h}{\lambda}\right)^2\left(\dfrac{1}{\lambda}\right) & \text{when }(\gamma_d h < 0.4) \end{cases} \quad (4.67)$$

Similarly, for vertically polarized radiator

$$E_{\psi v} = j\eta \dfrac{\gamma_d I_o l e^{-j\gamma_d r}}{4\pi r}\sin\theta[2j\cos(\gamma_d h\cos\theta)] \quad (4.68)$$

$$R_{\psi v} = 2\eta\pi\left(\dfrac{1}{\lambda}\right)^2\left[\dfrac{1}{3} - \dfrac{\cos(2\gamma_d h)}{(2\gamma_d h)^2} + \dfrac{\sin 2\gamma_d h}{(2\gamma_d h)^3}\right] \quad (4.69)$$

By roughly approximating available in the literature expressions, we can obtain the ratio of the radiation resistances of a radiator in free space to that over the perfectly conducting ground plane. Thus, for example for a horizontal half-wavelength dipole:

$$\psi_{hg} = \dfrac{R_{hg}}{R_r} \approx \begin{cases} 0.35 \bullet (h/\lambda) & \text{when } h/\lambda < 0.4 \\ 1 & \text{when } h/\lambda > 0.4 \end{cases} \quad (4.70)$$

Of course, the presented formulas do not cover all the possible applications. They are given mainly to illustrate the general principle, and the reader is encouraged to use the original literature sources on antennas.

✦ Why not approach the line and shield radiation coupling problem directly? Since the use of Maxwell's equations in the integral form is possible only for the simplest field and system configurations, in general, to determine radiation from transmission line and scattering by transmission line, we must operate with the Maxwell's equations in the differential form. Despite the large volume of theoretical and experimental work done to date (e.g., see Refs. [P.5, 4.2,4.19-22]), this old problem is an "unfinished business." For example, it is especially attractive to obtain an analytical solution, but usually this is done at the price of severe limitations. Respective derivations can be found in the literature for both near- and far-field zones. For example, a case of a plane wave illuminating a perfectly conducting uniform two-wire transmission line of finite length (susceptibility coupling) is treated in Ref. [4.21], and analytical expressions are derived in both the time and frequency domains. Because of theoretical difficulties, numerical methods gain wide acceptance for the field-to wire coupling analysis. In Ref. [4.20], the moment method is used to consider some typical problems of "super-high frequency" field-to-wire coupling to a transmission line of finite length. Both the radiation from transmission line and the scattering by transmission line problems are addressed. In this particular case of centimeter-long and millimeter-long waves, it appears more convenient to consider the scattering problem, which permits the use of antenna theory. In Ref. [4.22], a finite element analysis is applied to susceptibility coupling between an externally incident plane wave and an infinitely long multiconductor line. The homogeneous Helmholtz equation is solved using two-dimensional formulations of the partial differential equations for horizontal and vertical polariza-

Emi Environment and Electromagnetic Coupling in Shielding 311

tions of the incident wave, in the absence of reflections. As a result, the authors present the maps of the scattered and total electric and magnetic field distributions for the TE and TM waves in the frequency range up to 10 GHz.✦

4.5.5 Aperture Coupling in Radiation Field

The electrical size and shape of an aperture may impact the energy penetration through the shield in the most profound ways. Not only it determines the amount of EMI "leakage", but also the penetrated field configuration and coupling of the shield to the shielded circuits. It also dictates the approaches and models to study the phenomena. Here a parallel can be drawn with the effects of the electrical size of radiators and sensors on the electromagnetic signal coupling and propagation characteristics.

Electrically Small Aperture Coupling
In sections 3.4 and 3.5 of chapter 3 and in the section 4.4.7 of this chapter we have discussed the role, physical processes, and numerical evaluations of electric and magnetic coupling related to the *induction* field penetration through *small* apertures in the shield. As we saw (see expression (3.95)), the mutual capacitance and mutual inductance introduced by small apertures depends on the size and shape of the apertures. Kaden modeled such small apertures with elementary dipoles situated in the center of the aperture. For an electric dipole (electric field coupling) the model dipoles have the axis perpendicular to the shield surface, and for magnetic dipole (magnetic coupling) - the dipole axis lays in the plane of the shield surface. In cylindrical shields, both the inductive and capacitive coupling coefficients are proportional to the ratio d_0^3/D^2, where d_0 = the aperture diameter and D = shield diameter.

You notice, that cube and square of geometrical dimensions parallel similar dependence, $1/r^3$ and $1/r^2$ of the field fall off with the distance r from the source in the induction fields. This should come as no surprize, because within an *electrically small aperture*, any point meets the *induction field* conditions (4.15): $\lambda \gg \Lambda$. But this condition is *independent* upon the induction or radiation nature of the field. It follows then, that as long as the aperture dimension meets condition (4.15), we can expect the same d_0^3/D^2 law for field coupling through the aperture *in both* induction *and* radiation fields. This is the main reason, why in chapter 3 we were able to incorporate the small aperture effects in the transfer function.

But you remember (see analysis in section 4.1.2), that in order to achieve the transfer function generality, we either "by-passed" or significantly simplified the "mechanism" of coupling to both surfaces of the shield. Therefore, with regard to small apertures, we, sort of, included their effects in the shield transfer function and "disconnected" these effects from coupling . Thus we conclude that no special discussion about the effects of small aperture on coupling is necessary, unless the aperture size is comparable to the wavelength. But then, these apertures are no longer small and we must treat them as *electrically large apertures*!

Electrically Large Aperture Coupling

Within the area of an electrically large aperture, the $\lambda \gg \Lambda$ condition does not apply any more. Suppose, the aperture is large enough to meet the far field conditions. How does the far radiating field coupling through the large depend on the aperture size and shape?

To be sure, radiation through an aperture is an extremely complex process. First of all, in the case of electrically *and* physically "extremely large" apertures, we should discriminate between *two* energy penetration effects: *direct* energy flow through the "free" space within the aperture and the effects of aperture boundaries. The direct energy flow is no different from the "infinite free" space wave propagation laws, as far as the signal wave frequency and polarization are concerned. The only effect will be on the reduction of the "visible" area of the aperture — *"virtual"* aperture cross-section A_v, which determines the *amount* of energy penetrating through the aperture. Applying the laws of geometrical optics, the virtual aperture cross-section can be related to actual physical cross-section A_s as

$$A_v = A_s \cdot \cos\Omega \tag{4.71}$$

where Ω is the wave incident angle at the aperture (i.e., between the direction of the wave propagation and the normal to the "plane" of the aperture — "plane" is quoted because the aperture will not necessarily lay in a "flat" surface").

Obviously, the previous case is not of great significance to us. One reason is that if the aperture is that big as to make the direct energy flow through it the *main* factor, then there is not too much shielding "left there to even talk about". The second reason is that, as a rule in EMC, we look at the *worse* case emissions / immunity, and thus we should assume $\cos\Omega = 1$. And finally, you don't have to be a "rocket scientist" to come up with or use the formula (4.71).

By far, the most important case is when the effects of aperture boundaries should be accounted for. To model it we can use *directly* the principles developed for aperture-based antennas of different shapes. We can also reduce the aperture study to an investigation of *reciprocal* "patch" antennas, using the *Babinet's principle*. Whatever is the method, it should account for three factors:
- Source field configuration ("EMI environment")
- Aperture geometry: size, shape, "depth", i.e., respective shielding wall thickness
- Surrounding shielding wall size and conductivity.

In this chapter, we will get a "flavor" of the problem, while certain respective other important details will be reviewed in Chapters 5 and 7.

Consider an infinitely large, infinitely conductive ground plane S of *"infinitesimal thickness"*, as shown in Fig. 4.23. Mounted at S, is a rectangular aperture with dimensions a and b, respectively, along the axes x and y of a rectangular coordinate system. Suppose, the aperture is illuminated by a uniform electromagnetic plane wave described by components $(E_y = \hat{y} E_o)$, H_x.

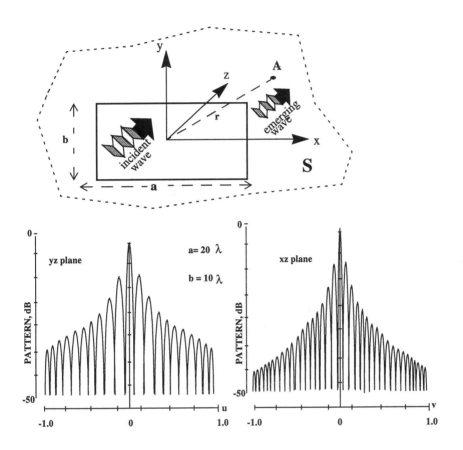

FIGURE 4.23 Principal plane far field patterns through rectangular aperture

The "emerging" field at the other side of the ground plane can be described in a spherical coordinate system by a set of pretty complex equations:

$$E_\theta = j\beta \cdot \frac{e^{-j\beta r}}{2\pi r} \cdot E_o ab \cdot \left[\sin\phi \left(\frac{\sin X}{X}\right)\left(\frac{\sin Y}{Y}\right)\right] \tag{4.72}$$

$$E_\phi = j\beta \cdot \frac{e^{-j\beta r}}{2\pi r} \cdot E_o ab \cdot \left[\cos\theta \cos\phi \left(\frac{\sin X}{X}\right)\left(\frac{\sin Y}{Y}\right)\right] \tag{4.73}$$

$$E_r = H_r = 0; \quad H_\phi = \frac{E_\theta}{Z_o}; \quad H_\theta = -\frac{E_\phi}{Z_o} \tag{4.74}$$

where

$$X = (\beta a/2)\sin\theta\cos\phi; \quad Y = (\beta b/2)\sin\theta\sin\phi; \quad \beta = (2\pi)/\lambda \qquad (4.75)$$

and $Z_0 = \sqrt{\mu/\varepsilon}$ - wave impedance of the dielectric.

Just a simple inspection of (4.72) - (4.75) leads to important conclusions:
- The radiation is *proportional* to the aperture area *(ab)*
- There are multiple lobes and nulls at the radiation pattern, generated along both axes, x and y, according to the laws a/λ and b/λ.
- The pattern in the plane y=0 has nulls of the field at mλ/a, m=1,2,3, ...
- The pattern in the plane x=0 has nulls of the field at mλ/b, m=1,2,3, ...

To simplify, consider the field only in the principal planes. For example, for a uniform (within the aperture area) incident field $E_o = E_y$, the E-plane (yz-plane, $\phi = 90^0$ and H-plane (xz-plane, $\phi = 0^0$). Then, $\cos\phi = 1$ and $\sin\phi = 1$, respectively.

Since $\beta = (2\pi)/\lambda$, we can express the respective radiation patterns as a function of ratios a/λ and b/λ in radians. The resulting field patterns are plotted in Fig. 4.23. As shown, the radiating field penetrating through the aperture indeed has multiple lobes in both principal planes, with the major lobe directly opposite to the aperture. Perhaps, this sheds additional light at the boxed-in slot antenna model in Fig. 1.24. Also, this information may suggest an *explanation* of the experimental data in Fig. 1.28.

✦ The most characteristic feature of the equations (4.72) through (4.75) is the dependence of the field on the function of the type sinX/X. We have already encountered this function in chapter 1 (section 1.2.4), when discussing the electrical signal spectra in the frequency domain. There, we were interested in the "extreme" behavior of the function: at very small and very large values of X. Here though, the argument X describes the radiation pattern, and we would like to view the *continuous* behavior of the function (or rather its modulus). This function is well known, so we can just "borrow" its graph from any mathematical textbook (see Fig. 4.24). Because practical radiation pattern is related to the field amplitude or power, it should be expressed by the *modulus* of the sinX/X function. This is shown in Fig. 4.23 in logarithmic units (dB).

Babinet Principle

✦ Consider Maxwell's equations in a source-free region:

$$\nabla \times E = -j\omega\mu H$$

$$\nabla \times H = j\omega\varepsilon E$$

Except for the sign in the left part of the equations, the equations' structure is identical, while the variables "trade places". These equations express the so called *duality* property of electromagnetic field: the solutions for electric and magnetic field components should have identical structure either and thus can be obtained by corresponding interchange of variables in any of the equations. The difference in signs is an expression of the *conjugal* (complementary) properties of the electric and magnetic fields: a source s_1 is conjugate to source s_2 when the distribution of *electric* and *magnetic* source currents and charges in one is replaced by the corresponding distribution of *magnetic* and

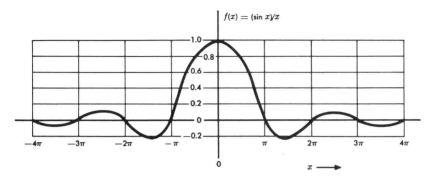

FIGURE 4.24 The (sinx)/x function

electric currents and charges, in the other. We can see examples of respective solutions, say, by comparing equations (4.72) and (4.73) with 4.74). Of course, we should remember the fundamental condition for Maxwell's equations duality: the source-free region

The Maxwell's equation duality, applied to the electromagnetic wave propagation in free space, is the basis for the Babinet's principle, first formulated with regard to optics. Consider two parallel infinite size plane shields (see Fig. 4.25).

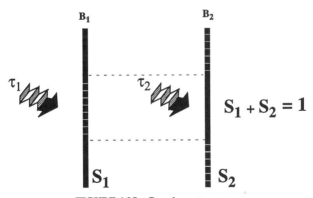

FIGURE 4.25 Complementary screens

For discussion sake, let's define the shielding effectiveness as a ratio of the field behind the shield to the field which would exist in the same point of space with the shield absent, but the same source in place. Let a source τ_1 to the left of an infinite shield B_1 produce a field which penetrates into the shielding space with the shielding coefficient S_1. Similarly, let S_2 be the shielding coefficient of a complementary shield B_2 with the source τ_2 to the left of the shield, which is conjugate to the source τ1. Then the Babinet's principle states that

$$S_1 + S_2 = 1 \tag{4.76}$$

316 Electromagnetic Shielding Handbook

Now, consider a field source illuminating a structure consisting of an infinitely large shield with an aperture and of a placed in free space, opposite to the aperture, conducting " patch" identical in shape to the aperture in the shield (see Fig. 4.26).

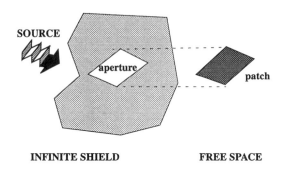

FIGURE 4.26 Aperture in the shield and complementary patch

The result of the Babinet's principle is, that the sum of the field penetrating through an aperture, with the field diffracted over a complementary patch is equal to the field with no shield at all. That is, such a structure will produce no shielding effect, whatsoever.

As "sad" this conclusion is in itself, it does facilitate the determination of radiating field penetration through apertures. Indeed, easy to see that using (4.76), we can *reduce* the problem of aperture field penetration to the problem radiation of a complementary antenna. The mathematical apparatus and basic formulas to do this can be found in the referenced literature.✦

4.6 TRANSIENT RESPONSE OF SHIELDING

This section belongs to this chapter and to chapter 3, as well. We have defered its discussion until this point, because by now the implications for both the transfer and coupling functions should be better understood.

As we discussed in Chapter 1, there are two main approaches to the study of transient effects. The "brute force" approach investigates the circuit behavior under the impact of transient excitation "as is;" that is, in the time domain. For example, using this approach, we were able to study near- and far-end crosstalk between the line lines by comparing the pulse amplitude in the disturbing and disturbed lines. Although we were able to arrive at some important practical conclusions, in general this approach is not all that convenient because of the many and varied difficulties in identifying the transient parameters and their interaction with the EMI environment at large (and the line lines and shields, in particular). On the other hand, investigation in time domain is often useful in order to separate different effects (e.g., electric and magnetic coupling, contributions of interconnection wiring to the measurement results, and so on). Hence, the numerous attempts to study shielding phenomena using the time domain, especially in such fields as electromagnetic pulse (EMP), lightning, crosstalk, and even in studying the effect of pigtails (e.g., see Refs. [4.23–27]).

A general approach to the study of transient effects is to reduce them to the frequency domain. For example, using the Fourier transform, it can be written in time domain [4.25]:

$$V_{se}(z, t) = \frac{1}{2\pi}\int_{-\infty}^{\infty} I_{is}(z, \omega)Z_t(\omega)e^{j\omega t}d\omega \qquad (4.77)$$

Although the solution of the integral (4.78) in the general case of arbitrary current and transfer impedance characteristics may present serious analytical and/or computing difficulties, for some relatively simple specific cases it may be obtained. For example, the transfer impedance of a thin solid homogeneous shield [see Eq. (3.49)] can be represented in the time domain as [4.25- 26]:

$$z_t(t) = \begin{cases} \dfrac{\pi}{a\sigma^2 d^3 \mu}\sum_{n=1}^{\infty}(-1)^n n^2 e^{-n^2\pi^2 t/(\sigma\mu d^2)} & \text{when } t > 0 \\ 0 & \text{when } t \leq 0 \end{cases} \qquad (4.78)$$

Using the fact that the Fourier transform of a product is the convolution of the Fourier transforms of the factors, and the fact that $z_t(t) = 0$ when $t \leq 0$, the electric field intensity at the shield surface opposite to the return current path (for solid shield it is equal to the voltage drop at this shield surface per unit length) can be written as

$$E(z, t) = \int_0^t I_{is}(z, \tau)z_t(t-\tau)d\omega\tau \qquad (4.79)$$

where

a = the shield radius

d = thickness

$\tau_s = \sigma\mu d^2$ = scaled time (denoting the time necessary for magnetic flux to diffuse through the shield)

Substituting the specific signal and shield parameters into those equations, the electric (and magnetic) fields can be calculated at the shield surfaces illuminated either by induction or radiating fields expressed in time domain.

5

Shielding Effectiveness for EMI Protection

Rationale

5.1 THE TECHNICAL BOTTOM LINE OF SHIELDING PERFORMANCE

The shield transfer function and shielding coupling functions may serve, as a rule, only as *indirect* measures of the shielding performance. The reason for this limitation is that these functions describe partial, "ingredient mechanisms" of the EMI energy transfer from the source to the victim, but not the "whole" process. Therefore, while in the preceding chapters we had multiple opportunities to appreciate the use of transfer impedance and shielding coupling parameters for shield design comparisons and shielding evaluations, from a system point of view, the most convenient characteristic seems to be a complete, *direct* evaluation of the *shielding effectiveness* of an electronic system, as a whole. We expect, that the direct evaluation of the shielding and the respective figure of merit could be *immediately* applicable to resolve the electronic system EMC performance problems. Is such figure of merit possible and if yes, what should it be based on?

For an answer, we will look into the comprehensive shielding model, suggested in chapter 2 of this book. In chapters 3 and 4, we have used this model as a *guide* to identify and understand complex shielding phenomena and as a *methodological* tool to develop roadmaps through shield transfer and coupling "jungles". Thus, the previous chapters have provided the insight into the subjects and problems of the EMI environment, shield transfer function, and shielding coupling, respectively. However, until we knew more specifics about the model element interactions, we had to limit the discussion to *isolated* effects of those elements, without bringing them together.

At the first sight, the direct, total measure of the shield performance can be easily obtained analytically, as well as experimentally. Analytically, using our comprehensive shielding model as a "template", we can try to assemble together all five identified "ingredient" shielding processes into the desired shielding effectiveness parameter. Experimentally, we may try to *measure* the same *desired* shielding effectiveness parameter by comparing EMC performance of the electronic system for the shielding options under study. Yet practical implementation of such analytical and experimental techniques may not always be that "straight forward".

The involved issues include the identification of appropriate shielding effectiveness criterion (or criteria), obtaining analytically and/or experimentally the corresponding figure(s) of merit, relating the figure of merit to the system shielding model, and expressing it via the ingredient components, by assembling ("bridging") these components together. We have already discussed many of the corresponding issues in chapters 1 and 2 (e.g., see section 2.6.2). The purpose of the discussion here is not to repeat the previous material, but to emphasize some salient points important for practical shielding applications. Further in this chapter, these will be illustrated with specific examples.

Consider the notion of the *desired* shielding effectiveness parameter. We use electromagnetic shielding to meet the electronic system EMC performance objectives. These objectives may be based on respective *EMC* standards and regulations, as well on the system *functional performance* standards and regulations. One problem that we have identified previously is that the system EMC performance objectives are not always clear, nor the relationships between the EMC and functional requirements and between the shielding evaluation criteria based on them, are always explicit. Take "commercial" EMC. Until recently, the emphasis was mainly on emissions. There, *at least two* goals must be met, which results in *two* respective sets of *desired* shielding effectiveness parameters: to limit electromagnetic pollution of the environment and to limit EMI within the system itself.

The electromagnetic pollution problem is addressed in the EMC regulations, which impose strict limits at the system emissions. Even if meeting these limits is not always easy, at least the evaluation criteria are clear (see chapter 1 in this book). On the other hand, the emissions effects on the same system functional performance are usually perceived as the manufacturer-relegated "system quality". Such functional performance "link to EMI" is even more emphasized in immunity regulations. In this case, it often appears necessary to evaluate the effectiveness of shielding with regard to the *system performance objectives*. Using such shielding effectiveness parameter, it may be possible to formulate the shielding effectiveness in terms of system functional performance. For example, we can measure the shielding effectiveness in units related to bit error rate (BER) of a communications system under the impact of EMI fields, currents, and voltages. The corresponding figure of merit may be a ratio of BER measured (or calculated) in the presence and the absence of the shield. There is nothing wrong to evaluate the shielding effectiveness in the terms of communication system errors, control system failures, number of damaged biological cells, etc. Moreover, this presents an almost ideal way to "automatically bridge" all the separate shielding processes in the comprehensive model into one unified figure of merit.

But let's not get "carried away", because such "ultimately" system approach is not always convenient, practical, and even possible, from the implementation and evaluation points of view. One problem arises because such implementation and evaluation (analytical or experimental) must be performed on a *specific system as a whole*. But we have already seen the "cons" and "pros" of this (e.g., see Fig. 2.26). Indeed, the fact that one system improved its performance when shielding was applied, does not guarantee that identical improvement, if any at all, will occur in another system, when the same shielding is applied. Another problem arises because in a complex system it is not always possible to "assign" the particular EMI manifestation to a particular source. Remember the discussion of the primary and secondary EMI sources in sections 1.1.3 and 1.2.2 of chapter 1? There we noted that shielding will have a *visible* effect on the system performance *only* when it is applied to the *dominant* radiator in emission case or dominant sensor - in immunity case. When the emissions or immunity of *secondary* radiators and sensors are improved, e.g., by shielding, the useful effect is obscured by other, dominant elements, and may not be detectable at a system level. So, in general case, by evaluating shielding effectiveness at the *specific* system level, we may just not know, *if* and *how much* is effective the shielding. Of course, you can say that such shielding does no harm either, but do you want to pay for something that you are not sure does "any good" as well? So much for the system shielding effectiveness based *completely* on "specific" system functional performance.

On the other extreme, we can "strip" the system to the "bare" shielded elements, and then measure the shield performance "unobscured" by other other parts of the system. An example of such *completely* (or "almost completely" - see discussion in chapter 4) *"general"* system performance is transfer impedance parameter. But while such parameter is repeatable and relatively easy to measure, it also have serious drawback: it does not reflect the system specifics. Talk about "catch 22"!

As we see, with regard to shielding evaluation there are manifold "intertwined" conflicts between the need to determine the shielding effectiveness for EMI protection in terms of system EMC performance and functional performance, as well as in terms of the specific and general system performance evaluation. To this we have to add the need to account for multiple physical and mechanical shielding chracteristics, reliability, manufacturing "logistics", system economics. The specifics depend, often in conflicting ways, on a multiplicity of factors: frequency and signal propagation regimes, system and shielded circuit configurations, line and shield transmission parameters and loads, shielded circuit and shield mutual placement, shield construction, termination, and grounding conditions, potential ground loops.

How to resolve all these conflicts?

To start with, because of the multiple involved factors, complex relationships between them, and constant need for trade-offs between conflicting requirements, no "cookbook" approach is possible here! If so, let's get ready for the "hard work", which "as always in shielding", should be based on the adequate background. Also, even if we know the underlying theory, we still will need a good "dose" of creativity and "non-conventional" thinking in finding solutions. As an illustration, consider the conflict between the shielding effectiveness figure of merit "generality" and "specificity". Usually, such conflicts are resolved by the means of a compromise: the system must be specific *"enough"* not to "lose the sight" of its performance objectives, but, on the other hand, general *"enough"* to permit "non-obscured" evaluation of the effectiveness of shielding. But how much is enough?

If this sounds "tricky", it is! This is what makes electromagnetic shielding discipline a science *and* an art, as opposed to extremes running from a "mechanistic" compilation of recipies to "black magic"! Only an in-depth knowledge combined with hands-on experience and engineering ingenuity may lead to (but even then, not guarantee!) the right solutions to practical shielding problems. We thus necessarily have to moderate our "ambitions" and scale the expectations down from the "global" shielding solution to tackling separately certain practical problems.

In this chapter, we will consider several typical electronic systems or subsytems which employ shielding protection. For each of the systems we will consider one or several appropriate shielding effectiveness figures of merit. We will look for a shielding effectiveness figure of merit based on the system EMC performance. Our objective is the *" technical bottom line"* of electromagnetic shielding: what does the shield contribute to the EMI protection from the system point of view.

From the methodological point, we use the general shielding models developed in chapter 2, in which we "plug in" respective "ingredients". In the previous chapters, we have identified and researched the five main elements of electromagnetic shielding model: external and internal EMI environment, external and internal coupling, and the shield transfer function. Our models relate the EMI environment adjacent to *two opposite surfaces* of the shield to the characteristics of these environments, electromagnetic coupling, and shield transfer function.

"Bridging-in" the ingredients in our models to bring all these processes together in a single parameter — shielding effectiveness — is a complex process in itself. We will have to "match" all five ingredients (or only some of them), so they could "fit" together in the models. For example, we have included in the shield transfer function definitions certain simplified EMI elements of the environment and coupling at both surfaces of the shield. In its turn, the coupling function was also formulated with regard to certain EMI environment. And the EMI environment itself was considered at some quite general level. The question is, how a change in environment and possible respective change of the coupling function will affect each other and the behavior of the transfer function?

The answer can be obtained by simplifying the task down to a set of predetermined conditions, and investigating each of them using convenient assumptions. Thus we come to a need for certain classification of the shielding effectiveness evaluation problems. Unfortunately (or fortunately?!), there could be many criteria to guide the process of splitting one big problem into a set of smaller, better manageable tasks. These will result in multiple ways to classify the shielding problems, based on the study objectives and problem specifics.

In this study, we will consider a two-fold "orthogonal" classification of shielding problems, matching practical applications: one, based on EMI signal propagation regimes (expressed in terms of respective frequency bands), and the second — based on the shield geometrical "dimensionality". We have already discussed both these concepts. Indeed, we identified and analyzed in sufficient details in chapter 4 the roadmap of EMI propagation regimes: static and stationary, quasi-stationary, radiation. In chapter 3 (section 3.3.4), we have also "touched" upon the respective shield "dimensionality" and mentioned *one-dimensional cable* shields, *two-dimensional "parallel"* shields - ground planes, and *three-dimensional* product enclosures - *"boxes"*.

Here, we will additionally introduce the concepts of *closed* and *open* shields and consider shielding effectiveness solutions for the following important classes of electronic systems and / or their elements:

closed one-dimensional shielding enclosures, as exemplified by cable shields

closed three-dimensional shielding enclosures — shielding "boxes", cabinets, rooms / test chambers, cars

open one-dimensional and two-dimensional "parallel" shielding formed by grounded lines, traces, and ground planes, respectively, running parallel to the protected line; the parallel shielding applications range from integrated circuits ("chips"), to single- and multi-layer printed circuit boards to EMC test sites.

For these classes of systems, we will discuss the respective figure of merit of the shielding effectiveness and the effects of the shieding design features in the EMI signal propagation regimes. We will then address an important subject of shield grounding and its effects on the shielding performance. Needless to say, that both the number of considered problems and the extent of their analysis are necessarily limited. Our goal is not to review all potential shielding problems - this is impossible - but to create and illustrate a general framework which permits to address such problems and specifics of their solutions.

Background

5.2 SHIELDING EFFECTIVENESS IN STATIC AND STATIONARY FIELDS

5.2.1 Problems of Statics

While engineers often tend to concentrate on high-frequency (RF and microwave) shielding, on the other edge of the utilized frequency spectrum the static and stationary phenomena may also present serious problems "asking" for shielding solutions. Just to name a dozen of such EMC-related natural and man-made sources and related problems:

- geomagnetic (and cosmic, in general) fields which affect the navigation, geophysical studies, space vehicle control, etc (have you ever heard of a *magnetic compass*? -of course, it's just a joke, but it emphasizes the importance and scope of the discussed problems: literally from elementary magnetic pointer to Global Positioning System - GPS)
- high voltage and current power transmission lines generate DC or industrial frequency strong non-compensated electric and magnetic fields, corona, etc
- low voltage power distribution network generating noise ("hum") in audio electronics or, in the opinion of some journalists, even may be dangerous to your health[*]
- power transformers generate industrial frequency strong non-compensated magnetic fields
- large electric potentials associated with lightning (from Hz - to MHz range)
- the same, due to the electrostatic discharge)
- magnetic fields developed by on-board or separate power supplies (several hundred kHz typical)
- magnetic fields developed in audio speakers
- anode voltages at a CRT ("DC")
- magnetic fields developed at the deflection system of the same CRT
- EMI generated due to primary-to-secondary coupling in transformers
- EMI generated due to crosstalk coupling between PCB traces and cable lines

In these and many similar cases, electromagnetic shielding is one of the main mitigation techniques. For example, multi-media computer speakers often must be shielded to prevent audio noise and to protect the monitor, magnetic memory disks, and read/write circuitry and control.

Even some "innocent" consumer products may generate surprisingly large fields. To get "a feel" for "real life" field intensity values, Fig. 5.1 presents maximum magnetic and electric emissions at close distance from a computer VDT (brand withheld), measured in a VLF frequency band: 2 - 400 kHz.

Note the high field intensity values close to the terminal and rapid field fall off with distance from the source. This should come as no surprise, since we deal with low frequency fields which obey cubic and quadratic fall off laws (as opposed to RF and microwave frequencies obeing the linear fall off law - e.g., see. Fig. 5.). In such fields, the rapid fall off continues even at larger distances than shown in Fig. 5.1.

As an example, Fig. 5.2 presents the measurement results of magnetic field intensity fall off at several hundred meters from a 30 kHz radiating product.

---

[*]In the opinion of this author, the danger referred to in the last statement is not proven as yet and is sometimes blown out of proportion in numerous, and often quite conflicting, reports. But then, "inconclusiveness" may work both ways! Anyway, for those interested in the subject, abundant literature is available.

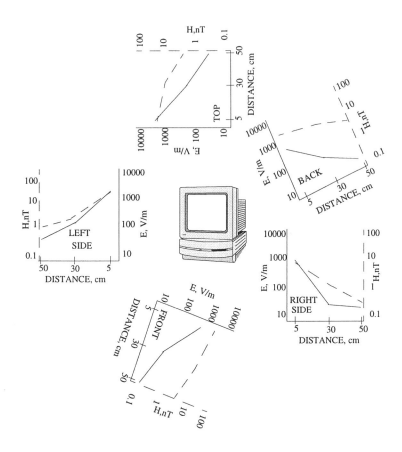

FIGURE 5.1 Maximum emissions from industrial computer terminal VDT measured in the VLF frequency band

5.2.2 Electrostatic ~~Metallic~~ Conductive Shielding

The substitute in the title of this section is not a correction of a "typo". Just several years ago, the only practical material for conductive shield should have been metal. Since then, modern technology have created advanced non-metallic conducting materials. Among them are *composites* and *inherently* conductive plastics. We will discuss such materials (albeit, quite briefly) in chapter 8. But here, point is, that we cannot any more limit electrostatic shielding to metals! However, because the metallic shields are still prevailing in the industry, the majority of our examples here will still refer to them.

326 Electromagnetic Shielding Handbook

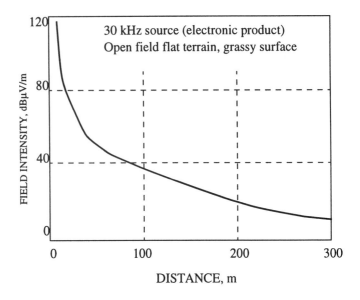

FIGURE 5.2 Low frequency field fall off with distance

Consider an important practical example - a shielded transformer. To be sure, such a device is not only, and even not mainly, associated with a transformer metallic container (e.g., 3D enclosure - box or cylinder), but more with a thin layer of copper tape (2D shield) separating the primary from secondary windings. However, each of these applications has different objectives, calls for its own shielding effectiveness figure of merit, and may involve different shield effectiveness evaluation techniques.

Anyway, while the need for a 3D transformer enclosure is obvious, what is the utility of a transformer separation shield? Here's the problem. In Chapter I, we have introduced two transmission modes: common and differential. One of the main means to fight the dangers of the common mode, is the use of a transformer (balanced-to-balanced, or balanced-to-unbalanced - "balun"), which stops the EMI-generating common mode, while being transparent to the useful differential mode transmission signal.

Theoretically, *any* transformer is supposed to stop the common mode penetrating from the part of the circuit connected to the transformer primary to the secondary winding. However, because of the interwinding capacitance, the common mode signal couples, or "leaks", into the secondary winding. By using an electrostatic separation shield between the windings (at primary winding side, or at secondary winding side, or both), the common mode signal is effectively "shorted" to the ground and coupling is reduced, as shown in Fig. 5.3. When properly designed, such shielding can dramatically reduce the system conducted and radiated emissions and susceptibility.

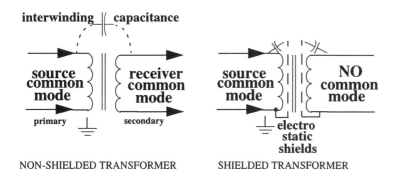

FIGURE 5.3 Use of electrostatic shield in the common mode suppression transformer

Now, some "physics" background.

Of course, the interwinding capacitance is a manifestation of electric field coupling. To determine the coupling functions in static (and quasi-static) electric fields, we operate with the field Eqs. (4.19) and (4.22). Consider an infinitely long conductor with a uniformly distributed positive electric charge q_1, Q/m. The charge has its electric field lines radially distributed in all directions around the conductor. This means that the electric field is transverse with regard to the conductor axis. At an observation point at a distance r from an infinitely long charged line, the field intensity is

$$E_r = \frac{q_1}{2\pi\varepsilon r}\hat{a}_r \qquad (5.1)$$

Note the *inverse linear* field intensity dependence on the distance from an infinitely long line.

If now another conductor, b, is placed parallel to the first one and illuminated by the field E_r, a negative charge, $-q_1$, will be induced at its surface. "Mathematically", this follows directly from the boundary condition of Eq. (A.19c). "Physically", the negative charge at metallic surface b is created by free electrons which were attracted by the positive charge in a.

To protect receiving conductor b from the field generated by source charge a, we can separate the conductors by a metallic shield. This can be a complete metallic enclosure ("closed" shield) of the source conductor a or of the receiver conductor b, or even an infinitely large metallic sheet separating the source from the receiver. The first option is shown in Fig. 5.4 a. Then, as before, the positive charge at a attracts the electrons to the inner surface of the shield. But now the external surface of the shield is charged positively, because it is "missing" the electrons that were moved to the shield's inner surface. As a result, the positive charge at a and the negative charge at the shield's inner surface compensate for each other, while the external shield's surface possesses the charge q_1! But in Eq. (5.1), r is the distance from the *center* of the charge, which means that the effect of q_1 will be the same, whether it is concentrated at axis a or uniformly distributed at the shield surface (as long as the symmetry is observed). So our shield is rendered useless. (Remember the previous section?)

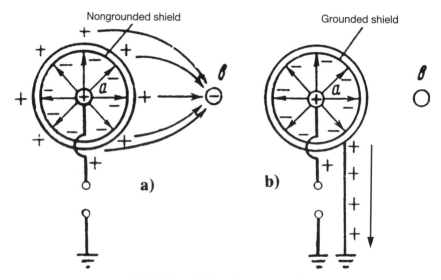

FIGURE 5.4 Shielding in electrostatic field

To fix the problem, we must compensate for the positive charges at the external shield's surface. One way to do it is to supply a charge $-q_1$ to the shield from an external source. In static regime, a conductive surface is equipotential, so this additional charge will be distributed uniformly at the shield surface, compensating for the original positive charge. A more "elegant" way to supply the negative charges is to ground the shield (Fig. 5.4 b). *Allegedly* (see more on this in section 5.7), the ground stores an "unlimited" stock of negative (as well as positive) charges which, given a conducting path (i.e., the grounding connection), can compensate for the positive charges at the shield's external surface. Now the whole system, consisting of the source charge at a and the shield, is neutral and, theoretically, the shield provides "100 percent protection."

We conducted the previous derivation for a linearly distributed charge - one-dimensional problem. We could have done similar derivations for a point, or a finite size 3D source charge. For instance, in the case of a point source, the only difference with (5.1) would have been the inverse *square distance* dependence of the field. All other derivations and considerations are identical.

As we can see, field theory gives a physically persuasive and mathematically simple explanation of electrostatic shielding. But what happens in a stationary field when the charges in a conductor move, albeit with a constant speed? Since the electric field of stationary currents is described by the same Eqs. (4.19) and (4.22), in principle, the uniformly moving charges produce a transverse electric field similar to the static ones, and the electric field shielding mechanism should be identical to that. Therefore, the same shield will protect from both static and transverse stationary electric fields. Moreover, as we will see later in considering the role of electric coupling in crosstalk between shielded lines, this can be true up to relatively high frequencies, with quite dramatic and sometimes surprizing effects, e.g., see later the discussion related to the Fig 5.11.

However, compared to the statics, the stationary regime produces some additional effects. One such major effect, the stationary magnetic field, constitutes the whole discipline of magnetostatics, and it will be discussed in the next section. But there are also additional phenomena associated with electric fields. Indeed, now we cannot disregard the resistance of the conductors. Because of the conductor resistance, the dc current in the conductor can produce a voltage drop in the grounding system and in the interfering conductors and the shield. The voltage drop in the shield can be the cause of the common-mode interference (see Chapter 1). The voltage drop in conductors and the shield produces the longitudinal component of the electric field, which is at the "roots" of galvanic coupling and coupling via the "third circuits." These two types of coupling are primary causes of crosstalk between and emissions from shielded cables, which we will address later.

Another way to look at electrostatic, and especially, stationary electric field shielding phenomena is by using circuit theory. In this case, we will operate with the concepts of potential and capacitance. Using Maxwell's equations in differential or integral form for electrostatic conditions (time derivatives = 0) and Coulomb's law, it easy to show that the electrostatic field and stationary field (in the areas containing no charges) are potential fields (that is, the work produced to move an electric charge against this field forces does not depend on the path of that charge) and therefore can be described by a scalar potential. The electric field potential at a distance r from an infinitely long line carrying a charge, q_1, is

$$U_r = -\frac{q_1}{4\pi\varepsilon}\ln r + U_0 \qquad (5.2)$$

Potential U_0 is the reference potential, which can be selected arbitrarily.

Capacitance is a measure of the ability of a system of conductors to store charge. We define the capacitance between two conductors carrying equal but opposite charges, Q and −Q, as a ratio of the charge to the potential difference between the conductors:

$$C = \frac{Q}{U_a - U_b} \quad , F \tag{5.3}$$

·When two electrically coupled parallel conductors a and b carry charges q_1 and $-q_1$, respectively, the potential difference between them depends on the conductor dimensions and distance between them (that is, geometry) and the dielectric constant. Thus, capacitance of a coaxial line with the external radius of the inner conductor r_i and internal radius of the outer conductor r_e, is

$$C_{coax} = \frac{2\pi\varepsilon}{\ln\frac{r_e}{r_i}} \quad , F/m \tag{5.4}$$

Similarly, capacitance in free space of a balanced transmission line with conductors of radius r_i placed at a distance a between their axes is

$$C_{bal} \approx \frac{\pi\varepsilon}{\ln\frac{a}{r_i}} \quad , F/m \tag{5.5}$$

Equation (5.5) is approximate because it does not account for the charge re-distribution at the conductor surfaces, due to their mutual coupling, as well as coupling to other conductors, including cable shielding.

✦ Three comments may be helpful in answering some of potential questions:

1. With regard to Eq. (5.1). Statics is the "ultimate" of the near field, no matter what's the geometrical, or geographical, distance. While the near field, as we know, falls off by square or cube distance law, the fields generated by the elements of an infinitely long line add up in such a way, as to produce the linear field intensity dependence on the distance.

2. Metallic sheet shield. While a *finite* size sheet presents an "open" enclosure, the *infinite* sheet may be viewed as a closed shield with the "closing" taking place in "infinity".

3. Note, that we have never even mentioned the shield thickness and/or conductivity: in static regime these parameters are of no consequence. So on one hand, the title of this section emphasizes *conductive* shield, on the other hand we claim that the shield conductivity is of no conse-

quence! Which statement is right, anyhow? Both!

This is true because in the static regime the charges are not moving; therefore, they don't have to overcome the conductors' resistance. However, with the just described physical model, it is essential that the shield were, "in principle", conductive to prevent the *isolation* and *concentration* of charges in separated shield areas and lack of charges in other areas and to facilitate the equipotentiality of the shield surface. Really, we understand that the higher is the shield resistance, the longer it takes for the charge to "spread" uniformly over the shield surface, so that in the shield with *no conductivity at all* the charge "levelling off" may never occur. But don't forget, that we are dealing with static fields, which have "all the time in the world" to do this "job". So, as long as there is *some* conductivity, "in principle", the equipotential surface will take place.

A little later, we will also discuss the *dielectric electrostatic* shield, which has, again "in principle", no conductivity, whatsoever. Of course, in "real life" there are no ideal shields and fields, but such approximations often help to understand the "nature of the beast" and arrive at practical solutions. Just be careful with and check your assumptions. Isn't this"fun"? ✦

5.2.3 Magnetostatic Shielding

Cylindrical Magnetostatic Shield of Infinite Length
Consider a conductor carrying dc current. Then per our definition, a static magnetic field will be generated in the space around the conductor. To study the stationary magnetic fields, we operate with field Eqs. (4.20), (4.21), and (4.23), or their equivalents in integral form (e.g., Ampere's and Gauss's laws—see Appendix). The way we apply and analyze these equations depends on the system geometry and magnetic properties. For example, applying Ampere's law to an infinitely long linear conductor, we found that current I in this conductor produces circular magnetic field lines in the planes perpendicular to the conductor axis. Equation (3.23) yielded the magnetic field intensity at the distance r from the conductor: $H^\phi = I/(2\pi r)$. (Again, as in the case of electrostatics, compare this with the inverse square distance law, as follows from the Biot and Savart law for an elementary length of current.) Note, that in Eq. (3.23) we did not even require the stationarity conditions!

How can we shield a conductor from such a magnetic field? The theoretical analysis answers the question. Here we will apply the same equations to study problems related to the coupling functions and shielding in magnetostatic fields (magnetic shielding from time-varying fields will be considered later). For example, in a relatively simple case of the magnetostatic field penetration from a current carrying conductor into a ferromagnetic half space, the analytical solution is given in [A.3] — by the way, a highly recommended reading!

Consider now an infinitely long cylindrical shield placed along the axis z in cylindrical coordinate system and illuminated by an external stationary magnetic field, H_0. Figure 5.5 shows a planar cut in such a cylinder perpendicular to the x axis. Variables are the shield's internal radius, r_i; external radius, r_e; and magnetic permeability, μ_s. The shield separates the space areas I, II, and III. Areas I and III are filled with the dielectric ($\mu = 1$, $\varepsilon = 1$). The objective is to determine the magnetic field inside the shielding enclosure (area II).

332 Electromagnetic Shielding Handbook

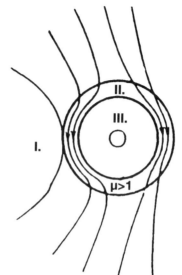

FIGURE 5.5 Shielding in magnetostatic field

Because in all three areas there is no current, the magnetic field in any point of space with coordinates r and ϕ is described by the Laplace equation in terms of scalar magnetic potential φ:

$$\nabla^2 \varphi = \frac{1}{r}\frac{\partial}{\partial r}\left(r\frac{\partial \varphi}{\partial r}\right) + \frac{1}{r^2}\frac{\partial^2 \varphi}{\partial \phi^2} = 0 \qquad (5.6)$$

where scalar magnetic potential is defined as

$$H = -\text{grad}\,\varphi \qquad (5.7)$$

✦ To solve (5.6) we need to separate the functions dependent on different variables - an example of a general problem when solving Maxwell's equations. In the previous section, we have identified the electric field potential as the work necessary move the electric charge against the field forces. Although no magnetic charges are known in nature (at least so far), we can formally introduce the scalar potential of magnetic field at frequency regimes and in the areas in space not containing currents, where $\nabla \times H = 0$, which is formally identical to (4.19). The equation (5.6) can be solved expressing the potential $\varphi = M(r)N(\phi$ as a product of two functions, each of a single variable (Fourier method). Substituting into Eq. (4.6) and after simple mathematical transformations we obtain:

$$\frac{r}{M(r)}\frac{\partial}{\partial r}\left(r\frac{\partial}{\partial r}M(r)\right) + \frac{1}{N(\phi)}\frac{\partial^2 \varphi}{\partial \phi^2} = 0 \qquad (5.8)$$

Unlike Eq. (5.6), each item in Eq. (5.8) is a function of only one variable, r or ϕ. Obviously, the sum of two functions of different variables can be equal to zero only when each of the functions is equal to the same constant with opposite signs. This results in:

$$\frac{r}{M(r)} \frac{d}{dr}\left\{ r \frac{d}{dr}[M(r)] \right\} = k^2 \qquad (5.9a)$$

or

$$\frac{r}{M(r)}\left[\frac{d^2}{dr^2} M(r) + r \frac{d}{dr} M(r) \right] = k^2 \qquad (5.9b)$$

and

$$\frac{1}{N(\phi)} \frac{d^2}{d\phi^2}(N(\phi)) = -k^2 \qquad (5.9c)$$

The solution to Eqs. (5.9) is well known. From it, we can obtain the solution to Eq. (5.6a) in general form as

$$\varphi^x = \left(C_{1x} r + \frac{C_{2x}}{r} \right) \cos\phi \qquad (5.10)$$

where x corresponds to the areas I, II, or III, respectively, and the coefficients C are the integration constants for these areas.

Thus, we need to determine six integration constants. As usual, these are determined from the boundary conditions:

1. Field at infinity (at $r = \infty$ the field $H_r = H_0$, which results in $C_{13} = H_0$)
2. In area I at $r = 0$, the field must remain finite, which results in $C_{21} = 0$
3. $\varphi^I = \varphi^{II}$ at $r = r_i$
4. $\varphi^{II} = \varphi^{III}$ at $r = r_e$
5. Boundary condition for normal components of magnetic induction at the interface between the areas I and II at $r = r_i$: $B_{di} = B_{mi}$
6. Boundary condition for normal components of magnetic induction at the interface between the areas I and II at $r = r_e$: $B_{de} = B_{me}$ ◆

Solving jointly all the equations with regard to the integration constants, we finally obtain*:

---

*Don't be surprised, if you see in the literature somewhat different expressions: the difference is caused by varied assumptions and, as a rule, all these expressions yield similar results

334 Electromagnetic Shielding Handbook

$$S = \frac{H^I}{H_0} = \frac{4\mu_s r_e^2}{r_e^2(1+\mu_s)^2 - r_i^2(1-\mu_s)^2} \approx \frac{4}{\mu_s} \frac{1}{1 - \frac{r_i^2}{r_e^2}} \quad (5.11a)$$

Equation (5.11a) indicates that magnetostatic shielding effectiveness depends on magnetic permeability of the shield and the its radius and thickness. Suppose, we have a shield with magnetic permeability $\mu_s = 1$ (that is, a nonmagnetic material). Then, from the full expression in Eq. (5.11a), we obtain S = 1; i.e., no shielding effect! On the other hand, even if the shield's magnetic permeability $\mu_s > 1$ but is not very large, and the shield thickness is small ($r_e \approx r_i$), then the shield effectiveness is not very good. Only thicker shields with large magnetic permeability provide protection from static and low-frequency magnetic fields.

The dependence of magnetostatic shielding effectiveness on the shield magnetic permeability and thickness (ratio of the thickness t to the radius r) is illustrated in Fig. 5.6 a by a contour plot and in Fig. 5.6 b as a 3D diagram, graphed per Eq. (5.11a).

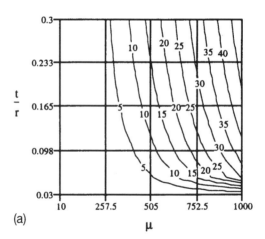

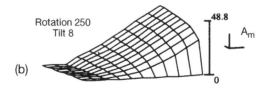

FIGURE 5.6 Shielding effectiveness of magnetostatic shields

- ✦ The *finite* length cylindrical shields are extremely difficult to analyze theoretically. Fortunately, the problem yields itself to numerical analysis, and solutions are available for different practically important situations. For example, the ref.[4.6] makes use of the so called *relaxation method* to solve several *axisymmetric* problems for variable shield magnetic permeabilities, ratios of the cylinder length to its diameter, with and without the "butt" lids, and parallel or perpendicular to the shield axis field incidence angles. Briefly, the findings are as follows:
  - A finite length shield performance is inferior to the infinite length, given all other conditions the same. The longer is the shell, the larger is shielding effect. With permeability in the range 100-1000, for the shield to be effective, the ratio of its length to the radius should be >10
  - When the length-to-radius ratio is around 2 and shield thickness to radius ratio is around 10-20%, the shielding effectivness up to 20 dB still can be achieved for permeabilities in the range 100-1000
  - There are considerable differences in the mode of shielding when the incident field is parallel or perpendicular to the shield's axis. ✦

- ✦ A similar difficult problem is related to the magnetostatic shielding effectiveness of a *finite width* planar shield. This problem is addressed in [5.49]. The proposed theoretical model is developed for multi-layer shields of open topology, constructed of perfect electric and perfect magnetic materials. It is based on conformal transformation between the fields of a line source over a finite width and infinite size planes. The perfect shield approximation can be used when the leakage around the conductor edges is significantly larger than that of direct penetration through the shield. In this case, the addition of layers beyond the second, does not improve the shielding effectiveness. ✦

Case Studies: Iron Pipe vs Conetic Braid, Audio Transformer Enclosure, Computer Monitor EMI

- ✦ Just to provide "a feel" for practical applications, consider three "real life" EMC case studies.
  1.The following measurement results were obtained using the Helmholtz coils (see Chapter 6). A one-sixteenth inch *iron* pipe provided around 20 to 25 dB shielding effectiveness in the frequency range from 50 Hz to 5 kHz. By comparison, an RG59 line, shielded by relatively thin but high magnetic permeability single- and double-layer *conetic* braids, yielded approximately 14 to 20 dB and 16 to 23 dB, respectively. As a rule, it is not simple to obtain large electromagnetic shielding effectiveness in magnetostatic fields! More details can be found in original literature (e.g., see Refs. [3.14, 4.6]).

  2.This is an EMC design case study, when a set of output audio transformers, mounted on the same single-layer PCB, would "crosstalk with each other". First, the transformes were shielded by "donning" on them high-magnetic permeability caps made from mumetal sheets - with almost no effect on crosstalk! Then, instead of the caps, the transformer were completetly enclosed in boxes, made of the same material. Now, the crosstalk was reduced by 8-30 dB.
  An illustration in Fig. 5.7 underscores the situation: an open shielding enclosure presents high resistance to the magnetic flux. The resulting "stray" magnetic field lines of force are the source of crosstak. Of course, the same is true for emissions and immunity cases.
  While the physical picture is pretty clear (e.g., compare the two situations in Fig. 5.7), theoretical analysis of this problem is quite complex and can be done only for a limited number of simplified geometries [P.1]. Using modern computers, the numerical methods can be applied to resolve numerous practical problems, as a rule with the required accuracy, but then, of course, the difficulties of general analysis become the trade off.
  3. Make an experiment: place two computer monitors side-to-side at the same desk. Chances are, you may notice the distortion of the screen image in one or both of them. Especially with the older monitor designs, these distortions could be quite severe, and often accompanied with

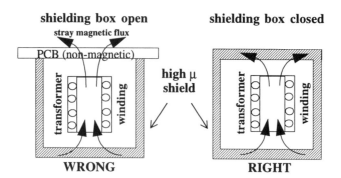

FIGURE 5.7 Transformer winding within open and closed magnetic shield

the frame and line synchronization glitches. Such interference is of course caused by the magnetic fields generated in the CRT deflection systems and circuits utilizing relatively high AC 3.currents. The CRT vertical scan circuitry generates ELF (extremely low frequencies) in the range 5 Hz - 2 kHz. The horizontal scan circuitry generates VLF (very low frequencies) in the range 2 kHz - 400 kHz.

Having several side-to-side computers at your desk is not exactly typical for a "regular" user, but in the preceding decade the problem of monitor emissions was given special attention because of the health concerns: effects on the eyes' fatigue and RF biological safety. While the evaluations of the health effects are still largely inconclusive, the implications for the industry are dramatic and expensive. One such implication bears directly with the problem at hand: lots of high magnetic permeability shielding around the CRT. Usually, the shielding is applied along with other EMI mitigation techniques, such as copper cancellation coils or cancellation bands (tapes) around the CRT deflection yokes and flyback transformers.

As a result, the ELF and VLF emissions from the CRT could be reduced down to 2.5 mG and 0.25 mG, respectively in these frequency ranges, to the "tune" of about 10% monitor price increase. For this "price" your monitor may also carry a label "Low Radiation". Mine does! ✦

5.2.4 Non-Conductive Statics Shielding

Non-Conductive Electrostatic Shields

Note, that while we required "at least some" conductivity of an electrostatic shield, there is no such requirement for conductivity of a magnetostatic shield. Just opposite: because of the skin effect, the larger is the shield conductivity, the smaller is magnetostatic shielding effect. We will address this problem in the next section. For now, let's just state that all the theory reviewed in the previous section treats magnetostatic shields as *non-conductive.* Thus similar considerations, as in the previous section, apply here. To arrive at spherical shell magnetostatic shielding performance formulas, the same method can be applied, as for cylindrical magnetostatic shielding reviewed in the previous section. The only difference is in the used coordinate system.

But what about non-conductive electrostatic shielding, is it possible? The answer is positive and easy to come by: that's when the electromagnetics theory shines in all its glory! Indeed, just by reviewing the equations (4.19) through (4.23) for our conditions (static fields, no currents or stored charges in dielectrics), we can see that there is really *no difference* between the electrostatic and magnetostatic shielding. For this reason, we can use the same equations (5.6a) through (5.11a) while substituting E for H and ε for μ. Thus by analogy, we can write the shielding effectiveness of an infinitely long dielectric shield in electrostatic field:

$$S_E = \frac{E^I}{E_0} = \frac{4\varepsilon_s r_e^2}{r_e^2(1+\varepsilon_s)^2 - r_i^2(1-\varepsilon_s)^2} \approx \frac{4}{\varepsilon_s} \frac{1}{1 - \frac{r_i^2}{r_e^2}} \quad (5.11b)$$

Spherical Static Shielding Shells

Expressions (5.11) can be used to evaluate, say, cable shielding. But what about 3D shielding, e.g., a computer enclosure? While accurate analysis of actual enclosures with quite complex geometry will, as a rule, require the use of numerical techniques, often an approximation of the enclosure by a spherical shell may be sufficient. The only difference between the evaluation of non-conductive cylinders and spherical shells is in the utilized coordinate system: cylindrical coordinate system in the first, and spherical coordinate system - in the second case. Otherwise, the same methodology can be applied to solve the equations for these two cases.

Thus, we place a spherical shield in a fixed, uniform magnetostatic field H_0, apply the same conditions as considered in the previous section, but express the Laplace equation in spherical coordinates (see Appendix) in terms of electric or scalar magnetic potentials. Again, we can use Fourier method to separate the variables and obtain three ordinary equations, each a function of only one variable: r, φ, and θ. The mathematics is somewhat more complicated, because now we have three variables, instead of only two variables in infinite cylindrical shield case. Anyway, the solution to this problem is well known (e.g., see [A.1], although in this source it is given as the end result of a problem to be solved - page 265):

$$S_{Hspher} = \frac{H^I}{H_0} = \left[1 - \frac{1 - \left(\frac{r_i}{r_e}\right)^3}{\frac{(\mu_s+2)\cdot(2\mu_s+1)}{2(\mu_s-1)^2} - \left(\frac{r_i}{r_e}\right)^3}\right] \quad (5.11c)$$

Similarly to equations (5.11, a and b), we can write the electrostatic shielding effectiveness of a dielectric non-conducting sphere, by substituting in (5.11c) H and μ for E and ε.

Note, the *cubic* dependence of the internal-to-external sphere radii, in contrast to the square *dependence in the cylindrical shield case.*

◆ For the sake of clarity, we have significantly simplified the shielding effectiveness formulas. An accurate solution, even in our idealised cases of infinitely long cylindrical shells and spherical shells, will be formulated in terms of the Bessel and Legendre functions — see the referenced sources.◆

5.2.5 Complexities and Specifics of Static Shield Design

Several concluding remarks, important for static shield design.

As we have just seen, *any closed conductive* enclosure (no matter how "thin" or "resistive") without apertures can provide "ideal" electrostatic shielding. Of course, practical enclosures will often contain apertures. Note, that since we deal with statics, *any* aperture can be considered *"small"*. In chapters 3 and 4, we addressed the issue of electrostatic field penetration through such apertures. It is also often beneficial to apply the circuit theory concepts to this problem. Thus, the capacitive coupling through an aperture can often be determined analytically and/or measured, especially for simple surface configurations. For complex geometries, numerical techniques can be applied.

The picture is much more complicated for magnetostatic shielding. While Eqs. (5.11) are simple, the magnetostatic shielding problem is not! When designing magnetic shielding, four factors should be considered: signal characteristics, shielding material properties, shield design, and shielding effectiveness evaluations. Which make the design and evaluation of magnetostatic shielding a "tricky" business, sometimes more an art than a skill. There is no need to delve into the details here, since abundant sources of information are available: from books to papers and articles to manufacturer catalogs and application notes. Some of the references can be found in the bibliography to this book (e.g., see 3.14, 4.6]. For these reasons, we will limit the following discussion to "only" half-dozen important considerations:

- There exists a large choice of magnetic shielding materials with magnetic permeabilities up to 400000 and in wide variety of brand names and price ranges. From "plumbing pipe" steel to sophisticated magnetic alloys - they come in all sorts of rigid and flexible sheets, tapes, boxes, cylinders, and any "fancy" shapes, called for by applications.
- High magnetic permeability materials are nonlinear; with the change of the field intensity, the magnetic permeability varies, reaching saturation at certain induction values. Unfortunately, the general trend is the μ reduction with magnetic field intensity increase. As a rule, the higher is magnetic permeability, the lower is the saturating field intensity. Obviously, this leaves "lots of room" for design choices, tradeoffs between the material permeability and the shield thickness (or multilayer applications), economical research, and optimization. Theoretical (analytical or numerical) evaluation of non-linear shields is pretty cumbersome and is usually based on iterative methods .
- The shielding effectiveness is also affected by several parameters that are specific to magnetic materials, including hysteresis, magnetostriction, and core loss. All these factors must be accounted for not only in shield design but also in shield performance measurements (e.g., in selecting the amplitude of the test field).

- Material properties vary broadly with respect to fabrication and handling conditions, temperature, flexing, and impact shock. As a rule, the higher the magnetic permeability of the material, the larger will be the shielding effectiveness instability, with experimental variances observed as large as tens of decibels.
- You have probably noticed that Eqs. (5.11) do not contain the frequency—and rightfully so, for magnetostatics. So, strictly speaking, it is applicable only at dc or low enough frequencies where equivalent penetration depth $\delta = \sqrt{2/(\omega\mu\sigma)}$ is large {see Eq. (A.58)}. But because of high magnetic permeability, the skin effect becomes significant at very low frequencies, leading to a reduction in the effective thickness of the shield. This, in turn, leads to deterioration of the magnetostatic shielding effectiveness. However, the good news is that with the frequency rise, other shielding mechanisms enter the picture, based on the quasi-stationary regime.

◆ For example, Kaden [P.2] gives the following expression for the effective value μ_{ef} of magnetic permeability of a metallic plate immersed in magnetic field:

$$\frac{\mu_{ef}}{\mu} \approx \begin{cases} 1 & \text{for } (t < 2\delta) \\ (1-j) \cdot \frac{\delta}{t} & \text{for } (t > 2\delta) \end{cases} \quad (5.12)$$

where t is the plate thickness and δ - the penetration depth (see chapter 3). Easy to see, that with the raise of frequency, the effective magnetic permeability quickly diminishes at the rate of the skin depth variation. All in all, the expressions in chapter 3 of this book permit to evaluate the shielding performance of shields with high magnetic permeability in the large frequency range. ◆

- The shielding enclosure shape, seams, gaps, apertures - and their arrangement with regard to the shielded object - may have a dramatic effect opn the shield performance. For example, the calculations indicate that shields with *smooth* edges perform better than shields with *sharp* edges at the surface. Thus, given the same crossection, a cylindrical shield is better than a prism shape.

- Multi-layer magnetostatic shielding holds a surprize. Indeed, after all that was said, we tend to think that the thicker is the shield, the better is shielding effectiveness. And this is true. However, let us compare two spherical shell shields of the *same thickness t:* one single-layer t_m, another - 3-layer with high magnetic permeability layer thicknesses $t_{1m} = t_{2m} = t_{3m} = t_m/6$, and non-magnetic dielectric (e.g., air) gaps between the magnetic layers $t_{1d} = t_{2d} = t_{3d} = t_m/6$. Which one would you expect to perform better?
  The numerical modeling performed in ref. [4.6] shows, that in the second case shielding performance is much better, and improves with the increase of the magnetic permeability. For instance, with $\mu = 1000$, there is about 60 dB to be gained. Also, the larger is the number of the layers, the better performance can achieved. This looks quite strange: we *reduce* the amount of magnetic material in the shield

(which makes the shield also lighter and, as a rule, cheaper), and *gain* in shielding performance! Why does this happen? The "secret" is in the variations of the incident angle and density of magnetic lines of force entering the shielding enclosure. It appears, that the magnetic lines of force deflect much stronger at the inner surface of the shield (from magnetic to non-magnetic media), than at the outer surface (from non-magnetic to magnetic media). Incidentally, the ref.[4.6] contains excellent graphical illustrations of these effects.

Looking at the previous phenomenon otherwise, to gain additional understanding it is now about time to "revive" the old notion of the energy *reflection* at the boundary between two media: as you see, this old theory may not always be wrong!

5.3 SHIELDING FOR CROSSTALK PROTECTION: FROM MILS TO MILES

5.3.1 "Compensation" Shielding from Low-Frequency Crosstalk

Consider a low resistance shield. This can one-, two-, or three-dimensional shield, respectively, just a linear conductor or a ground plane running parallel to a source or victim line or placed between the interfering lines, or an enclosure containing them. The signal frequency is low enough and the propagation regime can be qualified as quasi-stationary. In the quasi-stationary regime, the skin depth in the signal-carrying lines and the shield is much larger than the conductor thickness, so that current distribution in all conductors is similar to DC. The transfer impedance of such a shield is equal to its DC resistance. Thus, the shielding properties of such a shield are determined only by the coupling between it and the source and victim lines, respectively. In quasi-stationary regime, the electric and magnetic fields in dielectrics are reactive transversal. The electric and magnetic coupling coefficients may be considered independent on each other and addressed separately.

Consider parallel shield in quasi-stationary regime. Such a shield can be open or closed. The source is coupled to both the victim circuit and to the shield, with the shield, in turn, producing charges and currents in the victim line, which compensate the effect of direct coupling. In a quasi-stationary (as well as stationary) electric field, the coupling between the source / receiver conductors and the shield can be modeled with capacitors (see Fig. 5.8). Consider first a case when the shield is not grounded (both switches across C_{sg} are open). After V_1 is applied and the process is stabilized, then, neglecting the shunting effect of C_{sg}, the potential V_1 is applied to the load R_2 at the susceptible circuit via capacitors C_{1s} and C_{s2} connected in series. This is, of course, the same capacitance between these two conductors that also would have been exhibited in the absence of the shield (remember the behavior of ungrounded shield in Fig. 5.4 a?). However, when the shield is grounded (a switch is closed), the potential V_1 is applied to the ground directly via C_{1s}, bypassing the shorted series connection C_{s2} R_2. The respective values of the shielding coefficient

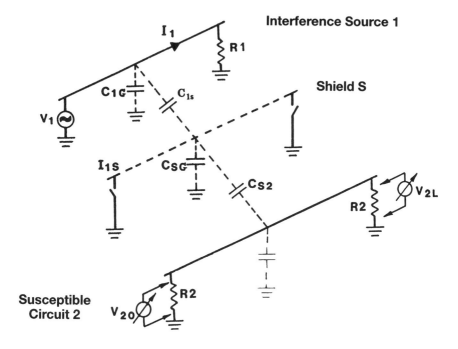

FIGURE 5.8 Capacitive coupling between shielded circuits

are 1 and 0. An important conclusion follows: a grounded shield completely protects the line circuits from electric field coupling. Food for thought: does it make a difference if both switches are closed?

Now, suppose the shield is not a complete enclosure. Such a shield can be formed by just a conductor or a finite size ground plane placed between the "crosstalking" lines, or it may be even a complete enclosure but with gaps and holes. Then, even if the shield is grounded, the electric field penetrates into the shielded space, resulting in electrostatic (a.k.a. capacitive) coupling. Obviously, the larger the gaps and holes, the greater this coupling will be. It may be worthwhile now to revisit the concepts of the transfer admittance developed in Chapter 3. The amount of electric coupling depends on the geometry and size of the shield imperfections. We have addressed some of such problems in chapter 3, when discussing the effects of the apertures, gaps, and seams on the shield transfer function. Generally, the solution to such problems isn't simple (e.g., see [P1]).

342 Electromagnetic Shielding Handbook

With regard to the magnetic quasi-stationary field coupling between the shielded circuits, we have to operate with a transverse magnetic field per Eq. (4.34). The source and receiver conductors and the shield are coupled via mutual inductances M_s (see Fig. 5.9).

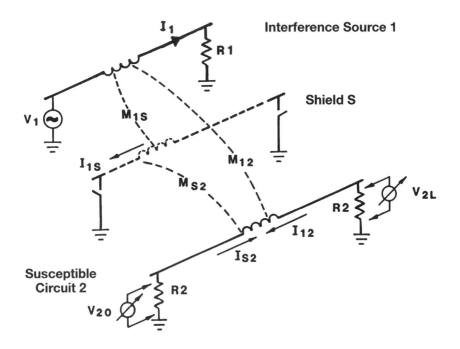

FIGURE 5.9 Inductive coupling between shielded circuits

When the current is flowing in circuit 1, electromotive forces are induced in both the shield and the disturbed circuit. In the latter, the induced electromotive force results in the EMI current I_{12} through the circuit loads at both ends ("bad news"). However, the effects in the shield depend on its terminations. Consider first the case when the shield is not grounded (both switches are open). In this case, no current can flow through the shield, and the shield is useless. (Don't forget, we are discussing an ideal situation, where the only grounding points are at the end of the shield. As a rule, at higher frequencies, this is not the case.) If we close one of the switches, the circuit is still open. Only when we close both switches at the line ends, is the circuit is closed, and the current I_{1S} can flow in the shield. In turn, this current induces an electromotive force in the disturbed circuit, resulting in the current I_{S2}, which is directed opposite to the current I_{12} and compensates for it.

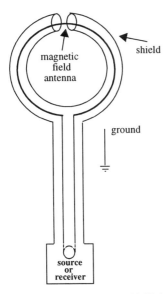

FIGURE 5.10 Low frequency shielded magnetic field antenna

◆ This difference in grounding requirements for electric and magnetic fields has permitted us to create effective designs for shielded magnetic loop antennas and magnetic field probes. For example see Fig. 5.10, By making the loop antenna shield from two isolated parts, and grounding each of them separately, the shield effectively suppresses the antenna sensitivity to electric fields without affecting its sensitivity to magnetic fields.

Moreover, as we will see later in this chapter, the magnetic shielding effect should be credited not to the grounding per se, but to creation of the closed current loop by providing proper current return. It just so happens in situation presented in Fig. 5.9, that by grounding the shield at both ends results in closing the shield current loop. Therefore, for the antenna shield in Fig. 5.10 to perform properly in the higher frequencies, the grounding may be not necessary at all. At these higher frequencies the main function of the loop antenna shield, is to provide for the antenna symmetry and protect it from common mode interference. More details on this subject can be found in literature, e.g., see Ref. [4.3]. ◆

How well does the shield work for magnetic coupling? Unfortunately, not as well, as it was in the case with the electric field coupling. For example, a detailed analysis of the involved phenomena is given in Ref. [4.1]. Consider magnetic coupling between two circuits in the presence of the shield. The line length is L. The voltage applied to the source line is U_1. The loss impedances of the circuits and shield are, respectively, z_{11}, z_{22}, and z_{ss}; the coupling impedances are z_{12}, z_{1s}, and z_{2s}; and the load impedances are z_{c1}, z_{c2}, and z_{cs}. The loads are assumed to be identical at both ends of each line. The currents in the source, victim, and shield are, respectively, I_1, I_2, and I_s. For the coupling between the circuits 12, 1s, and 2s, a simple set of three mesh equations can be developed:

$$(z_{11}L + 2Z_{c1})I_1 + z_{12}LI_2 + z_{1s}LI_s = U_1$$

$$z_{21}LI_1 + (z_{22}L + 2Z_{c2})I_2 + z_{2s}LI_s = 0$$

$$z_{s1}LI_1 + z_{s2}LI_{2(1)} + (z_{ss}L + 2Z_{cs})I_{2s} = 0 \qquad (5.13)$$

The longitudinal electromotive force, emf_2, in the victim line is composed of the sum of the electromotive forces induced in this line by the source (circuit 1) and the shield (circuits):

$$\text{emf}_2 = \text{emf}_{21} + \text{emf}_{2s} = z_{21}LI_1 + z_{2s}LI_{1s} = L\left[z_{21}L + \frac{z_{s2}z_{21}L}{(z_{ss}L + 2Z_{cs})}\right]I_1$$

(5.14)

The shielding coefficient is

$$S = \frac{\text{emf}_2}{\text{emf}_1} = \left|1 - \frac{z_{s2}z_{s1}}{z_{ss}z_{21}}\right|$$

(5.15)

The presented approach can be used to model and analyze important specific applications. As an example, Table 5.1 presents a summary of coupling models between shielded and nonshielded wires, taken from Ref. [4.9].

To gain additional understanding, let us review the experimental data shown in Fig. 5.11. The near-end (NEXT) and far-end (FEXT) crosstalk attenuation was measured between line pairs, either shielded and nonshielded, in otherwise identical cables. Consider first the nonshielded cables. As predicted by theory, the FEXT attenuation exceeds (that is, is better than) that at the near end by about 25 dB. As the frequency rises, both the NEXT and FEXT experience steady degradation.

For the shielded cables, the curves for NEXT and FEXT are almost identical—again, as predicted by our theory. However, while the crosstalk attenuation is pretty good, it is not as large as could be expected if both electric and magnetic coupling were compensated by the shield. Is it the electric or magnetic coupling that degrades the shield performance? Let us compare the FEXT attenuation of a nonshielded line to the FEXT of the shielded line. Aren't you surprised that the nonshielded line performs better than the shielded one? This is a real "eye-opener." Indeed, if we assume that the shield almost completely "stops" the electric coupling and "lets through" a significant enough fraction of magnetic coupling, then the difference between the electric and magnetic coupling at the far end of the line may be smaller in a nonshielded line (although, it depends on just how much magnetic coupling was reduced by the shield)! The last curve in Fig. 5.11 was FEXT measured with the shield connected to the ground only with a pigtail, while the other curves were obtained with the shield 360° terminated. At this stage of your reading, the result shouldn't need any comments.

5.3.2 Crosstalk between Shielded Lines in a Wide Frequency Band

This subject is not at all new and has been addressed in numerous works, some of which are enumerated below.

1. Crosstalk between two isolated coaxial pairs for arbitrary load of the tertiary, and in the absence of other conductors in the line was addressed as early as in the 1930s by Kaden [5.1] and Schelkunoff and Odarenko [5.2].

Shielding Effectiveness for EMI Protection 345

Table 5.1 SUMMARY: OPEN-WIRE AND SHIELDED WIRE INDUCED INTERFERENCE

INTERFERENCE CASE	SCHEMATIC	AC EQUIVALENT CIRCUIT	AC EQUIVALENT CIRCUIT OF VICTIM LOOP	OUTPUT VOLTAGE
(1) OPEN WIRE-TO-OPEN WIRE			$e_2 = j\omega M i_1$	AC RMS: $V_d = 2\pi f M_i \frac{R_d}{R_c+R_d} e_{L2}$ PEAK TRANSIENT: $V_{dMAX} = \frac{MI_1}{\tau} \frac{R_d}{R_c+R_d} A_{L2}$
(2) SHIELDED WIRE-TO-OPEN WIRE			$e_2 = j\omega M i_1 \cdot \frac{1}{1+j\frac{\omega L_{S1}}{R_{S1}}}$	AC RMS: $V_d = 2\pi f M_i \frac{R_d}{R_c+R_d} e_{S1} e_{L2}$ PEAK TRANSIENT: $V_{dMAX} = \frac{MI_1}{\tau} \frac{R_d}{R_c+R_d} A_{S1}$
(3) OPEN WIRE TO-SHIELDED WIRE			$e_2 = j\omega M i_1 \cdot \frac{1}{1+j\frac{\omega L_{S2}}{R_{S2}}}$	AC RMS: $V_d = 2\pi f M_i \frac{R_d}{R_c+R_d} e_{S2} e_{C2}$ PEAK TRANSIENT: $V_{dMAX} = \frac{MI_1}{\tau} \frac{R_d}{R_c+R_d} A_{S2}$
(4) SHIELDED WIRE-TO-SHIELDED WIRE			$e_2 = j\omega M i_1 \cdot \frac{1}{1+j\frac{\omega L_{S1}}{R_{S1}}} \cdot \frac{1}{1+j\frac{\omega L_{S2}}{R_{S2}}}$	AC RMS: $V_d = 2\pi f M_i \frac{R_d}{R_c+R_d} e_{S1} e_{S2} e_{C2}$ PEAK TRANSIENT: $V_{dMAX} = \frac{MI_1}{\tau} \frac{R_d}{R_c+R_d} A_{SS}$

Crosstalk between any number of coaxial pairs with outer conductors in contact was evaluated in the 1960s by this author [3.12, 5.3].
2. The effect of the conductors of the balanced circuits present in the line and of the overall line shield on crosstalk between two coaxial pairs for matched and short-circuit loads of the tertiaries was investigated by Schelkunoff and Odarenko [5.2] and Gould [5.4].
3. The most general study of crosstalk between the coaxial cables of arbitrary design can be found in References [3.12, 5.5, and 5.6], wherein reasonably similar results were obtained.

It was confirmed that the crosstalk between the shielded circuits in general, and the coaxial cables in particular, depends on the transfer impedance of the coupled shields and on the parameters of the third circuits (i.e., "coupling" characteristic). Since we have already dedicated the whole Chapter 3 to the study of transfer impedance, we will concentrate now on the coupling aspects, that is, the properties of the tertiaries. The EMI current generated by the longitudinal emf in a tertiary depends on the tertiary parameters: characteristic impedance and propagation constant (the attenuation and phase constant), and loads at the ends of it. For instance, if the outer conductors or shields are in contact at the ends of the line, the third circuit loads can be considered as "shorts," and if there is no contact they can be considered as "opens" (or rather, as a small capacitive load). When the considered tertiary element is far (electrically) from the line ends and the tertiary attenuation is high, the element can be viewed as loaded at the tertiary characteristic impedance.

In the following text, we will illustrate the basic considerations in calculating crosstalk between shielded cables by developing a relatively general mathematical model of crosstalk between coaxial lines. Since the model will account (in quite general form) for important line and line design factors, specific applications may be described by substituting the desired values of the variables in the model.

✦ Consider a system consisting of two coupled coaxial pairs ("P1" and "P2") with galvanically coupled outer conductors in the presence of other conductors and dielectrics. For example, a coaxial or shielded line may consist of a number of coaxial pairs (or shielded conductors) with grounded outer conductors (shields) in galvanic contact, other shielded entities which are isolated from the first group, separate isolated conductors forming balanced and nonbalanced open circuits, overall line shields, and so forth. Let P1 be the source and P2 be the victim. We will name "T3" the tertiary formed by the outer conductors of the pairs "1" and "2." With regard to P1 and P2, we will divide all other conductors in the line into two groups: the first group is in galvanic contact with the outer conductors (shields) of the "crosstalking" P1 and P2, and the second group is all conductors isolated from the outer conductors of P1 and P2. Obviously, the first group directly shunts the T3, and for this reason it can be considered as a part of T3. The elements of the second group form with the conductors of T3 a kind of "super" tertiary, T4.

First, consider how the longitudinal voltage generated in the shield is applied to the line loads (external or internal). Figure 5.12 illustrates the situation.

Shielding Effectiveness for EMI Protection 347

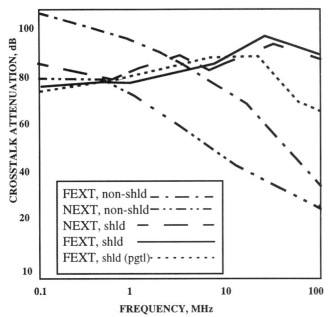

FIGURE 5.11 Crosstalk attenuation: shielded vs non-shielded
200 ft of twisted pair, NEXT and FEXT

The corresponding crosstalk model in a line is represented by an equivalent circuit of a small ("infinitely small") line element of length dx, as presented in Fig. 5.12.

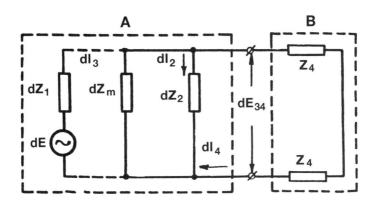

FIGURE 5.12 Model of crosstalk between coaxial or shielded (common mode) lines

Here dE is the emf at the outer surface of the outer conductor of the source. Let the line P1 has characteristic impedance Z_{c1}, propagation constant $\gamma_1 = \alpha_1 + j\beta_1$ (for convenience of notation, we will use the natural logarithmic measure of line losses α in nepers (1 nep = 10 ln $|U_1/U_2|$ = 8.67 dB), and transfer impedance Z_{t1}. The voltage U_{01}, applied to the near end of P1, generates a current in this line (with internal return path), which at the distance x from the beginning of the line is:

$$I_1(x) = \frac{U_{10}}{Z_{c1}} e^{-\gamma_1 x} \qquad (5.16)$$

Due to this current, the voltage drop at the external surface (i.e., for the external return path—see Chapter 3) of the line P1 element dx at a distance x from the beginning of the line is:

$$dE(x) = U_{10} \frac{Z_{t1}}{Z_{c1}} e^{-\gamma_1 x} dx \qquad (5.17)$$

Per the Thevenin theorem (see Chapter 2), voltage drop $dE(x)$ can be considered as an electromotive force applied to the combination of T3 and T4, that is, in the circuit formed between the external surfaces of the outer conductors of P1 and P2, shunted by the other galvanically connected elements of T3, and "looking" at the input impedances Z_4 in the direction of both ends of the supertertiary T4. (For simplicity, the input impedances in the direction of the near and the far ends of T4 are assumed to be identical, but this is not necessarily so.)

In T3, the $dE(x)$ generates a current I_3 and a current I_4 in T4. Since in T3 the current flows in an external circuit formed by the outer conductors of the P1 and P2, the resistance it encounters is the external loss impedance of these outer conductors (see Chapter 3). We will designate the loss impedances of P1 and P2 as $dz_1 = Z_{s1}dx$ and $dz_2 = Z_{s2}dx$. In this circuit, the resistance $dz_2 = Z_{s2}dx$ is shunted by the resistance of other galvanically connected elements $dz_m = Z_{sm}dx$.

For theoretical analysis, it is convenient to separate the currents I_3 and I_4. This can be done by applying the compensation theorem. (Check with the "survival kit" in Appendix). Thus, with regard to the two-port network B (corresponding to T4), the two-port network A (corresponding to T3) can be considered as an active element with the emf dE_{34}, equal to the voltage at the network T3 terminals in the open-circuit regime (that is, in the absence of T4). Thus, assuming that the T4 is disconnected, and using Eq. (5.17), we can find the current I_3 in T3 originated by the line element dx at distance x from the line beginning:

$$dI_3 = \frac{dE(x)}{(Z_{s1} + \eta Z_{s2})dx} = \frac{U_{10} Z_{t1} e^{-\gamma_1 x}}{Z_c (Z_{s1} + \eta Z_{s2})} \qquad (5.18)$$

Remember that parameters Z_s and Z_t are specified per line unit length (i.e., meter). The coefficient η in the denominator (5.18) accounts for the shunting effect of other conductors in the LINE with loss impedance Z_{sm}, which are in continuous contact with the outer conductors of P1 and P2. From Fig. 5.12,

$$\eta = \frac{Z_{sm}}{Z_{s2} + Z_{sm}} \qquad (5.19)$$

In fact, the assumption of the disconnected T4 not only has theoretical utility but also practical significance. Indeed, it corresponds to an important case of crosstalk between two shielded or coaxial circuits in "free space," i.e., in the absence of, or large enough distance from, other conductors which are isolated from the first two. For this reason, it is worthwhile to evaluate the EMI signals arriving to the near and far ends of the victim P2, in the absence of T4.

Because of the presence of the shunting elements in T3, the current dI_{23} flowing in the outer conductor (shield) P2 with external return path is only a fraction of the tertiary current dI_3:

Shielding Effectiveness for EMI Protection 349

$$dI_2 = \eta dI_3 = \frac{U_{10} Z_{t1} e^{-\gamma_1 x}}{Z_c\left(\frac{1}{\eta} Z_{s1} + Z_{s2}\right)} \tag{5.20}$$

At each element dx with internal return of the circuit P2, current dI_{23} develops voltage $dE_2 = dI_2 Z_{t2}$ dx:

$$dE_2(x) = \frac{U_{10} Z_{t2} Z_{t1} e^{-\gamma_1 x}}{Z_c\left(\frac{1}{\eta} Z_{s1} + Z_{s2}\right)} dx \tag{5.21}$$

Inside P2, voltage dE_2 can be considered as longitudinal emf which is applied to the input impedances toward the near and far ends of P2, just as it was in T3. However, unlike in the tertiary, in this case line P2 generally can be considered matched, loaded on its characteristic impedance Z_{c2}, and identical at both the near and far ends. Therefore, both near and far end "looking" voltages dU_{20x} and dU_{21x} are equal:

$$dU_{20x} = dU_{21x} = \frac{dE_2(x)}{2} = \frac{U_{10} Z_{t2} Z_{t1} e^{-\gamma_1 x}}{2Z_c\left(\frac{1}{\eta} Z_{s1} + Z_{s2}\right)} dx \tag{5.22}$$

Obviously, before arriving at the near and far ends of P2, the signal from the element dx will change by $-\gamma x$ and $-\gamma(l-x)$, respectively:

$$dU_{20} = dU_{20x} e^{-\gamma_2 x} = \frac{U_{10} Z_{t2} Z_{t1} e^{-(\gamma_1 + \gamma_2)x}}{2Z_c\left(\frac{1}{\eta} Z_{s1} + Z_{s2}\right)} dx$$

$$dU_{21} = dU_{20x} e^{-\gamma_2(l-x)} = \frac{U_{10} Z_{t2} Z_{t1} e^{-(\gamma_1 - \gamma_2)x} e^{-\gamma_2 l}}{2Z_c\left(\frac{1}{\eta} Z_{s1} + Z_{s2}\right)} dx \tag{5.23}$$

By integrating the currents from all the elements dx, we obtain the near- and far-end crosstalk voltage:

$$U_{20} = \int_0^1 dU_{20} = \frac{U_{10} Z_{t2} Z_{t1}}{2Z_c\left(\frac{1}{\eta}Z_{s1} + Z_{s2}\right)} \int_0^1 e^{-(\gamma_1+\gamma_2)x} dx$$

$$= \frac{U_{10} Z_{t2} Z_{t1}}{2(\gamma_1+\gamma_2)Z_c\left(\frac{1}{\eta}Z_{s1} + Z_{s2}\right)} \left(1 - e^{-(\gamma_1+\gamma_2)l}\right)$$

(5.24)

$$U_{21} = \int_0^1 dU_{21} = \frac{U_{10} Z_{t2} Z_{t1} e^{-\gamma_2 l}}{2Z_c\left(\frac{1}{\eta}Z_{s1} + Z_{s2}\right)} \int_0^1 e^{-(\gamma_1-\gamma_2)x} dx$$

$$= \frac{U_{10} Z_{t2} Z_{t1}}{2(\gamma_1-\gamma_2)Z_c\left(\frac{1}{\eta}Z_{s1} + Z_{s2}\right)} \left(1 - e^{-(\gamma_1-\gamma_2)l}\right)$$

(5.25)

Dividing U_{20} and U_{21} by the right-hand parts of Eqs. (5.24) and (5.25), respectively, we obtain crosstalk; or taking a logarithm of the inverse crosstalk, we obtain crosstalk attenuation. For instance, assuming crosstalk between identical coaxial pairs or shielded common-mode circuits ($\gamma_1 = \gamma_2 = \gamma$, $Z_{t1} = Z_{t2} = Z_t$, $Z_{s1} = Z_{s2} = Z_s$), and in the absence of other conductors ($\eta = 1$), we obtain crosstalk attenuation at the near and far ends, respectively of:

$$A_o = \ln\left|\frac{8\gamma Z_c Z_s}{Z_t^2 \left(1 - e^{-2\gamma_2 l}\right)}\right|; \quad A_1 = \ln\left|\frac{4Z_c Z_s}{Z_t^2 l}\right| \quad (5.26)$$

Equations (5.26) are quite familiar from the theory of crosstalk between coaxial cables. Also, when the line length and losses are electrically small, Eqs. (5.26) yield:

$$A_o = A_1 = \ln\left|\frac{4Z_c Z_s}{Z_t^2 l}\right| \quad (5.27)$$

Equations (5.26) and (5.27) are simple and easy to use. But remember, that to ensure this convenience, we imposed a quite restrictive limitation: we assumed that the shields of the interfering circuits are in continuous contact. Owing to this, the only characteristic of the tertiary T3 we needed was Z_s, which could be viewed as a characteristic of P1 or P2 as well. However, in the general case of arbitrary shield arrangements, they may or not be in touch. There, the EMI signal will propagate in the T3, "regulated" by its parameters γ_3 and Z_{c3}, and also by the "loads" Z_{03} and Z_{13} at the near and far ends of the tertiary.

To determine the currents in the disturbed line, we can use the same approach as before. However, now the integrals {Eq. (5.24) and (5.25)} become double integrals because the current at each element in T3 is determined by contributions of all the elements of P1, and the current at each element in P2 is determined by contributions of all the elements of T3. In the process of these integrations, the T3 terminations at the near and far end affect the input impedances that the currents in T3 "face" in respective directions [see Eqs. (A.68) through (A.71)]. The respective derivations can be found in the literature (e.g., see Ref. [5.4]). For reference purposes, we will present the final result of these derivations for a somewhat simplified case of a line length l, with parameters $\gamma_1 = \gamma_2 = \gamma$, $Z_{c1} = Z_{c2} = Z_c$, $Z_{03} = Z_{13} = Z_3 \neq Z_{c3}$, and $\kappa = Z_3/Z_{c3} = \tanh r$:

$$U_{20} = \frac{U_{10} Z_t^2}{2 Z_c Z_{c3} (\gamma_3^2 - \gamma^2)} T_{03}; \quad U_{21} = \frac{U_{10} Z_t^2}{2 Z_c Z_{c3} (\gamma_3^2 - \gamma^2)} e^{-2\gamma l} T_{13} \tag{5.28}$$

where

$$T_{03} = \frac{\gamma_3}{2\gamma}(1 - e^{-2\gamma l}) + \frac{(\gamma_3^2 \kappa^2 - \gamma^2)\left[2 e^{-\gamma l} - (1 + e^{-2\gamma l})\cosh \gamma_3 l\right]}{(\gamma_3^2 - \gamma^2)\left[2\kappa \cosh \gamma_3 l + (1 + \kappa^2)\sinh \gamma_3 l\right]}$$

$$+ \frac{-\left[\kappa(\gamma_3^2 - \gamma^2)(1 + e^{-2\gamma l}) + \gamma \gamma_3 (1 - \kappa^2)(1 - e^{-2\gamma l})\right]\sinh \gamma_3 l}{(\gamma_3^2 - \gamma^2)\left[2\kappa \cosh \gamma_3 l + (1 + \kappa^2)\sinh \gamma_3 l\right]}$$

$$T_{13} = \gamma_3 l - \frac{(\gamma_3^2 + \gamma^2)[(1 + \kappa^2) e^{\gamma_3 l} + (1 - \kappa^2) e^{-\gamma_3 l}]}{(\gamma_3^2 - \gamma^2)[(1 + \kappa^2) e^{\gamma_3 l} - (1 - \kappa^2) e^{-\gamma_3 l}]}$$

$$+ \frac{-2(1 - \kappa^2)(\gamma_3^2 - \gamma^2)\cosh \gamma_3 l - 2(\gamma_3 \kappa + \gamma)^2 e^{\gamma l} - 2(\gamma_3 \kappa - \gamma)^2 e^{-\gamma l}}{(\gamma_3^2 - \gamma^2)[(1 + \kappa^2) e^{\gamma_3 l} - (1 - \kappa^2) e^{-\gamma_3 l}]}$$

By substituting in (5.28) the respective values of parameters, they can be reduced to (5.26) and (5.27). We'll return to these equations when discussing the grounding effects on shielding.

Now, returning to the model in Fig. 5.12, we can use (5.28) to incorporate the super-tertiary T4. Thus we obtain the near- and far-end crosstalk by integrating EMI currents from each P1 element dx twice: in the super-tertiary, and in the victim. Leaving the derivations as an exercise (they also can be found in the quoted references), we will present the final equations:

$$A_0 = -\ln \left| \frac{Z_{t2} Z_{t1}}{4 Z_c} \left\{ \left[\frac{1}{Z_{s1} + \eta Z_{s2}} - \frac{1}{8 Z_{44}} \frac{\gamma_4^2 (\gamma_4^2 + \gamma^2)}{(\gamma_4^2 - \gamma^2)^2} \right] \frac{1 - e^{-2\gamma l}}{2\gamma} - \frac{\gamma_4^2}{8 Z_{44}} \frac{\gamma_4^2}{(\gamma_4^2 - \gamma^2)^2} \frac{2 e^{-2\gamma l} - (1 + e^{-2\gamma l})(\cosh \gamma_4 l)}{\sinh \gamma_4 l} \right\} \right| \tag{5.29}$$

$$A_1 = -\ln \left| \frac{Z_{t2} Z_{t1}}{4 Z_c} \left\{ \left[\frac{1}{Z_{s1} + \eta Z_{s2}} - \frac{1}{8 Z_{44}} \frac{\gamma_4^2}{\gamma_4^2 - \gamma^2} \right] l + \frac{\gamma_4^2}{8 Z_{44}} \frac{4 \gamma_4^2}{(\gamma_4^2 - \gamma^2)^2} \frac{\sinh \frac{(\gamma_4 + \gamma)}{2} l \sinh \frac{(\gamma_4 - \gamma)}{2} l}{\sinh \gamma_4 l} \right\} \right| \tag{5.30}$$

In Eqs. (5.29) and (5.30), the parameters with indices "4" are those of the "super-tertiary." It was assumed that the super-tertiary is matched at the ends, so $Z_{44} = \gamma_4 Z_{c4}$. In practical terms, it more expedient to measure Z_{c4} and γ_4. As we mentioned, such "matched" conditions can be assumed if the super-tertiary is electrically long and/or its losses are relatively large. Formulas for the general case, and their analysis for arbitrary loads at the ends of the super-tertiary, are given in Refs. [3.12, 5.6]. It can be shown that when $\gamma_4 = Z_{s1} = \infty$, Eqs. (5.29) and (5.30) transform into Eqs. (5.26) through (5.28).

Again, we remind you that the crosstalk values these expressions are in *nepers* - see comments to (5.16).✦

5.3.3 Shielding for Crosstalk Protection in Printed Circuit Boards and Integrated Circuits

Shielding Traces and Ground Planes

There are numerous applications shielding techniques to IC and PCB EMI protection, even if they are not always recognized as shielding problems. One typical application is parallel shielding of clock distribution lines. As we have discussed in chapter 1, a clock line presents one of the most EMI-troublesome circuit element.

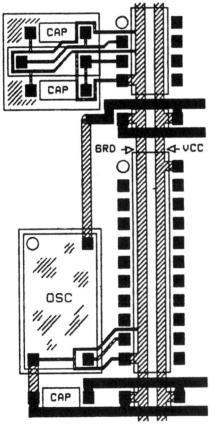

FIGURE 5.13 Shielding clock traces

To protect the clock trace on a PCB or backplane, it is often surrounded by shielding traces (see Fig. 5.13), which reduce the crosstalk (and radiation). Here, a parallel may be drawn with a coaxial cable which has slits in the external conductor. When the design incorporates a ground plane, the shielding traces are connected to the ground plane. In longer lines, such connection is done in a number of groundings points. (in coaxial cable analogy, this will be equivalent to reducing the lengths of the slits, or like substituting the slits with apertures).

Some important issues to keep in mind:

• *separation shielding mode* for interfering lines or traces located at the *opposite* sides of the ground plane

• *parallel open* shielding mode when shielding lines or traces are used along the direct path line or trace, or when the interfering lines or traces traces located at the *same* side of the ground plane

• relatively small length or area; of course, we are talking "*electrical* size" - so at higher frequencies and larger objects (say, backplanes) this may not the case

• often only *partial* mutual run of shielding and shielded conductors or finite size coverage of the area by the shield

Shielding Effectiveness for EMI Protection 353

- *not* necessarily *parallel* run of the crosstalking circuits, as well as the shield

In general, there are many similarities between the parallel shielding processes in PCBs or ICs vs flat cables with ground plane, as well as any non-shielded wire or conductor which runs in proximity to other metallic structures. We have already discussed and illustrated with experimental data the effects of an open configuration "parallel" shielding in cables (see section 4.4.7). For example in Fig. 4.17, the near end crosstalk in ribbon cables, 3 m long with ground plane, presents a typical case of two-dimensional open shielding, in comparison with non-shielded lines, and with closed cable shielding. The testing was done in time domain with the pulse rise times of 1 ns and larger. Examples of a one-dimensional open-configuration shielding (SG and GSG) crosstalk effectiveness are given in Fig. 1.39, in comparison with other types of non-shielded and shielded configurations. In both cases, the use of parallel shielding resulted in approximately 2-3 times crosstalk reduction (in linear units).

These results are equally applicable to the flat ("ribbon") cables and printed circuit board (PCB) traces. The previously reviewed theory of crosstalk between shielded lines and the presented formulas can be used to address the respective effects as well as the particular application specifics on the crosstalk performance, permits to understand them, as well as to derive appropriate mathematical expressions. The only thing to watch, is the "electrical geometry of scale", that is the "electrical size" of the considered systems, which is determined by the ratio of the geometrical dimensions to signal wavelength. Then our derivations are correct, no matter what are the actual geometrical dimensions and signal wavelength, taken separately (of course, within identified limits separating the crosstalk in reactive fields and radiation regimes).

Consider for example, how effective is parallel shielding at the integrated circuit (IC, or "chip") level. Fig. 5.14 presents the crosstalk measurement results between microstrip lines on a chip substrate (see Ref. [5.7]), powered by 50 ps pulses. Although the line length is in mm, for a 50 ps risetime, it is comparable to the wavelength. As shown, the insertion of shielded traces results in 2 to 5 times crosstalk reduction, similar to what we had in the PCB or flat cable cases. Another important factor is the trace proximity to the ground plane.

Non-Parallel Lines / Partial Shielding

So far, we have discussed crosstalk between parallel lines of infinite length. However, in the PCB and IC layout, this may not always be the case. Thus, in the adjacent PCB layers, or even in the same layer, the crosstalking lines may be placed at different angles with regard to each other. Other conductors, which may be placed between the crosstalking lines and play the role of parallel shields, may not run the whole length along these lines. Thus we have non-parallel and partial runs of shielding with regard to the protected lines.

Again, the general crosstalk theory can be applied, and the results are available in the literature. For example, when the lines are crossing at some angle, θ, and / or the shield covers only a length, L, of a total line length, l, then the shielding effectiveness (defined in Ref. [5.8] as the difference in crosstalk in this partially shielded line and in a nonshielded line intersecting at the same angle) can be found as:

354 Electromagnetic Shielding Handbook

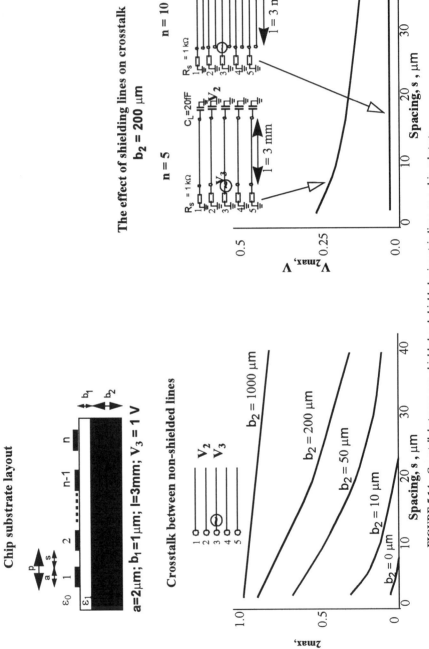

FIGURE 5.14 Crosstalk between non-shielded and shielded microstrip lines on a chip substrate

$$\zeta = -20\log\left|\frac{1 + (\eta - 1)\tan^{-1}\left(\frac{L}{2d}\sin\theta\right)}{\tan^{-1}\frac{1}{2d}\sin\theta}\right| \quad , \text{dB} \quad (5.31)$$

where d is the closest distance between the interfering lines, and η is the shielding coefficient of an infinitely long shield of the same design.

When $\theta = 0$,

$$\zeta = -20\log\left|1 + (\eta - 1)\frac{L}{l}\right| \quad (5.32)$$

These results explain one of the fundamental rules of the PCB layout: 90° line crossing in adjacent layers. They also explain *some* of the problems with running the microstrip close to the ground plane edges, e.g., see Fig. 4.19 (top configuration) and related text (however, for more details on the ground plane edge effects see later, section 5.4.4).

Eqs. (5.31) and (5.32) are applicable only under the assumption of constant current distribution along the lines and, also, we are addressing the differential-mode coupling in a transverse reactive field. Likewise, in Fig. 5.11, with a further increase of frequency and/or line length beyond those shown in graph, the losses in the line increase. This results in the reduced effect of the contributions to the NEXT from the line lengths located far enough from the near end of the line. On the other hand, the contributions to the FEXT from all elementary lengths of line remain almost constant. Figure 4.7 presents a good illustration of these effects. Also, a simplified description of the involved processes can be found, e.g., in Ref. [5.8].

For non-parallel lines, to appreciate the scope of crosstalk problems and to obtain further insight into the involved mechanisms, consider crosstalk between bi-level crossing interconnections, embedded in GaAs IC substrate. This problem was treated in time domain in [5.7] on the basis of general crosstalk theory, as well as using computer simulations based on approximations of inverse Laplace transformations. In supporting experiments, the crosstalk was measured with pulses ranging from several picoseconds to hundred picoseconds, the line lengths of several hundred microns, and the values of crosstalk were in order of several percents. While the reader is referred to the original text for details, we will emphasize here the relative *independence* of crosstalk on interconnection length, inter-level distance, and crossing angle.

Grounding Shielding Traces for Crosstalk Protection
Should or shouldn't the shielding traces (or conductors) be grounded for crosstalk protection? The positive answer to this question follows from the discussion in section 5.3.1. As we saw, it was sufficient to ground the shielding trace only in two points, the line ends, to protect from both, electric and magnetic coupling. This is "theory", and as we saw, it is correct when dealing with *closed* shields.

But what about open shields, e.g., such as parallel shielding traces, or even ground planes? As could be expected, the answer is not simple! To see the problem, consider first a microstrip line over a ground plane. This is a typical example of an *open* line.

The effect of the ground plane can be modeled by an image conductor, as shown in Fig. 4.20, with both the "real" conductor and its image forming a balanced line. Now, instead of our microstrip, we have a *balanced pair*. The configuration of the fields generated by this line is shown in Fig. 4.5 a (with the symmetry axis A representing the ground plane).

If another microstrip line is placed at the *same side* of the ground plane with the first, but *outside* of it and in the "action zone" of the first one, the two lines will be electrically and magnetically coupled. Because the fields generated by the conductor and its image are directed opposite *outside* of the area occupied by the line, the coupling is reduced. Of course, without the ground plane, the image does not exist and the coupling reduction would not occur — that's what parallel shielding is all about.

The electric and magnetic fluxes, which act as "carriers" of this coupling, occupy the space "all around" the source line. However, in the microstrip and striplines, when the ground plane is present, there is a tendency to attribute all the coupling flux only to fields that are generated in space *over* the crosstalking conductors. And indeed, if you compare the cross section area *over* the conductors (theoretically, infinity in a microstrip) with the space *under* the conductors (say, 10 mils to the ground plane) it seems that such an assumption might be right. However, one should keep in mind that the distance between crosstalking lines on the IC and, very often, PCB may be extremely small, so that field intensity is high since its fall off haven't yet been significant.

Consider now Fig. 5.15.

Fig 5.15 a, shows two coupled microstrip lines, a and b. The arrows can be interpreted as coupling EMI energy fluxes, or electric and magnetic coupling fields. In the latter case, the arrows may be substituted by respective field lines of force. If a circuit theory interpretation is applied, the arrows may be substituted by respective mutual capacitances and inductances. In any case, we see that the coupling occurs *above* (designated as "A" in the figure), *underneath* ("C"), and *in the plane* ("B") of the conductor strips.

In Fig.5.15 b, a *shielding microstrip* s is shown inserted between the lines a and b. The microstrip s is not connected to the ground plane. Checking with Figs. 5.8 and 5.9 and related text, confirms that, given the assumptions*, the coupling models in Fig. 5.15 a and 5.15 b are equivalent (that is, yield the same results) at least theoretically. Using our designations, we can write: B = D+E, while A and C are left unchanged (probably, "almost" unchanged).

In Fig. 5.15 c, a shielding microstrip s is *grounded* at one or two ends (the far end grounding trace is shown dotted). It follows from Figs. 5.8 and 5.9 and related text that this should "take care" of coupling.

---

*As you remember, the "main" assumption was that of the quasi-stationary fields. Also, at high enough frequencies, the shielding microstrip can be considered as "grounded" via its partial capacitances and inductances to the ground plane. For these reasons, the presented model should be always verified before application.

Shielding Effectiveness for EMI Protection 357

a) Coupled microstrips

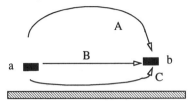

b) Shielding trace between coupled microstrips

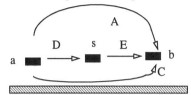

c) Shielding microstrip, 1- or 2-point grounding

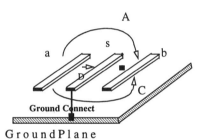

d) Shielding microstrip, multi-point grounding

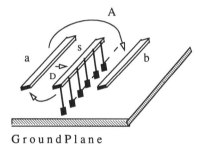

e) Shielding stripline, multi-point grounding

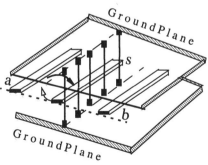

FIGURE 5.15 Coupling between microstrips and striplines with and without shielding trace

Unfortunately, in our case of *parallel open* shielding this is true only partially: as the arrows in Fig. 5.15 c show, only *"in-plane"* EMI flux E is shorted to ground, and, probably, fluxes A and C are somewhat reduced, but not eliminated. Why? Because even if the shielding microstrip is grounded, but *only* at the ends or at some large enough intervals, the "free space" above the strips, as well as "free space" in the "apertures" formed *between* the grounding interconnections *under* the strips, provide paths for coupling!

For experimental confirmation of this statement, revisit Fig. 1.40. The GS and GSG circuit configurations have grounded shielding conductors between the lines - typical example of open parallel shielding. Nevertheless, the crosstalk between the single-ended lines separated by these shielding conductors is large, much larger than the crosstalk between coaxial, or shielded balanced lines which represent closed separation shielding. Obviously, in coaxial and balanced shielded cables, which feature *closed one-dimensional* shields, all electrica and magnetic coupling paths between the source and victim lines can be bypassed.

This explains the EMI coupling flux above the microstrip, flux A in Fig. 5.15 c. But what about flux C, *under* the microstrip, that is between the strip and ground plane, when the latter is present? Consider experimental data, this time taken from available literature [5.9]. This particular PCB was specially designed to perform experimental investigation of crosstalk between microstrip lines in wireless applications. The PCB was double sided, with the back side completely plated with copper and the top side also plated with copper, except around the top traces, which were also copper microstrips. The microstrip lines were 83 mm long and 2.8 mm wide, with adjacent line separation 20 mm center to center, of which 14.7 mm was plated by copper.

The diagram in Fig. 5.16 presents the minimum, that is, the worse case, crosstalk attenuation between microstrip lines on a PCB, in the frequency band 3 MHz through 6 GHz (actually, the test was performed by measuring scattering parameter S_{12} — see Appendix). By selectively using shielding enclosures, two components of the coupling were separated and identified: air path coupling in free space over the microstrips and substrate coupling between the microstrips and ground plane.

As you can see, in this particular setup the substrate coupling is about 30 dB worse, than the airpath coupling. It follows from the diagram that coupling via the substrate path is the determining factor, when no shielding is present. Therefore it can be expected that when both air and substrate coupling is present, the *total* coupling is almost equal to the substrate coupling. The described experiment proves that this is really the case. Closing both the air path and the substrate coupling improved the crosstalk attenuation to over 90 dB. Revisiting now Fig. 5.7, one can see that this is actually an extension of the same principle of creating, or at least approaching a closed shielding enclosure.

It is relatively easy to shield the microstrip in the experimental, laboratory setup. But how can we shield *practical* PCBs and ICs, especially in multi-layer structures? One practical way to shield microstrip lines is suggested in [5.9]: to use a special set of vias.

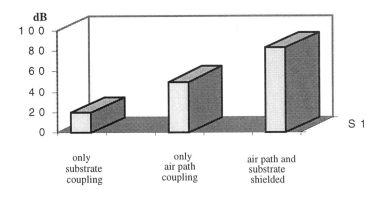

FIGURE 5.16 Crosstalk Attenuation Between Microstrip Lines

We will expand on this idea, by proposing its expansion shielding microstrip traces, and to striplines, as well as to IC, or other PCB component (inductors, capacitors, transformers), footprints. Vias are metal plated cylindrical passages (apertures) used to interconnect lines (or ground/power planes) placed in different PCB layers. Although practical via diameters are small, by placing them densely in a required pattern, a shield can be formed around the object to be shielded (e.g., a line, or an IC footprint).

An example of such a separation shield, placed in the path of substrate coupling, is shown in Fig. 5.15 d.

✦ Of course, this leaves still untreated the air path A. One way to expand the idea is to use a stripline configuration, as shown in Fig. 5.15 e. Of course, we will need to "sandwich" our lines between two ground planes, which may be economically non-desirable. But when we deal with, say, clock distribution in a sensitive application and at very high frequencies, that may be a viable option comparing to the alternatives — see chapters 2 and 7. ✦

We can now return to the important question: how densely should the vias be placed to provide for effective shielding microstrip or stripline? Theoretically, this question can be answered on the basis of the coupling through apertures. As we indicated earlier in this section, the "air passages" formed by adjacent vias, the ground plane and the shielding strip, can be treated as apertures. In chapter 3 of this book, we have reviewed the effects of small apertures on coupling. For instance, formulas (3.95) determine the dependence of inductive and capacitive coupling coefficients on the aperture size and the number of apertures. The same approach can be used in evaluation the high density via patterns for PCB and IC level shielding.

✦ For example, let us apply this theory to the problem at hand. According to the experimental data given in [5.9] for a described laboratory setup, by reducing the via density in steps from 500 mil to 250 mil to 100 mil separation, about 15 dB in crosstalk improvement was achieved with each step

360 Electromagnetic Shielding Handbook

density change, respectively. A further 20 dB improvement was gained by implementing two rows of via holes with 100 mil separation. Now, noting per formulas (3.95) that the coupling coefficients are proportional to the aperture diameter in cube power and to the number of apertures, it is easy to calculate that the experimental data is about right! ✦

It is proper to repeat here a "word of caution": don't forget that vias violate the integrity of ground planes, preventing in certain circumstances for the return current to take the "optimal" route with the smallest possible area. So, while gaining on crosstak between certain lines, we may increase crosstalk, as well as radiation and susceptibility, of other interconnections. We have already discussed this subject in chapter 4 (e.g., see Fig. 4.18, illustrating the "via congestion" on the ground plane).

5.3.4 Crosstalk in Coaxial and Shielded Transmission Lines

Using experimental data (some of this data is new, while other was originally published in Refs. [3.12, 4.7, 5.3, 5.10-11]; also abundant information on the subject is available in the literature), we will illustrate the theory developed in the previous section with examples of crosstalk between shielded lines. Among our objectives is to emphasize the dependence of crosstalk on the shield transfer parameters and coupling mechanisms, as well as to review typical crosstalk data. Consider the following "case studies":

1. We start with reference to the data in Fig. 5.11. It was reviewed in Section 5.3.1, and the only reason we mention it here is that this was low frequency differential-mode crosstalk between shielded and nonshielded balanced pairs in electrically short line.
2. As we saw, the crosstalk between reasonably good (with respect to the shielding) coaxial pairs is caused mainly by the common-mode currents in the line outer conductors. We will compare now crosstalk at relatively high frequencies between two short coaxial lines ($l = 2$ m), each consisting of two identical pairs: one line with single-braid outer conductors, and the second line with double-braid outer conductors. The test setup is shown in Fig. 5.17. In each line, one line was powered by a sweep generator (disturbing line); the crosstalk at the second line (disturbed line) was measured by a spectrum analyzer. Figure 5.18 shows the frequency characteristics of crosstalk attenuation in these two lines. Also plotted are the transfer impedance frequency characteristics of a single and a double braid shields.
   There is a notable correlation between the transfer impedance and crosstalk frequency characteristics. For an electrically short line, Eqs. (5.27) predict that the near- and far-end crosstalk attenuations should be equal and, for the identical lines, proportional to the square of transfer impedance. In this particular case, the transfer impedance increases with the frequency rise, but the difference between the single and double braid shields is about two orders. This leads to the corresponding difference in crosstalk attenuation of about 60–70 dB.

Shielding Effectiveness for EMI Protection 361

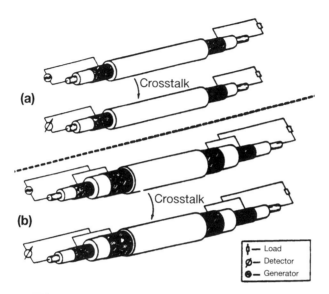

FIGURE 5.17 Coaxial cable setup for crosstalk measurements

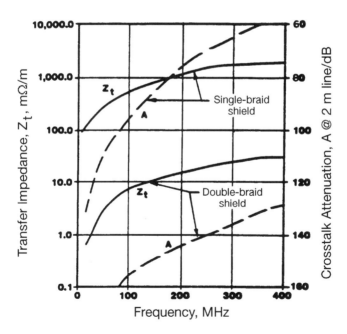

FIGURE 5.18 Crosstalk attenuation vs transfer impedance (CATV cables)

3. The effects of the different shield design at the low and high frequencies are compared. A special coaxial line (L = 1.5 m) was assembled, consisting of three coaxial pairs in one plane: two pairs with identical single-layer braided shields, and

362 Electromagnetic Shielding Handbook

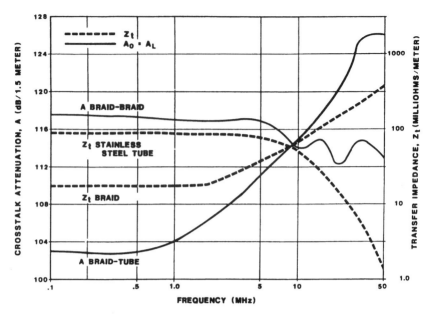

FIGURE 5.19 Crosstalk attenuation vs transfer impedance (artificial coaxial line, L=15 m)

one pair with a stainless steel tube solid homogeneous outer conductor. One of the braided pairs was located in the center of the assembly and acted as a disturbing circuit, while the second braided and solid tube pairs were located on both sides of the first pair symmetrically and acted as the disturbed circuits. The test procedure was identical to that in previous example

The test results are presented in Fig. 5.19. In the low-frequency region, the transfer impedance of both shield types (braided and solid tube) remains constant, and so does the crosstalk attenuation. While the disturbing circuit is the same for both cases, the difference in crosstalk magnitude depends on the transfer impedance of the disturbed circuits, which in the low-frequency range is larger (e.g., worse) for the steel tube. Correspondingly, the crosstalk attenuation is less (worse) for this line.

With the frequency rise, the transfer impedance of the tube diminishes, leading to the increase in crosstalk attenuation between the braid-tube pairs. The reverse takes place for the braid-braid pair combination.It is interesting to note that, at the frequency where $Z_{tbraid} = Z_{ttube}$ (the proper curves are intersecting), the crosstalk attenuations are also equal.

4. Here, we will "get a feel" of the effect of isolated conductors in the line (super-tertiaries) on crosstalk between two coaxial pairs in this line. We will consider a relatively long line at low frequency. As we will see, the presence in the line of balanced pairs brings about quite "fancy" crosstalk performance between coaxial pairs.

The measurements were performed on two carrier lines containing four 1.2/4.4 mm coaxial pairs with outer conductors (solid copper tube + two opposite direction spiral steel tapes) in continuous contact. Additionally, one line contained five balanced pairs. The line length was 6 km. The testing was done at the frequency 60 kHz. (The crosstalk attenuation in this heavy, shielded line quickly increases with frequency so that the highest magnitude of the transfer impedance in the frequency band of interest was observed at 60 kHz). The line configuration with regard to the tertiaries is illustrated in Fig. 5.20.

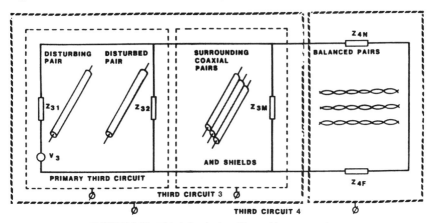

FIGURE 5.20 Third circuits in cables and interconnections

The line-length dependence of crosstalk between two coaxial pairs in lines without and with balanced pairs is shown in Figs. 5.21 and 5.22. When the line length is relatively small and frequencies are low (that is, the line length is electrically short—approximately, up to 1 km in Fig. 5.21 and 5.22), only the primary tertiary (T3) and the shunting effect of two other coaxial pairs have to be taken into account. In this case, Eqs. (5.26) can be used. The crosstalk attenuation decreases in inverse proportion to the logarithm of the line length, while crosstalk frequency dependence follows that of the shield transfer impedance.

With the line length increase, the super-tertiary formed by the balanced pairs (if present) must be accounted for. Now, in the line without balanced pairs, the NEXT line length characteristic oscillates with a fading amplitude, according to the phase shift and attenuation of the currents coming to the near end from different line elements. The far-end crosstalk characteristic slides down, almost with the logarithmic law.

On the other hand, the presence of the balanced pairs in the line drastically changes both the patterns and magnitudes of NEXT and FEXT. In general, an improvement of crosstalk attenuation is observed, reaching 10 to 20 dB on the length of a repeater span (6 km).

The presented experimental data was used to validate theoretical Eqs. (5.29) and (5.30). The disparity between the experiment and theory never exceed several decibels, which validates the developed formulas

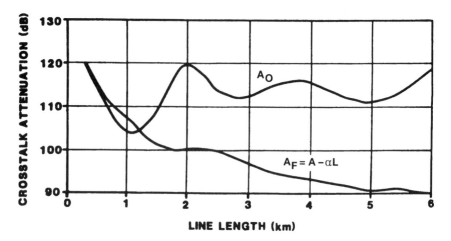

FIGURE 5.21 Crosstalk line-length dependence in coaxial line without balanced pairs (1.2 / 4/4 mm cables, 60 kHz)

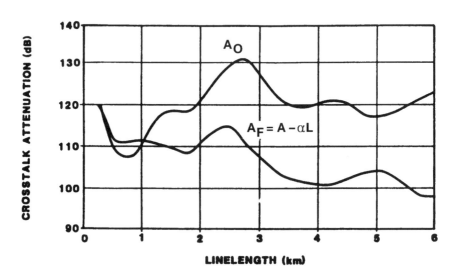

FIGURE 5.22 Crosstalk (NEXT and FEXT) line-length dependence in coaxial line with balanced pairs (1.2 / 4/4 mm cables, 60 kHz)

5.4 SHIELDING IN RADIATING FIELDS

5.4.1 Shielding from Radiation vs Shielding from Crosstalk

Indeed, what is the difference, "as far as the shield is concerned"? We will approach the analysis using the shield functions in the system. Previously, we have formulated the three functions of the shield in a system, which must be considered in their interaction:

- shield as a line conductor
- shield as an energy barrier
- shield as a coupling element - both, to 'internal' and 'external' fields.

We will try to gain a further insight into the shielding problems by comparing these functions in the crosstalk and radiation EMI regimes.

The function of the shield as a line conductor in the radiating regime is no different from the static or quasi-stationary regimes: as long as quasi-stationarity conditions are met and there is a closed circuit, we can always determine the line behavior using circuit theory. Also no different is the barrier (transfer) function — the same transfer parameters apply. Thus, it is possible to draw a parallel with crosstalk and, before we proceed, the reader may want to review Sections 1.3, A.5, and the second portion of 4.2.1. In addition, Ref. [5.11] presents some details related to the material presented herein.

The fundamental difference between the shielding performance in crosstalk and radiation conditions is determined by the external environment and coupling to external fields. First, different laws govern the field character, distribution, and propagation: the crosstalk phenomena are associated with transverse reactive induction fields, and the radiating fields are associated with active radiated losses on energy leaving the line (far-zone radiation), or a combination of radiated losses and induction processes (near zone radiation). Second, whereas crosstalk implies the presence of *two* mutually interfering systems, in radiating fields one of the systems is absent (the source system in the susceptibility case, and the victim system, in the emissions case) or, in principle, is replaced by a field probe or field structure, respectively.

Consider now the shielding problem in a far-field radiating zone. Adopting the shielding coefficient (shielding effectiveness) definition as a ratio (difference of logarithms) of the electric field intensities generated by the shielded and nonshielded conductors, and expressing the radiating electric field in the form of Eq. (4.57), we can define the shielding performance in typical antenna terms, as a function of the radiator/sensor (the line, or the shield, or any other conductor, for that matter) element factor and other applicable factors:

$$S = \frac{E_{\Sigma s}}{E_{\Sigma i}} = \frac{E_{es}}{E_{ei}} \frac{\int_{-1/2}^{1/2} A_{es} A_{eg} A_{et} dl}{\int_{-1/2}^{1/2} A_{is} A_{ig} A_{it} dl} = \Upsilon \frac{E_{es}}{E_{ei}} \qquad (5.33)$$

Of course, all the conditions and limitations adopted in the formulation of Eq. (4.57) apply. Such definition may often be adequate in the far fields *deterministic* environment, which can be reduced to a single EMI source and single receiver, while it may be incorrect (or at least incomplete) in a more complex *statistical* environment, as well as near zone fields.

As you remember, the criteria for a radiating near field can be defined per Eq. (4.17), which relates two parameters: the boundary between the far and near radiating field zones, and the critical electrical size of the radiator/sensor (Section 4.2.5). The near-field configuration depends on both the system geometry and the distance of the system from the observation point. For these reasons, it is impossible to theoretically predict the field configuration "at large." However, using numerical techniques, excellent results were achieved in modeling specific problems, e.g., human exposure-related problems (see more about this problem in section 5.8).

When dealing with near fields, only the most general considerations can be spelled out, and any specific system must be considered separately, usually using numerical methods. Even when the solution can be obtained for some simplified configurations, they are good only for the particular adopted conditions {e.g., see the note in Chapter 2 (Section 2.4) and Fig. 4.8}. For these, the solution of practical problems is often relegated to numerical modeling.

On the other hand, the analogy between the shielding for crosstalk protection and shielding from a near-zone radiating field is closer than in the case of far radiating field. For this reason, the crosstalk study methods may be quite applicable for evaluating shield performance in near fields. In essence, near- and far-end crosstalk can be viewed as the figures of merit for the shielding effectiveness in the quasi-stationary field! Then, the respective Eqs. (5.26) through (5.30) are applicable and sufficient to calculate this "crosstalk" shielding effectiveness.

The structure of these equations is pretty much similar even to equations describing the shielding from far radiating fields. The most important analogy is the dependence of crosstalk on the shield transfer impedance. However, since two line shields are involved in crosstalk, there are two transfer impedance parameters. But if, instead of one of the cables, an electromagnetic environment at large is considered, then only one shield will affect the shielding properties, and the similarity between the crosstalk and radiation shielding becomes even closer. Even the near-field configuration dependence on the observation point and the system's relative position in space can be modeled with proper variations of the third-circuit parameters. We will return to this question again in section 5.4.3.

◆ Although much research has been done for the near field of antennas, extensions of these results to radiating lines in the EMI environment are scarce (e.g., see Refs. [4.19, 5.12]). For example, the Ref. [4.19] considers the near field generated by a radiating rectangular plate which is placed over and normal to an infinite horizontal ground plane. The computational models are based on the representation of the elementary currents in the plate by filamentary dipoles with sinusoidal current distribution or an array of Hertzian dipoles. Applying the moment method to the computational models, the plots are obtained representing the dependence of the electric field on the electrical distance r/λ from and electrical size Λ/λ of the plate. Reference [5.12] presents useful experimental data.◆

5.4.2 Deterministic vs Statistical Nature of EMI Environment: Indoor Propagation of PCS Signal

◆ The purpose of the following exersize is to alert to the complexities of the EMI environment which may need to be addressed while solving electromagnetic shielding problems, and to indicate some ways to simplify the problem solutions. The EMI source location with regard to the evaluated shielded system, the test site configuration, the signal averaging algorithms - these and many other factors may contribute to the evaluation result interpretation, repeatability, as well as calculation strategy asnd test logistics.

We have already analyzed the role and certain specifics of the EMI environment and field configuration in the electromagnetic shielding problems. As a rule, we dealt with a *singular* correpondence between the field characteristics and respective independent variables, e.g., see Figs. 4.5, 5.1, and 5.2. While such *deterministic* analysis often (although, not always) fit the purpose in static and quasi-stationary fields, it appears not sufficient in the context of the microwave (and even lower frequencies) radiating field environment. For example, it is usually accepted that the field generated by a point source, falls off with the distance r from the source by the law $1/r^2$ in the near field zone and with the law $1/r$ — in the far field zone. Also, the field stays constant or stationary in time. Therefore, given an EMI source, it could be assumed that at the shield location the EMI field variations and configuration were similar to those at the source. The only difference was that the amplitude and phase were modified according to the signal propagation channel, per Fig. 1.10.

With regard to the high frequency signals, this may be correct only for *isolated* sources in *free space* or in certain idealized conditions during mathematical derivations and in corresponding test set up. But in "real life" applications, including testing, the EMI environment is complicated by the presence of multiple sources as well as multiple objects which reflect, scatter, and absorb the propagating signal. The result is an electromagnetic environment which varies both in the space and time domains. This is true for near, as well as for far fields.

In practice, the complexity of the near field EMI environment is usually emphasized and attempt is made to more or less account for it, in contrast to the relative over-simplification of the far field EMI environment. For instance, to avoid the field calculation complications, the circuit theory approaches are often used. Then, the induction and even near radiating field EMC performance is formulated in terms of crosstalk, where the interfering systems, their configuration, and mutual arrangement are pretty much defined and "refined". However, while the near field configuration is indeed comlex due to the electric and magnetic field independence and mutual coupling between the interfering objects and related active losses, the over-simplification of the far field may not always be justified. The point is, that the far field distribution in space and in time is often anything, but uniform and simple!

One illustration of this fact, the so called Rayleygh fading, is shown in Fig. 4.5 e. To give a "taste" of the "Lord Rayleigh Universe" ensuing problems, consider a microwave signal propagation environment, such as dealt with in wireless (e.g., cellular or PCS) applications. When a radiated signal is generated in this environment, a multipath channel is formed by fixed and randomly moving signal reflectors and scatterers. As an illustration, Fig. 5.23 presents a simulation of a 5 GHz signal power distribution in the time- space (distance) environment for a certain topography. The distance between the transmit and receive antennas is assumed 44.7 m. As we see, at any given space point,

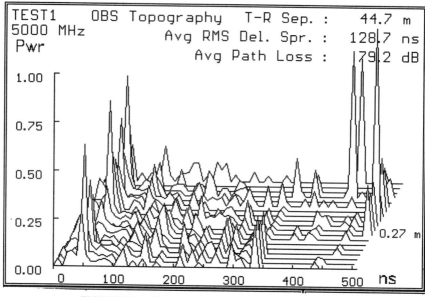

FIGURE 5.23 Radio signal distribution in time-space domain

or at any given time moment, the electromagnetic environment is anything, but deterministic! Such environment must be treated statistically. At the best, in deterministic terms this environment can be characterized by average path loss and average RMS delay spread.

In mathematical terms, if a signal s(t) is transmitted in such channel, the received signal may be expressed as

$$r(t) = \sum_n \alpha_n(t) s[t - \tau_n(t)] \quad (5.34)$$

where $\alpha_n(t)$ and $\tau_n(t)$ are the attenuation and the propagation delay for the signal received on the nth path.

Because both the line of sigght (LOS) and scattered signals are received, statistically such multipath propagation can be modeled by a Ricean-distributed random variable

$$R = \sqrt{\sum_{i=0}^{n} x_i^2}$$

with a pdf:

$$p_R(r) = \frac{r^{n/2}}{\sigma^2 \cdot s^{(n-2)/2}} \cdot e^{-(r^2 + s^2)/(2\sigma^2)} \cdot I_{n/2-1}\left(\frac{rs}{\sigma^2}\right) \quad (5.35)$$

In general, the variable parameters depend on such factors as Doppler effect, delay spread, Rayleigh fading (random, with zero moment distribution), shadowing, etc. Obviously, it isn't simple to "plug

Shielding Effectiveness for EMI Protection 369

in" this kind of expressions in the EMI and shielding analysis problems, which are complex enough without these complications. Just imagine, using (5.34) and (5.35) to calculate the shielding effectiveness, when each of the items would have to modified by complex and different coupling and transfer functions!

To facilitate the evaluation of the in-building environment, simplified models can be utilized, which emphasize the necessary factors, depending on particular applications. For example, for slower moving objects and limited distances, it may be sufficient to account only for the *mean* power loss and slow signal *shadowing statistics*. But even if we average in some way the environmental factors (say, take the signal *means* - which is quite often done in certain wireless applications), or limit them to LOS only, there are still enough problems left.

Consider a transmit antenna in free space, radiating power P_T, W, at frequency f, GHz. The *mean* LOS signal power at the receive antenna located at a separation distance r (in m) from the transmit antenna, can be expressed as

$$P_R = P_T \cdot G_T \cdot G_R \cdot \frac{Q}{f^2} \cdot r^{-\alpha} \qquad (5.36)$$

where G_T and G_R - are the transmit and receive antenna gains, respectively; constant $Q \approx 6 \times 10^{-4}$. Exponent α determines the mean propagation path loss. In free space $\alpha = 2$. In the simplest case for in-building environment similar to (5.36), propagation models can be adopted (as for free space), but with different propagation loss exponents. Depending on the building construction materials, floor plans, and even on the installed furniture, the typical values of α vary from 1.8 to 4. For example, Fig. 5.24a illustrates the *averaged* field intensity variations at 6 GHz, LOS along an aisle in a large building, with the radiating source installed at the beginning of the aisle (point 0). A definite trend is visible, which, when averaged, can be expressed in terms of path loss or, equivalently, in terms of the coefficient α. However, when diffraction is involved, the relatively monotonous path loss dependence on the distance from the source can be abruptly violated, as shown in Fig. 5.24b. Of interest, all these obstacles in the signal propagation path can be viewed in the context of shielding. In this respect, the coefficient α itself may serve as a figure of merit for shielding effectiveness of the respective construction elements - see section 5.8.2.✦

5.5 FROM TRANSFER IMPEDANCE TO SHIELDING EFFECTIVENESS AND CROSSTALK ATTENUATION

Consider now a line oriented in free space, along the axis z, in which a common-mode current is excited. This could be a single-ended circuit or a multiconductor nonbalanced or balanced circuit. Unless this is a special case of a line with an arbitrary load at the end (e.g., a CMOS gate which has a low capacitance input), the line has an adequate matched return for the transmission signal (that is, differential-mode current) and produces induction fields. But, in this case, we are interested in the common-mode, or antenna current in the line, for which no close enough return can be identified. As we have already discussed, common-mode currents appear on the direct and return path conductors for *inherent* reasons due to the finite distance between the conductors, as well as because of the line asymmetry, as a result of crosstalk, of ambient field coupling, and for other "external" reasons often beyond control.

370 Electromagnetic Shielding Handbook

(a) Line of sight (LOS)

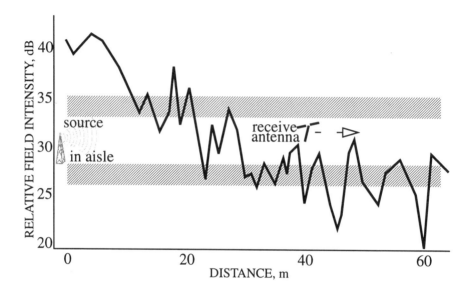

(b) LOS vs diffraction)

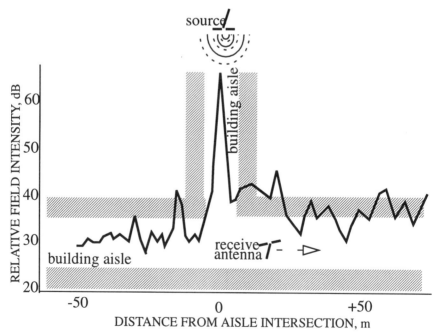

FIGURE 5.24 A 6 GHz radiating source field variations along a building aisle

Shielding Effectiveness for EMI Protection 371

The antenna current distribution along the considered line depends on the presence, configuration, and electromagnetic properties of other conductors that are mutually coupled to the considered line. Our line ends (one end or both, depending on the location of the feeding point) are assumed to be open-circuit for the common-mode currents. From antenna theory, we know that in such a configuration the antenna current on the surface of the thin conductor is distributed in a standing wave pattern with zero current at the open end. The electromagnetic fields of such an antenna are determined by Eqs. (4.52) through (4.54) in general, or by Eqs. (4.41) through (4.45) in the case of a short line. In any case, we can present the radiated field as a function of the current at some reference point of the antenna:

$$E_i = \zeta_i I_o \tag{5.37}$$

Equation (5.37) holds for many line configurations, with the difference between them being accounted in the expressions for ζ_i.

To determine current I_o, we will assume that a voltage, V_{io}, is applied to the beginning of the line. Then I_o is determined by the radiation resistance, R_{ri}, of the line. It can be shown that, at least for not very long lines (this, of course, means electrical length, and the condition is often not very severe: $< \lambda/2$), the radiation resistance of a thin straight wire is independent on the location of the voltage feed along the line and can be described by Eq. (4.64) in general, and by Eq. (4.58) for a short radiator. (Actually, this independence of the radiation characteristic on the location of the point at which the voltage is applied to the radiator is not fundamental and is made only to simplify the discussion. It results in the equal contributions to the radiation from each line element, wherever it is located in the line. When necessary, for instance to evaluate longer lines, the different contributions can be integrated). Then,

$$E_i = \zeta_i I_o = \zeta_i \frac{V_{io}}{R_{ri}} \tag{5.38}$$

where we introduced the radiation factor ζ_i to account for the rest of Eqs. (4.58) and (4.64).

When the line is enclosed in a cylindrical concentric conducting shield which is connected to the signal returns at the ends, then, with regard to common mode current, the shield acts as an outer conductor of a coaxial pair. Now, the current $I_i = V_{io}/Z_{ci}$ returns in the shield as $I_{si} = -I_i$ in the internal circuit, both direct and return currents are at a close distance, and the fields radiated by them are compensated in any point of space in the far radiation zone (all the words here are important). Here, Z_{ci} is the characteristic impedance of the shielded line (for the common-mode signal). Assume that, with the addition of the shield, all the current returns in the shield and the current's value did not change. Outside of the line, the tangential magnetic field due to direct path current I_i in the line conductor is com-

372 Electromagnetic Shielding Handbook

pletely compensated by the field of the return current $I_{si} = -I_i$ in the shield (see Fig. 4.8). Thus, we can expect perfect shielding (i.e., shielding coefficient 0, or shielding effectiveness attenuation ∞). And this really takes place, as far as current I_i in the line is concerned!

However, with an imperfect shield (nonhomogeneous shield with gaps and seams, or even a homogeneous shield with finite conductivity—see Chapter 3), crosstalk and radiation still occur due to the non-compensated current, I_{se}, in the shield with external return, which is generated by the voltage drop at the shield's external surface. The measure of the shield imperfection that compromises its electromagnetic barrier quality is the transfer impedance (and transfer admittance, in certain conditions). As we saw in Chapter 3, the voltage drop at the element dm of opposite to the direct path shield's surface is:

$$dV_{se} = I_{si}Z_t dm = \left(-\frac{V_{io}}{Z_{ci}}\right) Z_t dm \tag{5.39}$$

Now, instead of the current I_0, the EMI is produced by the voltage dV_{se}, which generates a current I_{se}. Of course, inside the emf source dV_{se} this current is opposite to its direction in the load. We have already studied how current I_{se} was generated in the process of crosstalk (Section 5.3). In the crosstalk model, the voltage dV_{se} acted as an emf in the external circuits (tertiaries) composed of the source and victim line shields and other conductors which provided the current I_{se} return. The electromotive force was applied to the return via two impedances in series: external line input impedance in the direction toward the near end of the line, and external line input impedance in the direction toward the far end of the line (e.g., see Fig. 4.10). The EMI signal in the external circuit thus "looked" in these two directions. Comparing the voltage at the beginning of the source circuit to the voltages induced at the near and far ends of the victim circuit, we came up with the near- and far-end crosstalk between the shielded lines.

In the radiation (antenna) model, the current I_{se} "return" is provided by the radiation resistance R_{rs}, while the voltage V_{io} is substituted with the equal in magnitude and opposite in sign emf. Similar to Eqs. (5.36) through (5.39), we can obtain the radiating field generated by the shielded line (under the assumed conditions):

$$E_s = \zeta_s I_{se} = \zeta_s \frac{V_{io}}{Z_{ci} R_{rs}} Z_t dm \tag{5.40}$$

✦ There are fundamental differences between the EMI in shielded and nonshielded lines. In a nonshielded line, the sources of EMI are the common-mode current, I_0, and the transverse components of the electric and magnetic fields (the electric field radial component and the magnetic field tangential component determined the electric and magnetic coupling in crosstalk, but did not affect the radiation in the far zone). In a shielded (coaxial) line, the transverse electric field is terminated inside the shielded space, and I_0 becomes a differential-mode current. In general, the magnetic field is determined by three currents: (1) direct path current in the shielded conductor I_0, (2) return (for

Shielding Effectiveness for EMI Protection   373

I_o) path current in the shield ($-I_o$ in the internal circuit), and (3) the antenna current $-I_{se}$ at the shield's external surface. We can present the total current I_s in the shield as $I_s = -I_o + I_{se}$. The current I_{se} in the shield has the same direction as current I_o in the shielded conductor (but opposite to the return current in the shield associated with the internal circuit). Since the I_o and I_{si} are concentric and directed opposite, the related field is determined by the residual difference $I_{ix} - I_{sx}$. Although with proper design, this difference can often be made negligible ($I_{ix} - I_{sx} = 0$), this should not be "taken for granted." For example, if the shield is disconnected from the shielded line, no current will return through it, which is equivalent to little or no shielding effect. In general, the ratio $I_{sx}/I_{ix} = \xi \leq 1$.

Of course, the radiation is also present, and in principle it must be considered in the crosstalk study. But as we found in the quasi-stationary regime, the radiating field contribution to the losses of energy in the "crosstalking" line was negligible. Moreover, since the lines as a rule *are not designed* to radiate the energy and are inefficient radiators (there are exceptions, e.g., "leaky" coax), we will neglect the radiation loss effect on the total energy propagating in the line. This assumption is extremely important to simplify the study of the field radiating from the line shield.✦

Taking the ratio of Eq. (5.40) to Eq. (5.38), we can obtain the shielding coefficient in free space in terms of the shield and shielded conductor radiation resistances and radiation array factors:

$$S = \frac{E_s}{E_i} = \frac{\zeta_s R_{ri}}{\zeta_i R_{rs}} \frac{1}{Z_{ci}} Z_t dm = \frac{\xi_s}{\xi_i} \frac{1}{Z_{ci}} Z_t dm \tag{5.41}$$

Or, in terms of shielding effectiveness ($A = 20 \log |1/S|$),

$$A = -20\log\frac{\zeta_s}{\zeta_i} - 20\log\frac{R_{ri}}{R_{rs}} + 20\log Z_{ci} - 20\log Z_t dm$$

$$= -20\log\frac{\xi_s}{\xi_i} \frac{1}{Z_{ci}} - 20\log Z_t dm = A_r - 20\log Z_t dm \tag{5.42}$$

Expressions (5.41) and (5.42) lead to a fundamental conclusion: the shielding effectiveness is proportional to the shield's transfer impedance! Although the transfer impedance is not the only factor affecting the shielding effectiveness, it is difficult to overestimate the importance of this conclusion for practical applications. For example, consider such an important case of the shield design comparisons for the same line and in same electronic system. Then, the system- and line-specific item A_r often (but not always) can be assumed to be identical for different shields, and the relative shielding effectiveness of shields 1 and 2 can be obtained from Eq. (5.41) as

$$S_{rel} = \frac{S_1}{S_2} = \frac{Z_{t1}}{Z_{t2}} \tag{5.43}$$

What can be simpler?

Comparing now (5.42) for shielding effectiveness from radiating fields to (5.26) through (5.30) for crosstalk shielding effectiveness, we see that with regard to the transfer impedance parameter, their structure is similar. The only difference is in the presence of the two transfer impedance parameters — Z_{t1} and Z_{t2} — in the case of crosstalk, vs only one transfer impedance parameters — Z_t — in the case of the radiation. Of course, this reflects the presence of two shields in the crosstalk process (although, if one of the "crosstalking" lines is not shielded, the respective Z_t will be absent), vs only one shield - in the radiation process.

Applications

5.6 EMC AND SHIELDING PERFORMANCE OF TYPICAL ELECTRONIC CABLES

We will use experimental data to illustrate the EMC performance of shielded electronic cables in the radiating field. Actually, in the course of the book so far, we have quoted many practical examples applicable to this subject. (For instance, you may want to revisit the Section 2.2 in Chapter 2.) For this reason, here we will concentrate mainly on the radiated emissions and/or susceptibility and shielding effectiveness of some typical electronic cables. We will also try to draw a parallel between the EMC performance of these cables and the transfer impedance of their shields. According to Eqs. (5.41) and (5.42), we can expect the shielding effectiveness to be proportional to the transfer impedance of the shield.

Figure 5.25 is a plot of the frequency versus shielding effectiveness characteristics of a flat line with three often used "cigarette wrap" shields: solid aluminum foil, braided aluminum foil, and copper mesh. Also, for comparison, it presents the performance of a shielded connector. We need this latter curve here only for reference, while a more detailed discussion of connector performance will be presented in the next section. The measurements were performed in the 30 MHz to 1 GHz frequency range. (If this frequency range does not "ring a bell," you may want to revisit Chapter 1). To obtain this data, all the conductors of a line were connected together and used as a single-ended circuit. The radiated emissions from this circuit were then measured in a line, with and without the shield.

As we can see, the shielding effectiveness degrades with the frequency—from 30 to 60 dB at lower frequencies to about 10 dB at higher frequencies. While the copper mesh is the best at lower frequencies, it becomes the worst at higher frequencies.

Shielding Effectiveness for EMI Protection 375

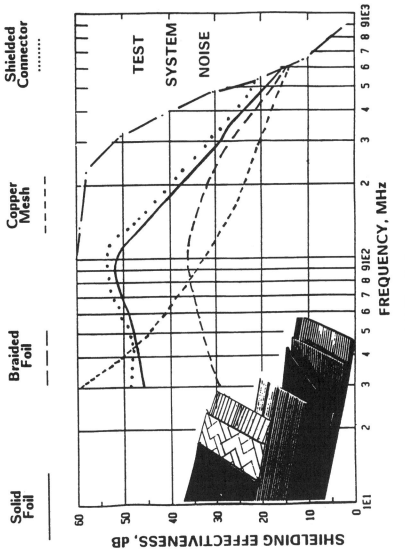

FIGURE 5.25 Shielded flat cables' and connectors' shielding effectiveness

376 Electromagnetic Shielding Handbook

This relative behavior is easily explained by the properties of the transfer functions of the shield. As a matter of fact, we could have obtained equivalent relative data by using the frequency characteristic of the transfer impedance of these shields. However, comparing with the transfer impedance, the shielding effectiveness is a less general and more system-specific parameter (see Chapter 2).

In principle, the shielding effectiveness parameter can be used to evaluate the absolute EMC performance of the cables. Indeed, is the shield "good" or "bad"? To answer this question, we need to refer to the more system-oriented information. For example, we could measure the field intensity at a specified distance from the line powered with a known drive signal. Figure 5.26 presents the field strength at 3 m distance from the same cables, driven by a 0 dBm signal. Although Fig. 5.26 results are identical to those Fig. 5.25 with regard to the relative performance of the shields, it also provides insight into the actual field strength of emissions from the line under certain conditions (that is, those of the test environment).

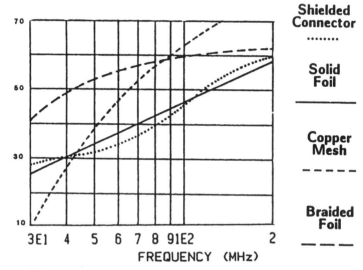

FIGURE 5.26 Peak emissions from shielded flat cables and connectors
(field strength in dBμV/m at drive signal = 0 dBm)

Similarly, the effects of the shield nonuniformities can be identified and explored with reference to Fig. 5.27.

Figure 5.27 presents the frequency characteristics versus comparative EMC performance of four popular CATV drop cables. Three different techniques were used: transfer impedance (Fig. 5.28a), shielding effectiveness measured in the radiating emissions regime (Fig. 5.28b), the shielding effectiveness measured in the radiating susceptibility regime (Fig. 5.28c), with respect to the line under test. While the transfer impedance and "emission" shielding characteristics were measured at continuous frequency sweep, the "susceptibility" shielding was measured at three discrete frequencies only: 34, 102, and 148 MHz.

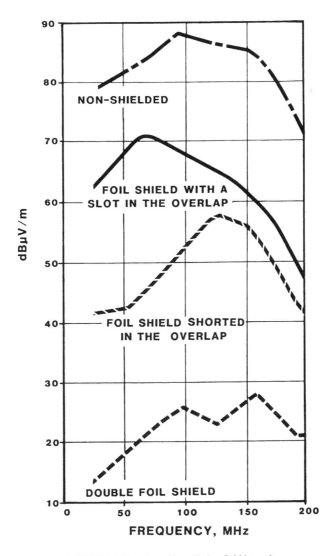

FIGURE 5.27 Flat cable radiation field intensity

As shown in Fig. 5.28b, the addition of every extra layer increases the shielding effectiveness (and reduces the line radiated emissions) by 15 to 25 dB. In the frequency range over 100 MHz, the 95 percent copper braid performs much worse than a thin aluminum foil. To explain the cables' performance, consider their transfer impedance frequency characteristics. The comparison of Fig. 5.28a and Fig. 5.28b proves their remarkable correspondence.

Now, referring to Chapter 3, we can explain the relative performance of the aluminum foil and copper braid shields. At lower frequencies, the transfer impedance of the shield is determined by its dc resistance which, of course, is much lower in the copper braid. However, at higher frequencies, the field penetration through the aper-

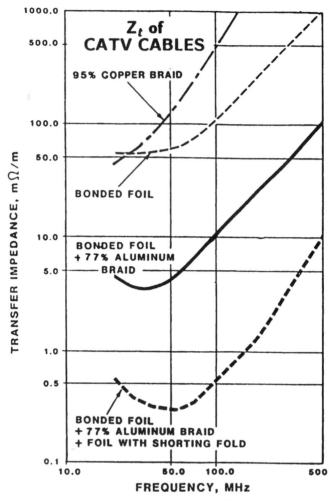

FIGURE 5.28a EMC performance of CATV cables expressed in terms of transfer impedance

tures and braid spirality degrade its performance, while the seam in the foil shield does not produce such drastic degradation. As might have been expected, susceptibility performance of the same cables obeys the same patterns.

5.7 GROUND PLANES AND RADIATION: FROM EMC TEST SITE TO PCB

✦ Consider an experimental immunity test described in literature [5.13]. Two PCB were manufactured with identical, microprocessor-based circuitry: one with PCB trace return, another - with ground plane return. The PCBs were illuminated by RF fields with intensity varying from 10 to 100

Shielding Effectiveness for EMI Protection 379

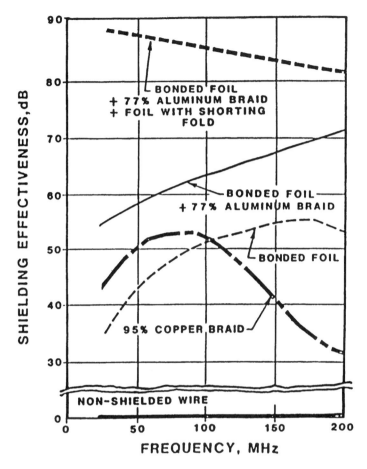

FIGURE 5.28b EMC performance expressed in terms of shielding effectiveness (emissions regime)

V/m in the frequency range 10 - 500 MHz.

When monitoring the microprocessor output, two types of malfunctions were observed: a) the microprocessor would cease to operate on application of the field but would continue working when the field was reduced, b) the microprocessor would lock up and require a reset to continue working. During the experimental test, significantly less malfunction occurences were observed in the PCB with the ground plane. Although at the PCBs with and without ground plane the malfunctions would occur at different frequencies, the lower field intensity thresholds for malfunctions were 50 V/m and 30 V/m, respectively. ✦

PCB with a ground plane is a typical example of a 2D open shielding configuration. We have already discussed the effects of parallel shielding in crosstalk. But how does the ground plane affect the emissions/susceptibility? One way to understand, is to combine two models: the generation of electrical images of emission / susceptibility objects by a metallic surface and the representation of a combination of the actual and image radiators as an active antenna array. Both models have been analyzed pre-

380 Electromagnetic Shielding Handbook

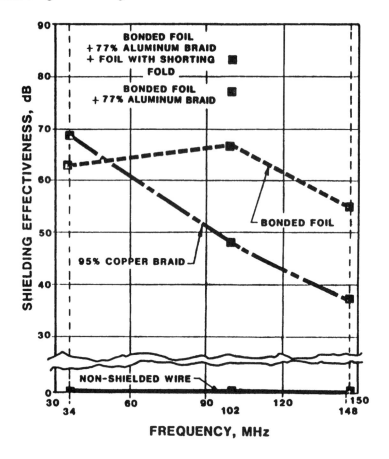

FIGURE 5.28c EMC performance expressed in terms of shielding effectiveness (susceptibility regime)

viously in sufficient detail, so we will just illustrate the effects with radiation sources over a ground plane at an open area test site.

As we saw, common mode radiators can be modeled with electric elementary sources, and differential mode radiators - with magnetic sources. Consider now simple models of vertical and horizontal electric and magnetic linear sources over ground plane (see Fig. 5.29)

This corresponds to an EMC radiated emissions test setup (identical cases can be made for immunity/susceptibility, so the general nature of the following discussion is preserved). The effect of the ground plane can be accounted for by the images of "real" sources. The electric source images have the same as the real source direction for vertically polarized objects, and the opposite directions for horizontally polarized objects. For magnetic sources - vice versa. At the figure, these images are shown with dashed lines.

The electromagnetic theory proves that under certain conditions (including ideal infinite size ground plane of large conductivity), the emissions of the "real" source over

Shielding Effectiveness for EMI Protection 381

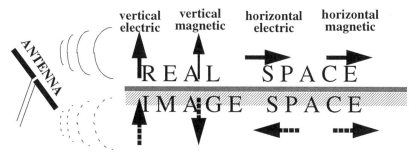

FIGURE 5.29 Radiation sources over a perfect ground plane

ground plane are identical to the emissions generated by an *array* of two sources: real and "imaginary", but with ground plane absent. .

This situation was analyzed in many literature sources, e.g., see Ref. [1.8]. The effects of such an array, as "seen" by a receive antenna, depend on the relative distance between the source and ground plane on one hand, and to the receive antenna on the other hand. An EMC test site, for example, deals with far field measurements, where these distances are large enough, comparable to the signal wavelength. In this case, instead of the image model, a reflection model can be used (e.g., see Fig. 5.30) - both give identical results.

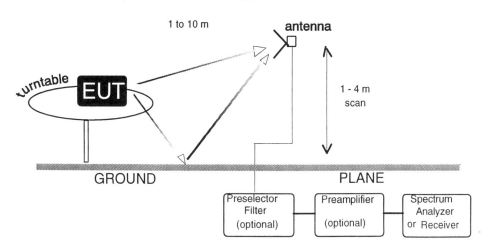

FIGURE 5.30 Radiated test setup (OATS, semi-anechoic chamber)

382 Electromagnetic Shielding Handbook

In particular, from the representation in Fig. 5.30, it follows that the direct and reflected waves sensed by the receive antenna, have to travel different distances and therefore, have different time delays. Since at the source, the phases of these two waves were the same, the difference in phase at the receive antenna position will be determined by this delay. If the receive antenna (or the source) change their position (e.g., receive antenna height scan), the phase difference of the direct and reflected rays arriving at the receive antenna, also changes. Obviously, when the phases of the direct and reflected wave coinside (every 2π of the wavelength λ), the receive antenna will register maximum, and when the phases are opposite - minimum. For instance, an OATS height scan of a radiating source may look, as shown in Fig. 5.31.

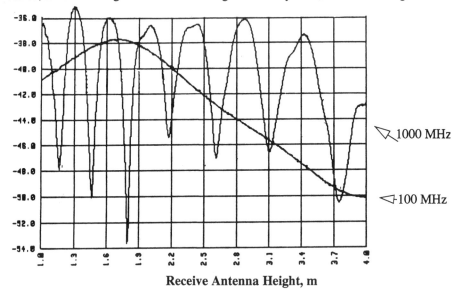

FIGURE 5.31 Receive antenna level variations @ 100 MHz and 1000 MHz, when height-scanned 1 - 4 m at OATS

There is nothing new in this theory of an open area test site [2.1]. We have also looked into the respective formulas in chapter 4, (e.g., see Fig. 4.20 and related text). But what does it all say about open parallel shielding using an EMC test site ground plane? As you can see from Fig. 5.31, at different heights over the ground plane the receive antenna levels may be smaller or larger than those in free space. Thus, the presence of the ground plane may reduce or increase emissions (and susceptibility) of any particular source, which can be interpreted in terms of positive or negative "shielding effectiveness" of the ground plane at the OATS. Obviously, we would hardly want to use the EMC test site ground plane for shielding purposes.

So much for the "large scale" ground plane open parallel shielding!

But what's the difference between the OATS ground plane and the ground planes in printed circuit boards (PCB) or flat cables? We did see (e.g., Fig. 2.13 in chapter 2) that the presence of ground plane has proven so useful for EMC! What makes a shield a shield? We are looking for such conditions which result in desirable ("positive") shielding effect of the ground plane.

To arrive at the answer, consider Fig. 5.30 and 5.29 with reference to each other. We see that the "secret" of the shielding effect is in the relative "perception" by the receive antenna of the waves from the real and imaginary sources. For the ground plane to act as an *ideal* shield, the signals from both the source and its image *at the receive antenna location* must have opposite phases and identical amplitudes. This happens when both the direct and reflected waves experience identical delay before arriving to the receive antenna. In practice, this condition depends on the relationship between the signal wavelength and mutual positions of the source and image with regard to the receive antenna. It can be met in several ways, e.g., we may require that the distance D from the receive antenna to the source is large, relative to the source height h over the ground plane, *and* the ratio D/λ doesn't vary significantly (in terms of λ), with the change of the system geometry.

Returning to the Fig. 5.29, we see that just such conditions usually have place in the multi-layer PCB and in the flat (ribbon) cables with ground plane. These contain "horizontal" electric and "vertical" magnetic sources* with conductors (traces) parallel to the shield. Each source and its image are directed opposite. The distance between the actual radiator and its image is a miniscule fraction of the wavelength. The receive antenna is in the far field, but *not as far* as to accumulate large difference for the reaching it actual and image receive signals. Open parallel shielding indeed!

How effective is ground plane?

Consider a practical "case study". A single-layer PCB (see Fig. 5.32) was tested at the EMC test site and failed the radiated emission test (Fig. 5.33 - top graph). The addition of a ground plane as a second layer, resulted in the radiated emission drastic reduction over the whole investigated frequency band, and meeting the FCC limits (Fig. 5.33 - bottom graph).

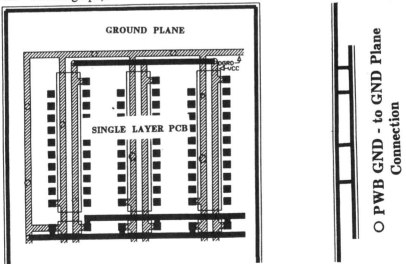

FIGURE 5.32   Use of external ground plane to reduce radiated emissions

---

*The "horizontal" and "vertical" are quoted, because it is not the "horizontality" or "verticality" per se which is important, but rather the source direction with regard to the ground plane: parallel or perpendicular (obviously, all other directions can be reduced to these two).

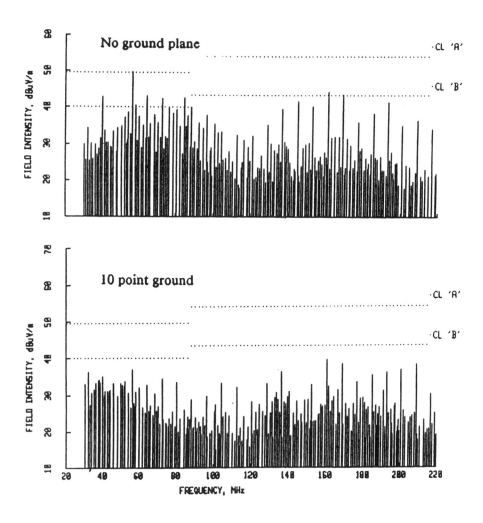

FIGURE 5.33 Ground plane effectiveness in reducing radiated emissions

As they say - no need for further comments. Nevertheless, we will return to the same test case in chapter 7, when discussing the system grounding. At that time we will also clarify the so far "cryptic" note at the bottom graph with regard to "10 ground points".

No discussion of the use of the ground plane can be complete, without touching upon the subject of the ground plane size. Indeed, this is an important practical subject. As a rule, when the direct path trace over ground plane runs close to the ground plane edges, this results in an increase of crosstalk and radiation. This prompted the formulation of specific PWB layout rules limiting the distance from the trace to the

Shielding Effectiveness for EMI Protection 385

ground plane edge (e.g., the so called nH-rules [5.14], where H is the distance from the direct path to the return ground plane and $n=3$ through 100 — coefficient).

The usual explanation invokes the generation of stray fringing fields which couple into the neighboring PCB interconnection layers, generating crosstalk as well as couple into the "environment" at large, generating radiated emissions. But why would such fields appear, to start with? Revisit the Fig. 3.15. Just how will the high frequency current return, when part of the respective area of the ground plane will be partially, or completely, absent? Of course, it will take another route. But then the mirror image nature of such current return (with regard to direct path) will be violated. Now, both the "original" current and the return are not any more compensating each other, which results in the "stray", non-compensated fields. The consequences are clear: emissions / susceptibility and crosstalk increase!

Fig. 5.34 presents experimental data [5.15], comparing the radiated emissions from a stand-alone wire to those from a wire with ground plane strip returns in three configurations: the radiated wire is centered over the strip, the radiated wire shifted to the edge of the strip, and the radiated wire shifted to the edge of the strip but with the second strip at the other side of the wire (see Fig. 5.34).

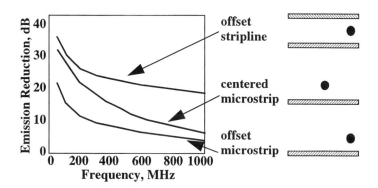

FIGURE 5.34 Ground Plane Return Effects

Using the accepted terminology, the first two configurations correspond to the microstrip geometries, while the third configuration - to the strip line. Several conclusions can be drawn:
- A ground plane return is better than *"no ground plane"*, even when the direct path is traced close to the edge of the ground plane
- Tracing the direct path close to the ground plane edge resulted in up to 10 dB increase in radiated emissions
- The strip line configuration provides for lower radiated emissions than the microstrip configuration, even when the direct path is traced close to the edge of the ground planes.

The "edge" effect is also related to the height of the signal trace over the ground plane. The reason is, that although the *bulk* of the return current in the ground plane concentrated "right under" the direct trace, a certain amount of return current "spreads out", occupying a larger area (e.g., see simulation results in Fig. 3.15). And the higher is the signal trace over the ground plane, the larger is the "spread". The current density quickly falls off with the distance from the line of maximum current concentration. The available research indicates that practically the return current in the ground plane can be assumed as spread in the strip of width equal 2-3 times the height of the direct signal trace over the ground plane*.

The increase of the signal trace height over the ground plane results in larger emissions (and susceptibility) of the respective line. For example per experimental data in [5.22], the trace height increase from 0.2 mm to 1.6 mm, resulted in 10-15 dB radiated emissions *increase*, almost constant in the frequency range 900 - 300 MHz. However, with the frequency reduction from 300 MHz to 100 MHz, this increase was *tapering off*, so that below 100 MHz there was *no difference* between the emissions corrseponding the two different direct trace heights over the ground plane.

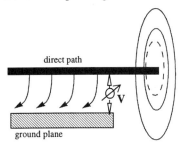

FIGURE 5.35 Partial return in ground plane

Of course, the worse ever situation occurs when the direct path trace runs only partially over ground plane and *extends over its edge* (see Fig. 5.35). Any way you look, this is an antenna: a microstrip antenna if the shown direct path is a trace on a PCB with an interrupted ground plane, or a "macro" antenna formed by a cable attached to the direct path. Now, if there exist potential difference V between the direct path and the ground plane (at the edge), it will act as an EMI voltage applied to the antenna terminals. Such antenna will radiate into the environment, or pick up *any* signal which "happens to be" transmitted via, or in the vicinity (because of crosstalk), of such trace. And of course, the effectiveness of such a radiator is determined by its electrical length, as well as magnitude of V.

Using circuit or antenna theories as outlined in the previous material, and relatively simple experimental techiques, it is not too difficult to numerically evaluate such ground plane effects. Also, abundant literature is available on the subject, presenting experimental data, as well as theoretical analysis, e.g., see [5.18-22]. Although the authors do not always treat the problems in shielding context, the results can usually be applied to those.

5.8 SHIELDING ENCLOSURES FOR RADIATING FIELD PROTECTION

In chapter 2, we have already looked at several shielding enclosures' performance, but we never attempted to explain the underlying processes. Now we will try to

---

*For instance, similar considerations make the rationale for the so-called 3H "rule of thumb" [5.14]

"make sense" of this old, as well as new, "business"*. We will also continue this discussion in chapter 7.

5.8.1 Critical Parameters of Shielding Enclosures

First, let's ask ourselves, what are the characteristics of an enclosure that determine its shielding performance? What most often comes to mind, is wall[†] material and thickness. However, we have already seen the problems with such "common sense" approach: while *in principle* such answer is correct, it is *incomplete:* practically the wall material and thickness are not the only, and often not the most critical, parameters (e.g., see section 2.3.3 in chapter 2). We suggest the following list of enclosure parameters — not necessarily a complete list and not necessarily in order of parameter importance:
- Enclosure integrity
- Cable penetrations
- Apertures and their treatment (e.g., use and performance of gaskets)
- EMI source and receiver position within and outside of the enclosure
- Enclosure wall reflectivity
- Enclosure size and shape
- Wall material and thickness (yes, they *are* important parameters, though not always critical in practical applications)

Some of the named characteristics may seem not as important as others, or even irrelevant. But that's often only at the first sight! Take for example the enclosure size and shape. What do they have to do with shielding performance? Such questions could have been asked by those who limit the shielding performance and shielding effectiveness just to the shield transfer function (and indeed, as we have seen in chapter 3, the effect of the enclosure size and shape on transfer function is often minimal). Hopefully, by now we don't have this kind of confusion.

We know better by now! Such characteristics as the enclosure size and shape, the wall reflectivity, the EMI source and receiver position within and outside of the enclosure — they affect the EMI environment inside and outside the enclosure. And we already know, how important is the EMI environment to the shielding process at large, to the external and internal electromagnetic coupling, in particular, and to the very definition and figure of merit of shielding effectiveness.

5.8.2 Radiated EMI Environment Issues

Shielding Enclosure Internal Resonances and Shielding Effectiveness

If a shielding enclosure with conducting confines is electrically small, the field configuration within it will be determined mainly by the near field effects. And in the

---

* However, we need again to summon our "usual caveat": because of the extremely large body of available theoretical and experimental research material, we will be able to consider only selective data, and those interested in more detail, are encouraged not to limit themselves to this material.

[†] By enclosure *walls* we will understand also the ceiling and the floor, unless indicated otherwise.

388 Electromagnetic Shielding Handbook

If a shielding enclosure with conducting confines is electrically small, the field configuration within it will be determined mainly by the near field effects. And in the electrically large enclosure — the far field effects and multi-moding may be dominant. Of course, we underscore here *electrical* size (see chapter 4). The "electrical" size of the enclosure is emphasized for a good reason: a geometrically large EMC test shielding room accomodating airplanes and tanks must be considered *electrically small* at low enough frequencies (say, at 10 kHz, the wavelength is 30 km!), while a microwave PCS terminal could be deemed "large" at 5.7 GHz (λ = 5.2 cm)! In both situations, these effects can be investigated using the "usual" apparatus of Maxwell's equations and / or the circuit theory.

- If an *isotropic* radiation source is placed inside a conductive enclosure, the physical processes and mechanisms governing the energy distribution and propagation depend on the electrical size of the enclosure. The same mechanisms, as discussed in chapter 4, are at work here. When the enclosure is electrically small, no effective radiation can occur within it — thus all the processes are reduced to static or quasistatic interactions. The effects of walls (as well as other objects within the enclosure) can be accounted for by using either the near zone reactive field models, or equivalent to them circuit models, e.g., the concepts of *partial capacitances* and *inductances* or *crosstalk coupling*. Within an electrically large enclosure, the radiating waves will propagate in *all* directions illuminating the walls and experiencing multiple reflections, as shown in Fig. 5.36.

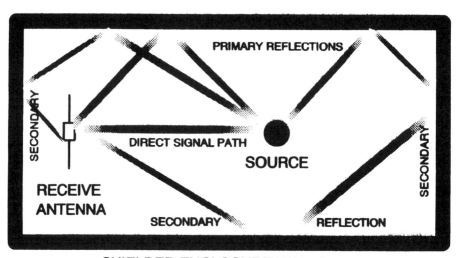

FIGURE 5.36 Signal reflections in shielding enclosure

When the enclosure is "extremely" large or its walls are lined with RF absorbers, the reflections may often be disregarded. Then, the signal at large enough distance from the source can be considered a plane wave, and far field radiation concepts and techniques can be used. However, when the reflections reaching the receiver are significant, multiple internal resonances are generated. The signal received by an antenna placed at some distance from the source *within* the shielding enclosure, will experience large variations — like illustrated in Fig. 5.37 for an EMC shielding room test site. As shown, the field intensity variations with frequency and / or location may exceed 60 dB! These variations create such serious problems for EMC measurements, that in many EMC regulations throughout the world the *untreated* shielding room was all but prohibited from use for RF radiated emissions testing! As a rule, internal resonances are required to be suppressed either by using open area test site or by lining the shielding room confines with RF absorbers.

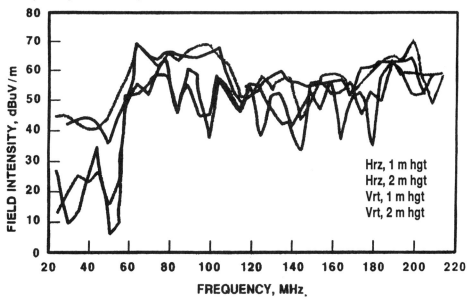

FIGURE 5.37 Infinitesimal dipole emissions in shielding room (no absorbers)

All these phenomena are well known and researched, e.g., see [1.12-17, 2.1, 2.3, A.1-5]. The only reason to mention them here is their impact at shielding effectiveness evaluation of such enclosures. Here the problem stems from the nature of the EMI environment, which affects the methodology as well as results of the shielding evaluation, both theoretically and experimentally. Suppose, we want to measure the shielding effectiveness of a large enough reflective enclosure. We install a source antenna *inside* the enclosure, and the receive antenna - *outside* (this is a typical emissions case, but similar problems can be considered also for immunity). Taking the *ratio* of the field intensities *within* and *outside* the enclosure, we obtain a *measure* of *shielding effectiveness* for this particular enclosure.

Easy? Yes.

Right? Probably, not!

If you analyze Figs. 5.36 and 5.37, it appears that not only the field intensity is *different* in various parts of this enclosure, but when we place the source antenna in different point within the enclosure, the field distribution also changes. Additionally, with the variations in antenna position, we introduce such effects like near/far field

390 Electromagnetic Shielding Handbook

transformations, variations of source and receiver coupling into the enclosure walls which is different for different shapes of enclosures, different wall reflectivities (due to wave incidence angle dependence of wall material and RF absorber properties). Another complication is due to multi-moding. To study the shielding enclosures, multi-moding effects are important and must be accounted for when defining the EMI environment within the enclosure. Practically, these effects are not an exception, but rather a rule*.

✦ Indeed, "everybody knows" that in the far field we are dealing with a plane wave, which is essentially a TEM field structure where the electric E and magnetic H field vectors are located in the plane *perpendicular* to the direction of the wave propagation. But a similar TEM H-E field configuration is also formed in quasi-stationary reactive fields. Thus, there is much in common between the fields configurations in radiated plane wave and transmission line, e.g., coaxial cable. Practically this analogy is often used in EMC to great advantage. For example, this permits to correlate the test results obtained in such fundamentally different EMC test facilities as open area test site and TEM cell. Another useful application of plane wave / transmission analogy is to reduce the complexity of theoretical analysis as well as numerical modeling and simulation of extremel;y elaborate problems of EMC electromagnetics. Some of such approaches were discussed in sufficient detail in chapter 4.
Things get much more involved in *resonating* enclosure! Along with the TEM propagation mode, a resonating cavity may support transverse-electric (TE) and transverse-magnetic (TM) propagation modes of different orders, where the magnetic or electric field vectors, respectively, are located in the wave incidence plane. In general, a resonant mode in a cavity with conducting confines can be viewed as a standing wave pattern for incident and reflected waveguide modes [A.2]. Then, any of the infinite number of *possible waveguide waves* might be formed, with any integral number of half waves between shorting ends. Since a waveguide is a media with *one-dimensional* direction of energy propagation (viewed as infinite length), to apply waveguide principles and waveguide theory, we need to *select* the "direction of wave propagation". For example consider a rectangular cavity with dimensions a, b, and d, along the x, y, and z axes, respectively, of a rectangular coordinate system. Equating the cavity to a rectangular waveguide and selecting the direction of propagation along the z-axis, the following two sets of field components can be obtained [A.2].
For the TE_{mnp} Mode we have:

$$H_x = -\frac{C}{k_c^2} \cdot \left(\frac{p\pi}{d}\right) \cdot \left(\frac{m\pi}{a}\right) \cdot \sin\frac{m\pi x}{a} \cdot \cos\left(\frac{n\pi y}{b}\right) \cdot \cos\left(\frac{p\pi z}{d}\right) \qquad (5.44)$$

$$H_y = -\frac{C}{k_c^2} \cdot \left(\frac{p\pi}{d}\right) \cdot \left(\frac{n\pi}{b}\right) \cdot \cos\frac{m\pi x}{a} \cdot \sin\left(\frac{n\pi y}{b}\right) \cdot \cos\left(\frac{p\pi z}{d}\right) \qquad (5.45)$$

$$E_x = \frac{j\omega\mu C}{k_c^2} \cdot \left(\frac{n\pi}{b}\right) \cdot \cos\frac{m\pi x}{a} \cdot \sin\left(\frac{n\pi y}{b}\right) \cdot \sin\left(\frac{p\pi z}{d}\right) \qquad (5.46)$$

---

*For instance, exploiting this principle, a whole class of EMC test facilities were developed - stirred- or tuned-mode chambers (see [2.1], also in chapters 1 and 6)

$$E_y = -\frac{j\omega\mu C}{k_c^2} \cdot \left(\frac{m\pi}{a}\right) \cdot \sin\frac{m\pi x}{a} \cdot \cos\left(\frac{n\pi y}{b}\right) \cdot \sin\left(\frac{p\pi z}{d}\right) \quad (5.47)$$

Analogously, for the TM$_{mnp}$ Mode

$$E_z = D \cdot \sin\frac{m\pi x}{a} \cdot \sin\left(\frac{n\pi y}{b}\right) \cdot \cos\left(\frac{p\pi z}{d}\right) \quad (5.48)$$

$$E_x = \frac{D}{k_c^2} \cdot \left(\frac{p\pi}{d}\right) \cdot \left(\frac{m\pi}{a}\right) \cdot \cos\frac{m\pi x}{a} \cdot \sin\left(\frac{n\pi y}{b}\right) \cdot \sin\left(\frac{p\pi z}{d}\right) \quad (5.49)$$

$$E_y = \frac{D}{k_c^2} \cdot \left(\frac{p\pi}{d}\right) \cdot \left(\frac{n\pi}{b}\right) \cdot \sin\frac{m\pi x}{a} \cdot \cos\left(\frac{n\pi y}{b}\right) \cdot \sin\left(\frac{p\pi z}{d}\right) \quad (5.50)$$

$$H_x = \frac{j\omega\varepsilon D}{k_c^2} \cdot \left(\frac{n\pi}{b}\right) \cdot \sin\frac{m\pi x}{b} \cdot \cos\left(\frac{n\pi y}{b}\right) \cdot \cos\left(\frac{p\pi z}{d}\right) \quad (5.51)$$

$$H_y = -\frac{j\omega\varepsilon D}{k_c^2} \cdot \left(\frac{m\pi}{a}\right) \cdot \cos\frac{m\pi x}{b} \cdot \sin\left(\frac{n\pi y}{b}\right) \cdot \cos\left(\frac{p\pi z}{d}\right) \quad (5.52)$$

Here,

$$k_c^2 = \left(\frac{m\pi}{a}\right)^2 + \left(\frac{n\pi}{b}\right)^2 \quad (5.53)$$

$$\beta = \left[\left(\frac{2\pi}{\lambda}\right)^2 - k_c^2\right]^{1/2} = \frac{p\pi}{d} \quad (5.54)$$

$$\beta d = p\pi, \text{ with p an integer} \quad (5.55)$$

λ is resonant wavelength. Then,

$$k = \frac{2\pi}{\lambda} = \left[\left(\frac{m\pi}{a}\right)^2 + \left(\frac{n\pi}{b}\right)^2 + \left(\frac{p\pi}{d}\right)^2\right]^{1/2} \quad (5.56)$$

Integration constants C and D are determined from the source characteristics and boundary conditions.

In the above text, the words "direction of wave propagation" are quoted because in a cavity the selected direction is *arbitrary*, while it has a definite meaning in a waveguide. It's important to realize that the obtained mathematical description (5.44) - (5.56) of a particular field pattern in a cavity will depend on the *selected* direction of propagation, and the formulas will change, if another propagation direction is selected. However, this difference is "artificial" and one set of formulas can be reduced to another one.

Note two facts:

- The TM and TE modes of the same order m, n, p have identical frequencies (they are degenerate modes)
- As the mode order becomes higher, the resonant frequency increases. Thus, to support the higher mode resonance at the same frequency, the enclosure size must be larger, than to support the lower order mode at the same frequency. ✦

With regard to shielding, the importance of the above considerations is two-fold:
- Obviously, the EMI environment within the shielding enclosure affects this enclosure shielding effectiveness. For instance, by varying the source position within a shielding enclosure, the different field modes can be emphasized and the field intensity maximums or minimums can be "strategically" placed (consciously or not) with regard to the critical areas of the enclosure (e.g., seams, windows, etc).
- When a shielding enclosure is used as an EMI test facility, the shielding effectiveness tests performed inside this facility are affected by this environment. For instance, for an illustration of the potential differences in measuring shielding effectiveness in anechoic room and a stirred mode chamber see [5.23-24].

Radio Signal Penetration Into (and Out of) Buildings

This problem became timely with the proliferation of wireless telecommunications: cordless telephones, PCS, wireless utility billing and control, even garage openers, etc. Business and residential buildings, as well as other large scale structures can be viewed as a special kind of shielding enclosures. Their shielding performance can range from an almost ideal to "virtual". Presently, quite an extensive literature is available on RF and microwave penetration into and propagation within the buildings, e.g., see [5.25-31].

✦ An example of an almost ideal enclosure is an underground mine, which is pretty much isolated from the "surface" signals. Such "enclosure" was even recommended as an EMC test site, e.g., see [2.1]. At the "opposite end", a plastic or wooden enclosure over an OATS, effectively protects this test site from the weather elements, while having very little effect on the measured signals. Of course, both "ideal" and "virtual" performance is achieved within a specified frequency band and other constraints. Thus, the same weather enclosure over an OATS can significantly absorb, reflect, and distort the upper end microwave signals (or even RF signals), especially when the enclosure is covered by rain or ice. By the same token, while a thick enough layer of ocean water prevents the radio signal propagation, a submarine communications channel can be established at frequencies of several Hz. ✦

Unless special EMI hardening measures are taken (e.g., like lining the enclosure with shielding material), the shielding effectiveness of majority of buildings falls between the ideal and "virtual". But then, given the complexities of electromagnetic wave penetration into and propagation within a building structure, the very figure of merit of this shielding effectiveness becomes non-trivial. Very often, because of the signal statistical nature and propagation complexities, the definition of shielding effectiveness as a ratio of field intensities "in front" and "beyond" the building becomes impossible or meaningless.

But there are "ways out"!. For example, in section 5.4.2 we have discussed some of the related problems and presented formula (5.36) which expressed the mean LOS received signal power as function of frequency, distance between the source and receive radiators, and the mean propagation path loss exponent coefficient α. In (5.36) coefficient α was *applied* to a *distributed path loss*, but it was also indicated that, under certain conditions, α may serve as a figure of merit of the shielding effectiveness of the obstacles in the signal propagation path. Because although a parameter of distributed path loss, α itself is not tied to the *path length* r. Therefore, if we *normalize* the receive power to a certain distance, the coefficient α may serve as a measure of attenuation for "whatever" is placed in the signal propagation path.

Indeed, taking the ratio of the receive to transmit power per (5.36), obtain *transmission power loss ratio*:

$$\frac{P_R}{P_T} = \aleph \cdot r^{-\alpha} \quad (5.57)$$

where $\aleph = G_T \cdot G_R \cdot Q/f^2$

"Fixing" $r = r_0$ and taking logarithm of transmission power loss ratio, obtain the *transmission loss attenuation* parameter:

$$A_P = \alpha \cdot 10\log r_0 - 10\log \aleph \quad (5.58)$$

As we see, parameter A_P is *directly proportional* to coefficient α. In section 5.4.2, we have indicated that in *free space* $\alpha = 2$, which yields

$$A_{Pair} = 2 \cdot 10\log r_0 - 10\log \aleph \quad (5.59)$$

. Now, if with the *same setup*, we *insert* a building wall with $\alpha = \alpha_w$ in the signal propagation path, measure transmission power loss ratio A_{Pw}, and deduct from it A_{Pair}, obtain a figure of merit of the shielding effectiveness:

$$A_w = (\alpha_w - 2) \cdot 10\log r_0 \qquad (5.60)$$

Such figure of merit permits a simple comparison between the shielding properties of different structures, and is a useful tool in determining the link budget and other characteristics, necessary for the wireless communications and other applications planning. As a matter of fact, the values of α have been measured for many important practical wall materials, and are available in literature.

As an example, Table. 5.1 presents the coefficient α (called distance/power law gradient) for several important construction wall materials at 900 MHz [5.25].

Note, that the indoor propagation conditions may be similar to those in a waveguide, in which case the losses will be *smaller*, than in free space.

In using the just described approach to shielding effectiveness, care should be taken *not to mix* different propagation modes. Indeed, as shown in Fig. 5.24b, while in general the diffracted signal obeys a certain loss increase law with the separation distance from the source, this law abruptly changes when the LOS conditions are taking place.

In average, depending on the wall material and thickness and signal frequency, the shielding loss of building materials range from 2 to 40 dB, at the windows — 2-7 dB. Of course, by lining the building with special shielding materials and controlling leakage through the seams, slots, and windows, the buildings may be significantly hardened against EMI.

Table 5.2 **DISTANCE/POWER LAW GRADIENT OF WALL CONSTRUCTION MATERIALS AT 900 MHz**

CONSTRUCTION	α
AIR	2
PLASTER BOARD	3.7
BRICK	3.9
REINFORCED CONCRETE FLOOR	5.1
STEEL	5.7
METAL SUPPORT STUDDING+METAL FACED PARTITIONING	6.5
INDOOR WAVEGUIDE KIND PROPAGATION	1.4 - 1.8

5.8.3 Radio Wave Penetration Through Apertures, Seams, and Slots: Navigating Through Literature

About the Subject

This is an extremely important and extremely complex subject. Despite the age-old history and urgent practical needs, many problems still remain unsolved, or "half-solved" (that is, mathematical equations are derived but cannot be solved, or vice versa, only "purely" numerical solutions are available without any generalization and analysis). For this reason, it is a favorite topic with academic "serious" and "commercial" research, as well as of PhD dissertations. As a result of such research efforts, many literature sources are presently available on numerous related subjects, with different degree of accuracy and "utility". The latter term is coined with regard to the ability of the engineers in the field to apply the results to specific solutions of real life problems.

Like in case of many other problems reviewed in this book, the subject is so complex on one hand, and the available sources are in such abundance — on the other, that it is impossible, impractical, and "no need" to go into every detail and every source and to repeat all this research here. Besides, in being too detailed, we risk to lose the sight of "forest behind the trees", which will be a deviation from this book's intent.

Therefore in this section, we will heavy rely on available literature sources. Our objectives are many-fold:
- Indicate at possible solutions
- Explain and "bring closer" to practical applications some solutions, where it is desirable and possible to do
- Attempt to "make sense" of some of conflicting recommendations
- Use the research results to explain some "tricky" and "unexplainable black magic" issues in shielding
- Direct the readers to at least some appropriate sources, so that from that point they could continue the research on their own*.

Rationale

Even with simplifying assumption, the solution to the aperture wave penetration problems is not simple and tightly related to the EMI environment issues. In the previous section, we have addressed the issues of the EMI environment within a shielding enclosure. There, we mentioned that in electrically small enclosure, the environment is defined by the near zone reactive fields, but as the cavity becomes electrically large, the radiation mechanisms are involved, including multi-moding. In this, latter case, the cavity resonance should be accounted for. However, when dealing with aperture radiation, the other kind of resonances should also be introduced: those associated with the aperture itself. Since there is no reason to expect that both the aperture and the enclosure resonances depend on the same factors and / or occur at the same frequencies, the picture becomes quite complicated.

The available experience proves that analytical solution of aperture, slot, and seam shielding performance is possible only for quite simplified sets of parameters, while as a rule, comprehensive treatment is impossible. However, the modern meth-

---

*Again, it should be repeated that this review is necessarily limited. I apologize for not even mentioning many sources and not giving credit to their authors, which they certainly deserve. The readers are strongly encouraged not to limit themselves *only* to the following discussion or to the quoted sources, for that matter. Although we have repeated this recommendation many times through the book, there is never too much of a good thing.

ods of numerical analysis employing powerful computers, permit to evaluate just about any shielding problem. By now, numerous methods were suggested for such solutions and numerous literature is available (see Appendix). For these reasons, we will limit our discussion to basic principles and certain illustrations, while leaving the detailed resolution of specific problems to numerical techniques.

Usually three alternative approaches are used, and the existing literature sources make good use of all of them:

- Direct solution of Maxwell's equations under certain approximations. This approach is potentially comprehensive, however often unsurmountable difficulties may be met both in mathematical formulation and numerical calculation of the problems. Of course many numerical techniques are presently available to facilitate the solution
- Reduction of the problems to circuit theory, and especially transmission line theory. This permits relatively simple and elegant formulation and interpretation of the problems, however inherent linitations of the circuit theory must be attentively watched. One way to "bypass" these limitations is to reduce the task to a set of elementary steps and then use the numerical techniques in conjunction with modern computers
- Reduction of the problem to a wire or strip (patch) antenna cases, using the Babinet principle (see chapter 4). Throughout the book, we have "liberally" applied the antenna theory principles to shielding problems. As you will see later in the analysis of specific problems in this section, even if the direct electromagnetics or circuit solution are sought, the use of antenna analogies often permits to keep the "common sense" through the most complex problems and derivations, as well as suggest convincing interpretations. In this respect, in case of aperture penetration, it is important to pay attention to inherent specifics of aperture antennas. Indeed, when dealing with wire or patch antennas, we have related their radiation characteristics to the current distribution in the antenna. Obviously, such approach does not make sense with regard to the apertures.

We will consider the aperture, slot, and seam resonances (their interaction with the cavity resonances will be addressed in chapter 7, when discussing the shielding ewnclosure design). Geometrically, a slot and a seam can be viewed as "degenerated" apertures, in which one dimension becomes small (slot) or "infinitely" small (seam). Therefore, it could be expected that the analysis of apertures, slots, and seams might be performed using identical methods. The problerm is, though, that in order to obtain palatable solutions of shielding problems, simplifying assumptions must be made. It appears more convenient, and often the only practical way, to have these assumptions different for each type of the shielding singularity!

Another issue is with the driving EMI source, illuminating the aperture. Historically, three approaches were used, as illustrated in Fig. 5.38:

- Illumination of the aperture with a plane wave
- Illumination of the aperture with an electrical dipole source
- Illumination of the aperture with a magnetic loop source

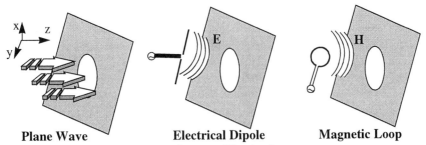

FIGURE 5.38 Aperture illumination sources

Each of these approaches is used in the literature and will be illustrated here.

Aperture Resonances

As we have established in Chapters 3 and 4, the aperture electromagnetic wave penetration effects depend on the source field distribution and the aperture size and shape. In chapter 3, we have considered electrically small apertures in the shield, and modeled them with electric and magnetic dipoles. This permitted to obtain solution for small apertures both in induction radiation fields. We have successfully applied these solutions to the study of braided and mesh shields. In Chapter 4 we have also considered the solution for large apertures in radiating fields. We have determined that even if the illuminating field is uniform plane wave, the field appearing at the other side of the aperture has three-dimensional multi-lobe radiating patterns.

Of course, three-dimensional radiation field patterns can be built for different sets of aperture parameters per (4.72) - (4.75). We have also obtained a useful insight by conveniently visualizing the two-dimensional patterns in the principal E- and H-planes. Indeed, in this case obtain from (4.72) - (4.75):

$$E_\theta = j\beta \cdot \frac{e^{-j\beta r}}{2\pi r} \cdot E_o ab \cdot \frac{\sin[(\beta b/2)\sin\theta]}{(\beta b/2)\sin\theta} \quad \text{(E-plane)} \qquad (5.61)$$

$$E_\phi = j\beta \cdot \frac{e^{-j\beta r}}{2\pi r} \cdot E_o ab \cos\theta \cdot \frac{\sin[(a/2)\sin\theta]}{(\beta a/2)\sin\theta} \quad \text{(H-plane)} \qquad (5.62)$$

As an illustration, the yz and xz plane patterns for a = 20λ and b = 10λ were shown in Fig. 4.23. As we saw, the radiation field intensity is *the strongest* over the center of the aperture (x=0, y=0).

Role of Non-Conductive Seams in Shielding Enclosure

Let's have no doubt, the *non-conductive* seam role in shielding enclosure is extremely *negative*: seams impair shielding effectiveness. This said, we are still left with questions: in what manner, how much, what to do about it? Actually, we have already considered the effects of the seams in chapter 3 (e.g., see section 3.4.1 and Fig. 3.16). There, we identified the problems with the shield transfer function due to

398 Electromagnetic Shielding Handbook

the necessity for the shield currents to "bypass" nonuniformities in the shield conductivity. The same "current problems" are at the roots of the radiation from the seams.

In principle, the problem can be solved by using a combination of the circuit theory concepts and Maxwells equations, and applying boundary conditions at both sides of the seam to determine the integration constants. However in the radiation case we also have an option to apply the Babinet's principle (see chapter 4) to reduce the task to antenna problems. Indeed, considering a seam as an infinitely narrow dielectric (e.g., air) gap, we can substitute the seam with an *equivalent* infinitely thin antenna.

Now, we can apply all the familiar formulas for electrically small and long antennas to evaluate the effects of the seams. For example, if the seam is linear, formulas in section 1.3.2 of chapter 1, and (4.41) - (4.45) and (4.53) of chapter 4 can be used.

The Enigma of Slot Radiation

Why enigma? Because in the existing literature you find two conflicting statements:

Statement # 1. "A slot is just like a seam: its radiation is determined (some go even further and claim *proportionality*) *only* by a slot *length*, and *does not* depend on the slot *width* (or *area*, for that matter)

Statement # 2. "A slot is just like an aperture: its radiation is determined by the slot area (and shape, and electrical size, etc. - see previous subsection).

So, which is right? Allegedly, you can find arguments for both statements. But as you will see in a moment, such ambiguity is quite superficial, and it is caused either by wrong assumptions or improper applications of basic principles.

For example, applying the Babinet's principle which reduces the seam problem to an (infinitely "thin") antenna problem, and using formulas (4.41) - (4.45), we can indeed claim that the field intensity radiated by a seam is proportional to the seam length. And this is the guideline given in some literature sources (e.g., see [1.20]). But don't forget two (at least) important limitations of this approach! The *first* limitation of (4.41) - (4.45): these equations are valid only for electrically small lines (read: "seams", in our case). For longer lines (seams), much more compex formulas, e.g., like (4.53), must be used, where linear radiation/length law does not hold. Also all other problems with characterizing electrically long antennas apply, as outlined previously in chapters 1 and 4. The *second* limitation stems from the assumption of an "infinitely small" antenna radius (read: "seam cross-section = width × depth), compared to the wavelength and antenna length. There goes the claim that these formulas are applicable also to *wider seams*, that is, *"narrow slots"*!

So, between the "extremely narrow" seams and "wide enough" apertures, lies the "enigmatic country" of many *practical* radiating shield slots. Obviously, to outline the borders and study this "territory" is of great practical importance.

Consider first calculation results. Fig. 5.39. presents a set of curves obtained by simulating (TLM numerical techniques were used) the effects of different electrical length and width seams in an otherwise perfect shield [5.32]. As shown, both the slot length and width affect the radiation. Note two facts:

- The effect of the length variations on the radiated emissions is much stronger than the effect of the width variations.
- At high frequencies (at the end of the simulated frequency range) in the 10 cm long slots, the difference between the radiation from different width slot is quite small - at least, much smaller than at the lower frequencies.

But there is more! Per [5.32], when a 10 and 6 cm long slots, both 1/2 cm wide, were excited at their half-wavelength *resonant* frequencies (1500 MHz and 2500 MHz, respectively), both yielded identical field intensity @ 1 m distance (far field) from the slot. However, when in the example in [5.32] the frequency was reduced well *off* the resonance, the emissions were *proportional* to the slot length at the same frequency, or to the frequency — at the slots of the same length.

✦ Comparing these results to the data in Fig. 1.27 where the maximum emission amplitudes from a cable at resonant frequencies were identical for all resonant frequencies, it is easy to come to a conclusion that such behavior is a manifestation of a more general law! Of course, this fact has serious consequences for the engineering and design of shielding enclosures and interconnections, especially with regard to EMC.✦

After reading through a large part of this book, the explanation of these results shouldn't be difficult. Indeed, let's apply the Babinet's principle to this problem. Now the problem reduces to a comparison between "infinitely" thin and relatively "thick" antennas. What's the difference between such antennas? Very simple — its their bandwidth: a thin antenna is a very narrowband antenna, while a thick antenna - has a much wider band (there is no need for references here — any one will do). Also, we can expect the thin antenna to exhibit a "sharper" polarization than a thick one (remember Fig. 1.25?)

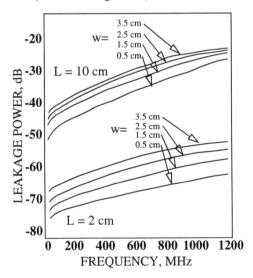

FIGURE 5.39 Leakage power for 2 cm and 10 cm long slots of different widths, per [5.32]

As we know, there is little difference between the radiation properties of narrowband and wideband antennas *at resonant frequencies*. Returning back to our slots from the "trip to Babinet's domain", this explains why close to resonant frequencies the difference between the narrow- and wide slots was small.

At *other than resonant* frequencies, the effectiveness of a "thin" resonant antenna (in our case, read *narrow seam*) quickly reduces resulting in significant decrease of radiating fields, while a "thick" antenna (read, "wider slot") retains its effectiveness in a much wider frequency band. Indeed, from Fig. 5.39 follows, that a five times variation of the slot width resulted in less than 10 dB change of the emissions, while a similar ratio in the slot length — resulted in about 35 dB difference. Note, that 2 cm long slots were well below their resonant frequencies, as well as 10 cm long slots — at a large part of the investigated frequency range.

Using the just outlined "antenna-kind" approach, you can easily analyze the limits of the suggested analogy (consider it a "home assignment").

Slots in the Shields of Finite Thickness

✦ Why not view a slot as a *narrow* aperture and use the same formulas, e.g., (5.61) and (5.62), for its evaluation? While in principle this is possible, now we have to critically review at least one basic assumption from the previous analysis: the *infinitely thin* shielding wall. In a large aperture, this assumption could be "tolerated", and it permitted us to arrive at relatively simple expressions. But when the shielding wall thickness becomes comparable to the aperture dimensions, the infinitely small thickness assumption may be not true any more. To be sure, this is not a new problem, and information is widely available, e.g., see [5.33-37].

A defining assumption is that the slot length and the signal wavelength are much larger than the slot width and depth. Also, usually the conductivity of the shield (i.e., of slot "walls") is considered either infinite, or very large (of course, implying a metal, but not necessarily so - see chapter 8 for shielding material alternatives). The result of this latter assumption is that we must compare the slot depth (shielding wall thickness) to the signal skin depth, which, of course, *reduces* with the signal frequency.

The theory behind the finite depth slots is quite complex, the solutions usually requiring different kinds of assumptions. For example, [5.34] suggests general formulas for an *infinitely long narrow slot* in *perfectly conducting* shield. For such slots in thick conducting wall only the transmission line TE mode can propagate through the slot region. Using the method of moments, the electromagnetic energy transfer through the slot can be characterized by general admittances (matrices), generalized voltage vectors, and generalized source current vectors. By neglecting the higher order TE modes and all TM modes, an approximate *one term moment* solution to the general problem can be obtained, which leads to a relatively simple equivalent circuit.

Basically, this equivalent circuit presents a parallel-plate transmission line of length d (equal to the shielding wall thickness) which "direct and return paths" are formed by the slot walls. The line can be characterized by a transmission line admittance matrix (see Appendix). The input Y^a and output Y^c admittances of this line are the aperture admittances which are determined mainly by the properties of the media on both sides of the aperture (that is, in two half-spaces separated by the shield). The transfer admittamce of the line can be approximated by admittance $Y_o = 1/wZ_o$, wher w - is the slot width, and Z_o — wave impedance (aka, "intrinsic impedance") of the dielectric *inside* the slot.

From there, important characteristics of the *slot transmission line* are deduced. Related formulas are suggested in [5.34]. A narrow slot of width w and depth d will have multiple resonances when d is slightly less than an integer number of half wavelengths of the transmitted signal (λ is the signal wavelength within the aperture, therefore it depends on the characteristics of the dielectric within the slot):

$$d_{res} \approx \frac{n}{2}\lambda, \quad n = 1,2,3, ... \tag{5.63}$$

Before we explain the meaning of the slot resonance, let's define the transmission coefficient of the slot $T = T_{trans} / T_{inc}$, as a ratio of the power T_{trans} transmitted through the slot and *dissipated* at the

Shielding Effectiveness for EMI Protection 401

admittance Y^c at the output of the slot the "c" half-space to the power T_{inc} incident from the "a" half-space at the *unit length* of the slot with *width* d. At a resonance, the transmission coefficient T_{res} of the slot can be estimated as

$$T_{res} \to \lim_{w \to 0} \left[\frac{4\lambda v}{w\pi(1+v)^2} \right] \qquad (5.64)$$

Here, $v = \varepsilon_c / \varepsilon_a$, where ε_c and ε_a are dielectric constants in the regions *in front* and *behind* the slot
The plots in Fig. 5.40 are based on data from [5.34] and present transmission coefficient as a function of the ratio of the slot depth to the signal wavelength. They provide insight in the performance of infinitely long narrow slots of different width filled with air or lossy dielectric:

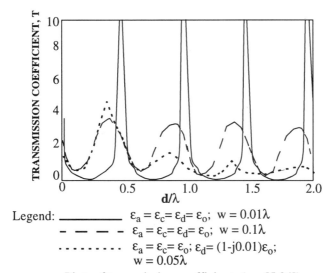

Legend:
— $\varepsilon_a = \varepsilon_c = \varepsilon_d = \varepsilon_o$; $w = 0.01\lambda$
– – – $\varepsilon_a = \varepsilon_c = \varepsilon_d = \varepsilon_o$; $w = 0.1\lambda$
· · · · · $\varepsilon_a = \varepsilon_c = \varepsilon_o$; $\varepsilon_d = (1-j0.01)\varepsilon_o$; $w = 0.05\lambda$

FIGURE 5.40 Plots of transmission coefficients (per [5.34])

- Note, that the resonant phenomena become significant, when the slot depth d constitute a "sensible" fraction of the signal wavelength. For example, using a 2 GHz signal and assuming we are looking at the ratio $d > \lambda / 20$, the limiting slot depth will be 7.5 mm - pretty thick shield, indeed.
- With non-lossy dielectric, the transmission coefficient T > 1. That should come as no surprise: what would you expect from a non-lossy resonator?
- Of course, when the losses are introduced, they affect the transmission coefficient in a "normal" for tranmission line way. By the way, even if the data on the Fig. 5.40 relates to lossy dielectric *inside* the slot, a similar effect will be produced if the losses were introduced by lossy shielding wall material, which will act as lossy conductors of the equivalent transmission line model.

Well, how to use this theory? Unless you want to go through a rigorous analysis, a "rule of thumb" estimate is pretty simple:
1. Determine the ratio d/λ

2. If d/λ is *small* enough value below the first resonance frequency, just disregard the depth of the slot
3. If d/λ is *large* enough value, but such that at *any* frequency of interest is *far enough* from *any* of the resonant frequencies, just disregard the depth of the slot
4. If d/λ is *large* enough value, such that at *any* frequency it is *close enough* to *any* of the resonant frequencies, compute T_{res} for the slot. The power transmitted through the slot is

$$T_{trans} = (Tw)_{res} T_{inc} \qquad (5.65)$$

where T_{inc} is the incident power density.

5. If the losses either in the dielectric inside the slot, or in the shield walls of the slot are significant, use more accurate formula (5.64) to calculate T_{res}.

A different approach was taken in [5.35-36]. There, a transmission line model was introduced, which is capable of modeling the resonances of *finite length* narrow slots in *final thickness* shield, including a lossy slot. Another fruitful approach [5.37] relies on time-domain representation of the unknown electric field, or an equivalent magnetic surface current, to derive a respective integral equation. Numerical methods are suggested for the solution of the equation, leading to determination of the field penetrating through a slot of *arbitrary* shape, but for an *infinitely thin shielding wall*. ✦

✦ To "bypass" the complexity of analytical approach and dependent on it expressions, a host of empirical formulas were suggested, based on a "mix" of several models: equivalent to aperture magnetic and electric dipoles, waveguide propagation considerations, and some generalization of experimental data. As an example, the following formula presents the "worse case" shielding effectiveness of rectangular aperture placed perpendicularly to the propagation signal of wavelength λ>>2L with crossection L × H (L is the larger dimension: L > H) and thickness t:

$$A = -20\log\left(\frac{\sqrt{LH}}{0.24\lambda}\right) + 27.3\left(\frac{t}{L}\right) \text{ , dB} \qquad (5.66)$$

When using such empirical formulas, one should keep in mind the limitations of the respective approximations and" inject" a good deal of"common sense". For example, expression (5.66) does not account for the resonant phenomena in the aperture, while you can see from the data in Fig. 5.40 how significant they can be.✦

Aperture Leakage in a Large and Lossy Reverberating Chamber

✦ In reverberating chamber, the integrated over a certain period of time (related to rotation of the mode stirrer) the amplitudes of all modes are maximized. All the generated modes are resonating. This permits to operate with the average field strength, assumed uniform throughout the enclosure interior. This creates an opportunity to create a theory based on the power balance for aperture excitation. In [5.38], a mathematical model is developed for this problem based on four loss mechanisms: wall loss, absorption by lossy objects inside the enclosure, aperture leakage, and power dissipated at the load of the receive antenna. The model is solved for a large enough reverberating chamber with lossy loading.

Due to the field averaging the model does not yield fine details of the field distribution and shield effectiveness, but on the positive side, it does not require the knowledge of detailed geometry, of the cavity, RF energy absorbers, and apertures. Nevertheless it presents a simple and robust method to evaluate the aperture field penetration and respective shielding effectiveness "at large" for both CW and pulseinto the enclosure through an aperture, strongly depends on the Q-factor of the enclosure (remember how much time we spent describing the dependence of shielding effectiveness upon the EMI environment?). ✦

Periodically Perforated Shields

✦ Another area of interest deals with shields with multiple perforations. While in general case, the theory behind this problem is extremely complex, several specific cases were addressed to date, including the near and far field problems, use of advanced mathematical methods and numerical techniques (e.g., spectral-domain analysis), time domain vs frequency domain. Since this is quite extensive and complex subject, and to a large degree, "unfinished business", the reader is referred to original literature for details. To start with, we can refer to works done by H. Kaden [P.1], who considered periodical arrays of apertures in the shield. Several other sources can be named, which appeared in later years, e.g. [5.39-42].

One important conclusion follows from the comparative analysis of near and far field radiation caused by field penetration through an array of apertures in a flat shield:

The predicted degrading effect of the apertures on the near field shielding effectiveness is significantly worse than that for far field shielding effectiveness.

As an illustration, the results of the shielding effectiveness simulations were presented [5.41] in the frequency band 1 through 1000 MHz for an infinitely large perfectly conducting shield of zero thickness. A periodic in two dimensions array of rectangular apertures is mounted at the shield, with (5×6) mm dimensions of each aperture measured in x and y directions, respectively, and spaced with a period 20 and 60 mm in the x and y directions, respectively. The shield was illuminated by a radiating current loop with a radius of 15 mm and placed parallel to the shield at a distance 15 mm (the distance measured along the axis z). At the shadow side, the receiving loop had a radius 5 mm and was scanning in plane parallel to the shield at a distance 10 mm.

Defining the shielding effectiveness as a ratio of the maximum (during the scan) magnetic field intensity with the shield placed between the transmit and receive loops, to the field intensity at the same position when the shield is absent, a shielding effectiveness of about 10 dB was obtained, *constant* in the whole frequency range 1 - 1000 MHz. The same shield, when illuminated by a plane wave, exhibited a far field shielding effectiveness almost *linearly* reducing with the frequency from 70 dB @ 1 MHz to 10 dB (that is, identical with the near field shielding effectiveness) @ 1000 MHz.

While some of the premises of the presented analysis might be questionable (e.g., the adopted definition of the shielding effectiveness, comparison of the fields just at 10 mm distance from the shield with a much larger far field distance, near field coupling of both, transmit and receive, loops into the shield), this is an extremely important conclusion. ✦

Joining Panels in Shielding Enclosures

We would like to finish this section on the "experimental note", to emphasize the importance of testing. Indeed, while the theory can explain and predict the *general trends* of radio wave penetration through the shield imperfections, the last word in *specific* applications frequently belongs to the experiment. The reason is of course in the numerous "real life" factors which "interfere" with even the best of theories and may be difficult to quantify.

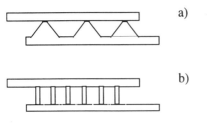

FIGURE 5.41 Contact at the shield panel joints

Thus, with regard to the problem at hand, try to predict the actual current paths at the panel joints in Fig. 5.41 for two (hypothetical) situations. Obviously, the current path depends on the number and area of the contacts, the number, shape, and size of the formed "inductive" slots, volume and surface conductivity of the formed contacts, and many other factors — leave alone the environmental and ageing variability of the same (see also chapter 7). However, these are the parameters that determine the joint "transparency" to the electromagnetic signals and, in the final count, the shielding effectiveness. An extremely practical matter, indeed.

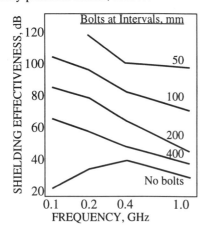

FIGURE 5.42 Shielding effectiveness of bolted metal joints

By the meantime, experimental data is not that difficult to obtain, either for the specific designs, or their analogs and models. For example, Fig. 5.42 illustrates the frequency characteristics of shielding effectiveness (MIL-STD-285 test procedure — see Chapter 6) measured on test samples. The test samples were formed by joining with M6 bolts two 35 μm thick electrolytic copper foil plates (affixed to plywood cuts) with a 100 mm - wide perpendicular edge section [5.43]. As shown, there is a direct dependence of shielding effectiveness on the distance between the bolt connections at the joint. Of course, with different metals, and under different environmental and even test conditions in general, the data may significantly vary.

5.8.4 "Mobile" Shielding Enclosures

Cars and trucks, ships and submarines, airplanes and satellites - as a rule they all have metallic confines. Comparing to "stationary" shielding enclosures, the *mobile* shielding specific has to do with EMI environment: the necessity to account for Doppler effect when the mobile enclosure is moving with high enough speed. Obviously, this is not of major consequence to the shielding properties of mobiles. Otherwise, the features of the mobile enclosures of interest from the shielding point of view, are determined by their intended application.

For example, the in-vehicle environment is an important factor determining the area of coverage of cellular communications and PCS. The following vehicle characteristics (but, sure, not only those!) influence the in-vehicle environment and affect its electromagnetic shielding performance:
- Large window and windshield apertures, comparing to the enclosure size
- Multiple enclosure parts with non-conducting seams, e.g., doors, hood, trunk
- "Cramped " interior, with varying degree to the "ambient" exposure of different areas within the enclosure
- Presence of driver and passengers, as well as seats, screens, etc.
- Presence of "internal" EMI sources

Any of these, and other, mobil shielding specifics can be addressed using the approaches and techniques that we have already discussed, as well as numerical methods. For this reason, we will limit the discussion here to certain important illustrations.

Consider a car salon. To start with, it may be viewed as a cavity confined by metallic walls. In their turn, the walls are not solid, they contain large apertures for windows and windshields. Moreover, the walls on different sides of the car are not really in contact (at least, in the RF sense). The car salon cavity is lined by non-conducting materials with diverse absorptive and reflective properties. It also contains inside numerous absorptive and reflective objects: vehicle controls, seats, passengers. The boundaries of this cavity are not uniform geometrically, nor electrically. Thus the vehicle salon can be separated in three distinctly different large areas: bottom — open at the top metal box between the car floor and the bottom of the door windows and windshield, top — metallic roof, and middle — air space limited by the side windows and windshield. The windows and windshield may be metallized. There are also many smaller areas, which are or are not a part of the named large areas, e.g., foot space at the driver and passenger side in front bottom of the salon under the dashboard. Depending on particular car design and investigated problems, you may come up with a different subdivision of the salon areas.

We can expect, that the in-vehicle electromagnetic environment is determined by the following phenomena:
- Electromagnetic wave penetration through the windows and windshield
- Cavity resonances
- Ground plane effects in the near field, e.g., the effect of the roof, floor, doors on the near field coupling

It can be expected that the bulk of the signal energy penetrates inside the vehicle cabin (receive, or immunity, regime) or is emitted in free space (transmit regime) through the windows and windshield. Then the radiation pattern is a function of the size and shape of these openings. If now we identify a car window with the aperture ab per Fig. 4.23, the picture will correspond to a situation when a signal is excited within a car cabin. Now we can apply to the evaluation of the car cabin radiating and shielding characteristics formulas (5.61), as well as the principal plane patterns in Fig. 4.23. As illustrated in Fig. 4.23, the pattern directivity

$$D = \frac{4\pi}{\lambda^2}ab \tag{5.67}$$

Of course, this formula and the corresponding patterns are true when the same conditions apply as in Fig. 4.23: rectangular aperture, uniform amplitude and phase signal wave distribution, large aperture size with regard to the wavelength. Fortunately, such assumptions can often (by no means, always!) be justified, with different degree of accuracy, in mobile shielding applications. And when they cannot be justified, usually the only recourse is to use numerical techniques.

In a similar manner, the problem of multi-moding can be addressed. The energy penetrating inside the vehicle cabin from the outside (receiving mode) or generated inside the cabin by the handset (transmitting mode), will excite local oscillations within the cabin cavity. As a first approximation, the car cabin may be considered as a rectangular prism cavity with metallic walls and dielectric lining. Then, the different modes excited within the cavity can be determined by (5.44)-(5.56) as a function of the cavity shape and dimensions, absorption/reflection properties of the car interior lining, and the objects present in the car. Because of internal resonances, the cavity may enhance or suppress the signal at different frequencies and different propagation modes. This will result in the vehicle cabin either shielding or "amplifying" the signal, respectively, in comparison to the free space propagation. Experimental illustration of this phenomena was presented in chapter 2 (see Fig. 2.9).

These considerations are supported by experimental data. The data illustrates the theoretical analysis, and also can be used for optimal placement of portable in mobile.

✦ In Fig. 5.43, the receive signal patterns are shown of a portable coaxial dipole antenna at 800 MHz [5.44]. The comparative measurements were performed in the vehicle and in the anechoic chamber ("outside" power, as per Fig. 5.43 legend), with the antenna at the head level and at the belt loop side level.

As shown, the field generated by unobstructed source in the vehicle exhibits a multi-lobe pattern, while the pattern is relatively smooth in free space (compare a) and b) in Fig. 5.43). Also, in certain directions, the signal in the vehicle cabin is lower ("shielding") or higher ("gain") than in the anechoic chamber, as discussed above. From the same graphs follows that inside the vehicle, the pattern is distorted. This is due to internal resonances. Additionally, when the antenna is located at the belt level, the human body provides for additional shielding.

However, even in the direction where the human body does not obscure the signal from the source, the signal intensity is significantly lower than when the antenna is at the head level. This indicates that in the vehicle cabin the signal loss is larger at the seat level than at the window level. Also in the vehicle, the field pattern is distorted by the internal resonances. ✦

✦ To get "a feel" for the shielding parameters of a mobil, consider experimental data. Field distribution measurements were performed in a 4 door midsize sedan placed on a flat site and illuminated by a 6 GHz signal (see Fig. 5.44). This signal was generated by a standard horn antenna to achieve the desired angular resolution. The signal in the car was received by a biconical antenna connected to a power meter. Both transmit and receive antennas were vertically polarized and the height of the transmit antenna was fixed.

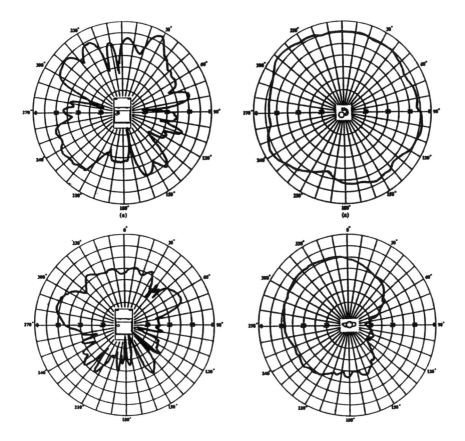

(Top) 800 MHz coaxial dipole, head level, in-vehicle signal, 0° elevation
(Bottom) 800 MHz coaxial dipole, belt loop side level, in-vehicle, 0° elevation

(Top) 800 MHz coaxial dipole, head level, outside, 0° elevation
(Bottom) 800 MHz coaxial dipole, belt loop side level, outside, 0° elevation

FIGURE 5.43 Receive signal patterns (power (dB) of a portable coaxial dipole antenna at 800 MHz (Hill and Kneisel, IEEE Trans. on VT, Nov. 91)

First, the signal power was measured in free space at the car location (with the car absent). Then the car was parked in this location. During the test, the car was placed in various positions with respect to the transmit antenna. Data was collected with the transmit antenna pointing in the following directions: driver side — Direction 1, front — Direction 2, back — Direction 3.

408 Electromagnetic Shielding Handbook

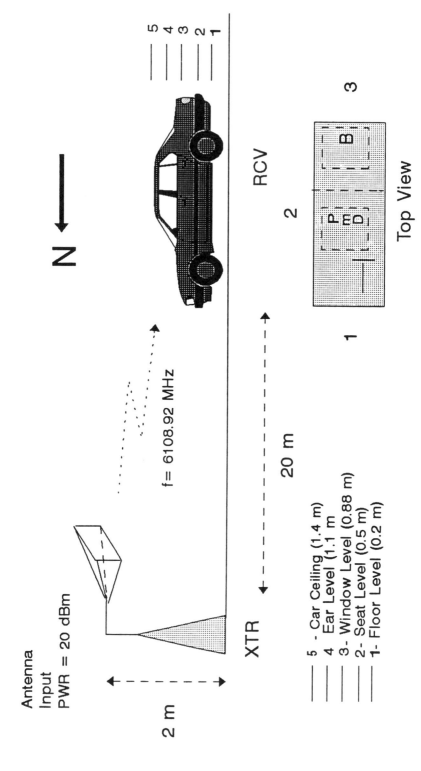

FIGURE 5.44 In-vehicle propagation: experiment layout

Shielding Effectiveness for EMI Protection 409

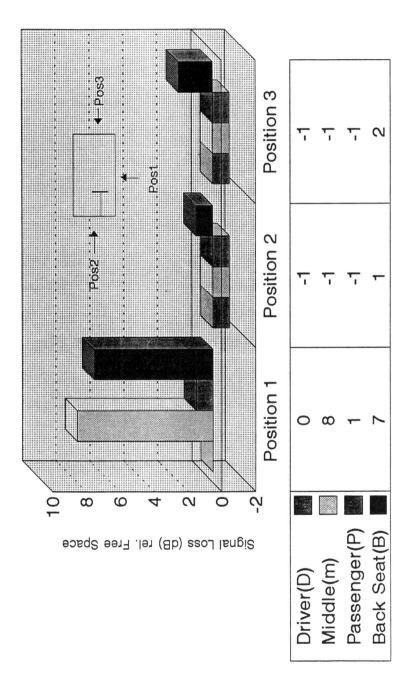

FIGURE 5.45 In-vehicle propagation: relative signal distribution in car cabin

With the transmit antenna pointing in the above directions, the signal power measurements were made inside the car at positions D — driver, P — passenger, B — back passenger. Five different antenna height levels were investigated (from 1 to 5) corresponding to the heights above the ground plane from .3 m to 1.6 m. Transmit antenna directions, in-car positions, and receive antenna height levels are shown in Fig. 5.44.

Each measurement was made at several locations within 5 cm (i.e., λ @ 6 GHz) from the nominal points, with the average entered in the log. Also, an attempt was made to evaluate the maximum and minimum receive antenna indications in some positions.

The results of the measurements are presented in Fig. 5.45 and 5.46. Following is a brief summary of the results.

6. Field distribution inside the vehicle environment is non-uniform, which results in respective variations of shielding effectiveness in specific areas of the cabin:
- shielding effectiveness (i.e., propagating signal loss) variations relative to free space ranged from -2 to 23 dB
- Minimum shielding effectiveness was observed at ear level when adjacent to windows, even when the passenger head was between the incoming wave and receive antenna
- Variations at the ear level did not exceed 8 dB
- Maximum shielding effectiveness was observed at the car floor level (23 dB)
- The shielding effectiveness at the seat level was higher than at the ear level.
2. Shielding effectiveness is larger when the wave penetrates through the car side, than through the front or back
3. Car window glass had negligible effect on shielding effectiveness (however, this was regular, non-treated glass)
4. Antenna position with respect to head had no effect on shielding effectiveness

✦

5.9 WIRELESS PRODUCT SHIELDING

Wireless Product Shielding Needs and Problems

Microwave wireless applications of the shielding technology underscore the problems with near and far field "collision". Cellular and PCS communications, medical and military wireless devices, "consumer" radar technology (e.g., microwave ovens, police speed monitors, etc), remote control products — all may be used in closed proximity to the human body. As we have discussed previously, in the microwave band the "useful wireless signal" propagation is governed by the *far field* laws, "*multiplied*" by the statistical complexities of the propagation environment. However, when a telephone handset is pressed to the ear of the user, the microwave oven "leaks" or a remote control device is held against the heart, the body around this area including the brain and other important organs, is irradiated by the *near fields* generated by the same products (as well as by lower frequencies generated within the device). That makes all these technologies dangerous near field EMI "offenders". The governments, the regulatory authorities, the occupational and public safety administrations, the manufacturing and service companies, the trade associations, the corporations, scientific & engineering societies, attorneys, and just people using cellular phone "in the street" — all have vested interest in the subject.

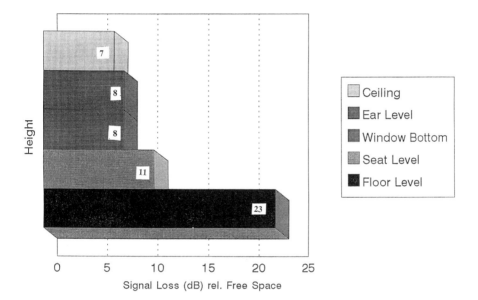

FIGURE 5.46 Loss distribution in vertical plane through the middle of the cablin
(loss vs height in front section)

While it is well beyond the scope of this book to discuss this topic at any depth, still, we cannot avoid commenting on its *technical* implications at the subject at hand: electromagnetic shielding. Take telecommunications: cellular and PCS. Presently, mobil communications occupy the frequency band, roughly, from 30 MHz to 6 GHz. Within this band, there are several microwave frequency sub-bands dedicated to the cellular and PCS transmission, the most widely used being 800 MHz, 1.9 GHz, 2.4 GHz, 5.7 GHz[*]. In designing the respective devices (handsets, base stations, amplifiers, antennas, etc) many "regular" shielding needs are identified, just as for other types of electronics. Thus, shielding product enclosures, ground planes and shielding

---

[*]With the popularity of the wireless technology, no doubt that new frequencies will be allocated to this use, and rather sooner than later.

traces on PCBs and ICs, shielded cables and other interconnections — all these techniques can be used as we have already discussed in-depth. However, there are "special" problems *inherent* to the wireless communications, which are of interest to us. Among such problems, two related near field effects captured in recent years the attention of the engineering community, as well as the public at large:
- Human exposure biological safety. This is a relatively old problem, which became important with the latest explosive growth of wireless technology
- Hearing aid electromagnetic compatibility with cellular and PCS devices. The study of this field is still in its infancy, although significant strides were made to date of this book publication.

Biological Safety

✦ Presently, the researchers concentrate on three fundamental biological effects of microwave radiation:
- thermal impact within a large tissue volume — we will call them "macro-effects"
- cellular membrane permeability variations — "micro-effects"

neuron "communications" disruptions — probably, of electromagnetic nature

To date, the most studied are the thermal macro-effects, due to the heat generation in randomly moving tissue molecules under the impact of the electromagnetic field. The amount of generated heat is frequency and tissue property dependent, with the penetration depth in the tissue from several millimeters to several centimeters. The existing standards (e.g., see ANSI C95.1) define the limits to electromagnetic radiation thermal effects in terms of the so called specific absorption rate (SAR). We have briefly mentioned the existence of such limits in section 1.1.2 of chapter 1. Basically, SAR presents the time derivative of the energy absorbed by a specified tissue mass, and is measured in units of W/kg (watts per kilogram). SAR can also be expressed as a function of electric field strength E_i in the tissue:

$$\text{SAR} = c_i \cdot \frac{dT}{dt} = \frac{\sigma \cdot E_i^2}{\rho} \quad (5.68)$$

where dT/dt is the time derivative of temperature in the tissue, K/s;
c_i - heat capacity of body tissue in J/kgK,
σ - conductivity of body tissue, in S/m,
ρ - density of body tissue in kg/m^3

From (5.68), using the established limits for SAR, some guidance limits for the human exposure power density and field intensity can be established. An example of such guidelines is shown in Fig. 5.47. As could be expected, there is abundant literature available on the subject, e.g., see [5.45-46].

For *illustration purposes*, let us use the limits in Fig. 5.47 to answer an important practical question: is computer dangerous to human help? Fig. 5.48 presents 2.4 GHz emissions from a 90 MHz Pentium computer and a wireless LAN adapter at 0.1 to 10.0 m distances from respective products [5.47]. As you see, there is "nothing to be afraid of", at least so far! (Or is it?)

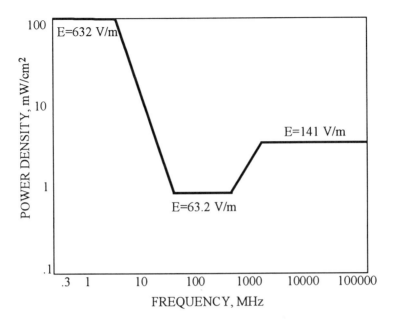

FIGURE 5.47 ANSI C95.1 RF field exposure safety guidelines
(Based on 0.40 W/kg SAR limit in exposed tissue)

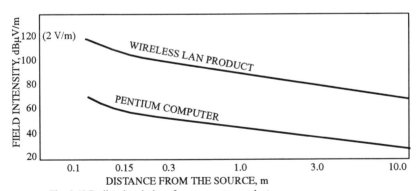

Fig. 5.48 Radiated emissions from computer products

✦

Hearing Aid EMI

✦ The limits to the hearing aid compatibility are determined by the noise at the device amplifier input, which appears as a resuolt of *audio rectification* of a microwave system. The existing or under development standards (e.g., ANSI C63.19 and ANSI C63.20) define these limits as well as repective test philolsophy. Briefly, to provide for hearing aid compatibility, there is a need to work at "both ends": to limit the wireless telecommunication devices emissions *and* improve immunity of hearing aid devices to the illuminating fields. As a result, several immunity categories of hearing aid devices (providing acceptable signal-to-noise ratio in EMI fields in the range 10 to 50 V/m), can be *matched* (for EMC) with several levels of telephones for E- or H-field emissions. ✦

Shield that Handset

So, how can electromagnetic shielding be used to improve the wireless microwave telecommunications devices compatibility with hearing aids and meet the human exposure standards? Due to many complicating factors in these problems, the shielding solutions can run from "trivial" to quite "creative". For instance, by applying a coat of electrostatic shielding, containing silver paint, inside or outside a hearing aid, 5 to 30 dB immunity improvements were achieved, depending on the hearing aid model [5.48]. We will return to this problem in chapter 8, when discussing the effectiveness of different coating shield techniques.

The use of electromagnetic shielding methods and technology for meeting human exposure standards presents a real challenge to the knowledge and creativity of designers. Indeed, here two conflicting requirements must be reconciled: *reduce* the *near* field exposure, while *not affecting far* field transmission! Technically, this can be achieved by selectively using shielding to achieve respective radiation pattern forming. In general, this problem still awaits it solution (and you thought that in the course of the past hundred years all shielding problems were solved!).

✦ Today the efforts in applying electromagnetic shielding to resolve biological problems are already under way in the industry. The applications range from serious scientific research to purely speculative, unfortunately sometimes driven by "biological scares" or even desire for publicity and/or money-making.

An example of the first is the application of shielding to a cellular phone antenna: such work is progressing as a part of R&D efforts by handset manufacturers and academia. For instance, as reported in Ref. [5.50], a shielded cellular transceiver antenna was proposed that significaly reduces the near electric field "facing" the user, while not harmfully distorting the antenna parameters in the useful, "opposite" directions: far field radiating pattern, input impedance, and the gain of the antenna. The antenna is shielded with PEC and the shielding is coated by a layer of a lossy magnetic material.

The second example deals with many types of the "shielding clothes" — aprons, shirts, jackets, skirts, dresses. Among the large number of absolutely "legitimate" and useful applications, one can find certain manufacturers' claims of "total biological protection". The problem with many such claims are that they often provide insufficient data with regard to the considered potential sources, field character and configuration, applicable frequency range, or the provided amount of shielding effectiveness.

For example, what is the meaning of a 100 dB shielding effectiveness of the clothes material as measured in a plane wave field using a specific test technique, when the outfit will be used in a strong nagnetic field of a power line? Then again, a very simple test can easily prove that the "biologically justified" apertures in the shielding outfit may seriously violate its integrity (e.g., for such vital body elements as "a head", which may be worth to have and use to evaluate the needs of "biologically safe" dress, if anything).

This is not to say, that a static-dissipative lab coat or a smock consisting say of a combination 63% polyester+35% cotton+ 2% carbon black will not do the static protection job in the cleanroom. But it cannot be recommend for protection from the radar microwave radiation, or from magnetic fields generated by utility power transformer! ✦

Human Body as a Shield: Pondering Over A Problem

Revisit Fig. 5.44. When reviewing the patterns with and without the head or the body present, it is obvious that the body acts as a shield by reducing the handset field in certain direction. But how to determine the respective shielding effectiveness? The problem appears not that simple.

First, we have to define the figure of merit for this shielding effectiveness. Allegedly, we could take a ratio of the field measured at some distance in the *direction of the body* to the field in just *opposite* direction at the same distance. However, this may not necessarily work! Indeed, the head or the body occupy a significant solid angle with regard to the handset antenna radiation. As a first choice, we could have selected the *center* of this angle as the direction of propagatiopn towards the body. But then, two problems arise:

1) different human individuals possess bodies of different shape and size, so which one we account for?

2) what do we do if this directions points toward a *null at the pattern?*

Another option would be to *integrate* the field, or radiated power, at certain angles *in* and *opposite* the direction of the body. But then, which angle to select and how to account for multi-lobe patterns?

It seems, that the answers to all these questions should come from the basic premises suggested in this book: the criteria for evaluation and figures of merit of shielding phenomena *must* be determined by the objectives of the analysis.

6

Shielding Measurement Techniques and Apparatus: The Tools of the Trade

6.1 THE PROOF OF THE SHIELDING

6.1.1 Measurement Objectives

The proof of the shielding is in testing. Recognizing this "eternal truth," the shielding industry has come up with a myriad of measurement techniques and procedures, hardware and software designs, literature descriptions, test standards, guides, and practices. Indeed, the bibliography to this chapter, as well as many general purpose references in other chapters of this book containing descriptions and / or explanations of respective test techniques, may look intimidating, and many more sources exist. Why not just refer to these sources and let the users select what they need? However, that's exactly the problem. First, many widely used test procedures often do not provide sufficient underlying background nor the application domains. Second, because the practical testing needs often cannot be defined in certain and unique terms, even if they *seem* obvious. A good illustration of this fact was presented in chapter 2, where we demonstrated the confusion with existing definitions of the shielding effectiveness. Similarly, not only may different evaluation criteria be applied to a single problem, but as a rule, the measurements to the same criterion can be performed in a number of ways. As you remember, in evaluating the shields and shielding systems, we found the way out of this predicament by relating the shielding objectives to the EMC objectives.

418 Electromagnetic Shielding Handbook

Anyway, an engineer in need of shielding measurements is faced with a vast body of information on available shielding measurement techniques. Our objective here is not to repeat this information (although a reference handbook on the shielding test methods might not have been such a bad idea), but to develop a systematic approach to its utilization. The initial step to approach this "information overpopulation" is in answering two "naïve" questions: what do we measure, and how do we measure it? With regard to these questions, common sense does suggest certain guidelines, similar to those for electromagnetic compatibility discipline at large (see chapter 1). First, the measurement results must constitute a true representation of the shielding performance. Second, the measurements must be repeatable. Finally, the implemented techniques should match the realities of the specific technical, economic, and regulatory environment.

Indeed, the specifics and objectives of measurements may substantially influence the scope, selection, and implementation of the shielding models, as well as test techniques and procedures. For example, compare the conditions that may be imposed on shielding test procedures in industrial and research applications.

- Industrial environment requirements:
  a. repeatability
  b. potential for test automation
  c. ease of interpretation
  d. simplicity of equipment setup and test sample preparation
  e. small time and labor consumption
  f. low equipment expenses
  g. specialized procedures and tooling
  h. easy physical interpretation of the test results and their connection with the product quality
- Research environment requirements:
  a. high accuracy
  b. close relevance to investigated phenomena
  c. versatility of application
  d. test stability
  e. expanded frequency band and dynamic range
  f. special and possibly unique requirements

Thus we come to a conclusion that it may be desirable and expedient to use different models to develop the shielding measurement procedures and test equipment setups for industrial and research environments, even if the shielding system and its working regime are the same in both applications. Just as it is the case with the shielding analytical evaluation, but with respectively different requirements and constraints.

Optimally (though not necessarily mandatory), to obtain a meaningful and practical figure of merit, and to facilitate the test result interpretation, the shielding measurement philosophy could parallel that of the shielding analytical evaluation. Then, the shield measurement will be based on the system approach to the same shielding models and coupling mechanisms as developed in previous chapters. For this reason, when we discussed the shielding models and coupling mechanisms, we did not ex-

Shielding Measurement Techniques and Apparatus: The Tools of the Trade 419

plicitly mention the distinction between shielding analysis and measurement. As you see, that was not an omission, on our side, but a deliberate "ploy".

If we adopt the just described philosophy, it transpires that the development of criteria for measurement technique selection, realization, and correlation should be based on the shielding application specifics and needs (see Chapters 1 and 2) and the shielding evaluation dimensions (Chapters 2 through 4). In particular, in Chapter 2 we have defined a pair of "polar" concepts for the shielding evaluation: generality and relevancy levels (see Fig. 2.26 in Section 2.6.2). A similar relationship is applicable with regard to the shield measurements. Indeed, in the course of the discussion in Section 2.6, we even addressed the practicality of the shield evaluation criteria by referring to the test realization. Thus, we can introduce test system levels (see Fig. 6.1) that are identical to the shield evaluation levels (Fig. 2.26). We also can utilized the respectively developed principles of the test generality and relevancy..

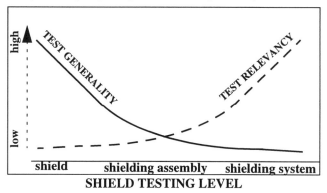

FIGURE 6.1 Shielding test generality and relevancy vs system levels

But along with *similarities*, the shielding system evaluation and testing tasks also have certain *different* objectives and "agendas" and may significantly diverge in certain important aspects. Indeed, while the main thrust of shielding analytical evaluation is on the physics of the phenomena, the measurement needs primarily accentuate hands-on factors that often conflict with the accurate theoretical modeling of the shielding system. For example, measurement repeatability is usually improved when the test system configuration and measurement procedures are simplified and standardized

6.1.2 Test Space Environment

Generally, there are tremendous diversities among shielding system requirements, numerous EMC and shielding performance parameters and limits, electromagnetic coupling and EMI transfer mechanisms, shield designs and working environments, frequency ranges, grounding conditions, measured values (dynamic range), and so on. Indeed, just to develop the shielding evaluation principles in previous chapters, we had to identify at least three shielding-related system levels (electronic system as a whole / subsystem, shielded assembly or finished product, and shield "itself") and three signal propagation regimes (static and stationary, quasi-stationary, and radiation).

But this is not all! Because of this complexity, we can hardly expect that a universal test method can exist that is equally applicable to all system levels and shield working conditions. And it is true that, at least so far, no single shield evaluation and testing method has been created by the industry to address all the complex problems, and no shielding test standard is comprehensive enough to become universal. Combining these with the differently formulated EMC and shielding limits and test implementation factors transforms shielding measurement into a multidimensional problem.

Thus conceptually, we can refer to the shielding measurement problem in terms of a multi-dimensional test space, which provides the test environment. While the *mathematical* representation of the test space environment yields many theoretical and practical benefits, it is highly convenient to have also a *geometrical* representation. Of course, this can be done for the number of factors not exceeding three. For example, we have just named three of these major dimensions: system level, shielding parameters (which may also reflect the shield working regime), and measurement realization technique.

The respective visualization in a 3D environment is shown in Fig. 6.2. Fig. 6.2 relates this shielding testing problem to a three-dimensional *testing space,* in which separate measurements procedures are represented as *data points*. Any specific test procedure (data point) can be mapped in this space as a *function* of all the space dimensions. Thus in Fig. 6.1, each measurement procedure is a function of 3 coordinates

$$P = \vartheta(C, L, M)$$

Keep in mind that we have limited this illustration to 3 dimensions *only* for the sake of graphic visuality. In general case, the testing space may have more than 3 dimensions[*].

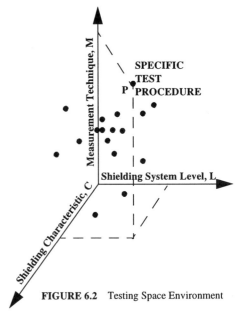

FIGURE 6.2 Testing Space Environment

Since there is a definite gravity between certain "values" at the different coordinate axes, there is a tendency to form *constellations* of data points in the testing space. For example, the system level tests will "gravitate" towards shielding effectiveness type of characteristics, while general shield tests will be "closer" to transfer

---

[*]Here, we have introduced the "multi-dimensional " approach with regard to the shielding measurements. However, its utility goes well beyond shielding or even EMC disciplines, for that matter. The same philosophy might be found useful in testing applications at large.

function types of measurements. Practically, not all the "points" of the test space can be filled, which reflects the non-availability of respective test procedures for combinations of given shielding system levels and shielding characteristics. For instance, the transfer impedance testing is often (but not always!) inconsisten with the higher electronic system levels.

Another problem stems from the necessity to correlate the test results obtained via different measurement procedures. If the same shield evaluation needs can be met using different techniques, which generally serve different purposes and are not equivalent, then we have to devise a way to bring all of them to some "common denominator." Therefore, we must add to our list of problems the test result correlation. This is necessary not only to compare different measurement methods but, even more important, to compare the measurement results with the shielding requirements and EMC limits.

6.1.3 Shielding Measurement Rationale

Three Definitions

In the following discussion we will "heavy" rely on the concepts introduced in chapters 1 and 2 of this book. We start with definitions of three terms[*]:

- *Shielding system global shielding effectiveness* denotes a measure of EMI protection quality of the *complete shielding system,* which *may* or *may not* be *linked* to the particular electronic electronic system (that is, the shielding system is evaluated and tested *in its entirety*, but may be considered *together with* or *separately from* the electronic system it protects.
- *Shield element global shielding effectiveness* denotes the contribution of a *specific shield* to the *complete electronic system* EMI protection quality measure (that is the element in question is evaluated and tested *as a part* of this complete electronic system).
- *Shield element local shield effectiveness* denotes a measure of EMI protection quality of a *specific shield, separately of any electronic system* (that is, the element is evaluated tested *in isolation* from a larger system).

Note the difference in terms *shielding* vs *shield* effectiveness. It follows from Fig. 6.1 that the latter measures are related to a more general test (that is, its results are applicable on a broader scale to different systems) than the first ones, which are, as a rule, are applicable *mainly* to the specific systems under test.

It is also worth to repeat that we bestow certain *flexibility* at the terms *system* and *element.* This flexibility is generated by the freedom to look at these terms in such a way that is *convenient* for the analysis and test objectives of any particular problem. On one hand, *any* system may be viewed as an *element* of a larger, "higher order" system — no matter how complex this particular element is *per se*. On the other hand, *any part* of a system — no matter how simple or small is that part — may be viewed itself as a " high order" system, even as a complete system .

---

[*]Here, we can all but repeat the usual "caveat": the given definitions are pretty much arbitrary; their main purpose is to identify the subject and avoid ambiguity in the discussion.

Shielding System Global Shielding Effectiveness

Consider first the *shielding system effectiveness* which, per our terminology, is a more relevant (with regard to the specific system under test — EUT) and less general (not necessarily applicable to other EUTs) test, than the element effectiveness. Important to remember that the electronic system itself can be considered at different levels: (e.g.,

LAN ⇒ specific node subsystems ⇒ specific product and interconnection subsystem ⇒ etc).

The *choice* of the system to be considered depends on the test objectives. We have discussed this subject in previous chapters. Therefore here we will just note that a correct choice can significantly facilitate the test and its result interpretation, while the consequences of a wrong choice can be disastrous. This is usually, where the experience counts! Since endless choices are possible, we will restrict the following discussion to only several important options:

- shielding system / subsystem as a whole
- shielding assembly: product enclosure / cable / connector / gasket
- shield "proper".

Anyway, it appears that the shielding system effectiveness *must* be measured at the *respective* system level*. Whereas this is true in general, the distinction should be made between the *shielding system* and an *electronic system* which this shielding system is protecting. The fact is, that we can test the *shielding system* performance *with or without* the protected *electronic system* or its elements present!

As an example of a shielding system / subsystem *as a whole*, consider a computer and a printer designed with shielding enclosures, and interconnected by a shielded cable. There, the two enclosures and the cable shield (together with proper connectors and gaskets, if present) constitute a shielding system, which protects a "computer+printer+cable" electronic system. As just discussed, we can imagine two 'polar' options of testing *the complete shielding system*:

a. The shielded electronic system is tested "as is": the EUT consists of the *electronic system with* the *shielding system*, as manufactured and / or assembled. Hopefully, at this stage of our discussion you don't wonder how to test (see Chapter 1): of course, by measuring and comparing emissions or immunity of our EUT *with* and *without* the shielding applied! This is a typical example of an "absolute" test.

b. *Only* the *shielding system, without* the "stuffing" — actual computer, printer, or even cable core — is measured. Then, in our example, the EUT consists of two shielding enclosures interconnected by the cable shield. The same "emissions / immunity" measurements principles can be applied. But now, to measure the emissions or immunity of such EUT, we need to introduce an EMI signal source or a field sensor, respectively, first *outside* (that is, in *"free space"*), then — *inside* our EUT (that is, within the shielding enclosure confines).

The same options are applicable also to shielding assemblies consisting of different *combinations* of one or both product enclosures, cable, connectors, and gaskets.

---

*This is true even for the shield-"proper". Even if the shield "itself" carries more of an element attribute than a system, but as we saw in Chapter 3, practically even in this case we *need* to consider some kind of a system, though a simplified one.

Even the *shield "proper"* (that is without system reference) can be treated this way. However as we found in chapter 3, it is much more practical to "strip" such elementary system almost to "bare bones", leaving only rudimentary system attached. That's how we came up with the shield transfer function parameters.

Which test, a. or b., should we prefer?

Both have merits and demerits. The "absolute" shielding system effectiveness (option a. — at any level) can be *directly* applied to the respective level electronic system EMC performance. For instance, if shielding is used to meet EMC emissions compliance limits, by applying the shielding system and performing the system emissions or immunity test (per applicable standard), you can see "right away" if the limits are met. By *deducting*, say, shielded system emissions from the unshielded system emissions, *some measure* of the shielding system effectiveness can be obtained. To apply this shielding system effectiveness to a second, but *similar*, system, you can simply *deduct* this *shielding system effectiveness* from the *second unshielded* system emissions (both in dB). And since the shielding test was done at *the same* system and using *the same* measurement procedure as in the EMC compliance test, there is a high degree of assurance that the shielding will work as predicted.

Easy to see, that when the *new* electronic system is *not* similar to the one used for shielding system effectiveness measurements, the shielding *may* perform differently. First, there are the interaction effects between the shield and the specific protected elements. Second, usually there are additional effects and interference due to the presence of *other* subsystems, which may be not a part of this shielding system, but a part of the electronic system under test (EUT). On one hand, the presence of these "other" elements facilitates the working shield environment which is close to the actual application — that's the "good news". On the other hand, in the presence of these, other subsystems, especially if they are not shielded, the shielded subsystem may *not* (and often will not— just *because* of the shielding protection) *be* the primary radiator — see chapters 1 and 4. Then such test results may be generated by the interference from the primary radiators, and will not necessarily indicate the value of shielding effectiveness produced by the shielding (the "bad news"). This is why we have italicized the words *"some measure"* in the previous paragraph. Therefore, the shielding performance "absolute" figure of merit obtained using one EUT, may be inapplicable to another EUT — that is, the test result *generality* is extremely low (it's a good illustration of this term utility, isn't it?).

Consider now the most specific shielding system test, *"detached"* from the electronic system itself . We can view such system as a *model* or *simulation* of an actual system. While the previous "absolute" test considerations apply, there are important specifics. On the positive side, now we *can* limit the test to *only* the shielding system in question, because we *don't need* the rest of the electronic system to provide for the system generators / receivers and radiators / sensors. Now, our test becomes more sensitive (due to the absence of interference from other system elements) and more general (that is, the results are applicable to a wider class of electronic systems, than in the case of the absolute test). However, by the same token, this test is less relevant with regard to a specific system.

424 Electromagnetic Shielding Handbook

Shield Element Global Shielding Effectiveness and Shield Element Local Shield Effectiveness

Based on the concepts of electronic system EMI synthesis and analysis developed in Chapter 1, each shielding system element can be treated as an element of the respective active antenna array. Then, following the shielding evaluation flowcharts developed in the previous chapters, we have *two options* of performing the shielding element test:

- The *element global shielding effectiveness* can be determined experimentally by measuring and comparing this element's contribution to the system radiated emissions or radiated susceptibility with this element unshielded and shielded, respectively.

  Since different electronic system levels can be employed in such a test, the *respectively* correponding levels of the element global shielding effectiveness will be obtained. The same considerations apply, as in the complete shielding system test. In the *complete shielding system* test, one *element* system test "stands out" — the "absolute" test, in which the *complete actual* system is used. But now, the *electronic system* emissions or immunity are compared when *only* the *tested element* is unshielded and shielded, while all the other system elements do not change (preferably, they are kept as designed for actual application — with or without the shield).

  If the EUT is a *shielding assembly* which includes, along with the shield under test, *all or any* other elements — electronic product shielding enclosures, shielded and non-shielded cables, connectors, gaskets — the measurements of an element shielding should also be considered a system test, resulting in the *element global shielding effectiveness*.

- In Chapter 3, we have introduced the shield transfer parameter set as a measure of the *element shield* (that is, local) *effectiveness*, as contrasted with the *element system* (that is, global) *shielding effectiveness*. There, we have concentrated on the shield transfer impedance as the most (almost "the most", remember?) general figure of merit of the shield EMC performance. As we saw, the transfer impedance test can also be considered as an extreme case of a glopbal system test — kind of "degenerated" system test.

Shield Element Global Shielding Effectiveness: Illustration

✦ If you still keep wondering about this "good news / bad news stuff" about the system shielding test, following is an illustration based on actual data.
  Consider a test performed in a semi-anechoic chamber (SAC) in the radio frequency band of 30 MHz to 220 MHz, as described in Chapter 1. Fig. 1.12 in Section 1.1.4 illustrates the test setup, and Figure 1.13 gives an idea of the data obtained. As a part of the test program, the test was performed with shielded and unshielded cables, and also with the cables disconnected alltogether. Let us determine the system shielding effectiveness of the cables (that is, the cables are the tested system element).
  There are several ways to use this data to derive the cable shielding effectiveness. For example, projecting the maximum bounds of the emission spikes in Fig. 1.13 (which were derived with the cable under test shielded and nonshielded) and deducting one curve from the other, we can obtain the frequency characteristic of the "absolute" system shielding effectiveness of that particular cable (see

Fig. 6.3). Moreover, if necessary, the effect of shielding of any group of cables or cable/connector assemblies can be determined in the same way. Because this is done on the actual system, with the actual signal generated by the system in the most realistic working conditions, we have accounted for all the factors affecting the radiation from both nonshielded and shielded EUT (see also the previous discussion in this chapter, as well as in Chapters 1, 2, and 4).

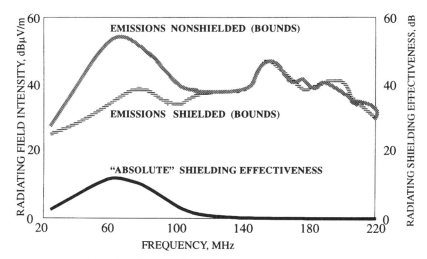

FIGURE 6.3 System radiated emissions and shielding effeciveness

However, along with these advantages, there are also serious drawbacks to such a shielding effectiveness test, some of which were mentioned in Section 1.1.4. Also, a brief analysis of using this "absolute" test for cable shielding effectiveness evaluation is given in Section 2.3.3. Generally, there are two groups of problems associated with the absolute system shielding effectiveness measurements. The first group can be traced down to the measurement method: the cable under test contribution to the system emissions can be registered only if the cable under test is a dominant radiator (see the second author's note in Sections 1.1.4 and the discussion in Section 2.3.3). Really, how would you otherwise explain the fact that the cable shielding *or* the cable complete disconection result in the same emissions?! But even if the nonshielded cable were a dominant radiator in the system, after it is shielded, it most probably will become a secondary radiator, and its contribution will not be detectable in the "absolute" test. (And isn't that the goal of the shielding?)

Consider Fig. 6.3. While it might be reasonable to expect that the emissions from the nonshielded cable in the frequency range below 100 MHz determine the maximum emissions from the system as a whole, the contribution of this cable to the system emissions at higher frequencies is not clear without additional investigation, because when we shield or disconnect this cable, the system emissions do not change. The situation is even more uncertain in the case of system emissions with the cable shielded, because we cannot be sure that the shielded cable is still a dominant radiator in the system. Now, what do we get by combining (that is, deducting) two "uncertain" curves? The only information we can obtain from the "absolute" element system shielding effectiveness curve in Fig. 6.3 is that, in the frequency range below 100 MHz, the actual shielding effectiveness is not worse than defined by the curve, while the 0 dB part of the curve is meaningless. Indeed, this is not extremely useful. ✦

6.1.4 Roadmap to Shielding Measurements

We can now introduce a certain classification of shielding testing prolems. Since the selected system level determines the applicable shielding definition, from there we can proceed to selecting the corresponding measurement techniques and consequently their practical realization. A respective "roadmap" to shielding measurements can appear as shown in Fig. 6.4. This roadmap suggests a shielding measurement classification along six criteria: system level, shielding characteristics, measurement technique, EMI signal carrier, test facility/jig, and test signal. As shown, the analysis is far from simple, especially in view of interdependence of different measurement attributes. (That's one of the reasons why we needed a roadmap, in the first place.) In the following sections we will review the basic principles of shielding measurements, based on this roadmap.

◆ It should be realized that the suggested "roadmap" is arbitrary and incomplete, since it is impossible to give a classification of all the real-life objectives, conditions, and realizations of shielding evaluations and measurements. For this reason, the classifications in Fig. 6.4 may be considered only as a guide. Departure from this system may be possible and *desirable* (or even *mandatory*), when justified by the problem specifics and test objectives.◆

6.2 GLOBAL SYSTEM SHIELDING EFFECTIVENESS MEASUREMENTS

To test the shielding effectiveness, it is the most expedient, whenever possible, to use the existing general purpose EMC test facilities and procedures. The bibliography to this and other chapters of this book contains many EMC test standards [6.1-11], reference and textbooks [2.1], and original papers and articles describing the EMC test facilities and the respective general EMC measurement techniques, their theory, design and use [too many to list]. Generally these details will be omitted from further discussion here. In the next sections, we will concentrate only on the specifics of their use for different kinds of shielding measurements.

6.2.1 System Shielding Effectiveness Measurements in RF Radiating Fields

There exist a number of test facilities which are the most often specified and used for general purpose EMC measurements:
- open area test site (OATS)
- anechoic (semi-anechoic) room
- absorber-lined OATS
- TEM (GTEM) cell
- shielding room

Shielding Measurement Techniques and Apparatus: The Tools of the Trade 427

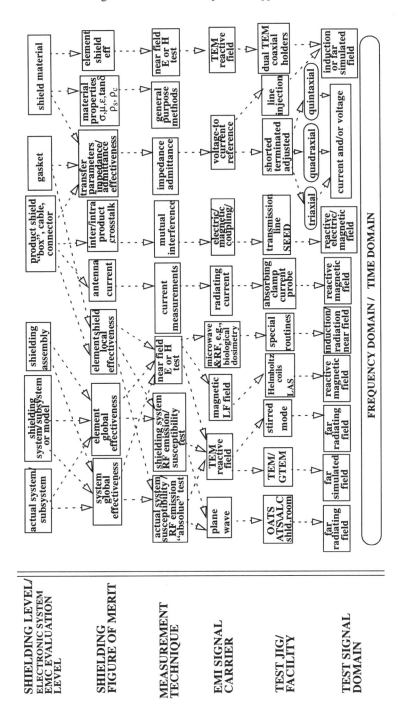

FIGURE 6.4 Roadmap to shielding measurements

Far Radiating Field Measurements

General purpose ("FCC", "EN", etc. type) EMC emissions measurements in the "radio frequency and microwave band" (30 MHz to 1 GHz and higher) are usually performed in an open area test site (OATS) or in an absorber-lined chamber (ALC) which contain a ground plane of large enough size. Probably, the simplest of all can be[*] OATS, which either has no confines, or may be enclosed only in a weatherproof non-conducting enclosure. On the contrary, the ALC has a conducting shielding enclosure, lined with electromagnetic energy absorbers, which dampen the reflections inside the facility, as shown in Fig. 1.12. The OATS is used to perform emission testing, while the ALC can be used for both emission and susceptibility test. As a rule, during the radiating emission test, the system under test (SUT) is installed on a turntable and rotated 360° at a specified height over the ground plane — usually about 1 m (e.g., see Fig. 1.12 for a system setup in an ALC). However, large equipment is installed at the ground-plane level. A receive antenna placed at 3, 10, or 30 m from the SUT is height-scanned (usually 1 to 4 m over the ground plane) until the maximum emissions from the system are measured. The maximum of the emissions is reached at such receive antenna heights and turntable azimuth angles where the direct and reflected wave amplitudes are maximal, and phase difference between them is minimal.

Each facility possesses certain specific qualities, plus certain merits and demerits that define its preferential application areas. For instance, although the plain shielding room (without absorbers) is recommended for emission testing by some of the standards (e.g., military standards), the signal reflections from the walls make a reliable test all but impossible, except for very low frequencies (e.g., see Chapter 5 of this book for illustrations). However, if proper "care" is exerted, the plain shielding room may be used for immunity testing. Two problems inherent in the OATS and ALC are EUT reactive coupling to and the radiating field distortion by the ground plane and the so-called "blind" angle — a solid angle in space "not seen" by the receive antenna during its height scan and EUT rotation (e.g., see Ref. [2.1, 2.3]). Both these problems can affect shielding measurements because coupling to the ground plane may distort the radiation patterns of the shielded and nonshielded system elements differently. Whether this will result in variations of the determined shielding effectiveness will depend on the character of the signal variations with the change of the position of the cable.

In principle, both these problems can be solved by administering "free-space" environment testing in conjunction with an equal separation distance path between the EUT and a circularly scanned (around the EUT) receive antenna. Such a test might be performed in a fully anechoic chamber or in an absorber-lined open area test site (ATS) [6.12 -13] — see Fig. 6.5. Although these test facilities often are not approved for "official" EMC testing, nothing prevents their use for shielding effectiveness measurements!

---

[*]But not necessarily is: the list of OATS "amenities" may include the weatherproof non-conductive enclosure, special shielded instrumentation room, TV monitors at the ground plane, airconditioning and ventilation, etc.

FIGURE 6.5 Evaluating radiating emissions and shielding effectiveness of a PCS handset at an absorber lined OATS (partial lining shown)

Near Radiating Field Measurements

In the near fields, the quality of "solid" (that is, without perforation) metallic shields is determined by magnetic field penetration. The basic configuration to measure the shielding effectiveness usually consists of a set of two loops — transmit and receive — placed at the opposite sides of the shielding barrier (e.g., see Fig. 2.19). Usually, CW testing is used and the two loops are placed parallel to the shielding wall and at the same axis. To account for the effect of apertures in a perforated shield, the method was modified, by incorporating the receive loop scan [5.41]. Several alternative near field testing methods were also suggested, based on the use of TEM cells, dual TEM cells, and other techniques. They are used mainly to test shielding materials — see section 6.5.3 later.

TEM Environment

The free-space test environment can be also simulated in the transverse electromagnetic (TEM) cell (see Fig. 6.6), which physically presents an expanded segment of coaxial cable [2.1]. Presently, a several TEM cell modifications are available. While they may differ in design, the principle of their operation is similar. Also, a number of the TEM "cousins" exists: parallel plate line, "G-strip", etc.

As we know, the electric and magnetic field vectors inside a coaxial cable are orthogonal. Thus, although this is inherently an induction field, its structure is just like the structure of a plane wave radiating field. Moreover, by having the coaxial cable center and outer conductors of rectangular cross sections of proper dimensions, and by tapering the cable at the ends (to avoid reflections and to fit standard coaxial cable connectors), the field in a certain volume of the cell can be made uniform, with both electric and magnetic lines of force linear.

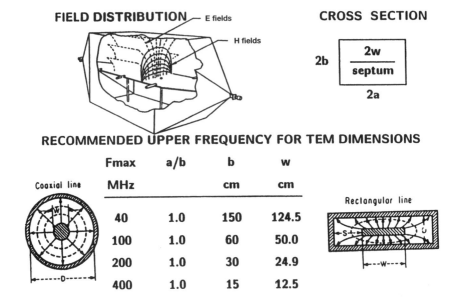

FIGURE 6.6 Transverse electromagnetic (TEM) cell

This TEM cell can be used to test small systems in a sufficiently low frequency range. At high frequencies, this facility has internal resonances and generates higher-order modes. Although this results in serious limitations with regard to the maximum size of the measured system and the maximum operating frequency, a TEM cell has been used successfully to measure the emissions and susceptibility of relatively small systems in the frequency band not exceeding 100 to 400 MHz (depending on the cell and system size). To "catch" all the energy radiated by the EUT (or to irradiate the EUT from all possible directions (in susceptibility testing), the EMC measurements are conducted for the EUT positions in three orthogonal planes, and the results are used to calculate the total field.

The drawbacks of the TEM cell have been addressed to a certain degree in the gigahertz transverse electromagnetic (GTEM) cell shown in Fig. 6.7a . By matching one end of a TEM cell with a distributed load and applying RF absorbers to this end, the internal resonances and the higher-order modes in a GTEM cell can be suppressed up to very high frequencies—some authors indicate 18 GHz as a limiting frequency. Also, unlike in the "regular" TEM cell, the septum (that is, the "center conductor") of the GTEM cell is shifted toward the roof of the cell, thus increasing the useful volume inside the cell. The test methodology is similar to that in the TEM cell and involves the system rotation in three orthogonal planes. There are several sizes and alternative designs of the GTEM cell available on the market. As an exam-

(a) EMCO model 5305 GTEM cell, used for emissions
and susceptibility testing from DC to 1 GHz

(b) Wireless product compliance test in the GTEM cell
with "hyper-rotation" capabilities

FIGURE 6.7 Gigahertz transverse electromagnetic (GTEM) cell

ple, an installed GTEM cell with hyper-rotation is shown in Fig. 6.7b. The hyper-rotation feature permits the GTEM cell to rotate around the EUT installed inside, while the EUT itself is immobile. This is especially convenient when the EUT maximum radiated emissions or radiateing patterns have to be determined. With some conditions attached, the GTEM is accepted by several regulatory authorities to perform "official" EMC testing and may be a useful instrument for shielding effectiveness evaluation.)

Shielding Room Reflections Put to Good Use

A mode-stirred (tuned) chamber is a large cavity with reflective (the higher the reflections — the better!) metallic surfaces whose boundary conditions are continuously and randomly perturbed by a rotating conductive tuner or stirrer. The field generated in such chamber is formed by randomly distributed plane waves coming in all directions, which are used to illuminate the electronic system under test (e.g., EUT measurements are illustrated in Fig. 6.8.

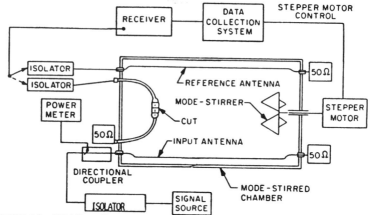

FIGURE 6.8 Shielding effectiveness measurement in a mode-stirred chamber per MIL-STD-1344A

When the number of such waves is large, their distribution becomes almost uniform, which results in the relatively smooth frequency characteristic of the field inside the chamber (see Fig. 6.9).

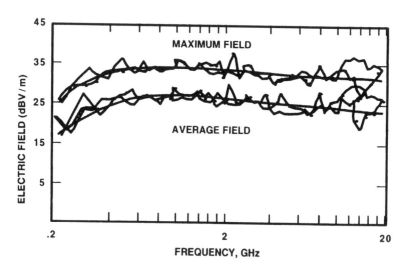

FIGURE 6.9 Electric fields generated inside a mode-stirred chamber

Such a facility can be used to test emissions and susceptibility of electronic systems. No rotation is required, due to the inherent test field characteristics. However, the directional information of the field is lost. In practice, this test is limited to frequencies not less than 200 MHz and requires complex processing of the test data; however, the maximum measurement frequency can be extremely high.

The mode-stirred chamber has proven itself as a very powerful tool for measuring small system, assembly, and element — a cable, connector, gasket, or material shielding effectiveness. It is especially useful for making measurements of objects with "irregular" geometry (e.g., connectors, or multi-branch cable assemblies) and testing in an extremely high frequency range — up to 40 GHz.

6.2.2 System Measurements in Magnetostatic Fields

Obviously, here we are discussing the near field measurements. To measure system shielding effectiveness in static and stationary magnetic fields, setups consisting of systems of magnetic loops are used. For example, the so-called Helmholtz coil sets [6.17] consist of a pair of *parallel* current loops (Fig. 6.10). The word *parallel* is related to geometry. Electrically, the coils may be connected either in parallel, or in series, with the latter preferable to facilitate the field uniformity and calibration. The Helmholtz coil sets are readily available with 12" to 48" coil diameters and also can be designed to other dimensions. In a properly designed system, the magnetic field in the space between these coils is uniform and can be easily controlled and/or measured. Depending on the applications, the *upper* frequency of the Helmholtz coil use may range range up to 20 - 30 kHz to tens of MHz.

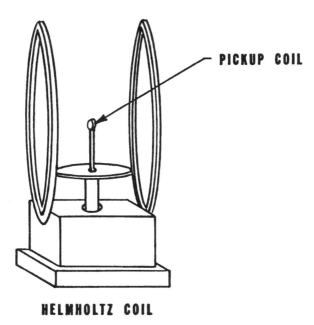

FIGURE 6.10 Shield effectiveness test set for magnetic fields (Helmholtz coils)

To measure magnetic emissions from larger systems in a wider frequency band (9 kHz to 30 MHz) the loop antenna system (LAS) was suggested [6.18]. An LAS consists of three mutually perpendicular large-loop antennas, each having a standardized diameter of 2 m. It measures three orthogonal components of the effective magnetic dipole moment generated by the EUT.

6.2.3 Near Field RF and Microwave Measurements

At the other end of the frequency spectrum are the near field RF and microwave shielding effectiveness measurements. Again, to test the shielding effectiveness, we want to make the best use of existing EMC test procedures. Then, by comparing the EMC performance in terms of the respective units for unshielded and shielded EUT the corresponding measures of system and / or element shielding effectiveness can be figured out. Except for the difference in the EMC test procedures, the same principles applied to determining the shielding performance, as described above.

Unlike in the far field RF and near field magnetostatic system testing, the near field RF and microwave measurements require, as a rule, *custom* EMC test facilities and test methods. This is due to the fact the in near RF and microwave fields, the EMC test conditions are difficult to predict and maintain. Two examples of such testing, considered in the previous chapter (section 5.8), dealt with microwave wireless and PCS issues: biological dosimetry and hearing aids EMI. Of course, for numerous important particulars and requirements of the custom near field EMC test methods and facilities, one should refer to the respective specifications. Thus, for the mentioned above examples, the detailed EMC test procedures are described in the respective standards or related literature, e.g., see [6.19-21].

6.3 SHIELDING ASSEMBLY MEASUREMENTS

6.3.1 Specifics of Shielding Assembly Measurements

It is just about time to ask an "insidious" question: if the "absolute" or other complete system shielding effectiveness tests are so problematic that they may render the measurements useless, why do we dedicate so much attention to their discussion? There are several reasons. First, in certain practical situations this *is* what we may be interested in. Truly, if the system and its environment are known and physically exist, a system test may present the most specific (for the particular system), the most reliable and relatively simple way to determine if the shielding is sufficient (within the indicated limitations).

Second, similar "absolute" system techniques can be used to measure the shielding effectiveness of separate subsystems, shielding assemblies, and elements of the base electronic system. We have already noted the *flexibility* of the terms *system* and *element* (see above, the end of sub-section "Three Definitions"): any of these items may be viewed as a system or as an element — whatever is convenient for the problem at hand. Using this flexibility and properly choosing the system, subsystem, or assembly under test, the limitations caused by the relations between the dominant and secondary radiators (sensors) in the system can be "by-passed", while to a large degree retaining the advantages of system test. Of course, this makes the EUT choice even more crucial.

Third, the problems encountered in the system shielding effectiveness measurements are not unique to the "absolute" testing. Similar problems may plague the shielding effectiveness measurements at any level. For example, even if our intention is to test *only* a *cable shield*, we will still need some means of connecting the cable and its shield to the generators, loads, receivers, and other related equipment. Thus, we will end up with a system anyway, and *always* have to deal with some kind of a system, whether we want to or not. Then, in the previous example, if the emissions from the setup interconnections are comparable to or exceed the emissions from the cable under test, we run into similar dominant / secondary radiator problem, as just described for the system test. Experience proves that such situations are responsible for many failed test plans,wrong test results and their interpretation.

6.3.2 Shielding Assembly Measurements in Radiating and Magnetostatic Fields

When testing shielding effectiveness, an important case of a system (or subsystem, or element — depending on our choice and need — see above) is the "box"/cable/connector assembly. By treating such an assembly as a system consisting of specific elements, we can design very useful shielding ananlysis and synthesis measurements. First, it permits us to evaluate the EMC performance and shielding effectiveness of a given assembly, as a "whole" at a "relatively realistic" level in the context of a larger system. But it can also be used to evaluate the performance of a single element — a "box", a cable, or a connector in an "assembly environment." For instance in a "box"/cable/connector assembly with "good" (in terms of shielding) boxes and connectors these elements become secondary radiators, then the shielding effectiveness of the cable (dominant radiator) can be determined. Or in a cable/connector assembly, a "good" cable (secondary radiator) permits us to determine the shielding effectiveness of the "not so good" connector(s). The key to such a test is correct selection of the dominant and secondary radiators (sensors).

In general, the methodology of the cable/connector assembly shielding effectiveness measurements may replicate that of the shielding system test or an element system test. For example, to measure an assembly, it can be installed at the turntable in the OATS or ALC, ATS, TEM/GTEM cell, or a mode-stirred chamber and powered with a given signal. The emission results are compared with the emissions from a nonshielded line or the power fed into the assembly (e.g., see Fig. 2.25).

An alternative arrangement is to measure the susceptibility of the assembly in a specified field (see Fig. 6.11). This susceptibility-based measurement technique is especially convenient when testing nonlinear shielding materials because then the EMI field can be accurately calibrated. It is especially simple in TEM or GTEM cells. Indeed, if the voltage applied to the cell is U, then in the absence of resonances the electric field intensity in a TEM facility with the distance D from septum to the ceiling (or floor) is:

$$E = U/D \qquad (6.1)$$

When testing in a GTEM cell, the EUT position along the cell's longitudinal axis can be selected so as to obtain a wide range of test field intensities)

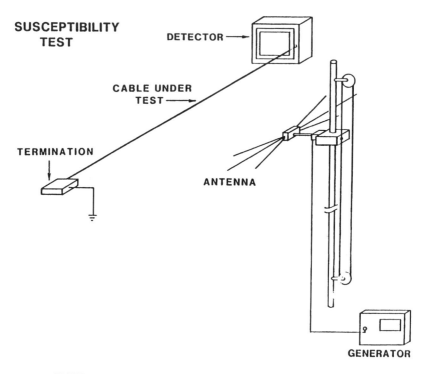

FIGURE 6.11 Cable EMC performance evaluation, susceptibility test

Testing isolated (from the system) assemblies has both positive and negative implications. The main advantage is that in the absence of other radiating system elements the emissions from the assembly are not obscured by other radiators. Also, unlike the "absolute" system test, which is based on the actual system-generated signal (which are usually digital in modern systems; see Chapter 1), the assembly testing can be performed using both digital and analog signals of a specified amplitude and wave shape parameters supplied by the signal generator. Usually, a convenient and sensitive measurement setup is based on the combination of a receiver or spectrum analyzer and a tracking generator, though noncoherent measurement routines can be also successful. An example of data obtained with such a test in a TEM cell is shown in Fig. 6.12 (compare this to Fig. 2.11).

However, by extracting the assembly from the system, the mutual coupling of it to other system elements is changed, which may produce a drastic effect on its performance. The modeling techniques referred to in Chapter 1 and further developed in

Chapters 4 and 5, at least permit an approximate evaluation of the introduced inaccuracies. Thus, applying the described methods of the system EMI analysis to the active antenna array modeling the system, the effects of the test conditions on the shielding effectiveness measurement can be estimated.. 2

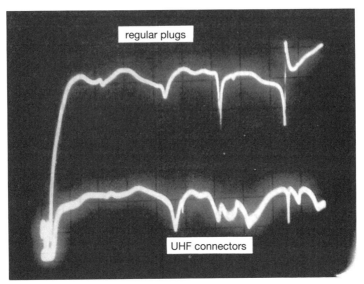

FIGURE 6.12 Shielded power supply cord emissions, 0–500 MHz, TEM cell test

In this respect, these practical questions must be answered with regard to the development of the assembly shielding effectiveness measurement procedure:

- Should we or should we not rotate the measured shielded assembly and height-scan the receive antenna when measuring in the OATS or ALC, or rotate the assembly in the orthogonal planes during the measurements in the TEM and GTEM cells?
- Should we look at the horizontal or vertical assembly polarization, or both?
- Do we need a ground plane and, if yes, then what configuration, size, and distance to it?

Allegedly, when the unshielded reference and the shielded assembly are almost geometrically identical, there is no need to do that. However, this may not always be the case! As we saw in Chapter 4, the antenna current propagation mode in the shielded and nonshielded line and the radiating field configuration (radiating pattern) depend on the grounding of the shield and its termination (e.g., pigtail), on the presence of and the distance to other conductors and ground planes within the near radiating zone. This produces differences in the number, magnitude, and direction of the radiating pattern lobes. As a result, the maximum field intensity in the shielded and non-shielded lines can occur at different azimuth and elevation angles, which mandates for the turntable rotation and receive antenna height-scan.

Considerations similar to those for radiating fields apply to measurements in low-frequency magnetic fields. For example, the experimental magnetostatic shielding effectiveness data quoted in Chapter 5 (see the beginning of Section 5.2.3) were obtained by inserting samples of nonshielded and shielded transmission line into the Helmholtz coils powered by specified current and then comparing the induced voltages in the line.

The "thing" to watch during such a test is the nonlinearity of high magnetic permeability materials. As a rule, with an increase in the field intensity, the magnitude of the magnetic permeability decreases so that, at saturation, no magnetostatic shielding will occur at all! For this reason, it is necessary to specify and generate a range of field intensities during measurement. As a rule, this does not present any major measurement problems. For example, a simple relationship exists between the magnetic field intensity, H, in the center of the area bounded by Helmholtz coils of radius, R, and the number of turns, N, and the current, I, in the coils:

$$H \approx 10^{-6} NI/R \qquad (6.2)$$

Thus, the desired magnetic field intensity can be controlled by varying the test current in the coils.

✦ Paralleling the "infinite" number of the shielding effectiveness definitions and respective evaluation parameters, there are "infinite" shielding effectiveness measurement techniques available and used by many engineers in different fields of shielded assembly and cable applications. Some of these techniques are "legitimate" extensions of sound engineering practices, while others may "defy the laws of physics" and generate intolerable errors.

In this respect, one shielding effectiveness measurement method is especially notorious. You have probably already figured out that we mean MIL-STD-285. Per this standard, the measurements are performed in a shielding room, which may have reflecting surfaces. The CUT is powered by a signal, and the radiating field intensity is measured at 1 m distance from the CUT. The near-field configuration, CUT and antenna coupling to the ground planes of the shielding room, and wave reflections from the shielding room confines are some of the problems that make this test almost useless.

6.3.3 Antenna Current Measurements

Since the field radiating from or received by an assembly is determined by the common-mode (antenna) currents in the nonshielded line conductors, or at the shielding enclosure surfaces — cable shield in shielded lines, connector shield in shielded connector, and product shield in shielded products — the measurement of the actual radiating fields can be substituted with the measurement of antenna currents in these conductors. Then, the shielding effectiveness—defined as a ratio of antenna currents in nonshielded conductors and the shield—can be determined.

In principle, this current can be measured (or induced in the assembly, in susceptibility testing) easily, using a *clamp-on* current probe over a cable (see Fig. 6.13 a) or a *surface* current probe over a shielding enclosure or connector surface.

Shielding Measurement Techniques and Apparatus: The Tools of the Trade 439

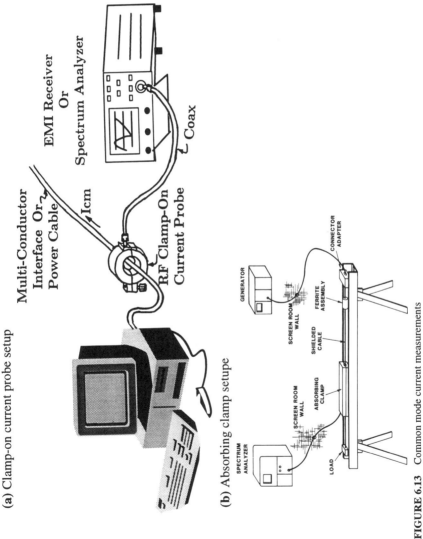

FIGURE 6.13 Common mode current measurements

Because the differential-mode currents are equal in amplitude and opposite in phase, such a current probes will not register (or create, in susceptibility testing) them and will indicate only the common-mode currents in the line or shield. Moreover, by combining the current probe with a ferrite sleeve over the radiating element (e.g., cable), which can be moved (slid) along the cable, different radiating lengths of cable can be set to resonate at different frequencies.

There are available numerous physical and electrical current probe designs. A simple and low-cost device to test the antenna current in cables and assemblies in the frequency range over 30 MHz is the absorbing clamp [6.14] (Fig. 6.13 b). In practice, the device consists of a set of split lossy ferrite rings attached to two halves of a hinged nonmetallic box, and a current probe implemented by capturing two or three of the bottom half-rings in a sense loop. The sense loop is formed by shorting the center conductor to the shield of the output signal cable at the clamp end, while the other cable end is terminated on a BNC or N-type connector to connect the cable to measuring device (usually, spectrum analyzer). To provide system immunity to ambient EMI, the measurement setup can be complemented by one or two additional ferrite tubes (they also can be made from half-rings) which are installed on the assembly under test close to its feeding point from the signal generator, and/or at the sensor cable. In the frequency range of interest (30 to 1000 MHz) the insertion loss of the ferrite assemblies is around 15 to 20 dB.

When the box is closed, the ferrites form a cylindrical channel through which a cable can be pulled. A 5 to 6 m cable or assembly under test (a nonshielded wire or a shielded cable) is inserted in the clamp, and the whole setup is installed on a horizontal nonmetallic support with a rail. The clamp has wheels and can slide smoothly over the measured cable length. The line in the cable is fed with a test current (a single frequency of interest, or a CW sweep) which generates an antenna current in the nonshielded line or the shield. This current is picked up by a probe and measured with a spectrum analyzer or receiver. The reference radiating power is determined by slowly moving the clamp over the cable and registering the maximum receiver readings at each frequency of interest. When performing a sweep test, the receiving spectrum analyzer can be set on "maxhold" function, and several runs of the clamp over the cable under test are used to obtain a steady curve. Comparing measured antenna current in the shield with the power supplied to the shielded line, or with the antenna current on a nonshielded wire, the shielding effectiveness is determined. Several techniques are possible and are used to perform the measurements and process the data.

To illustrate the mechanisms involved in the absorbing clamp and draw some useful information, consider measurement results. A tracking generator feeds a continuous scan of CW signal into a line. An absorbing clamp slides to several discrete positions along the line, and its output is measured with a spectrum analyzer. Fig. 6.14 presents a set of curves obtained for the same line: shielded (6.14a) and nonshielded (6.14b). Clearly, the magnitude of the measured current is 50 to 60 dB smaller in shielded line than in the nonshielded one. This proves that the absorbing clamp can be used to measure shielding effectiveness.

There is also another important attribute of the graphs in Fig. 6.14, namely, the relative independence of the shape of the curves (especially their maximums) with regard to the position of the absorbing clamp along radiating line, shielded or unshielded! It appears that, although the frequencies of the particular current maximums (resonant frequencies) along the line may shift, the maximum values of the antenna current at these resonant frequencies are quite close, differing no more than 5 to 10 dB.

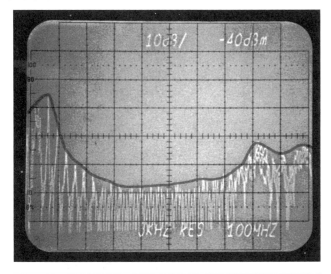

(a) with shield

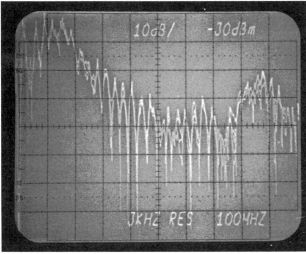

(b) without shield

FIGURE 6.14 Line output as seen on a spectrum analyzer screen

Since the radiation field intensity is proportional to the antenna current, the difference between the maximum emissions from such lines also should be close (although at different frequencies). Actually, we have already encountered a similar phenomenon in Chapter 1, when we considered the emissions from cables of different lengths (see Fig. 1.27). There, variations in the line length led to variations of the resonant frequency of the radiation; however the maximum magnitude of the emissions at the resonant frequency did not change significantly. Since the radiating emissions are defined by the antenna currents, and it is the electrical length of the radiator that determines the resonance frequency, it really does not matter what we change (frequency or length) or what we monitor (radiating field intensity or antenna current distribution), and in this respect the curves in Fig.6.14 and 1.27 are equivalent.

Then the conclusion should be that, although any particular line length has a set of resonant frequencies at which the emissions from the line reach their maximums, the maximum emissions from lines of different length are not all that different (although they occur at different frequencies). What does this mean in the context of EMC? Since the EMC measurements are conducted in a wide frequency band, and cables are manipulated individually to maximize the emissions (or susceptibility), this means that changing the current distribution at the radiator surface often will have only a limited effect on the result of the EMC testing.

We will continue the discussion of the Fig. 6.14 in Chapter 7, with more extremely useful conclusions.

Although absorbing clamp was originally meant to measure emissions from and shielding effectiveness of the power cords of domestic appliances, lately it found wide applications as a tool to measure electromagnetic emissions from electronic cables. Its popularity is caused by two advantages. First is the simplicity of making the measurements, process, and interpret the test results. It is really very logical to establish a link between the measured antenna current and the EMI potential of a cable or assembly. No special sample preparation or requirements to the test facility are required. The dynamic range is about 80 dB, and can be increased up to 120 dB by using amplifiers. Second, the use of antenna current measurements may be viewed as "a dream come true" from the system point of view. Indeed, to measure the antenna current at a cable assembly it does not have to be disconnected from the system. Thus, the system shielding effectiveness can be obtained. Moreover, even multi-branch assemblies can be tested this way, including the determination of contributions of each branch. Of course, on the negative side are the field distortions introduced by absorbing clamp itself (don't forget that ferrite assemblies have to limit the antenna current propagation along the cable under test).

The absorber clamp is used to measure the emissions from cables and cable assemblies. Similar results can be obtained on the basis of susceptibility measurements by using the so called current injection method [1.5, 6.15-16]. Although it was developed originally to test the emissions from consumer electronics and then expanded to susceptibility testing of military systems (e.g., see UK Defence Standard 59-41, 1986), it was also successfully used for measurements of the cable assembly shielding effectiveness. Both system-level and assembly-level shielding effectiveness can be measured. In the process of making the measurements, the EMI current is injected into the assembly under test with an injection probe (usually an inductive coil applied around the shielded cable). In this case, the shield transfer function can be determined by measuring and comparing the power supplied to the injection probe with the induced voltage between the shielded line and the shield.

While the use of current measuring and current injection techniques for general EMC testing is questionable and problematic (but there are strong opinions expressed on "both sides of the fence"), it is certainly an extremely useful tool for shielding measurements. We will compare the results of absorber clamp measurements to those of other techniques at the end of this chapter.

6.3.4 Shielding Effectiveness via Crosstalk Measurements

Coupling and crosstalk measurements present a convenient way to test the shielding effectiveness of cable assemblies working in near radiating fields and quasi-stationary regimes. Indeed, by comparing the crosstalk between the same lines and in identical positions, first nonshielded, then shielded, the shielding effectiveness of the shields can be determined. If only one of the lines is shielded, the effectiveness of its shield can be determined. When both lines are shielded, the combined effect of both shields is determined, which incorporates the measurement system geometry and third-circuit-related phenomena, along with the shields' transfer functions. These measurements can be done both in the frequency and time domains. Thus, significant advantages can be gained by using crosstalk-based and coupling-based measurements to determine the shielding effectiveness.

The setup for crosstalk measurements is clear from Fig. 1.38 (Chapter 1). The principles of crosstalk measurements are simple (this is not to say, that the realization of these principles is always easy!) The analog (CW) as well as digital test signals can be used, resulting in specific measurement data. However, the logistics of these measurements may require attention to detail. For example, at high frequencies, signal generators and receivers (spectrum analyzers) usually have a nonbalanced input/output. This should be kept in mind when measuring crosstalk between balanced pairs. In this case, to retain the cable under test symmetry, it may be necessary to use a balun or other balancing techniques (e.g., differential amplifiers).

But what shielding properties are measured in the crosstalk test? As we saw in Chapters 4 and 5, the differential-mode currents in cable lines generate transverse induction fields, while the common-mode currents generate longitudinal induction and radiating fields. In the absence of common-mode currents, the near- and far-end crosstalk between nonshielded lines in an electrically small cable length are formed by the sum and the difference of electric and magnetic coupling, respectively {see Eqs. (4.36) and (4.37), and Figs. 4.6 and 4.7}. Thus, by measuring and comparing the near- and far-end crosstalk, the electric (capacitive) and magnetic (inductive) coupling can be determined. When a solid metallic shield is placed between such lines, the transverse electric field does not penetrates it, whereas the quasi-stationary magnetic field is significantly attenuated, depending on the signal frequency and the shield characteristics (mainly conductivity and magnetic permeability—see Chapter 3). For example, Fig. 4.17 contained an illustration of the shield effect on the electric and magnetic coupling.

On the other hand, the common-mode currents result in longitudinal fields which also generate crosstalk, and also radiate. As far as the shielding performance is concerned, the most valuable information that can be extracted from crosstalk measurements between shielded lines is the shield transfer function. Expressions (5.26), (5.27), (5.29), and (5.30) in Chapter 5, illustrate the crosstalk dependence on the

444 Electromagnetic Shielding Handbook

shield transfer function (e.g., transfer impedance), the line electrical length, and the measurement setup.

This fact is exploited in many recognized and standardized shielding effectiveness measurement techniques, as well as in a host of "home-brewed" procedures used to solve specific shielding measurement problems. Consider the results of crosstalk measurement between two lines: one shielded, and the second nonshielded. These results can be used "as is" as a measure of shielding effectiveness or, after some mathematical processing, the shield transfer parameters can be determined. For example, comparison of these results to the crosstalk between similar lines, both nonshielded, readily yields a direct, convenient measure of the shielding effectiveness. Furthermore, by comparing the crosstalk between the identical lines shielded by different specified shields, their relative shielding effectiveness can be determined. On the other hand, using Eqs. (5.26), (5.27), (5.29), and (5.30), and additional data about the line electrical characteristics and the setup geometry, the transfer impedance (as well as the transfer admittance, loss impedance, and transmission effectiveness) can be calculated.

✦ When-using these principles for specific applications, a knowledge of the transmission line and shielding fundamentals is quite essential. As an example, consider a shielding effectiveness evaluation device that was extremely popular in the industry only about 10 years ago. To evaluate CATV coaxial drop cables, a *shielding effectiveness evaluation device* (S.E.E.D.) was introduced in 1972 by Belden Corp. This method proved to be fast and easy for shield performance measurements and thus was widely used in the industry.

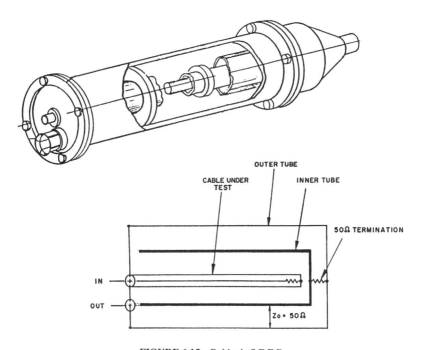

FIGURE 6.15   Belden's S.E.E.D.

A S.E.E.D. consists of a set of two copper (or brass) tubes. The cable under test (CUT) is inserted and supported coaxially in the inner tube, forming some kind of a quadraxial line (see Fig. 6.15) in which two circuits (one CUT, and the second formed by the S.E.E.D. tubes) are electromagnetically coupled. Usually, the CUT acts as a disturbing circuit, while the two S.E.E.D. tubes act as a disturbed circuit. Since the inner S.E.E.D. tube presents an excellent homogeneous shield with high conductivity, direct coupling between the circuits is precluded. The crosstalk signal transfers from one circuit to the other via the third circuit formed by the outer conductor (shield) of the sample cable and the inner S.E.E.D. tube.

As a result of double energy transfer (tested sample, third circuit; third circuit, S.E.E.D. line) the EMI related to the shielding effectiveness is attenuated 60 to 120 dB below initial level. This significantly reduces the device sensitivity: the better the shielding properties of the inner tube, the larger the signal attenuation and the less sensitive the device. It is easy to see that parameters of this third circuit depend on the diameter of the cable under test, its position inside the inner tube, and even on the material and thickness of the cable jacket. (This is certainly needed to preclude the shield direct contact with the inner tube.) The unmatched S.E.E.D. line (open at one end and shorted at the other end), as well as ground loops appearing when the unbalanced signal generator and receiver (spectrum analyzer) are connected to the system, result in a regularly oscillating with frequency signal that, as a rule, does not correspond to practical shield performance. This explains why a S.E.E.D. does not provide an acceptable frequency characteristic of shielding effectiveness.

A detailed theoretical analysis of this technique is provided in Ref. [6.22], and it was further developed by the author of this book at the company's request. The analyses were based on application of the circuit theory to crosstalk evaluation between the two lines formed by the CUT and the S.E.E.D. tubes in the presence of the third circuit. They confirmed the major problems with the use of this fixture.✦

6.4 TRANSFER IMPEDANCE AND CAPACITIVE COUPLING IMPEDANCE MEASUREMENTS

Without doubts, transfer impedance is the most popular cable (and connector, and assembly, and gasket, and even shielding material) shield characterization parameter. And rightly so. This parameter permits us to identify the most inherent shield "signature," which fits many practical applications. After the extended discussions in the previous chapters, we know its merits and demerits. Fortunately, there is no need to cover the details of the "infinite number" of existing transfer impedance measurement procedures, because they are pretty well documented. Shielding test standards, general reviews, and specific method descriptions are available of and about the underlying principles, fixture designs, test setups, and measurement results — the readers are strongly encouraged to peruse the bibliography to this chapter (how about "just" first fifty titles [6.1-50]), as well as numerous sources, not mentioned here. For this reason, in the following sections we will omit many details and concentrate on the fundamentals of the most important measurement techniques and their implementation.

6.4.1 Coaxial Structures: Is There a Sextaxial in the Cards?

Twinaxial, triaxial, quadraxial, quintaxial, ... ? This is not an exercise in Latin numerology, but just a listing of several transfer impedance measurement techniques. Historically, the basis for the earlier transfer parameter measurements was provided by the "bare-bones" graphical definitions per Fig. 3.2. In principle, these schematics

446 Electromagnetic Shielding Handbook

represent a specific case of crosstalk measurements between two circuits sharing a common element: the measured shield. One of these circuits (external or internal) provides the path for EMI current injection, while the other circuit is used to sense the coupled into it voltage. Then, the transfer impedance and capacitive coupling impedance can be determined from Eqs. (3.4) through (3.6). As evident from the schematic, the term "shorted" refers to the loads at the ends of the current injection and/or voltage measurement lines.

In practical measurement techniques, the current injection and voltage sensing circuits and their mutual arrangement are selected to facilitate the determination of the shield effect on the crosstalk between them. In this respect, one of the most widely used approaches is based on a triaxial line structure. There, the shielded cable sample is inserted in a circular tube which is a part of the test fixture. Now, two concentric coaxial circuits are formed: internal and external. In the internal circuit, the shield serves as the outer conductor while the center conductor is formed by the shielded line wires (for a coaxial line, it will be its own center conductor). In the external circuit, the shield acts as the center conductor, and the outer conductor is formed by the tube. The structure obtained looks like those illustrated by Figs. 3.3 and 3.5.

Over the years, numerous fixtures and jigs implementing such a procedure have been designed and built throughout the world under the auspices of different organizations and authorities. More often than not, they were created and used by the engineers faced with this task (including the author of this book, back in 1964). Although the principles associated with this test methodology seem straightforward, their implementation is often not simple. In particular, one of the most troublesome problems is associated with providing the "shorts" at the ends of both circuits. To illustrate this problem and provide an appreciation of the ingenuity required to solve it, Fig. 6.16 presents one of the practical designs used to perform such measurements [6.33].

An easy way to create a triaxial line is to pull a braided tube over the shielded cable. By 360° soldering this braid to the cable shield at one end, and to a coaxial connector at the other end, a triaxial structure is created that can be used in accordance

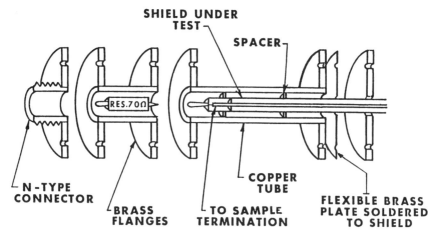

FIGURE 6.16 IEC triaxial

Shielding Measurement Techniques and Apparatus: The Tools of the Trade 447

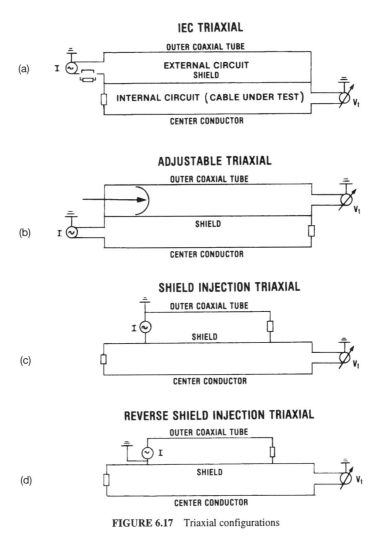

FIGURE 6.17 Triaxial configurations

with the schematic in Fig. 6.17 a (see Ref. [6.6]). Simplicity is the main advantage of such test, while its disadvantages are also obvious: uncertain and variable parameters of the external circuit, potential field leakage through the outer braid, and induction coupling of the external circuit through the outer braid to metallic surfaces outside of the fixture.

IEC Publication 96-1 recommends introducing the EMI current in the shield in the external circuit of the triaxial (i.e., external current return path). Two ways of introducing this interference current are suggested: without and with a resistor, inserted in series with the generator (see Fig. 6.17 a). The external circuit is shorted. (You saw in Fig. 6.16 what that takes, or it can be just 360° soldered to the fixture's outer tube if only one test is to be made.) The internal circuit is loaded on a matched impedance. (Since this is usually a cable under test, it will be 50 or 75 Ohm, but, in general, the internal circuit characteristic impedance must be determined). If the sample is electrically short (say, L < 0.1λ, which corresponds to the maximum measurement frequency of 30 MHz), the current distribution in the shield can be assumed uniform, and differentials in Eqs. (3.1) through (3.5) can be substituted for measured values of V_t and I. Moreover, owing to the short circuit, the action of the voltage in the external circuit, which is the source of the capacitive coupling, can be neglected, so the obtained result can be related to the transfer impedance only. To measure the capacitive coupling impedance, the IEC recommends a separate method (not presented here).

Another advantage of the short circuit method is that it partially (in the range higher than the outer tube cutoff frequency) neutralizes the ground loop between the generator and voltmeter. (For example, to visualize those ground loops, just imagine that all the grounded points in Fig. 6.17 are interconnected by conductors running some "fancy" routes!) As a rule, these devices will have unbalanced output and input, respectively. The location of the respective "grounds" is shown in all the schematics in Fig. 6.17. It can be shown that if the external circuit is not shorted, a part of the cable shield current will leak into the shield of voltmeter lead, and the measured voltage will be dependent on the shielding efficiency of this lead.

When the line is short, it is relatively simple to account for the loading conditions at the ends, because the line parameters may be considered as lumped components, determined by the loads at the terminations. In this case, the equivalent circuit 5.12 can be used. In this equivalent circuit, the input impedances Z_{ino} and Z_{inl} "looking," respectively, backward and forward in the circuit, can be assumed to be equal to the loads at the ends. (And they are much larger than the loss impedance dZ_1 at a small length of shield element.) The source dE in the equivalent circuit of Fig. 5.12 is applied to two impedances, Z_{ino} (near end) and Z_{inl} (far end), in series (dZ_1 is neglected). For example, if both impedances are identical, the voltage drop at each of them will be the half of the source voltage (more correctly, emf). But the source dE is formed by simultaneous contributions of the transfer impedance and the transfer admittance per Fig. 3.2 (read, magnetic and electric coupling, respectively). Also, the voltage generated at the near end of the disturbed line is determined by the sum of the contributions of electric and magnetic coupling, whereas the voltage generated at the disturbed line far end is determined by the difference of the contributions of electric and magnetic coupling (Fig. 4.6). As a result, the voltages V_0 and V_1 at the loads at the near and far end of the measurement system with length dl (we designated it dl to emphasize the electrically small length) can be presented as

$$V_o = -\frac{1}{2}(Z_t + Z_f)I \qquad (6.3)$$

$$V_1 = \frac{1}{2}(Z_t - Z_f)I \qquad (6.4)$$

These simple expressions can be used to determine the transfer impedance and capacitive coupling impedance of an electrically short shield from the measurements.

Shielding Measurement Techniques and Apparatus: The Tools of the Trade 449

✦ That could have been expected. When a signal generator introduces a signal current or voltage into one circuit of a triaxial line (internal or external), an interference voltage or current will be detected in the second circuit due to the leakage in the cable shield. This crosstalk between the internal and external circuits can be characterized by the respective coupling between them which, in turn, depends on the shield transfer function. The better the shield, the smaller this coupling will be. (With the "perfect" shield, there will be no coupling at all.) Thus, the shield transfer function can be found by measuring the crosstalk currents and voltages. But as we know, between two lines, two kinds of crosstalk can be identified: near-end crosstalk and far-end crosstalk. The potential difference between the outer tube and the cable shield is not zero, and the measured voltage V_t is dependent on the simultaneous action of tran [6.32, 6.47, 6.27]) reveals that, in this case, the measured response of the schematics in Fig. 6.17 a depends on the difference between the magnetic (Z_t) and electric (Z_f) coupling!

Which end should we measure while measuring shielding? For example, in Fig. 3.2 the measurements at the far ends are shown, but is it "right," and is it "enough"? We could have shown the "near end" measurements as well. The answer is provided by the theory.

The general theory of triaxial measurements can be built on the transmission line theory (e.g., see Refs. [P.5, 4.1, 5.2, 5.4–6, 6.27, 6.32, 6.47]). While the reader is referred to these and many other available works for a detailed analysis, we will limit ourselves here with simple physical considerations. Indeed, consider the "origins" of the transfer impedance and capacitive coupling impedance. The transfer impedance-related coupling is generated by and is proportional to (at least, in shields with linear characteristics) the current in the shield with the return path facing the shield surface that is opposite to the considered surface (see Chapter 3). Similarly, the coupling related to capacitive coupling impedance is generated by and is proportional to the voltage in the shield with the return path facing the shield surface that is opposite to the considered surface. Actually, even in linear materials, the proportionality of the fields to the currents and voltages takes place only in electrically small lines, but that's exactly how we introduced the shield transfer parameters.

Referring now to the Section 4.4, and in particular to Eqs. (4.33) and (4.35), we see that, at least formally, the transfer impedance is analogous to the magnetic coupling (by virtue of its relation to the EMI current and, thus, magnetic field), while the transfer admittance is analogous to the electric coupling (by virtue of its relation to the voltage and, thus, electric field). We emphasized the formal nature of this analogy because in Section 4.4 we introduced the concepts of electric and magnetic coupling for transverse fields while here, in our case of the triaxial line, both transverse and longitudinal fields may be generated. But it is easy to see that, in both cases, the EMI signals generated by the electric and magnetic fields in the disturbed circuit are in phase at the line near end, and in opposite phases at the line far end (as in Fig. 4.6). Thus, in both cases, the near- and far-end coupling and the near- and far-end crosstalk can be determined by Eqs. (4.36) and (4.37).✦

With the frequency rise, the line becomes electrically long, and its parameters no longer can be considered as lumped but as distributed along the line. The resonant properties of the short-circuited external circuit affect the shield current distribution, which no longer can be considered uniform. Especially important is the current phase nonuniformity. Also, voltage V_t now becomes dependent on the signal reflections in the external and internal circuit, as well as on the propagation constant difference in both circuits. As a result, the measured transfer impedance frequency characteristic exhibits a resonant character. In general, the behavior can be extremely complex. For example, Eqs. (5.24) and (5.25) and Fig. 5.21-22 relate the "flavor" of the problem. (The only difference is that the quoted equations and graphs account for two shields, whereas the test system contains only one shield—the cable under test.)

✦ It is worthwhile to look at these phenomena in some detail. Consider first the resonant effects. To separate these effects from the mismatch complications, we will assume that both the internal and external circuits are matched. In fact, the derivations here are no different from those in Section 4.4 and lead to expressions similar to Eqs. (5.24) and (5.25). Comparing these to the derivations in Section 4.4, the specifics in this case are:

- The presence of only one shield between the two "crosstalking" circuits: internal and external.
- In comparison with the terminating impedances, we will neglect the voltage drop at the loss resistances of the shield, outer tube, and the internal circuit center conductor.
- In Section 4.4, we disregarded the transfer admittance (electric) coupling, whereas here we will assume that in each element the transfer impedance (magnetic coupling) and capacitive coupling impedance (electric coupling) add at the near end and deduct at the far end, and so propagate to the line ends.

Assume that voltage V_{10} is applied at the near end of the fixture to circuit 1, which may be the external or internal circuit. That is, V_{10} is applied between the cable under test shield and outer tube or between the cable center conductor and the shield, respectively. Then, integrating the currents from all the elements dx, obtain the generated EMI crosstalk voltage at the near and far ends in the test fixture:

$$U_{20} = \int_0^l dU_{20} = \frac{U_{10}(Z_t + Z_f)}{2Z_c} \int_0^l e^{-(\gamma_1 + \gamma_2)x} dx = \frac{I(Z_t + Z_f)}{2(\gamma_1 + \gamma_2)}\left(1 - e^{-(\gamma_1 + \gamma_2)l}\right) \quad (6.5)$$

$$U_{21} = \int_0^l dU_{21} = \frac{U_{10}(Z_t - Z_f)e^{-\gamma_2 l}}{2Z_c} \int_0^l e^{-(\gamma_1 - \gamma_2)x} dx = \frac{I(Z_t - Z_f)e^{-\gamma_2 l}}{2(\gamma_1 - \gamma_2)}\left(1 - e^{-(\gamma_1 - \gamma_2)l}\right) \quad (6.6)$$

Usually even in a "long" enough measurement shield length the losses in the shield are small so that only the phase constants in (6.5) and (6.6) should be accounted for. Then we can further simplify these equations, assuming

$$\gamma \approx j\beta = j\frac{2\pi f}{v} = j\frac{2\pi f\sqrt{\varepsilon\mu}}{c} = j\frac{2\pi\sqrt{\varepsilon\mu}}{\lambda_o} \quad (6.7)$$

where the speed of light $c = 3 \cdot 10^8$, ε, μ is equivalent dielectric constant and magnetic permeability of the propagation media in the external and internal circuits respectively, and λ_o is the signal wavelength in free space. Now, we can express voltages V_{20} and V_{21} at the loads at the near and far end of the measurement system as functions of the cable length-to-signal wavelength ratio, $1/\lambda$. Note that the length variations of the voltages V_{20} and V_{21} are defined by the function

$$\varsigma = \frac{1 - e^{-j2\pi(\sqrt{\varepsilon_1\mu_1} \pm \sqrt{\varepsilon_2\mu_2})\frac{l}{\lambda_o}}}{j2\pi(\sqrt{\varepsilon_1\mu_1} \pm \sqrt{\varepsilon_2\mu_2})\frac{l}{\lambda_o}} \quad (6.8)$$

where the "+" sign corresponds to the near end voltage V_{20}, and the "–" sign corresponds to the far end voltage V_{21}. The far end voltage also contains an additional length-dependent item,

Shielding Measurement Techniques and Apparatus: The Tools of the Trade 451

$$e^{-\gamma_2 l}$$

Using simple mathematical transformations, it easy to show that the modulus of Eq. (6.8) reduces to

$$|\varsigma| = \frac{\left\{\sin\left[\pi(\sqrt{\varepsilon_1\mu_1} \pm \sqrt{\varepsilon_2\mu_2})\frac{l}{\lambda_0}\right]\right\}^2}{\pi(\sqrt{\varepsilon_1\mu_1} \pm \sqrt{\varepsilon_2\mu_2})\frac{l}{\lambda_0}} \qquad (6.9)$$

*(We had to go through this "exercise" with $\sqrt{\varepsilon\mu}$, because the shields and the cable jackets may have enhanced magnetic and dielectric characteristics. The thing to remember is that both the dielectric constant and magnetic permeability, in general, are equivalent characteristics exhibited by combinations of air, dielectrics, magnetodielectrics, and so on. The field configuration in such structures may be very complex, and the solution of the respective electromagnetics problems is often not trivial).

The oscillatory character of the $(\sin x)^2/x$ function is well known. Since the wavelength is inversely proportional to the frequency, similar resonant effects are observed with the corresponding signal frequency or line length variations. It is easy to see that, when the phase constants of internal and external circuits are equal and the lines are matched, such oscillatory variations take place only at the near end. This is illustrated by the experimentally measured voltages at the near and far ends of the matched triaxial setup (see Fig. 6.18).✦

✦ The resonances described by Eq. (6.9) are inherent to the line phase relationship. But there is more. When the current distribution along the shield is not uniform, voltage V_t is affected not only by the shield intrinsic transfer function property, but it also becomes dependent on the signal reflections in the external circuit, the reflections in the internal circuit, and the propagation constant (mainly, phase) difference in both circuits. In the general case of arbitrary loads, the voltages at the lines' ends can be calculated using Eqs. (A.68) - through (A.71). Now voltage V_t becomes dependent not only on the signal reflections in the external circuit but also on the reflections in the internal circuit, and on the propagation constant difference in both circuits. In general, the dependence can be extremely complex. Thus, in the shorted triaxial, the resonances described by Eq. (6.9) are overlaid by the reflections from the short-circuited ends. Now you can explain the experimental data which are often presented as "transfer impedance"; e.g., see Fig. 6.19.✦

Returning now to the triaxial measurement lines, we see that in general case of arbitrary line length and loads, the voltage generated at the "near end" of the triaxial fixture is determined by the geometrical sum (which accounts for phase variations along the line) of effects related to the transfer impedance and the transfer admittance, and the voltage generated at the "far end" of the triaxial fixture is determined by the geometrical difference of effects related to the transfer impedance and the transfer admittance. The measurement results are also affected by the mismatch in both internal and external lines. For these reasons, although the obtained data can still be used to characterize the shields, it cannot be viewed as the transfer impedance parameter, even if it is often labeled as such. (Sorry about that!) This is a matter of definition, and it may have far-reaching practical consequences. Indeed, the transfer impedance was defined on an infinitesimal shield length for good reasons.

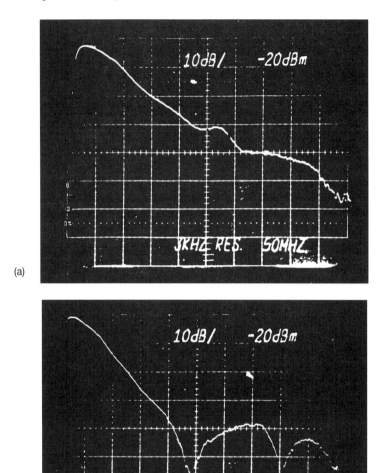

FIGURE 6.18 Experimentally measured voltages at (a) far and (b) near ends of matched triaxial test setup

However, if the measurement line is electrically short, Eqs. (6.3) and (6.4) indicate a potential method to determine the transfer impedance and transfer admittance (or capacitive coupling impedance) for the considered case of arbitrary line loads. Indeed, such techniques were suggested for measurement of cable samples, which use the same fixture to determine both parameters (e.g., see Ref. [6.36]). But to realize such a method in the general case of an arbitrary line length, we must find a way to "de-embed" the shield transfer parameters from pretty complex equations associated with the crosstalk between mismatched long lines.

Shielding Measurement Techniques and Apparatus: The Tools of the Trade

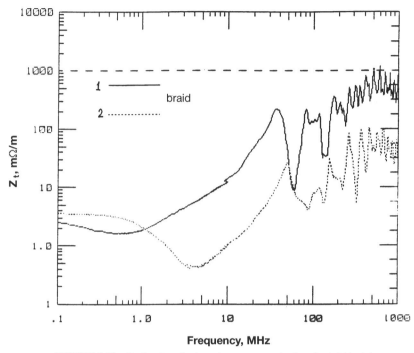

FIGURE 6.19 Surface transfer impedance measured using shorted triaxial

There have been suggested (and implemented) two main approaches to avoid complications originated by the mismatch in the external circuit. One approach was realized [6.34] as the adjustable triaxial (see Fig. 6.17 b). In the adjustable arrangement, the signal leaking through the cable shield into the external circuit generates two waves: a forward wave directed toward the voltmeter at the far end, and a reverse wave directed toward the near end. The analysis and mathematical expressions to calculate the voltage responses at the triaxials' near and far ends are given in Refs. [6.32, 6.47, P.5]. If the external circuit is shorted, the reverse wave reflects, and the voltmeter will indicate the geometrical sum of both waves. In the adjustable triaxial test method, the external circuit is provided with a short circuit plunger at the near end, which permits us to change the length of this circuit. When the length of a short-circuited line is adjusted to produce a maximum, voltage V_t is the arithmetic sum of two voltages: the forward wave for the matched case and the reverse wave for the matched case. In this way, the adjustable triaxial has been used up to 7.5 GHz. Notice that both the magnetic and capacitive coupling are present, and the adjustable triaxial does not permit us to distinguish between them. Another disadvantage is a relatively high minimum test frequency: for the reported fixture, it was found to be 500 MHz.

454 Electromagnetic Shielding Handbook

✦ A triaxial fixture with a sliding short, but to measure the cable shielding effectiveness (as opposed to transfer impedance), is described in [6.62]. The fixture was designed according to MIL-C-39012C standard. When used in the test setup employing a network analyzer (HP 8720), up to 110 dB shielding effectiveness was measured on 10 and 20 cm long cables in the frequency band 50 MHz - 18 GHz. ✦

The second approach to handle reflections is to match the external circuit. A relatively simple way to do that is realized in the shield injection triaxial (see Fig. 6.17 c), which provides for correct terminations with characteristic impedance at both ends of the circuit. It is possible to obtain the values of forward and reverse voltages, which can be used to calculate separately the transfer and capacitive coupling impedances, even if both external and internal circuits are not matched. Yet, if they are matched and have the same characteristic impedances and propagation velocities, the calculations become very simple and straightforward [P.1, 6.24]. There are two disadvantages, however, to the shield injection method. Since both generator and voltmeter have different reference points, a ground loop is formed between them. When the external circuit is not shorted, the ground loop distorts the measured signal at frequencies even much higher than the outer tube cutoff frequency, thus reducing the test range from below. On the other hand, since the outer tube and shield are not in direct contact at the end of the fixture, the system inherently cannot be "closed" (i.e., for radiation). Therefore, at higher frequencies, it radiates, reducing the test range from above.

A way to provide a common reference point to the generator and voltmeter, while having the circuits matched, is realized in a reverse shield injection triaxial, whose schematic is clear from Fig. 6.17 d. Yet such a system with a "live" outer conductor is extremely sensitive to external interference.

To "close" this system (with regard to radiation), a second outer tube can be added—a "guard tube," which leads to a quadraxial design (see Fig. 6.20a). Here, the cable shield and the driven tube form a reverse triaxial, while the guard tube is electrically connected to the shield (and to the ground) to provide the whole system shielding.

The quintaxial design (Fig. 6.20 b) is electrically identical to quadraxial. Mechanically, quintaxial is obtained if the quadraxial's driven tube is split on the outer driven tube and inner driven tube and electrically connected by fingerstock. The CUT and the inner driven tube form an assembly that can be detached from the fixture, thus providing easier access to the test sample.

One common disadvantage of the quadraxial and quintaxial configurations is their relative complexity. Another problems stems from the fixture designs. Obviously, the measured voltages V_t in Fig. 6.20 a and 5.20 b are the reverse wave voltage V_R, but to discriminate between the Z_t and Z_f both the reverse and forward V_F voltages are necessary.

To resolve the second problem, it can be proposed to provide the opposite to V_t CUT end with the second output to measure V_F. Such output can be provided in both quadraxial and quintaxial fixtures using isolated feedthrough connectors.

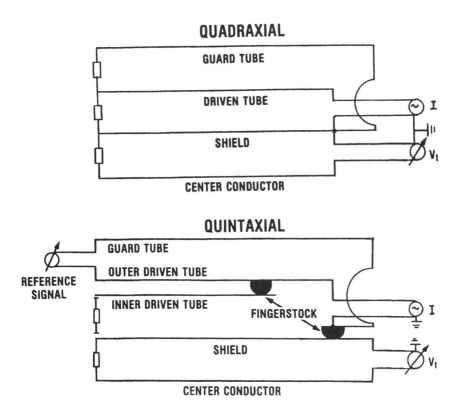

FIGURE 6.20 Quadraxial and quintaxial designs

6.4.2 Terminated Triaxial

Obviously, to avoid circuit termination problems, the test system must provide matched conditions for both generator and voltmeter circuits at frequencies where the cable under test sample cannot be considered electrically short ("high frequencies"), while short circuiting the outer tube with the cable shield at dc and low frequencies. So far, this goal has been achieved (albeit, not completely) in two test methods: terminated triaxial and line injection. In this section, we will consider the first method, and in the next section the second method.

The terminated triaxial [6.37] schematic is shown in Fig. 6.21 . In this method, the outer tube is electrically and mechanically shorted at the end of the test fixture with the cable under test shield. The generator and the matching load "injection" feedthroughs in the outer tube are separated from these shorts by ferrite core assemblies (usually, toroidal in shape) placed on the cable between the cable shield and the outer tube. They act as large inductive impedances at the end sections of the external circuit. Also, at high enough frequencies, the active losses in ferrites increase. Thus, at high frequencies, the impedance of the end sections is so large that the line loads at

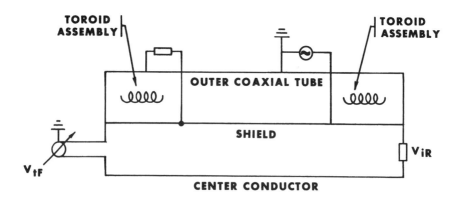

FIGURE 6.21 Terminated triaxial

the section ends can be considered as open, despite the physical shorts provided by the outer tube connection to the cable shield.

An example of a "typical" transfer impedance test terminated triaxial fixture [6.37] is shown in Fig. 6.22. To use the fixture, a cable sample is specially prepared

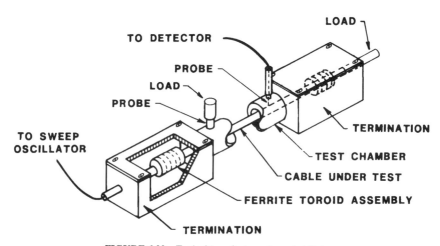

FIGURE 6.22 Typical transfer impedance test fixture

(Fig. 6.23). First, the jacket is cut out to expose the shield at two places, with the distance corresponding to the injection feedthrough (probe) spacing—usually 0.5 m or 1 m. The the cable under test sample is pulled through the test chamber (outer tube) and through the ferrite toroidal assemblies so that the exposed cable under test shield is just opposite the injection feedthroughs. Then, connectors are soldered to the cable under test sample ends, and the cable under test is connected to the feedthrough connectors at the ends of the termination boxes.

Now, if the external circuit is terminated at matched impedances (load and generator) at the "injection" feedthroughs, the "physical" shorts at the ends will not affect the matched conditions at high frequencies, where the ferrite assemblies are effective. However, these physical shorts at the ends will still act as shorts in the low frequencies and will "close" the system (for crosstalk and radiation from external circuit). This situation would be almost ideal if there were no "gap" between dc and the "high" frequencies.

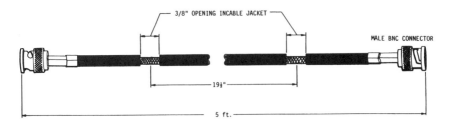

FIGURE 6.23 Sample preparation for transfer impedance test

Unfortunately, at these "low-to-moderate" frequencies, the ferrite assemblies provide neither very large nor very small impedance, thus introducing a frequency-dependent uncertainty in the measurements. It is clear from Fig. 6.21 that the lower the frequency, the smaller the impedances are at the ends of the external circuit. The generator "looks" into the matching load, connected in parallel with two shorted lines formed by the end sections of the external circuit. The frequency characteristic of the input impedance, Z_{in}, measured from the generator end (i.e., what the generator "sees") with external circuit matched is shown in Fig. 6.24 for two cases: with the toroid assemblies and without them.

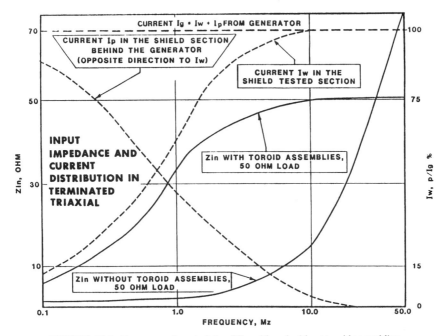

FIGURE 6.24 Frequency characteristic of Z_{in} with and without toroid assemblies

When the toroids are absent, the characteristic is typically that of the two short-circuited lines connected in parallel. Insertion of ferrite toroid assemblies (each 30 cm long, 3C8 Ferroxcube toroids) greatly improves the pattern but, in the range up to 10 MHz, the input impedance still is well below the 50 Ohm characteristic impedance of the line. This results not only in a mismatch but also in off-shooting the generator current to the parts of the shield outside the measured section of the cable under test sample. Of course, both effects distort the test results. This is one of the main drawbacks of the terminated triaxial test methods.

Another drawback is related to the different characteristic impedances and propagation constants (actually, predominantly, the phase constants) of the external and internal (that is, the cable under test) circuits. Indeed, while the cable under test transmission parameters usually can be controlled pretty well, for the external circuit these parameters depend on the relative diameters of the cable under test shield and external tube, and on the dielectric constant and losses in the cable under test jacket. Although there have been some suggestions to provide the test chambers of a variable size (by using a set of outer tubes of different diameters), their practicality is questionable.

And last, but not least, the process of pulling the cable through the test chamber, aligning the exposed shield with the probes and providing a good contact, and terminating the cable sample while inside the termination boxes presents a "hands-on" challenge.

Nevertheless, the terminated triaxial test method has been standardized by the IEC [6.7] and successfully used to obtain useful experimental data (e.g., see Refs. [6.40-41]).

Transfer Impedance Test Clamp

To provide for a full frequency range of coverage and simplify measurement performance, the transfer impedance test clamp was developed and built. It is based on the terminated triaxial schematics with important improvements of electrical circuitry and mechanical design [6.39]. From an electrical performance standpoint, it is desirable to create the capability to measure the whole frequency range of interest with the same jig, including low frequencies. Mechanically, it is the goal to facilitate and accelerate the test procedure, keeping in mind both industrial and research applications.

To achieve the first objective, the clamp actually incorporates two independent schematics: the terminated triaxial and the shorted triaxial. The terminated triaxial implementation, in principle, is no different from the one described above. However, the shorted triaxial schematic was specifically developed to be compatible with the terminated part of the test system (see Fig. 6.25). Here, the shorts at the ends of the external circuit are eliminated (points a–c and b–d are open), which essentially creates a shield injection schematic. The generator feeds the line via an unbalanced-to-balanced isolating transformer (balun), which physically breaks the ground loop. The short in the external circuit (short-circuit "cap") eliminates the distortive impact of the external circuit end sections (filled with the toroids).

However, to a certain degree, the presence of toroids prevents radiation from the open ends. To further reduce this radiation, the cable sample is mounted in the

Shielding Measurement Techniques and Apparatus: The Tools of the Trade 459

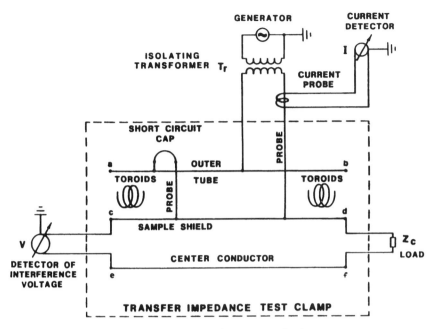

FIGURE 6.25 Low-frequency test circuitry

shielding enclosure (that is, the "outer tube") with special isolated feedthrough connectors c–e and d–f. This circuitry has proved to be effective from 10 kHz to 10 MHz. It can be used at even lower frequencies, with the proper choice of isolating transformer.

The second objective of the transfer impedance test clamp development was to facilitate practical measurements. Though different in embodiment, the commonly used coaxial transfer impedance test fixtures and jigs essentially consist of a cylindrical test chamber, which is usually made as a homogeneous or braided copper tube, provided with a proper set of terminations (e.g., see Fig. 6.21). Such a design presents significant inconveniences for practical measurements. This is due mainly to the need for pulling the CUT sample through the chamber tube and through the ferrite assemblies (if the terminated test method is used) in order to provide a triaxial construction. In this way, the connectors can be attached to the sample, as a rule, only after inserting the sample into the fixture. Hence, the sample, with connectors, cannot be prepared beforehand, which is important for production testing in industrial applications as well as for testing assemblies with permanently mounted connectors. The need for repetitive pulling and connector soldering makes it difficult for the operator to perform the necessary changes, adjustments, and mechanical manipulations without affecting the test conditions of the samples. This is especially important in performance stability measurements and research applications. Finally, the whole procedure of installing the samples in the fixture (and taking them out) and aligning the exposed shield "windows" right under the probes in the chamber consumes a significant amount of time.

460 Electromagnetic Shielding Handbook

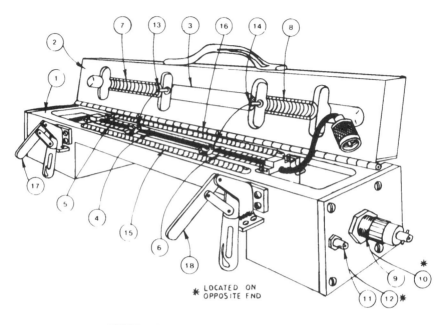

FIGURE 6.26 Transfer impedance test clamp (open)

FIGURE 6.27 Photo of a transfer impedance test clamp

To eliminate these inconveniences and shortcomings, the transfer impedance test clamp was designed in the form of a clamp (which is where the name comes from—see Figs. 6.26 and 6.27). The clamp is composed of the bottom (1) and the top (2) sections, containing semi-cylindrical grooves (3 and 4), and terminated at both ends with ferrite half-toroid assemblies (5, 6, 7, and 8). Bulkhead feedthrough connectors (9 through 12) are provided in special chambers at the ends of the clamp bottom section to terminate the CUT sample from the inside of the fixture. In the top section, feedthrough probes (13, 14) are built in with resilient (e.g., spring-loaded) contacts.

Now the connectors can be attached to both ends of the test sample, and the cable jacket can be removed under the resilient probes prior to installing the sample in the fixture. The sample then can be installed and terminated in the clamp's bottom section, and the clamp top section is closed, thus forming a terminated triaxial. Other options are to install the probes (13, 14) in the bottom section and/or substitute the resilient contacts with special conductive "holds" in which the exposed parts of the CUT shield are installed. In this case, there is no need to align the shield "windows" with the probe contacts.

In both cases, the probes serve as outer circuit current injection means. To eliminate the electromagnetic field leakage between the clamp's bottom and top sections, they are provided with RF gaskets (15, 16), while the fasteners (17, 18) press the whole system together. The transfer impedance test clamp was patented by Belden corporation [6.42].

Practical implementation of the transfer impedance measurements usually involves a sweep test under computer control. For example, Fig. 6.28 a illustrates a test based on the use of a spectrum analyzer with a tracking generator. Another option is to use a network analyzer, which has a built-in tracking generator and permits the determination of the transfer impedance modulus as well as phase (Fig. 6.28 b).

6.4.3 Line Injection Shielding Effectiveness Measurements

It follows from theory that, with proper design, the crosstalk measurements leading to the determination of the shield transfer function can be significantly simplified. The key to such simplification is to match the lines and their phase constants (including the tertiaries, if present). Indeed, we showed in Chapter 5 that, as "formidable" Eqs. (5.29) and (5.30) may appear, by properly selecting the line parameters, these formulas reduce to the much simpler Eqs. (5.26) and (5.27). We have also indicated that, in principle, the external circuit parameters can be changed by varying the dimensions of the outer tube—which is extremely problematic from the logistics point of view. Another technique based on a twinaxial principle has been suggested and implemented lately. This is the so-called line injection method [6.44-46].

Two transmission lines, the shielded cable under test (CUT) and the injection line (IL), are formed by a nonshielded wire and the shield of the CUT and are placed together to form a mutually coupled system (Fig. 6.29). The IL is fixed on the CUT shield by adhesive tape. While the characteristic impedance of the CUT is determined by its geometry (and dielectric), the characteristic impedance of the nonshielded IL is determined by the distance of the nonshielded wire from the shield. This distance is chosen so as to provide a matched transition from the coaxial cables feeding (at one end) and loading (at the other end) the IL. On the other hand, since both lines share the same common conductor (shield), this conductor itself can be viewed as a tertiary. The basic relations and "hands-on" details of the method are

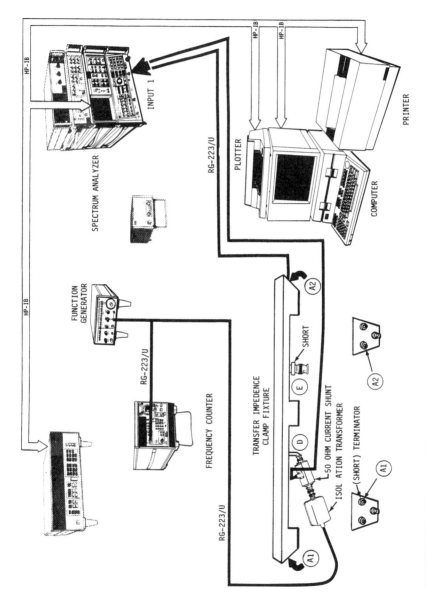

FIGURE 6.28a Test setup using spectrum analyzer and tracking generator

Shielding Measurement Techniques and Apparatus: The Tools of the Trade   463

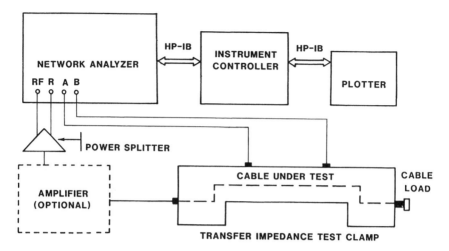

FIGURE 6.28b Computer-assisted transfer impedance measurement equipment test setup for frequency range > 10 MHz

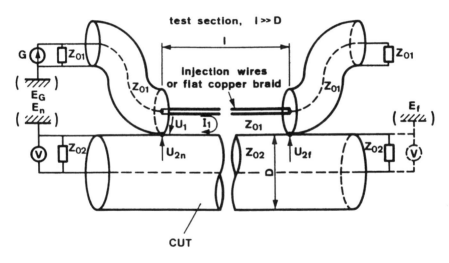

FIGURE 6.29 Mutually coupled system

given in Ref. [6.44], while other sources quoted in the bibliography provide additional information and illustrations of the method applications.

In the lower frequencies, the assembled line is electrically short, and thus there are no difficulties to relate the measured crosstalk to the CUT shield transfer function. But even at very high frequencies, when the assembly must be considered electrically long, the measurements yield realistic results because of the load and phase velocity matching in the CUT and IL. According to Eicher and Boillot [6.46], with a cable under test 50 cm long, the phase velocities in the shielded line and the injection line can be sufficiently well matched up to 3 GHz while, in a line 10 cm long, measurements up to 20 GHz are possible. The authors report that, for measurements at even higher frequencies, a fixed microstrip line setup with 5 cm length has been built.

Of course, the line injection setup is an "open" one. As such, it is prone to coupling to external near and far fields. For this reason, special care should be taken when designing and implementing this measurement setup.

Another potential correlation problem stems from the dependence of shielding performance on the test system configuration. As we have indicated in section 3.3.3 of Chapter 3, the transfer impedance *still is* a function of internal inductance, external coupling, and excentricity of the shielding system. But these parameters can be different in the line injection and triaxial test methods.

6.5 TESTING SHIELDING SYSTEM ELEMENTS

6.5.1 Shielding Enclosures and Building Structures

Shielding Effectiveness Measurements

Suppose, we need to test the shielding effectiveness of a large enclosure, like a shielding room, using a typical *insertion loss* measurement procedure. With all complications as described in the previous chapter (see sections 5.7.2 and 5.7.3), try to answer a very practical question: *where inside* the enclosure we want to place the source radiator and *where outside* the enclosure we want to measure this field?[*] And what type of the test antenna will fit the best when measuring such fields, including multiple modes? Also, it is clear that under certain circumstances, the shielding effectiveness may be *different* for emissions and immunity improvement use of the shielding enclosure. As a result, numerous shielding enclosure effectiveness evaluation procedures were created and are used in the field. Unfortunately, some of these procedures "fly in the face" of scientific evidence (just think of a certain "unnamed" military standard!). So care should be exersized in selecting respective evaluation procedures and their practical implementations. The neglect or misunderstanding of these problems are often the reasons why many otherwise fine experimental works failed to produce valid data, or to correctly interpret the measurements.

The existing enclosure shielding effectiveness evaluation procedures are usually based on modifications or combinations of one or several basic approaches:

- Look for an *"absolutely worse case"*. To obtain such figure of merit, we will have to investigate within the enclosure *all* possible positions of the EMI radiator (or sensor — if the immunity shielding effectiveness is tested). That's "a lot of hard work", to say the least. Another way to create the worst case field distribution, is to use the mode-stirred techniques. Indeed, such techniques are used in mode-

---

[*]By the way, does anybody still doubt the necessity of introducing the "EMI environment" factor into the shielding problems? For those "Thomases", now it is a good time to rethink your position and review the respective material of chapters 2 and 4, as well as the beginning of this chapter.

stirred (mode-tuned) EMC test chambers [2.1], and they permit to generate extremely large field intensities which could be relatively uniformly distributed within the enclosure, when sampled for a certain period of time and a number of tuner positions.
- Fix and standardize "once and forever" the radiator and sensor location points inside and outside the enclosure. This can be either a center point of the enclosure (e.g., see German standard VG 95370), or a pre-defined position and distance from the enclosure wall (e.g., MIL STD 285). Such tests are fast, but the results may not always make sense, because of ample opportunities to miss the most "leaking" areas and inherent non-repeatability. Indeed, following the MIL STD 285 recommendations, the source will be mostly in near field zone, with all the inherent problems. Nevertheless, many shielding enclosure evaluation standards are based on this philosophy. For example as you remember, the data in Fig. 2.8 in chapter 2 of this book was obtained per MIL-STD-285.
- Identify and evaluate (theoretically and/or experimentally) the "problematic" areas of the shielding enclosure and assume that the enclosure effectiveness is *determined* by the performance of these areas. Such problematic areas may include (but not limited to):
  — shielding panel joints (if not welded)
  — doors: electromagnetic seal of the seams, area, shape
  — wall material
  — windows
  — cable penetrations and cables
  — ventilation and lighting devices
  — fire protection equipment.

✦ To identify the "problematic" areas, different kinds of probes and RF "sniffers" can be used. However, this approach will only partially relieve the difficulties, because we will still have to move around the EMI source inside or outside the shielding enclosure to make sure that the EMI at the investigated area is maximized ✦

- Invoke and use a higher generality level model. As we saw in chapter 2, by using the models with higher generality, many specifics of particular task can be avoided (albeit, for a "price"!). As we indicated then, such strategy is especially expedient in comparative evaluations, say, of the effects of different wall materials, joint and electromagnetic gasket quality, apertures in the walls, and so on. For example, using the most general models, we came up with and successfully used for shield evaluation the transfer impedance parameter. This approach permits to investigate some particular *isolated* problems: performance and quality of enclosure wall, separate apertures (e.g., windows), separate cable penetrations, gaskets, joints, etc. But not the "whole picture"! This is especially convenient when comparison is made between the different options.

✦ Indeed, consider an allegedly simple problem: shielding effectiveness comparison of geometrically identical enclosures. For example, when two double-layer wall shielding rooms were comparatively tested — one lined with solid copper+steel walls and the other lined with two copper mesh layers — the advantages of the first room were obvious below several hundred kHz (magnetic shielding) and around and over 1 GHz (microwave region), while in the medium range of RF frequencies the shielding effectiveness of both rooms was over the measuring system dynamic range. Obviously, the same data could have been obtained just by measuring the transfer impedance of the wall material! ✦

Of course, such comparison is meaningful only when the studied parameter, e.g., wall penetration is the dominant EMI factor (remember our discussion about dominant radiators and sensors in section 1.1.3 of chapter 1?)

✦ Numerous existing enclosure shielding effectiveness standards and guides, including those referenced in the bibliography to this chapter, address the above problems with different degree of success. To date, probably one of the most "desperate" attempts to resolve all these issues was made in the IEEE Std 299-1991 [6.3]. This standard provides a uniform set measurement techniques and procedures to test room-sized, high-performance shielding enclosures from 14 kHz to 18 GHz, extendable to 50 Hz and 100 GHz, respectively. Recognizing the fundamental differences between the enclosure shielding performance in different areas of such extremely wide spectrum, the standard recommends to perform the measurements in seven fairly narrow bands, falling in three major test ranges, as indicated in the Table 6.1. Also referenced in the table, are recommended test antennas and test procedure (the Figs. numbers are per this book). The diagrams in Figs. 6..30-32 are presented "as is" from the standard to convey the "flavor" of the given recommendations.

Table 6.1 **Standard Measuring Frequencies for Shielding Enclosures per IEEE Std 299-1991**

STANDARD FREQUENCY	MEASUREMENT TYPE	ANTENNA	Fig. #
Low range	H Field	Small Loop	6.30
14-16 kHz			
140-160 kHz			
14-16 MHz			
Midrange	Plane Wave	Dipole	6.31
300-400 MHz			
850-1000 MHz			
High range	Plane Wave	Horn	6.32
8.5-10.5 GHz			
16-18 GHz			

The recommended procedures consist of two steps:

1. Preliminary scan. This shall be made along all accessible shielding faces to identify weak points of maximum electromagnetic leakage. Items that should be checked are doors, power-line filters, air vents, seams, and coaxial cable fittings.

2. Basic test, which is performed for standard locations recommended in IEEE Std 299-1991, as well as at the identified high leakage locations. It consists of a *free-field* setup calibration providing the *reference field intensity,* which is compared to the field penetrating through the enclosure.

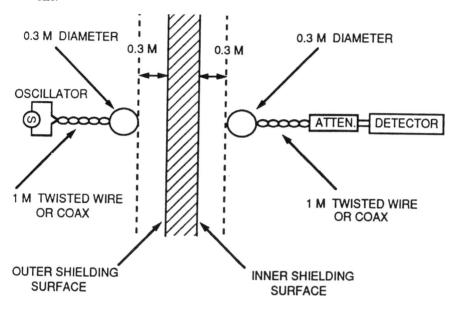

FIGURE 6.30 Small-loop test setup (per IEEE Std. 299-1991)

It is recommended to avoid testing at frequencies *close* to the lowest resonant frequency f_r of the enclosure ($0.8 f_r < f < 3 f_r$). Many additional recommendations are given dealing with instrumentation, setup, data acquisition and processing. ✦

As you can see, the shielding testing business is not for the "weak at heart and knowledge". But same could be said about many other topics in electromagnetic shielding.

Transfer Impedance Of Shielding Enclosure

Wouldn't it be less of a problem to measure the shielding enclosure transfer impedance, instead of an insertion loss type shielding effectiveness? For instance, in this way we can avoid the "nasty" problems of finding the positions of the signal source and sensor, dealing with radiating patterns, with reflections from extraneous objects, etc. And true, in Chapter 3 (sections 3.1.2 and 3.3.4, and respectively, Fig. 3.2d and Fig.3.14) we have addressed the conductor-to-shield coupling and transfer impedance issues in two- and tree-dimensional shielding surfaces and enclosures. Moreover, the problem is not new, and it was addressed both experimentally and theoretically, e.g., see [6.64-66].

Thus, it appears that the transfer impedance test on an enclosure is possible! How would one proceed with the measurements? Per general transfer impedance measurement procedure, there are two issues:

1. *Inject* the current into the shield
2. *Measure* the voltage between the shield and the shielded conductor

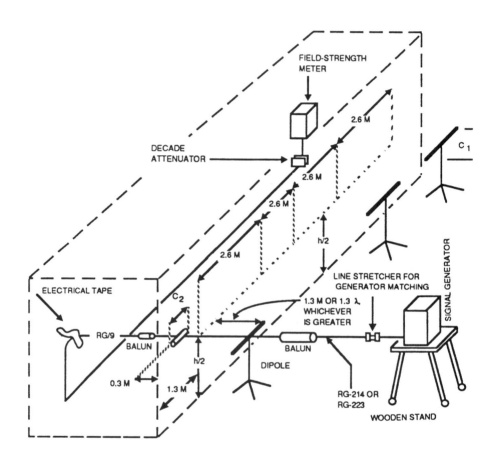

$C_1 = \lambda/2$
$C_2 = \lambda/8$
$C_3 = 2$ M or 2λ, whichever is greater

FIGURE 6.31 Midrange measurement setup (per IEEE Std. 299-1991)

Shielding Measurement Techniques and Apparatus: The Tools of the Trade

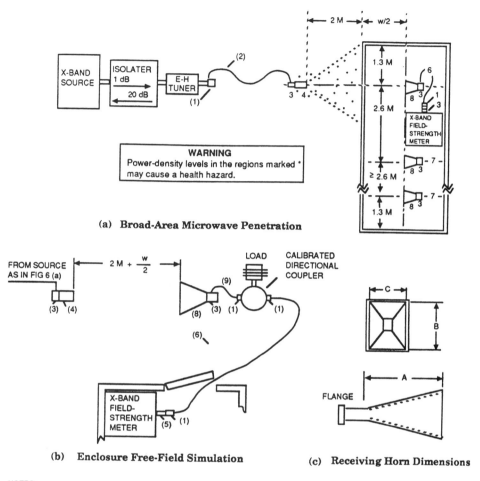

NOTES:
(1) Type N adapter coax to waveguide.
(2) Coaxial cable or waveguide.
(3) Adapter.
(4) Transmitter waveguide antenna, Table 5.
(5) Attenuator (if not within field-strength meter).
(6) Two-to-three meter low-loss coax, e.g., RG-214.
(7) Additional centerlines so that all areas are illuminated.
(8) Receiving horn antenna, Fig 6 (c) and Table 6; dimensions relate to standard EIA waveguides, flanges, and waveguide-to-coaxial transitions.
(9) Known low-loss coaxial cable about one meter long, calibrated in conjunction with (3) and (1).

FIGURE 6.32 High-range measurement setup (per IEEE Std. 299-1991)

470 Electromagnetic Shielding Handbook

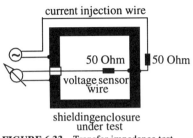

FIGURE 6.33 Transfer impedance test of a small enclosure

These issues are easier resolved for small enough shielding enclosures. In this case, the circuits can be used per Fig. 3.2, where the shielding enclosure under test "plays the role" of a coaxial cable external conductor, as shown in Fig. 6.33.

However, at larger enclosures such procedure may easily "lose common sense". It is still useful, when *local areas* of the enclosure are to be tested. For example, testing to the Fig. 6.33 paradigm may be useful to determine the effects of the shield non-uniformities: at the enclosure panel joints, apertures, cable penetrations, etc. If you revisit Fig. 3.22, you can see, how it "works".

Thus, just like in the case of shielded cables and assemblies, the transfer impedance may be an important shielding enclosure evaluation parameter which can be applied to whole enclosures or their "non-stationary" local areas. But don't forget the "price" we pay for such simplification — the constraints of the trasfer impedance test: a very general test model, in which many specifics of an actual EUT can be lost. For example, take the radiating patterns inside and outside of the enclosure. As difficult it is to deal with their problems, we should recognize that they carry important EMC information. This problem can be compared to one in the antenna studies: even if we know the power at the antenna terminals and / or current in the antenna radiating elements, the antenna gain may "work" to multiply many times the field intensity in certain directions!

6.5.2 How to Test Shielded Cables and Connectors?

You already know the answer: almost any of the discussed in this chapter techniques can be (and has been) applied to measure the shielding performance of these important shielding system elements. Just revisit Fig. 6.4 in section 6.1.4: system and element test, global and local, radiated and conducted, shielding effectiveness and transfer impedance, time and frequency domain, almost any routine and test facility — everything "goes". In this way, the shielded cable and connector testing options are all but endless — you can find detailed descriptions in the quoted (and not quoted) standards, papers, specifications. While discussing the testing roadmap, along with other shielding elements, we have often illustrated the use of the specific methods in the cable, connector, and cable / connector assembly context.

Traditionally, the transfer impedance measurement techniques used to be the main tool for "serious" cable and connector testing[*]. Other often applied methods are based on the use of antenna tests at the OATS or anechoic room, or TEM (GTEM)

---

[*] Leave alone dozens of other methods described in the literature — not that these are "non-serious", but they often lack the necessary justification and rigor, to rely upon them

cell measurements. However, at microwave and higher frequencies, the transfer impedance approach becomes difficult to implement. With the rapid expansion of the used frequency band in recent years, there was renewed interest to techniques based on shielding effectiveness approach, in particular, on the use of stirred-mode techniques. For example, EIA-364-66 standard [6.8] specifies the mode-stirred method of connector shielding effectiveness measurements over the frequency range of 200 MHz to 40 GHz. As a practical matter, the EUT in Fig. 6.8 could be very well a cable sample, or a connector, or an assembly, no matter how "irregular" in shape.

6.5.3 Testing Shielding Materials

Background

Per our classification, the shield material is not even a shielding system element, but belongs to a "sub-element" level, (although material is an important technological product). The more surprising it is then to see that sometimes a system shielding performance is reduced to *some* characteristic of the shield material. This is not to say that intrinsic shielding material properties are irrelevant to the system shielding and EMC performance. But revisit our *"parallel bypass"* model of the shielding performance in Fig. 2.21 (Chapter 2, section 2.5.2). There, the shield material is *only one* factor in the shielding performance. In fact, in chapters 2 and 3 we were able to quite persuasively demonstrate, that for "regular" *metallic* shields of *"average"* product or cable enclosure *thickness* this factor is often not even the most critical one, because the electromagnetic field penetration through *solid* shield (i.e., *diffusion* - see Chapter 3) is comparatively quite small. The *dominant* factors in such applications are those determined by shield imperfections (apertures, seams, slots) which are *always* present in real life products, as well as by system element interactions, grounding, etc. In our model, these effects resulted in "bypassing" the solid material "maximum" performance, as illustrated in Fig. 2.21. That's where the model's name came from!

As a "rule of thumb," a "good" *material* (to be sure, "solid" metal without imperfections and in idealized environment) shielding effectiveness or transfer impedance may determine *only* the *best possible* performance of the shielding system made from this material, given all other factors are *secondary* — a *potential* which is rarely reached in real life. Only when the material has outright *inferior intrinsic* shielding properties, its contribution to the" total" EMI flow *through* the shield may become significant, and even decisive. However, this is just the case with many *new shielding materials* and *composites*! As you will see in chapter 8, the shielding performance of these materials is indeed often marginal. And what's important practically, just such shield materials *need* to be tested to make sure that its performance is *better* than the performance of the *worse* "offender" in the shielding system.

Thus, the question of testing the shielding materials is not just "academic".

So, how to test the materials for shielding? As a rule, the *intrinsic* shield material performance data is supplied by the material manufacturers based on their own measurements. To meet this need, a whole test industry exists around shielding material measurements, based on respective test methods, techniques, and apparatus, some of them are quite "exotic". To date, this industry came up with more than "adequate" number of measurement techniques. Thus again (how many times in shielding?), the question is not so much in inventing new measurement techniques, but that of selecting and evaluating the existing methods, interpreting the obtained results, and determining how useful is this data.

Shield Material Test Classification

In the following, we will briefly review several important shield material testing concepts (see Refs. [6.52-61]). Refer to our shielding measurement roadmap of Fig. 6.4. The shield material box contains three groups of parameters:

1. intrinsic material electrical properties
2. material transfer parameters
3. material-related element shielding effectiveness

The intrinsic material electrical characteristics that are important from the shielding point of view are conductivity, magnetic permeability, electric permittivity, and dielectric loss tangent. Also included are the material surface resistance ρ_s and contact resistance parameter ρ_c. The corresponding methods for testing these parameters are well known and constitute an exciting and challenging field of material sciences and related disciplines — well beyond the scope of this book. Since these methods can be found in respective literature, there is no need, nor space, to present them here. Instead, we will concentrate on the interpretation of the material parameters in the context of shielding.

In the previous chapters we have addressed the effects associated with shield material conductivity (see discussion of the conductivity effects on transfer parameters in Chapter 3), magnetic permeability (see magnetostatic shielding effectiveness in Chapters 4 and 5), as well as contact resistance (see in Chapter 3 description of the role and importance of this parameter for overlapping foil shields). However, until now we have been concentrating mainly on metallic shields, while the role of *intrinsic dielectric* parameters ε and $\tan\delta$ in shielding performance was less obvious. But we will return to these subjects later in this Chapter 6 and Chapter 8, when discussing non-conducting shielding materials, e.g., chiral and magneto-dielectric shields.

With regard to shield transfer parameters, the associated measurement philosophies are similar to those of *shielding system element test*, with only certain features determined by the material specifics: measurements of the transfer impedance between two surfaces of the shielding material test sample, measurement of the flat shield or shielding enclosure insertion loss between two antennas, techniques based on the use of *simulated* far zone radiating fields generated in TEM/GTEM cell,.

Perhaps, you wonder whether these parameters aren't exactly what we need! However, the problem here is similar to that encountered in comparing different *shielding system element* evaluation levels: a *particularly* shaped sample (usually planar or cylindrical) in a *particular* test configuration is not necessarily representative of the actual shielding system performance in a more realistic environment. Thus, while the material shielding data provided by such measurements may be crucial for one kind of specific systems and applications, it also can be quite misleading in other systems and applications (e.g., see some examples in Chapter 2).

To find the way out of this maze, as often in shielding, we call "to the rescue" the test method *classification*.

First, let's make a distinction between the highly conductive shields, e.g., metallic or super-conductive ceramic (see Chapter 8) — but in principle not limited to those, and such newer materials like metal-coated dielectrics, dielectric-metal composites, inherently conductive plastics. In the previous chapters we have dedicated enough attention to the metallic shields and methods of their evaluation. The conclusion that we can draw from these previous discussions is that testing shielding performance of "solid" metallic materials is not really a big problem: any of the described techniques can be applied. As a rule, the transfer impedance test methods are quite adequate to test such materials. But you may not even need to test metallic shields. Because the theory of transfer impedance of *solid metallic* shields is so predictable and well developed, it is usually sufficient to *theoretically calculate* the shield transfer impedance.

Moreover, the excellent correlation between the solid material transfer impedance theoretical evaluations and experimental tests, results in an interesting "side" benefit: by combining calculations and measurements, effective methods were created [6.67] to measure such "difficult" parameter as magnetic permeability μ. The suggested technique encompassed forming a sampe (tape) of a flat ferromagnetic material into an outer conductor of a coaxial cable. By measuring the transfer impedance of this cable and knowing (or measuring - that's a "trivial test) the geometrical dimensions and conductivity of its external conductor, the magnetic permeability can be figured out. Also, by controlling the current in the outer conductor, electromagnetic, mechanical, chemical and other properties of the test environment, their effect on the magnetic permeability and its non-linearity can be determined, which in itself is usually a complex problem!

So, what is then left to discuss in the shield material testing?

A lot! When engineers (and sales people) speak about the *new* material shielding performance, more often than not, they mean materials with inherently *lower conductivity* (at least averaged in volume): metal-coated dielectrics, dielectric-metal composites, inherently conductive plastics. Such materials have been known and used in some "niche" applications for many years, but only recently they advanced to the "mainstream" practical shielding levels. But then, it's even more appropriate to know how they were tested (or how to test them yourself). "Caveat emptor"[*], and watch for "snake oil" — some of the advertised products might be that kind!

While we will discuss the "design" and performance of several such new materials in Chapter 8, here we are interested in their test methods. These methods are based on the induction field or "simulated" far field measurements. Probably, the most general analysis of several such basic shield material measurement techniques is given in [6.56-58]. The referenced sources also contain sample test results, obtained using these techniques. A somewhat modified summary of the far- and near-field shield material test methods per the quoted sources, is given in Table 6.2

---

[*]"Buyer beware" — lat.

Table 6.2 **Summary of shield material test methods**

Test Method	Frequency Low	High	Field Config.	Material Sample	Dynamic Range, dB
Continuous Conductor Coaxial Holder	1 MHz	1.4 GHz	Far: Plane Wave	Annular Disk	90-100
Flanged Coaxial Holder	1 MHz*	1.8 GHz	Far: Plane Wave	Circular+Reference (Ring&Disk)	90-100
Time Domain Technique	200 MHz	3.5 GHz	Far: Plane Wave	Large Sheet	50-60
				Mounted on Aperture	40-50
Dual TEM Cell	1 MHz	1 GHz	Near, or Glazing Plane Wave	Cover an Aperture	50-60
Apertured TEM Cell in Reverb Chamber	200 MHz	1 GHz	Near	Cover an Aperture	90-100

*Note some discrepancy with ASTM D 4935 recommended frequency range

Rationale

To arrive at the shield material test technique rationale, it is worth to analyze *another* method — one that is *not included* in the Table 6.2 summary, but probably would be the first to come to mind. Although, as you will see, that method doesn't fit our purpose, by analyzing it we will see the major problems with shield material measurements and using it as a "jumping-board", try to determine in several steps, what's to be done.

Indeed, why not just to install two antennas — transmit and receive — in "free" space and measure the *insertion loss* when an *infinitely* large sheet of shielding material is placed *between* the two antennas? But that's exactly why: while it is often *impractical* (i.e., difficult and expensive) to test an "infinitely large" metallic sheet in ambient free "free space", anything short of such ideal conditions is fraught with serious complications, distortions, and as a result, erroneous data. Thus, a "smaller" sheet illuminated by an electromagnetic field will be prone to edge effects and diffraction, while reflections, resonances, and multi-moding, as well as interference, will be generated if there were "extraneous" objects and ambient noise in space at both sides of the tested sample.

A "brute force" approach to avoid the "infinitely large size" and "ambient noise" complications, could be as follows. Suppose, instead of free space, we use a shielding room separated in two parts by a "very good" solid shield (at least, much better, than the tested material). Then we make an *aperture* in the shielding partition, place the transmit and receive antennas in the opposite sections of the shielded room. We then measure the insertion loss with the aperture *open* and *covered* with the testing material. The ratio of the two results can be viewed as a figure of merit for the shielding effectiveness of the tested material. In this way, we have eliminated two problems — ambient noise and diffraction. But then again, we have created two *new* problems: the separated sections of the shielding room (each of them can be treated as a smaller shielding room themselves) and the aperture generate their own resonances! Just as we described in Chapters 5 and 6.

However now, we are closer to the solution, because there are relatively simple and "cheap" ways to avoid these reflections and resonances. Think TEM cell: there we can create a TEM field configuration, at least at frequencies *below* the first resonant frequency for a given cell size. TEM cell supports reactive electromagnetic field configuration, in which both electric and magnetic field vectors are located in the transverse to the cell longitudinal axis plane. Along with the reactive induction near field, the TEM cell can be used to *simulate* far field. Although the field in the TEM cell is of inherently induction E-H nature, it's structure is identical to a plane wave in the far field. As you remember from previous discussion, the TEM cell is basically an *expanded* length of *coaxial cable*. The smaller are *cross-sectional* dimensions of the cell, the higher is the resonant frequency (e.g., see Fig. 6.6 in section 6.2.1).

But we really do not need a large TEM cell to perform the shield material measurements! Indeed, the *only* reason to use a large cell, is to be able to *insert* inside it (in the next two sub-sections, we will suggest two ways how to do that) a *large* sample of shielding material, so that the signal penetrating through this sample were *strong enough* to be detected. Since the environment inside the cell is shielded from the ambients, we can have a good signal/noise ratio with even a small signal (generated by a small material sample). Therefore, if we select a TEM structure, or a length of coaxial cable, of relatively small crossection, there will be no resonances in the cavity!

This leaves still open the question, how to place a material sample inside our our TEM cell / coaxial cable, so as to avoid diffraction, edge effects, and aperture resonances. We will consider below only *two* basic methods to do that, although there are many more available, and new methods still keep "popping up" in the literature.

Dual TEM Cell

The dual TEM cell uses two TEM cells mounted such as to enable a common wall — actually two adjacent walls, one of each cell, brought together side-by-side as shown in Fig. 6.34. The adjacent walls contain an aperture (usually, rectangular), which can be either "empty", or loaded with a tested material sample. One TEM cell is being driven by an appropriate signal. The amount of RF energy coupled into the second cell through the aperture, is measured at each end of the second cell, either separately, or simultaneously by mixing the two signal in a hybrid junction. Taking the ratio of the coupled energy as measured at the outputs of the dual TEM cell for the cases of unloaded and loaded apertures, the shielding effectiveness figure of merit can be obtained for the given sample.

Actually, there is a complex theory behind the "workings" of this device. Referring the readers to the original literature, e.g., [6.57, 6.59], we will underscore only the most general considerations. The dual TEM cell appears to be a typical near field device, in the respect that the ratio of electric and magnetic fields in the coupled sig-

Dual TEM cell in open position

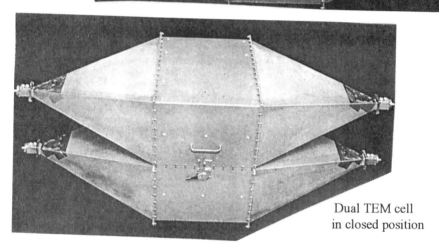

Dual TEM cell in closed position

FIGURE 6.34 Dual TEM cell test fixture

nal is not fixed (as in the far field), but depends on the material properties and aperture shape. Indeed, the TEM modes propagating in the "receiving cell" from the aperture in the forward and backward directions are determined by the sum and difference of the electric and magnetic polarizabilities. Now, the sum signal of the two outputs of receiving cell depends on the normal (with regard to the aperture) electric field coupling, while the difference of these outputs — on the tangential magnetic field coupling. Doesn't this remind you the addition and substraction of electric and magnetic coupling at the near and far end of crosstalking transmission lines (see Fig. 4.6 and equations 4.36)? And rightfully so, because there is a deep analogy between these two processes!

Practically, the polarizabilities of the aperture are not easy to determine, especially for the loaded (by the tested sample) aperture. Another complicating factor is the usually uncertain contact resistance between the tested sample and the TEM cell wall(s). In general, the contact impedance tends to degrade the measured shielding effectiveness.

◆ Just to give a "flavor" of this technique, here is "factual" information [6.57] of a dual TEM mode cell designed for a frequency band 1 MHz - 1 GHz:
- cross-section : 18 cm × 12 cm
- center plate width: 13.8 cm (gap between the center plate and cell walls is 2.2 cm)
- aperture: square with a side 5.08 cm
- dynamic range: 50-60 dB
- sweep time from 1 MHz to 1 GHz requires arond 30 minutes

◆

ASTM D 4935

One of the most known standards for the shield material testing is ASTM D 4935: Standard Test Method for Measuring the Electromagnetic Shielding Effectiveness of Planar Materials. This test method is based on the application of a coaxial holder. It applies to the measurement of shielding effectiveness under normal incidence, far field, plane wave conditions. The use of the standard is recommended within 30 MHz - 1.5 GHz frequency range.

Basically, the standard is written around a specimen holder, which is an enlarged coaxial transmission line with special taper sections and notched matching grooves to maintain a characteristic impedance 50 Ohm (see Fig. 6.35).

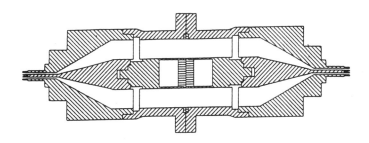

FIGURE 6.35 Cross section of ASTM D 4935 sample holder

A pair of flanges in the middle of the holder holds the *test specimen* of a *circular shape*. It allows *capacitive coupling* of energy into insulating materials through *displacement currents*. To account for the discontinuity caused by the test speciment in the transmission line, a *reference specimen* of a ring shape (actually a ring for the outer conductor plus a respectively sized smaller circle for the center conductor) is used of the same material and thickness as the tested specimen. To prevent *parasitic* (in this case) conduction currents to interfere with the *signal* displacement currents, all the means are used for electrically isolating the flanges from each other, including nylon screws joining the flanges.

The CW signal generator is connected to the input of the holder, and a corresponding receiver — to its output. By measuring first the received power with the reference specimen, then — the received power with the test specimen, and taking the ratio of the two, the tested material shielding effectiveness figure of merit is determined.

The standards specify the dimensions of the holder and specimens, preparation and calibration of apparatus, the test procedure, and related special considerations to insure the accuracy and repeatability of the obtained results.

6.5.4 Electromagnetic Gasket Test Specifics

First, "traditional" definitions.

An applicable part of the term entry in the IEEE Standard Dictionary of Electrical and Electronics Terms (ANSI/IEEE Std. 100) defines an electromagnetic gasket as "... a resilient insert usually between flanges intended to ... reduce power leakage and arcing". Although the Dictionary specifically applies this definition to waveguide (waveguide components), this is pretty much what this system element is supposed to do in any shielding application, just substitute the "flanges" for "adjacent surfaces" of shielding system elements.

How to test the electromagnetic gaskets? The rationale is as follows. Consider, per this definition, an electromagnetic gasket as an *extension* of the gasketed *shield*, that is, view the original shield + gasket as a *composite shield*. Then we can test such "amalgamated" shield as a shielded system element or even as shielding material (remember the concepts of element-vs-system definitions in section 6.1.3, as well as a further discussion in section 6.3.1?). If in our "composite shield" the gasket is attached to a "very good" original shield, the unit performance, *as a whole*, will be determined by the "worse ingredient", that is the gasket — just as our "parallel bypass" model predicts (see Chapter 2). Thus, the same techniques used to test the shield materials, can be applied to test the electromagnetic gaskets (e.g., see Refs. [6.68-75] for a literature "sampling")

To prove the point, Figs. 6.36 -38 present three alternative techniques to test electromagnetic gaskets: one based on the transfer impedance measurements, the second — based on the shielding effectiveness measurements using a flanged material coaxial holder similar to that described in ASTM D 4935, and the third — based on the shielding effectiveness measurements using stirred-mode techniques. The details should be clear from the diagrams. Many modifications of the presented, as well as other techniques were suggested and described in the literature.

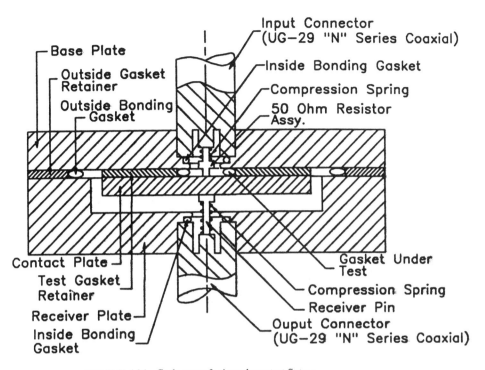

FIGURE 6.36 Gasket transfer impedance test fixture

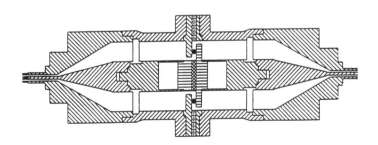

FIGURE 6.37 Cross section of modified D4935 sample holder to test gaskets

6.6 TESTING IN TIME DOMAIN

In the course of previous discussion, we have mainly focused on frequency-domain shielding analysis, evaluation, and measurements. There is no need to justify this choice, because there are many advantages to the frequency domain approach. However, the shielding discipline, and cable shielding in particular, presents ample oppor-

480 Electromagnetic Shielding Handbook

tunities to appreciate the related problems also. In particular it is often not a trivial problem to distinguish between the contributions of the CUT and the test support interconnections to the measured EMI signal. When testing connectorized assemblies, it is not always possible to discriminate between the effects of the cable and connectors, or between different branches of multi-branch assemblies. On the other hand, when using field intensity measurement-based test methods in the frequency domain, it is often impossible to separate the effects of the direct ray from the rays reflected from the facility confines. For this reason, the requirements to the test facility may be quite strict, leading to excessive costs. These and other problems are exacerbated at high frequencies and when several different coupling mechanisms act simultaneously.

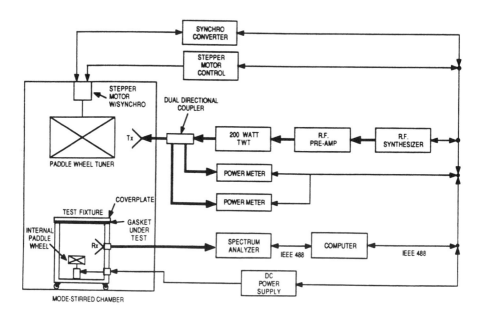

FIGURE 6.38 Schematic of stirred mode setup for gasket shielding effectiveness measurements

One effective answer to these problems is to conduct the measurements in the time domain [6.47–51]. In this case, instead of a monochromatic single or swept EMI signal, the cable or assembly under test is injected with a current or voltage step, and the system response is evaluated. In this case, the response of different elements of the system under test is delayed in time, and the delay corresponds to the distance the signal has to travel from these elements. When the pulse rise time is small enough to ensure the necessary resolution of the test system, the measurement results become independent of the line loads.

Shielding Measurement Techniques and Apparatus: The Tools of the Trade 481

It is interesting and important that, in principle, the same measurement setups ("hardware") can be used for testing in both the frequency and time domains. Also, when testing in both time and frequency domains, it is necessary to make measurements on a sample of short electrical length (although, in time domain, with the proper selection of the measurement parameters, this requirement is not as strict). Thus, to measure transfer impedance, the same triaxial or line injection methods and fixtures can be used. In this way, from the measurement logistics viewpoint, the main difference in hardware is that, instead of the synthesized generators and spectrum analyzers used in frequency domain measurements, the main tools in the time domain are a pulse generator and an oscilloscope.

Of course, the test signal selection and characterization, data processing, and test result interpretation in time-domain measurements ("software") present a number of specific problems. Fourier transformations provide the basis for understanding the relationships and mutual transitions between the measurement results in the frequency and time domains. Indeed, in Chapter 1, we reviewed digital signal fundamentals and looked at some applications. As we saw, there exists a unique correlation between the same EMI phenomena in the frequency and time domains. In particular, Fig. 1.39 illustrated the pulse crosstalk between cable lines, and in the subsequent discussion we were able to provide a parallel with the crosstalk in the frequency domain. But we also have seen in the present chapter that transfer impedance and transfer admittance measurement schematics can also be treated as crosstalk. Thus, it comes as no surprise that the response of the disturbed line in the process of transfer parameter measurements looks similar to that of Fig.1.39!

Does that mean that in the time domain we can measure the same shielding parameters as we addressed in the frequency domain, but without the identified problems? Yes, and no—because time domain measurement methods have problems of their own. One of the most serious drawbacks of measuring in the time domain is insufficient dynamic range. For example, when we specify 1 W power of the test signal in the frequency domain, it means that at each test frequency this is the power introduced into the disturbing line. However, when testing in the time domain, the 1 W test pulse energy is spread between the different harmonics all over the frequency band (see Chapter 1), so that only a fraction of this energy is attributed to each frequency of interest! Moreover, as we saw, the higher the harmonic order (that is, frequency), the smaller its amplitude. But higher frequencies are of primary interest! For these reasons, when testing in the time domain, there is often a need to have both the power amplifier at the source output and the low noise amplifier at the oscilloscope input. And we would want these amplifiers to have a wide bandwidth.

◆ With these "caveats" in mind, we will consider two time-domain transfer impedance measurement procedures. One technique is based on the line injection test method, and the other on a "classic" triaxial. Both utilize a test signal pulse in the shape of a unit step with a particular rise time: the amplitude of the pulse is gradually rising, then it stabilizes.

The time domain line injection method was suggested and implemented by Fowler [6.48]. When a step pulse current is injected into the disturbing line, the generated voltage impulses received at the ends of the disturbed line and measured by an oscilloscope are similar to those in Fig. 1.39. It follows from Fig. 1.39 that the received impulse amplitude presents a function of time delay, t. If we

482 Electromagnetic Shielding Handbook

relate the received voltage corresponding to different time delays to the current, we can treat the obtained curve as the transfer impedance time characteristic {it was named Z_t (time)}. The next step is to relate the time axis to the frequency axis. This is done using the relationship

$$f = \frac{1}{2\pi(t - t_0)} \tag{6.10}$$

where t_0 is the reference delay time.

Thus, the time transfer impedance characteristic is transformed into frequency transfer impedance characteristic. The reference delay time value significantly affects the accuracy and even validity of the results. As a rule, it must be somewhat larger than the time corresponding to the peak of the received spike. However, so far, no reliable and theoretically proven ways to select this parameter have been suggested.

The experimental data are presented in the frequency range up to 100 MHz, taken using frequency-domain procedures and time-domain procedures based on the suggested method. The obtained correlation strongly depends on the selection of the test setup parameters and t_0. For these reasons, the use of reference calibrators is strongly recommended.

The second technique was suggested and implemented by Demoulin et al. [6.50]. Here, the FFT procedures are used to determine the frequency characteristic of the receive signal from the time-domain measurements taken with a sampling oscilloscope. The main thrust of the utilized procedure is to make the measurements at a relatively small cable under test (CUT) sample — just 3 to 5 cm long. To obtain such a sample, a 1 m long CUT is measured in the shorted triaxial. Over the CUT shield, a "solid" copper shield layer is applied from both ends, leaving only 3 to 5 cm of the shield under test exposed.✦

✦ There is one more piece of related "unfinished" business: is there any use in the transfer impedance phase? In general, when dealing with the frequency domain, it did not seem like this information was necessary. However, when frequency-domain transfer impedance is used to characterize the time-domain (transient) performance of the shielded cable, that information may be important. There has not been very much research in this area. The available studies (e.g., see Ref. [143]) indicate that neglect of the transfer impedance phase may lead to errors in the evaluation of cable transient excitation, especially in the early terminal responses.

As of this writing, the IEC is planning to generate standards for time-domain transfer impedance measurement techniques along with the frequency-domain methods. ✦

6.7 TEST RESULT CORRELATION AND INTERPRETATION

6.7.1 "Apples and Oranges"

Practical measurements generate several types of correlation issues related to comparison between the testing performed at different conditions. The list of the most important such differences corresponds to the roadmap of shielding measurements of Fig.6.4:

- System levels: "absolute" system test versus assembly test versus product test versus shield material test

- Evaluation parameters at the same system level: shielding effectiveness versus crosstalk measurements versus transfer impedance
- Measurement techniques at the same level: system test (open field antenna test versus TEM simulation versus stirred mode chamber); assembly test (absorbing clamp versus TEM/GTEM versus stirred mode chamber); cable or connector (transfer impedance using the numerous techniques)
- Test signal: time domain versus frequency domain
- Correlation between identical techniques at identical facilities (which is a quality problem)

We start with system EMC performance evaluation diversity.

In the previous sections, we have devoted enough attention to the comparison of test relevancy and test generality at different system levels, so there is no need to repeat the conclusions. On the other hand, measurement result repeatability when identical techniques are implemented at identical types of test facilities is more of a quality problem, and it must be studied on the basis of general considerations of test uncertainty and statistics. There is also abundant literature available on the general problems of test result correlation obtained at different test facilities (and we have already quoted some of the sources). For example, Fig. 6.39 presents a PCS terminal radiation pattern and EIRP measurement results. The measurements were performed @ 2 GHz in four different test facilities. As shown, the variations were within 6 db. However, some literature sources report discrepancies in order of 10 dB and more! And all these tests are of the *same* nature — radiated test!

Even more discrepancy could be expected if *diverse* in nature measurements are performed. Take electromagnetic gasket measurements. To start with, discard the MIL-STD 285 type of test — it yields extremel;y unreliable results [6.73]. But even, when the measurements are performed in a special and quite precise sample holders — transfer impedance or ASTM D4935-type test — the discrepancies between different samples reach about 20 dB. A comparison between the two latter methods yields similar uncertainty [6.73].

Let us compare crosstalk and radiating shielding effectiveness in electrically short lines. The common-mode currents result in longitudinal fields which generate both quasi-stationary (crosstalk) and radiating (antenna) external fields. Since both common-mode crosstalk and antenna currents have the same source (longitudinal voltage drop at the shield external surface), we can expect a shield performance in these regimes to have certain common features. Thus we can expect that a comparison of formulas as well as measurement results will give an indication of both of them. On the other hand, not only are the measurement procedures different in the radiating field and crosstalk measurements, but even the shield working regimes are different.

Using Eqs. (5.27) and (4.57), we obtain the following:
Radiating field (emissions and susceptibility)

$$dE_{dr} = E_e Z_t \frac{dl}{\lambda} A_g A_t \qquad (6.11)$$

Quasi-stationary field (crosstalk)

$$A_{or} = A_{er} = 20\log\left|\frac{Z_c(Z_{31} + Z_{32})}{Z_{t1}Z_{t2}l}\right| \qquad (6.12)$$

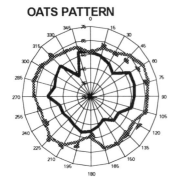

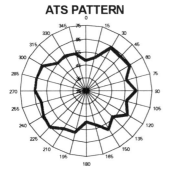

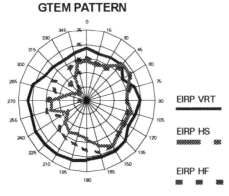

FIGURE 6.39a Radiating patterns of a PCS terminal @ 2 GHz

It is easy to see that, in both cases, the shielding effectiveness of a cable shield can be expressed as a function of the shield's transfer impedance:

$$S = 20\log|Z_t| + F \qquad (6.13)$$

Equation (6.13) is identical to (5.26-27) and (5.42) and consists of two items. The item F is test-specific. Given the diversity of system configurations, the available measurement procedures, and the shield working regimes, we can hardly expect to express the parameter F in equivalent terms for radiating field measurement and crosstalk measurement techniques.

✦ Moreover, as we just saw, the radiating shielding effectiveness measurements themselves can be performed in different ways, and these are not always easy to correlate. Techniques include the use of an OATS, ATS, TEM/GTEM cell, or an absorbing or current injection clamp, and some others. Thus, the absolute correlation even between the radiating fields and crosstalk-based techniques is all but excluded. Therefore, in our task of comparing shielding effectiveness radiating field measurement and crosstalk measurement techniques, we need to specify which radiating and which crosstalk techniques we mean ! ✦

On the other hand, because the Z_t-related items are common to all types of measurements (well, almost all—in some shield designs, the capacitive coupling impedance must be looked at), we can expect the relative correlation between different shield designs to be pretty good. And indeed, comparison of Figs. 5.18 and 5.19 on one hand, and 5.28a and 5.28b on the other, shows that essentially identical relative evaluations were obtained. Really, from Eqs. (6.12) and (6.13), it follows that, whatever test technique is used, the relative shielding effectiveness of two shields having different transfer impedances Z_{t1} and Z_{t2} but measured in identical conditions, is

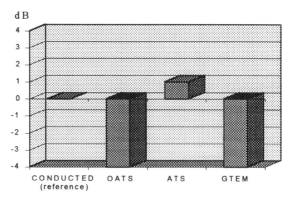

FIGURE 6.39b PCS terminal EIRP variations @ 2 GHz

$$S = 20\log\left|\frac{Z_{t1}}{Z_{t2}}\right| \tag{6.14}$$

But even when the same shielding evaluation parameter is selected, the accuracy and repeatability of test results is difficult to guarantee. It may seem that at least the transfer function is immune to the vagaries of different test methods. However, even this parameter is not invariant to the differences in field distribution at the shield external surface. Indeed, one cannot expect to create identical field distributions in such methods like shorted triaxial, terminated triaxial, or line injection (e.g., see Fig. 6.40)

The best correlation can be achieved between the methods of transfer function evaluation at electrically small EUTs. As the last example, consider the shielding effectiveness of three types of power supply cordage: (1) nonshielded (here, the "green" wire was treated as a shield, which it really is—remember the concept of "parallel" shield?) and shielded with (2) single-layer aluminum foil shield or (3) double-layer aluminum foil shield plus copper braid. The cordage was measured using

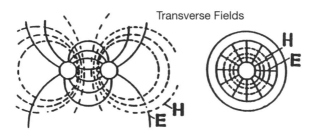

FIGURE 6.40 Electromagnetic field configuration in the shield test setup

three techniques: (1) transfer impedance (measured with a terminated triaxial) and shielding effectiveness measured with (2) an absorbing clamp as the ratio of antenna currents and (3) in an open site, as the ratio of electric field intensities. Figure 6.41 presents the experimental data. As shown, while the discrepancies between different measurement methods can reach 20 dB, the general trends of shield behavior, defined by Z_t and identified by all three methods, are similar.

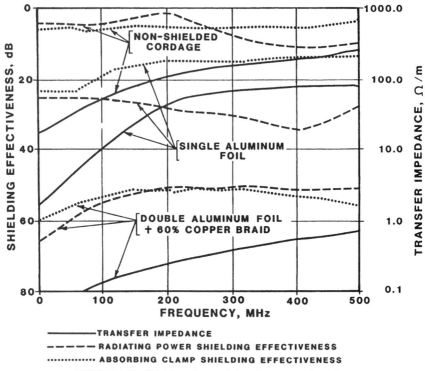

FIGURE 6.41 Shielding characteristics of power supply cordage

To summarize:
The absolute correlation between the shielding effectiveness parameters obtained by measurements at different system levels, as a rule, is limited and can be accurate enough only for certain relatively simple configurations. Within the same system level, the correlation between the shielding effectiveness values determined using different test methods depends on the similarity of the EUT geometry and field distribution at the shield surface.

6.7.2 Validating Test Procedures

How can we make sure that the test procedure setup is functioning right? It follows from Eq. (6.11-14) that correlation between different measurement methods, facilities, and techniques can be implemented on the basis of some reference shields. Using the reference shield as a "common denominator," the unknown constants of different methods can be determined and accounted for.

The idea of reference calibrators is not new: different type of the single frequency and comb generators to "check" and compare test facilities, "sample EUTs" with simple enough structures to calculate their potential response, simulation techniques — you name it!

The basic requirements for the reference calibrator are straightforward:

- It must simulate typical and mathematically predictable shield performance within a practical range of values and frequencies.
- It must be relatively simple but accurate.
- The parameters must not change with time, handling, and environmental variables.
- It must be installable in different test facilities and / or fixtures without their modification and in a manner similar to that of "real life" EUT.

Just as an example, we will consider here a calibrator for cable or connector measurements. In this case, a practical way to meet these requirements is to use a uniformly conductive pipe (e.g., see Refs. [6.43 - 48]).

The transfer impedance of a solid homogeneous pipe depends on it thickness, diameter, conductivity, and magnetic permeability (see Chapter 3):

$$Z_{ts} = R_o \frac{kt}{\sinh kt} \qquad (6.15)$$

A complicating factor is that the transfer impedance of a pipe with a thickness large enough to be mechanically practical quickly decreases with frequency and cannot be detected, even for metals with low conductivity (see Fig. 6.42, curve a). To fix this problem, two techniques can be implemented.

1. Use of extremely thin layer of metal over a cable core. When the thickness of this layer is small, the skin depth remains larger than the tube thickness up to very high frequencies, and the transfer impedance is either flat and equal to R_o, or at least reasonably large so as to be measurable (Fig. 6.42, curve b). Such a metallic layer can be applied using electroplating, electroless, vacuum deposition, magnetron sputtering, and so on. As an advantage, solid pipe reference standard exhibits distributed coupling properties that correspond to electronic cable shielding, so the calibrator is realistic.
2. Use of perforated pipes. The transfer impedance associated with n circular apertures of diameter a in a thin pipe (t « a) can be determined by a simple formula:

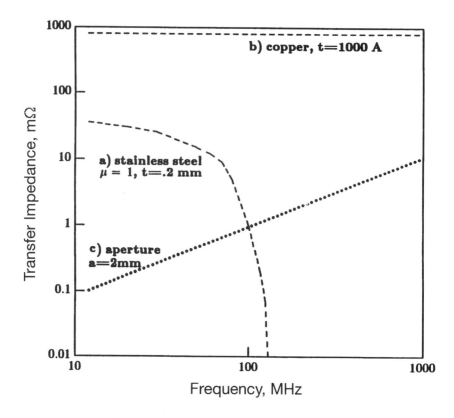

FIGURE 6.42 Transfer impedance reference standards

$$Z_{tm} = j\omega n \mu_o \frac{a^3}{6\pi^2 D^2} \quad (6.16)$$

A single aperture presents a lumped coupling that is proportional to frequency (Fig. 6.42, curve d). By placing multiple apertures on a pipe, the perforated standard can be designed to present distributed coupling as well. Of course, in a perforated shield, both the diffusion and aperture penetration effects (as well as capacitive coupling) must be considered simultaneously:

$$Z_{ts} = R_o \frac{kt}{\sinh kt} + j\omega n \mu_o \frac{a^3}{6\pi^2 D^2} \quad (6.17)$$

The shielding properties of such a calibration device are determined by the diffusion mechanism in the low-frequency range and by the energy penetration through the apertures in the high-frequency range. Since both items in Eq. (6.17) are complex values, the sum is also complex (see Fig. 3.38).

A practical version of such a calibrator was suggested by Madle [6.43] and is illustrated in Fig. 6.43. Other designs are possible and have been implemented (e.g., see Ref. [6.48])

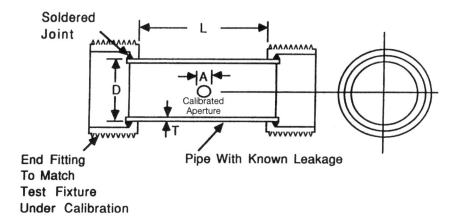

FIGURE 6.43   Generic transfer impedance standard

7
Shielding Engineering

7.1 SYSTEM APPROACH TO SHIELDING ENGINEERING

7.1.1 Shielding Engineering Problems

What is "shielding engineering", anyhow? Not an exception in the electromagnetic shielding discipline, the very term may itself be a subject of interpretation.

The American Heritage Dictionary defines "engineering" as

1. "The application of scientific and mathematical principles to practical ends such as design, construction, and operation of efficient and economical structures, equipment, and systems."

2. "The profession of or the work performed by engineer."

In such context, the shielding engineering is fraught with numerous and often overwhelming, problems involving different system levels, design objectives and requirements, physical mechanisms and process models, technical (and other) conflict resolution philosophies, and mathematical and experimental techniques. To *start* with, the goals of product manufacturers and system designers may *diverge*! Indeed, at a system-specific level, if the electronic system or its elements in question are not dominant radiators, there may be no need at all in shielding them or, for that matter, improving their EMC performance, at least, for the purpose of EMI reduction. Thus, while the goal of the specific product manufacturer is (hopefully) to design an economically and otherwise efficient shielded product, a computer system designer or a cable TV company may be not at all interested in using the products with improved EMI protection.

However, even at particular system levels where EMI considerations *are* essential, shielding may be neither the only nor the best alternative for EMI mitigation. In Chapter 2, we have identified a large number of alternatives to shielding.

Thus, the *first question* to answer in any particular shielding application is whether shielding is appropriate at all!

But suppose a decision to use shielded products has been made. Then the *next* task is to select and specify available "off the shelf" shielded products or add-on shields, or design such, if necessary. How do we proceed? First of all, in line with the definition at the start of this section, we may (or may not) define the shielding specification as the process (verb) and/or the result (noun) of deriving the shielding requirements from a given set of EMC and other system requirements. Similarly, we may define the shielding design as the process (verb) and/or the result (noun) of selecting the shield characteristics to meet the system shielding objectives.

Now, a whole new set of problems arise with regard to the system or product shielding *specifications* and *design*:

First comes all-too-familiar problem of which shielding performance definition to use (see Chapters 2 through 4).

Although we want to specify and design the system shielding for EMC performance, there are numerous other, "non-EMC" electrical and physical requirements to consider: transmission properties, mechanical strength, temperature stability, and so on.

Different shielding characteristics, functions, and objectives are also not necessarily coincident and may be in outright conflict.

Respectively, different shielding specifications are not necessarily coincident and may be in conflict.

As a rule, shielding specification and design are enacted by different people and different organizations or companies, with different goals and qualifications.

To resolve these problems, it is necessary to develop, analyze, and interpret the shielding models. Then the *second* question is about the adequate generality / relevancy of the needed shielding model, as discussed in the previous chapters. Of course, we must come up with the *model* itself — the *third* question to answer. The model should relate the shielding design objectives to the shielding system parameters and element shield characteristics. Using the model, the *fourth* problem so solve is to analyze the model and "solve" it for shielding system design specificswith and with regard to our design objectives. Now, that we have arrived at the shielding system and shield design, we are "all set" to determine the economics and particulars of using the shielding option.

The last step, is *once more* to compare the shielding option with other available options, that is, *"return back to the last question"* which is *the same* as the *original* question number one: *to shield or not to shield?*

Now, when we try to apply our "shielding design" definition, we are confronted by a host of practical but fundamental questions / problems:

Shielding Engineering 493

1. How do we define the shield and shielding system design characteristics and performance objectives?
2. What can we do about the interdependencies between different parameters and the often overlapping domains of the shield design characteristics?
3. Which shield design parameters and shield performance objectives are really important, and which of them can be disregarded for our analysis?
4. How do the different characteristics interact?

The answers to these questions are not at all trivial. However, the respective "roadmap" is available. It is called *system analysis*.

Practical experience convincingly proves that the rational process of specifying and designing a product or system, occurs through the application of the methods of system analysis, preferably in conjunction with the *optimization* techniques. Only then we can expect to reduce the number and mitigate the consequences of expensive failures. With regard to shielding design, we will formulate two *complementary* design problems: shield *synthesis* — determination of the shield design parameters for given shielding performance parameters, and shield analysis — determination of the shielding performance parameters for given shield design parameters. Note that *shield design* synthesis and analysis are *opposite* of *system EMI* synthesis and analysis, which result in the shielding design and specification objectives (see Chapter 1).

To illustrate the shield specification/design-associated problems, consider the task of designing two shields: an electronic product shielding enclosure and a cable braided shield. An impressive list of braided shield parameters are involved, with only several of them (far from a complete set) shown in Fig. 7.1.

The characteristics of the shielding enclosure in Fig. 7.1 a are selected on very general level, while the characteristics of the braid go in more detail to a quite "elementary" level. Even so, each of the parameters in Fig. 7.1 may itself present a complex and confusing problem.

✦ Consider two examples, one for each of the mentioned applications:
  1. The shielding can be applied at several levels: component, PCB, subsystem (e.g., shelf on a rack), system, multiple-system. Each of these levels will call for specific shielding means and design solutions, and must provide for the necessary levels of *shielding system integrity*. The latter is a direct function of the shielding system configuration, joining techniques, geometry and treatment of windows, vents, cable penetrations. Also, the other requirements shouldn't be passed by, which play the role of design constraints. Take for instance thermal management of the product enclosure. Practically, four different methods could be used: natural conduction/convection, forced air cooling (e.g., fans or blowers), heat exchangers / air conditioners (also can control humidity), heaters (to raise the temperature, when necessary). The use of each of the methods imposes certain requirements on the shielding enclosure affecting its EMC performance.
  2. Braided shield EMC performance. Consider a characteristic such as braid's surface coverage. Is it design or performance related? Surprisingly, this is a matter of one's point of view; that is, depending on the adopted definition, it can be looked at as either a design or a performance parameter. Indeed, the braid coverage is sometimes used as a measure of shield effectiveness (whether this is right or wrong is, in this context, immaterial). At the same time, the percent coverage is often

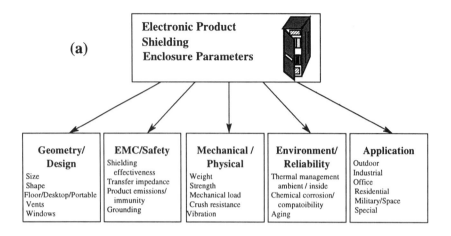

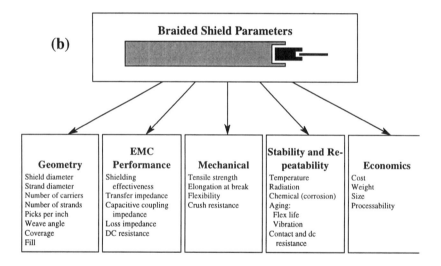

FIGURE 7.1 Partial list of shield parameters

specified as fixed for the particular shield design. Such usage is as "good" (or as "bad," for that matter) as the previous one since, for the given percent coverage, different shielding effectiveness values can be obtained by varying the strand diameter, the shield material, and other inherent shield characteristics.

By the way, braid coverage is in fact a poor performance criterion. The point is that there can be braids with different coverages that exhibit identical shielding effectiveness, and therecan be braids with identical coverages that exhibit different shielding effectiveness. Therefore, the higher coverage does not guarantee a better shielding effectiveness, and vice versa.

Shielding Engineering 495

In the preceding material, we had numerous opportunities to encounter this kind of predicament. Our usual approach was that, unless there were essential reasons to favor a certain definition, the choice was often pretty arbitrary, as long as a definition was given and agreed upon. Thus, returning to our example, we would rather attribute the braid coverage (if there is a requirement for this parameter, e.g., by a customer) to the design parameters, along with the shield's material, construction, and location within the cable geometry. As an opposite example, it is often convenient to attribute the foil shield surface resistivity (in ohms per square) to performance parameters, while the corresponding design parameters will be the foil's thickness and material (or the material's volume resistivity).◆

In summary,
we are left with an extremely complex shielding engineering questions/problems set, *starting* and *ending* with the *same* question: to use a shielded or non-shielded system or product?
- to shield or not to shield?
- to the development of the shielding performance guidelines;
- to the selection of specific shield material, construction, and geometry;
- to finalizing the shield termination and grounding.
- to shield or not to shield?

Each of these problems presents a separate shielding engineering step, while all the problems may be interrelated. The shielding specification/design process is aggravated by the difficulties in selecting the adequate generality and relevancy of the shielding models, as we discussed in previous chapters.

7.1.2 To Shield or Not to Shield: the First and the Last Questions

In real life, it is not always easy to make up one's mind with regard to the need for shielding—positively or negatively. As a rule, prior to making this decision, the system-level considerations and EMI mitigation optimization must be addressed. Indeed, sometimes a decoupling capacitor (5¢ cost) placed at the PWB can reduce the EMI signal and emissions to acceptable levels so that an expensive shielding enclosure or shielded cable wouldn't be necessary at all. Thus, it is often worthwhile to consider different EMI mitigation options and to look for "non-shielding" EMI mitigation alternatives (e.g., see Table 1.4 in Chapter 1, section 1.4)) *prior* to resorting to shielding.

But how do we choose among these options? Guidance is provided by the system approach to engineering. In the preceding chapters, we have dedicated much time and efforts to proving that the use of shielding is meaningful *only* within the context of the benefits to the whole system under consideration. The important feature of the system analysis is that it accounts for all the necessary aspects of system functionality. Thus, the selection of the best or just an appropriate EMI mitigation technique is based on custom- developed, and not necessarily EMI-related, criteria.

As an important practical example, consider two alternatives to cable shielding, i.e., the use of system filtering or fiber optic cables. The filters can be applied as discrete passive or active elements in series with the electronic cable line, as discrete common-mode chokes (e.g., a simple ferrite core over the cable), or as distributed EMI energy absorptive jackets over the cable core. While we will discuss the last option in little more detail later in Chapter 8, here we will qualitatively compare the first two options versus shielding. For simplicity, assume that only one set of filters is required, at the ends of the line. In this respect, the utilization of fiber optics is no different from the use of filters in electronic cables: its application will result in the use of two optoelectronic converters at the ends of the line.

Assuming that the necessary EMI mitigation level can be achieved using *either* (a) shielding or (b) filtering or fiber optics, the problem reduces to logistics (i.e., convenience, regulatory acceptance, mechanical strength, and even aesthetics) of, and economical considerations in, their applications. Relegating to "marketing" the logistics of the use of filters, ferrite cores, or optoelectronic converters, let us evaluate the economic consequences of these options (Fig. 7.2). While the EMI mitigation cost of the "non-shielding" options is the same one-time expense incurred at any line length, the cost of cable shielding is incremental and can be assumed (with certain caveats) to be proportional to the line length. Obviously, when the line length is small, this cost may be negligible compared with the cost of filters (plus a nonshielded cable) or fiber optics converters (plus fiber optic cable). Then the use of the shielded cable is cheaper than the other options. However, in relatively long lines, the cost of shielding may exceed that of the filters or converters.

In this case — "shielding vs filtering" — it is often possible to identify the line length below which the use of shielding is economically efficient, and above which filtering or fiber optics may make economic sense. Of course, this length is dependent on

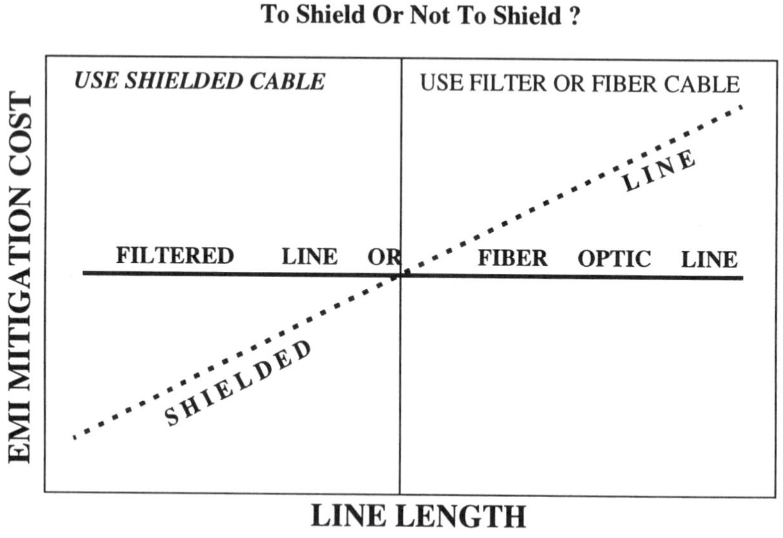

To Shield Or Not To Shield ?

FIGURE 7.2 Economic comparison of EMI mitigation techniques: shielding versus filtering and fiber optic transducers

the specifics of the cable and filter designs and on economics. Thus, to find this critical cable length, shielding and filter design problems still must be resolved in advance.

The considered problem may be viewed as just one (but extreme—to shield or not to shield?) specific case of a cable shielding system EMC specification. As we saw, it involved certain system considerations at levels higher than just the separate system element (e.g., cable) shield. Unfortunately, the system shielding specification may be far removed from the particular product shield performance per se, as in just considered case, where no shielding at all may have been required. In such situations, our best hope is to try to reduce the necessary system level upon which we must base the shielding specification and design solution. These system considerations must be resolved before the problem can be formulated in terms of the shielding design.

On the other hand, as we just saw, the very resolution of these system aspects may require preceding assumption or evaluation of certain shield design features; that is, the solution of a design problem (sounds like a "catch-22", doesn't it?). For example, the rate of growth of the "shielded line" "expense" characteristic in Fig. 7.2 depends on the shield design. That's why we referred to the shielding application decision in the title to this section as the first and the last question. This further complicates the shielding specification problem and may require the use of some kind of sequential approximation methods.

7.1.3 "Black Box" Model of Product and System Shielding

In the system approach to specification and design, the first step is to identify the sets of design characteristics and performance objectives, and to develop respective design/performance models which link these two sets together in the presence of certain general and system-specific constraints. In general, we can suggest a list of shield system and element design parameters, a.k.a. design options, which under certain conditions can be considered mutually independent. But ususally we will end up with a pretty large list of such parameters. For example, a limited (that is, incomplete) classification of major cable shield design options is presented in the form of a diagram in Fig. 7.3. As could be expected, it looks quite "busy." Moreover, different combinations of these design parameters are possible in the same shield. Thus, it is easy to figure out that the options in Fig. 7.3 may combine in more than 10^4 ways!

Should we consider *all* of these (and others, not indicated in this diagram) options, or only *some* of them? What criteria must be applied to their selection? The answers should be drawn from the shield functional objectives and physical and application constraints. There are multiple "driving forces" behind these objectives and constraints: from customer specifications to government regulations and standards, to adopted shield evaluation principles and available measurement techniques and instrumentation, to such "exotic" aspects as social and political realities to the environmental effects and laws of physics. With regard to shielding for electromagnetic compatibility, the potential "objective options" (see Fig. 7.4) can be as numerous as shield design

498 Electromagnetic Shielding Handbook

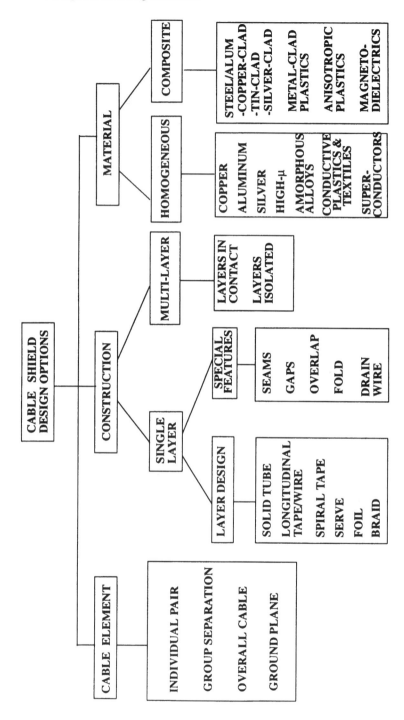

FIGURE 7.3 Cable shield design options

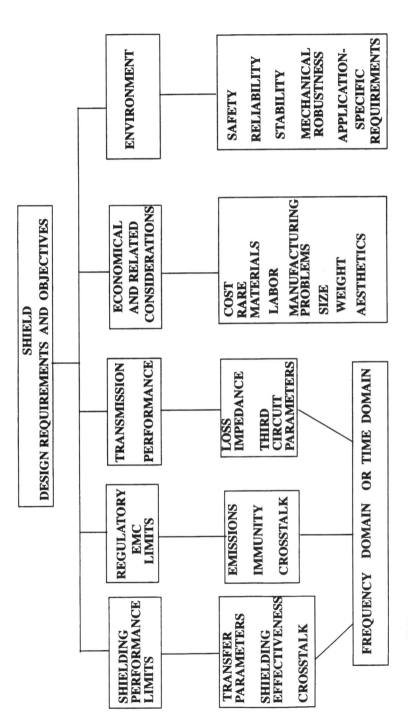

FIGURE 7.4 Shield design requirements and objectives

options. On the other hand, the constraints may range from the physical reality of the shield thickness being a positive number to the demand that the parameters in Fig. 7.4 cannot exceed (or must exceed, depending on the nature of the parameter), a specified value. So, the same parameters can be treated as objectives or constraints, depending on the specific problem.

Thus, again, we have to "take stock" of many interdependent parameters, often with overlapping functions, which cannot "automatically" be attributed to the category of objectives or constraints. And the shielding per se (that is, EMI protection) performance objectives are not the only ones to be studied. As shown in Fig. 7.4, there are also numerous "non-EMI" requirements, as well as physical and other constraints, which must be considered simultaneously with EMI. Also, these parameters can be interrelated. For example, we know that the shielding system-related emissions, immunity, and crosstalk can be expressed as functions (although not exclusive ones) of the shield transfer parameters or shielding effectiveness, while both latter characteristics are also related!

Returning now to our definition of the shielding system / product design, the next step will be to relate the shield design parameters to the shield's performance parameters; that is, to create design/performance models. These models can range from experimentally measured dependencies between specific design and performance parameters, to formulas and equations linking all or some of them, to sophisticated computer numerical algorithms. Again, consider an example. In Fig. 7.5, the shielding requirements are given in the form of the transfer impedance limit (as a matter of fact, this is an Ethernet cable specification). On the same graph, the transfer impedance frequency characteristics are plotted of several typical shields (experimental data). Assuming that the construction details of the measured cables are known, the graphs in Fig. 7.5 contain the necessary information to select or design the respective cable shields which meet the required limit (i.e., to solve the shield synthesis problem).

In this case, the shielding objective was directly related to the shield design. However, the limitations of an approach such as outlined in Fig. 7.5 are obvious: only a predetermined set of design parameters and only one performance objective are considered. Furthermore, even on such a limited basis, the design/performance relationships (experimental curves, or mathematical equations, or numerical algorithms) may be not known. Such a situation is typical for many problems of system engineering, and the system analysis discipline has developed its useful representation as a "black box" diagram.

An example of such diagram used for cable shielding design is shown in Fig. 7.6. Similar diagram can be created for the general case of shielding system or element design. In this model, the design options are shown as "inputs" (at the left-hand side) and the design objectives as "outputs" (at the right-hand side), and the constraints are incorporated inside the box. The design/performance relationships per se constitute the "contents" of the box. Of course, such a representation is largely arbitrary, and different models can be used. For example, the constraints could be represented as a different set of inputs to or outputs from the "black box," and so forth.

Shielding Engineering 501

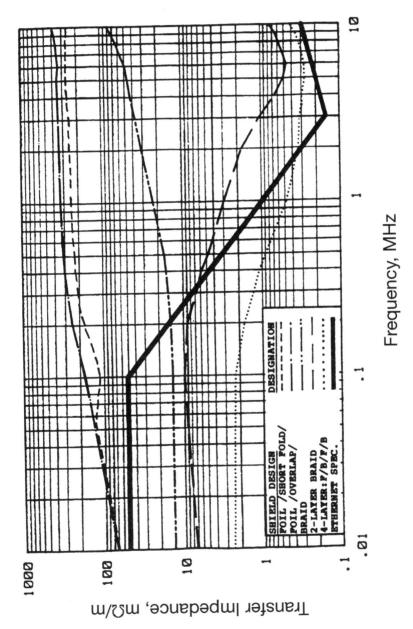

FIGURE 7.5 Coaxial cable shielding for Ethernet specification

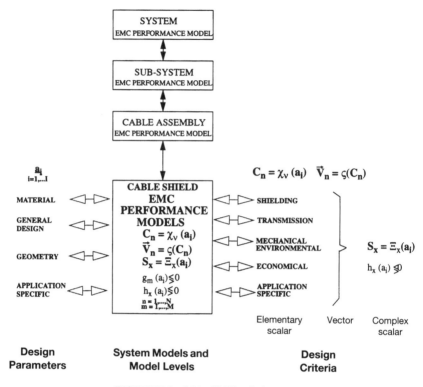

FIGURE 7.6 Cable shielding design process

The diagram in Fig. 7.6 can serve as a flow chart for cable shielding design, both synthesis and analysis, which is why the arrows are bidirectional. As we can see, there may be many design parameters (at the left-hand side of the model box in Fig. 7.6) and design objectives (at the right-hand of the model box). The relationships between the different groups of the shield design parameters and the shield's performance parameters are expected to be expressed by multiple *local* design/performance models which are additionally affected by the considered system levels. These shield design/performance models relate the shield design parameters to its performance objectives, both EMI-related and others, including mechanical, environmental, and economic. Mathematically, the design parameters play the role of independent variables, while the design objectives are functions of these variables.

As a rule, the shielding performance parameters of interest (e.g., the shielding effectiveness, shield transfer function, or other parameters, as appropriate) are not explicitly given but have to be deduced from other requirements. In case of designing for EMC, such requirements most often will be related to the emissions, susceptibility (immunity), and crosstalk (see Chapter 1). Now, before proceeding with the regular shield analysis and synthesis tasks we must define and determine the unknown EMC performance for the known shield design parameters or define and determine the unknown shield design parameters from the known EMC performance. To do that it is often necessary to define and determine one set of the EMC performance parameters or one set of the shield design parameters, respectively, when other parameters are given or known.

First consider the design parameters. Mathematically, they act as independent variables that can be varied within respective constraints. Obviously, a basic requirement of any given set of design parameters is their mutual independence. We called such independent parameters elementary. This does not mean that some "composite" design parameters could not be used that are functions of several elementary parameters. However, each of the elementary parameters must be represented in the adopted set only once, whether individually or as an "ingredient" part of a composite parameter. Otherwise, this may lead to several different models describing the same relationship, which makes the task ambiguous.

Similarly, each of the separate single shield performance objectives, or single design constraints, may be considered as an elementary criterion, or an elementary constraint, of the process. Mathematically, such criteria and constraints are associated with certain scalar variables that define a measure of the respective shield quality. This quality function may consist of a single (elementary) criterion which does not depend on other objectives, or it may be a composite function of several elementary criteria composed in accordance with certain rules. The only requirement for such composite criterion is its scalarity. As in the case of the design parameters, each of the elementary (independent) performance objectives (whether on its own or in a combination with others) must be used in the adopted set of design objectives (and based on the design criteria) only once.

For instance, the shielding effectiveness is a function of the shield transfer parameters—transfer impedance, capacitive coupling impedance, and loss impedance. Thus, any of these transfer parameters cannot be used in the same set of objectives simultaneously with the shielding effectiveness parameters. Moreover, since all the transfer parameters are functions of the shield material and geometry, they are not independent. If used individually or in combination with other objectives which do not depend on the same design parameters, any one of these parameters can be viewed as elementary and thus exclusively included in the set of objectives (criteria or constraints). However, these three parameters cannot be included simultaneously in a set of objectives, because they depend on the same design parameters and thus are not mutually independent. The problem can be resolved by presenting them as functions of the design parameters and then combining them in a composite noncontradictory criterion. In a way, this is equivalent to bringing them to a "common denominator." Of course, then the design objectives must be "translated" into this new environment by developing a new (as a rule composite) criterion based on some compromise between the involved elementary objectives.

♦ Actually, such parameters as the transfer impedance are not scalars but vectors because they have the magnitude and phase. Fortunately, as a rule, we are more concerned with their magnitude (i.e., modulus). However, in some situations, we need both the real and imaginary parts of the parameters, in which case these parameters cannot be treated as scalars. For example, when deriving the transfer impedance of a cable shield enclosing multiple wires, we considered this shield as an outer conductor of a coaxial cable and approximated all the shielded conductors with one conductor viewed as a center conductor of a coaxial cable. But to be accurate, we would have had to consider separately the transfer impedances associated with each shielded conductor, and then integrate the effects for all shielded conductors. These "partial" transfer impedance parameters may have different phases, which must be accounted for when summing the EMI contributions associated with different conductors. One such method was described in Ref. [7.1] (also, see the last author's comment to the section 6.6 of Chapter 6).♦

To design for elementary scalar criteria and constraints, the relationships between them and the design parameters must be identified; that is, the respective models must be known. Unfortunately, as we saw, the mathematical models of the shield performance are often quite cumbersome or, even worse, the expressions may be not available. Then, some reasonable approximations usually can be made and tested experimentally. Also, some problems can be solved on the basis of a qualitative evaluations, e.g., using ranging techniques. If everything else fails, physical models can be designed, manufactured, and tested, and then compared with the requirements. This opens possibilities for using empirical and statistical models. As a rule, it is advantageous to use several diverse techniques.

The problem is much more complex if several elementary scalar criteria must be met simultaneously. Some of the requirements may conflict with others, including the primary goal of providing shielding performance. Thus, it is intuitively clear that better shielding performance, or larger mechanical strength, will command an economic premium. (Although, as we will see when discussing shield optimization, our intuition is not always correct.)

Can several different scalar criteria be met simultaneously? The answer depends on the nature of and relationship between the criteria and the design goals. For the purposes of this analysis, different scalar design objectives, or criteria, can be separated into two major groups: *orthogonal* (or depending on *different* design parameters and, for that reason, *independent* on each other) and *non-orthogonal* parameters (depending on the same design parameters and, for that reason, *dependent* on each other). Within the non-orthogonal objectives, it is useful to identify the objectives with *compatible or conflicting* functions of (dependencies on) the same design parameters. By compatible function, we will understand the dependencies which have similar trends, although the rate of the function variations with the independent variable change can be different. The conflicting functions have opposite variation trends with the change of the independent variables.

Because different orthogonal objectives have "nothing in common," it is obvious that design for orthogonal criteria reduces to the design for single scalar criteria, as discussed above. It is also often possible, at least in principle and after respective analyses, to handle "more or less satisfactorily" the non-orthogonal objectives with coincident functions. Unfortunately, it is very rare that different design objectives are orthogonal or even non-orthogonal and non-conflicting. For example, the thicker the solid homogeneous shield, the smaller its transfer impedance, but the higher its cost will be.

Suppose now that different scalar performance parameters (e.g., transfer impedance and cost) are specified (required) at some levels that cannot be met simultaneously. The best we can do now is some kind of a "compromise" whereby neither objective is fully met but the objective "noncompliance" is somehow "balanced" (in the sense of the adopted rules). But how do we compare "apples and oranges?" That is, what kind of a compromise can exist between, say, the transfer impedance and the cost? This is a typical vector criterion optimization problem, and numerous approaches to its solution have been developed.

✦ One often used approach is to assign certain priorities, or weights, to different components of the vector criterion. (The guidelines for such assignments constitute another question.) Then some kind of a "fault function" can be developed that must be evaluated and compared to some limits during the design process. One simple way to develop such a function is to specify the ratio of different vector criterion components, e.g., to take the ratio of the transfer impedance to the cost. (How about specifying the units of such a parameter as ¾/m/$?). Of course, to do this we need additional information about the particular design environment, and it must be supplied by the customer. It is equivalent to comparing "apples to oranges," say, on the basis of amount of vitamins that each can provide. However, unless we have this "vitamin" information, "apples and oranges" cannot be compared! And so it works with the shield options.✦

7.2 METHODS AND TECHNIQUES FOR SHIELDING DESIGN

7.2.1 Basic Shielding Design Principles

In accordance with the preceding discussion, we will consider two methods that have been successfully used in the shielding design:
- design by constraints
- optimization.

In both methods, we will distinguish between local and global design problems.

In the *local* case, a single primary elementary or composite design criterion is used without regard to other requirements to the system. What particular criterion to use depends on the specific task which, itself, is generated by the EMC design objectives. For instance, this could be the transfer impedance or shielding effectiveness of a cable, singled out or within the system of various appropriate generality, and so on. (Don't forget that in this section we are dealing with the design for EMC performance although, in general, any of the characteristics of the shield can be mandated or selected as the primary criterion.)

506 Electromagnetic Shielding Handbook

In the *global* case, not only this primary design criterion, but also other system characteristics and requirements are accounted for. This can be done by complementing the primary criterion with other requirements, usually in the form of constraints, which must be considered simultaneously with the primary criterion.

7.2.2 Design by Constraints

When designing by constraints, the goal is just to meet the limit specified for the primary criterion which, of course, must be a scalar. In this case, for all practical purposes, this criterion itself can be viewed as a constraint because its purpose is to set the boundaries for the objective parameter variations (upper, lower, or "band") depending on the requirements. This is why we will call this approach design by constraint, or boundary design. The boundary design corresponds to a "pass/fail" process which is so common in EMC testing (e.g., testing to the radiated emission limits): as long as the limits are met, there is no difference whether they are met with large or small "margins."

Mathematically, the global boundary design problem is represented by a set of inequalities:

$$\begin{cases} \Re(D_i) < \text{const} \\ \Im_e(D_i) < \text{const} \\ \Im_e \in \Xi \\ D_i \in \Theta \end{cases} \quad (7.1)$$

where $\Re(D_i)$ is the selected criterion expressed as a function of the design parameters D_i (i = 1,2,...) and $\Im_e(D_i)$, and considered constraint (e = 1,2,...) selected within respective "eligible" domains Ξ and Θ.

The local boundary design problem is represented by the first f inequality in Expression (7.1):

$$\{ \Re(D_i) < \text{const} \quad (7.2)$$

For all practical purposes, there is not too much difference between the criterion $\Re(D_i)$ and constraints $\Im_e(D_i)$.

Another question is how to select the proper values of the constraint (const) in Expressions (7.1) and (7.2). This is the shield specification problem. Depending on the

Shielding Engineering 507

specifics of particular task, we can imagine three types of situations. First, the requirements of the particular shield performance characteristics can be specified in the standards or regulations, or by the customer requirements. This is the simplest case, because these specified value are "ready" to be used as const. When such a specification is not available, this const must be derived from other requirements. When the system in which the shielded cable will be used is known, then the corresponding system-specific models can be used to derive the requirements of the shield from the system EMC performance requirements. For example, the transfer impedance or shielding effectiveness const can be derived from the system emission requirements for given system configuration, currents and voltages. In a more general case, when the particular system is not known, the more general models should be investigated. One of the techniques that can be used in this situation is worst-case design, during which the "worst" system configurations and the largest currents and voltages in the shielded cables are assumed (e.g., see Ref. [7.2].

Also it is quite possible that the criterion and constraints requirements, or even different constraint requirements, will be in conflict (depending on the adopted values) so that the global problem in its entirety may not have a solution. Then, only the local solution will have to be implemented, which accounts for separate criteria having the highest priorities in the design project context. As a rule, the sign "<" may be substituted for ">" if the inverse values of the independent variables (design parameters) or the criterion (or constraints, in the second inequality) are taken.

✦ Figure 7.5 presents a typical illustration of the local boundary design process without the constraint requirements: different shield constructions are compared with regard to the transfer impedance limit. As shown, from the selection of shields in Fig. 7.5, only the four-layer shield (foil/braid/foil/braid) meets the limit. But experience has shown that more economical three- and even two-layer shields can be designed that do meet the limit. The "non-optimality" of the four-layer shield is a direct consequence of the limited shield selection in the considered set. However, even if the considered set of shields is expanded so that several shield designs meet the limit, it is still not clear which one to select. Such a selection could have been made by introducing additional criteria: shield cost, mechanical strength, corrosion resistance (chemical compatibility), and so on. Even larger benefits can be achieved when the shield is designed "from scratch" instead of considering "ready made" options. However, in this case, the criterion and constraint functions must be expressed as functions of the shield design parameters.

If in the previous example the transfer impedance and the shield cost are specified, this leads to a vector criterion consisting of two non-independent scalar components. This is because both the transfer impedance and the shield cost depend, at least partially, on the same design characteristics (e.g., shield thickness). Depending on the specific numbers associated with both requirements (i.e., Z_t and cost), a shield of a particular design (construction, geometry, material, and so forth) may or may not meet both requirements. In an attempt to find a design that meets both the technical and the economical objectives, we can attempt to change the shield design parameters according to the formulas that were described in detail in the previous chapters. These formulas can be viewed as mathematical models of the shield EMC performance which "cover" the EMC component (e.g., the transfer impedance) of our vector criterion. As a matter of fact, so far our models have been limited to the "technical" parameters. We also need the models of the shield's economic performance. We will illustrate some approaches to develop the models of the shield economic performance in the next section.

When both objectives cannot be met, the only recourse we have is to develop a composite scalar criterion based on the two given objectives. As we mentioned above, such composite scalar criterion represents a compromise between the different objectives.◆

7.2.3 Optimal Design

With the second approach, the design parameters are selected such that, while the specified limits are met (these limits act as constraints), a certain scalar criterion also (optimization criterion) must be maximized or minimized—whichever is the best from the adopted point of view. Naturally, we will call this process optimal design.

Mathematically, the global optimization process can be presented by a set of expressions similar to Expression (7.1), but with the criterion function optimized (minimized or maximized, depending on the nature of the criterion):

$$\begin{cases} \Re(D_i) \to \min \\ \Im_e(D_i) < \text{const} \\ \Im_e \in \Xi \\ D_i \in \Theta \end{cases} \quad (7.3)$$

where the designation of variables is similar to that of Expression (7.1).

Similarly, the local optimization process will account only for the first expression in (7.3):

$$\{ \Re(D_i) \to \min \quad (7.4)$$

There is no need to conduct a detailed discussion of the optimization principles here: a good library may have hundreds, even thousands, of titles dedicated to this subject. Several references are also included in the bibliography for this chapter (see Refs. [7.3-13]). For this reason, we will limit the following discussion to a review of the most general principles of shielding system design and optimization, illustrated by the analysis of several important practical examples.

We will use this discussion to illustrate both the boundary and optimal designs. The general principles are similar in both cases. As soon as a reliable enough design/performance model is developed, the shield design and optimization can be done using general telecommunication system optimization techniques (see Ref. [7.3]). According to those techniques, the shield is treated as a system in a hierarchical line of systems: shielded product (including cables), shielding assembly, shielding system. For example, if applied to electronic or telecommunications cable shielding, such "hierarchy" will reduce to cable as a whole, transmission line, network. The mathematical models provide the necessary quantitative relationships. The design and optimization procedure generally includes the following steps:

1. Design or optimization problem formulation. In this process, the main goals are realized, the sets of input (design) and output (performance objectives) parameters are defined, as are their desirable limits (constraints), and the system efficiency criteria are formed on the basis of the output parameters. The boundary design purpose is to achieve the specified values of the criteria, while the optimization process purpose is to achieve the maximal or minimal values of the optimization criteria by the proper choice of system (and shield, in particular) design characteristics, within the limitations of the adopted constraints.
2. Development of the design/optimization problem mathematical model. This model comprises the object function (relating the design or optimization criteria to the design characteristics) and constraints, which originate from the specific project conditions, production tolerances, and mutual dependence between the parameters.
3. Finding the acceptable or optimal solution. This is done on the design/optimization model by means of theoretical, graphical, or numerical techniques and/or optimal mathematical programming methods.
4. "Fine-tuning" of the model and solutions. This is of great importance because, as a rule, neither a completely adequate model nor its exact mathematical solutions will be available. Providing continuous analysis of the obtained results and a comparison with the measurements of some existing or specially designed and manufactured samples, the model can be adjusted to yield the necessary accuracy.
5. Application and interpretation of the design/optimization results. Results are related to real life constructions and conditions with consideration of factors and constraints that could not be included into the model because of their complexity or undefined impact.

7.3 CABLE SHIELDING DESIGN FOR EMC PERFORMANCE

7.3.1 Local Problems of Boundary Design

Because the primary purpose of the cable shield is to protect against EMI, we start with shield modeling and boundary design for EMC performance. The formulas in the pre-

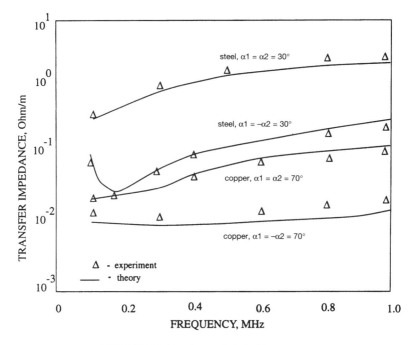

FIGURE 7.7 Two-layer spiral shield performance

ceding chapters can serve as models that permit us to conduct such an evaluation for many typical and practically important shields.

Only in the simplest cases, when the shielding evaluation parameter is defined and respective limits are established, is the solution to the design problems relatively simple. Thus we have reviewed Fig. 7.5, where the transfer impedance limits were compared with several typical shield designs. Unfortunately, such "straightforward" situations are "few and far between." As a rule, the shielding effectiveness (or the shield transfer function) limits are not explicitly given but must be deduced from other requirements. In case of design for EMC, such requirements most often will be related to the EMC parameters: emissions, susceptibility (immunity), and crosstalk (see Chapter 1). Now the shield designer is faced with the necessity to define and determine the EMC performance of the cable under consideration and only then using the concepts of the system EMI synthesis or analysis (depending on the problem at hand—again, see Chapter 1) to determine the shield performance limits related to the task.

In Chapter 3, we considered the transfer parameters of solid and spiral homogeneous and nonhomogeneous shields, single and multilayer, as well as braided shields and some special designs. Using these equations, we can determine the necessary values of the shield thickness, weave angle (in spiral shields), mesh density, desired foil overlaps, braid parameters, and so on, which are needed to meet the specified performance limits. For example, using Eq. (3.179), the curves in Fig. 7.7 were built for a

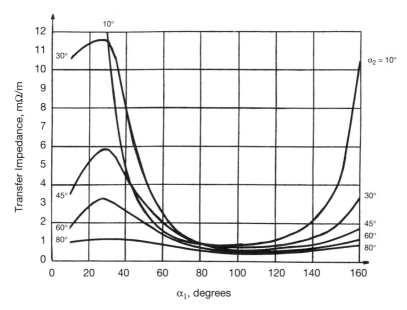

FIGURE 7.8 Dependence of transfer impedance on weave angles

two-layer spirally applied tape shield, and they show the dependence of its transfer impedance (for simplicity, the low-frequency case is considered here, but this is not necessary in general) on the tape material and weave angle (the thickness of these shields was from 0.05 mm to 0.1 mm).

Actually, the developed mathematical models of transfer impedance permit a systematic investigation of the dependencies of a selected EMC performance or shielding criterion, e.g., transfer impedance, on any of the "participating" design parameters (i.e., within the specified domain $D_i \in \Theta$). For instance, Fig. 7.8 illustrates the dependence of the transfer impedance of the same two-layer spiral shield (again, using the low-frequency case) on the weave angles of both layers. Obviously, if the transfer impedance limits are known (e.g., as shown in Fig. 7.5), these curves are sufficient to select the necessary weave angles, that is, for boundary design.

As another important case, we will consider in more detail the design/performance characteristics of a braided shield. We will base this analysis on the engineering model of a braided shield as derived in Chapter 3. Equations (3.118) through (3.120) relate the transfer impedance, loss impedance, and capacitive coupling impedance to design parameters, α, K (F), d, N, n, D, as well as the intrinsic properties of the used materials (σ, μ, ε) and the frequency ω (f) Some of the braided shield's most fundamental properties were highlighted in the Chapter 3. Here, we will illustrates the models of the braided shield transfer impedance as a function of its design parameters, which from a mathematical standpoint is equivalent to a multivariable function analysis.

Thus, considering the transfer impedance as a function (dependent variable) defined in several dimensions (design parameters—independent variables), we can use its spa-

512 Electromagnetic Shielding Handbook

tial sections to obtain 3-D views of the shield performance dependencies on selected design parameters. Each spatial section is formed by treating the basic expression as a function of only two variables (including frequency), while others are fixed. This permits very convincing geometrical interpretations, as shown in Figs. 7.9 through 7.13.

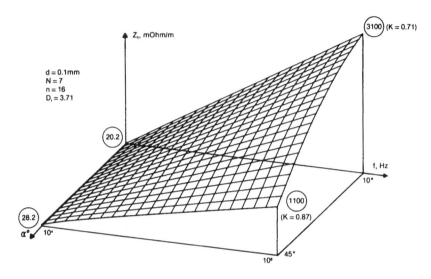

FIGURE 7.9 Braided shield transfer impedance vs. weave angle and frequency

Similar to the previously discussed examples, these graphics can be used to design by constraints (i.e., for boundary design).

Figure 7.9 presents the dependence of the transfer impedance on the weave angle and frequency. Also indicated are the fixed values of the other braid design parameters, the limits of the α and f variations, and the Z_t values at these extreme points (in circles). While the Z_t rise with frequency is anticipated, the dependence on the weave angle comes much as a surprise. It is widely accepted that with the weave angle increase, the transfer impedance becomes larger. Yet, according to Fig. 7.9, this is true only in the low-frequency range, while at the high frequencies the dependence is reversed. This is something to remember when designing braided shields.

✦ This can be explained as follows. In the low-frequency range, the transfer impedance value is defined mainly by the shield dc resistance, which increases with the weave angle. In the high-frequency range, along with the weave angle, the percent coverage has to be taken into account (which, in turn depends, on other, "more elementary" braid design characteristics including, but not limited to, the weave angles). Since parameters D, n, N, and D are fixed, the rise of the weave angle leads

to the percent coverage increase (in Fig. 7.9, the α change from 10° to 45° leads to K variations from 0.71 to 0.87) which reduces the transfer impedance magnitude. ✦

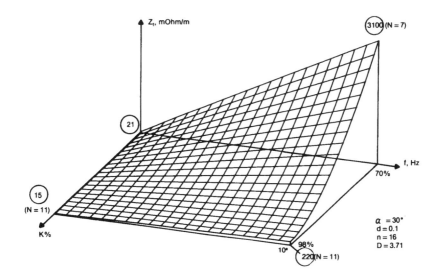

FIGURE 7.10 Braided shield transfer impedance vs. percent coverage and frequency

This is also confirmed by the plot in Fig. 7.10, which shows a transfer impedance decrease with the rise of the percent coverage. (Because α, d, n, and D are fixed, the change of K is being achieved by the variation of N.) Due to the proportionality to the frequency of the aperture penetration item in Eq. (3.119), this phenomenon is more manifested in the high-frequency range.

✦ And what about the transfer impedance increase with the weave angle? This happens when the percent coverage is fixed, which under the fixed d, n, and D can be achieved if N is being changed. As it follows from Fig. 7.11, when K is fixed, the increase of α really leads to the increase of Z_t. ✦

When the strand diameter becomes smaller, the transfer impedance increases faster, when the frequency is larger (see Fig.7.12). This is due to the increase of both field penetration through the apertures (because K decreases with other parameters fixed) and diffusion items. For a larger strand diameter, the coverage is also larger, which leads to a smaller high-frequency magnitude of Z_t. Therefore, less difference is observed between the high- and low-frequency transfer impedance.

It is interesting (and important in practice) that not all of the conceivable design combinations of the weave angle and strand diameter can be physically realized. These

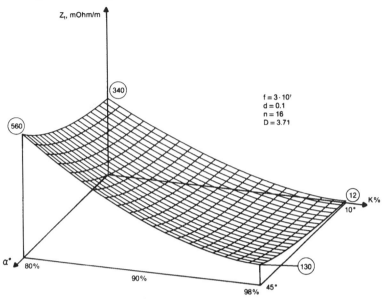

designs correspond to the white (non-filled) function surface on the α-d coordinate plane in Fig. 7.12.

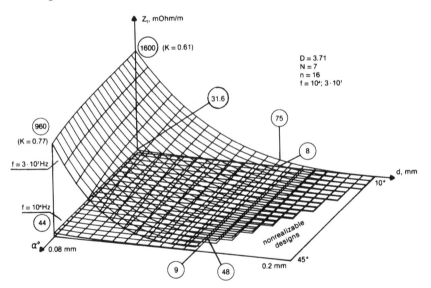

FIGURE 7.12 Braided shield transfer impedance vs. strand diameter and weave angle

When α, d, N, n are fixed, the increase of D leads to the reduction of K and therefore to a Z_t rise (see Fig. 7.13). , In Fig. 7.14 , comparative data is presented of several multilayer shield designs. Note that the transfer impedance of cable 4 is larger than the transfer impedance of cable 3, despite the fact that the braid coverage in the first one is larger, while all other parameters of these two designs are identical. This is caused

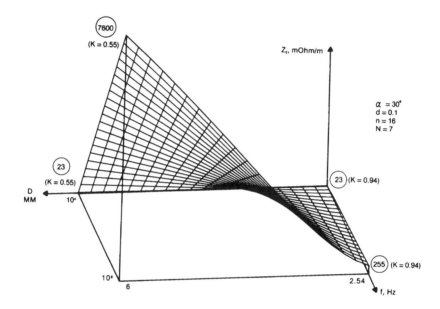

FIGURE 7.13 Braided shield transfer impedance vs frequency and shield diameter

by the difficulties in maintaining the "regular" geometry of the higher density braid (so much for the efficiency of higher density braids). We will return to this subject later.

The use of the models in Figs. 7.9 through 7.14 for boundary design by the transfer impedance criterion is self-explanatory.

7.3.2 Local Optimization

Optimization by local criterion is widely used in cable shielding. Thus, solutions are available of shields with minimum weight, transfer impedance, cost, physical strength, and so forth (e.g., see Refs. [P.2-3, 3.6, 3.12, 7.2-8]). In local optimization, as a rule, shield models are selected that express the shield characteristics of interest as a functions of the shield geometry, material, and so on (viewed as independent variables). Then the process is straightforward: the developed function is investigated on minimum or maximum (whichever is appropriate for a particular task) with regard to the independent variables. The obtained values of the independent variables (design parameters) are "declared" to be optimal.

Obviously, the validity and accuracy of the achieved results is determined by the validity and accuracy of the utilized models, and on the accuracy of mathematical pro-

516 Electromagnetic Shielding Handbook

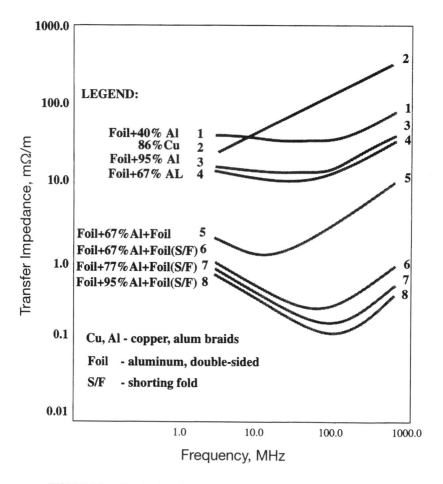

FIGURE 7.14 Transfer impedance frequency characteristics of multilayer shields

cessing techniques. But the main problem to watch for is the effect of requirements other than the adopted criterion. For this reason, the "last step" of the suggested design routine— result interpretation and application to real-life systems and environments— is vital for the success of such optimization.

Indeed, out of the context of the particular system, the adopted optimality definition may be defined quite loosely, or even incorrectly, which makes meaningless the very concept of optimization. For example, if we define "the best" shield as one having the smallest transfer impedance, we may end up with a silver (the highest conductivity) solid wall tube of "infinitely" large thickness! We may even suggest a golden shield to harden it against corrosion, or rare metal shields for high mechanical loads. Of course, the cost of such shields is also "infinite," let alone the "inconveniences" associated with logistics of manufacturing and installation. Then again, if we decide that "the

Shielding Engineering 517

best" shield is the one with the lowest cost, we may end up with an infinitely thin shield, which of course will have an "infinitely large" transfer impedance. Or if we want to "optimize" (in this sense) the shield flexibility, we will use a serve, but we have already seen that the serve has extremely poor shielding performance.

It appears like you cannot "have your cake and eat it too." Of course, the reason for these inconsistencies is that cost, transfer impedance, flexibility, and some other parameters can be in conflict. Nevertheless, the local optimization can be of great value when "you know what you are doing."

✦ It is important therefore to understand the reasons why optimization is at all possible.

As a rule, the process of optimization makes use of the *opposite trends* of the object function (that is, *criterion* function) with the variations of the participating *independent variables* (design parameters). For example, in composite multilayer shields, the reaction coefficients of ferromagnetic (e.g., steel) and non-ferromagnetic (e.g., copper) layers, have opposite signs. Then, by properly selecting the parameters of these layers, the shielding coefficient of a composite shield can be achieved, which is smaller (i.e., "better") than the product of the shielding coefficients of the ingredient layers (doesn't this remind you resonance in LC circuit?). It follows, that the optimality "at large" does not exist: it must *always* be considered in a certain "sense" defined by the optimization criterion (scalar or vector one)

Historically, one of the first such shield optimization models was formulated and resolved by Kaden [P.2] for multilayer solid wall shields with combination of ferromagnetic (iron) and high-conductivity (copper, aluminum) layers. He analyzed the expressions for the shielding effectiveness of two- and three-layer shields {actually per the terminology adopted in this book, it was the *transfer effectiveness*, e.g., similar to Eq. (3.143)}. For the fixed total thickness of the shield, the "optimal" (that is providing the smallest shielding effectiveness) shields required equal thickness in the layers. Similar (although numerically somewhat different) results were described in Ref. [P.3], and procedures are given in Ref. [7.4] for minimizing the weight and maximizing the shielding effectiveness of low-frequency shielding. The optimization models in these sources are based predominantly on transmission shielding theory (for this reason we are not going into the details here— remember in Chapter 2?). The braid optimization model suggested in Ref. [3.6] is based on the compensation of inductive coupling produced by the braid's apertures and the so-called *porpoising effect* (see the first comment in Section 3.5.2). A further improvement is achieved by inserting a magnetic layer between nonmagnetic braids. A more rigorous analysis of braid shielding performance optimization can be based on the vector diagram of the main coupling mechanisms (see Fig. 3.43). There, the shielding performance can be optimized systematically by mutually compensating the reactive parts of the coupling ingredients. Several practical examples follow below.

The problems of locally optimal (with regard to the transfer impedance) models of nonhomogeneous shields consisting of solid wall and spiral layers, nonmagnetic and ferromagnetic, have been theoretically investigated and experimentally tested as described in Ref. [3.12].✦

As an illustration of the results that can be achieved by such optimization, consider two-layer spiral shields. In Figs. 7.7 and 7.8, the transfer impedances of several such shields were plotted as a function of frequency and weave angles. This data was calculated per Eq. (3.179). Using the same formula, the optimal values of the weave angle α_{1opt} of one of the spiral tapes and the transfer impedance Z_{topt}/Z_{tsolid} (where Z_{tsolid} is the transfer impedance of a shield with solid layers of the same thickness) were cal-

518 Electromagnetic Shielding Handbook

culated and plotted in Fig. 7.15 as functions of the weave angle α_2 of the second spiral tape. Except for the weave angle, both tapes were assumed to be identical. As shown, the minimal (optimal) values of transfer impedance are observed when the tapes are wound in the opposite directions (i.e., α_{1opt} and α_2 are situated in different quadrants), while the values of α_{1opt} are shifted toward $\pi/2$. It is no surprise that the minimal optimum value of the transfer impedance is achieved at $\alpha_1 = \alpha_2 = \pi/2$, that is, when instead of spiral tapes, solid tubes are used. For the considered shield design, these results are representative for frequencies up to 500 kHz.

✦ It is interesting to follow the derivations of the expressions for optimal design used to plot the graphs in Fig. 7.15, from the general Eq. (3.179). To simplify the task, the case of relatively low frequencies was considered (up to 500 kHz), when the tape resistance can be approximated by their dc resistance ($z_{t1} = z_{s1} = R_1$) and ($z_{t2} = z_{s2} = R_2$). Because of non-ferromagnetic tapes, the axial magnetic flux through them can be neglected. Then, Eqs. (3.179) and (3.180) can be approximated as

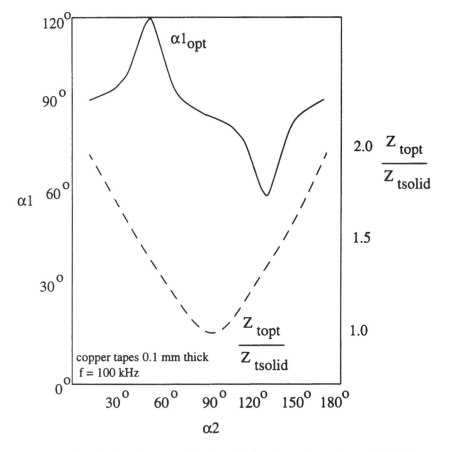

FIGURE 7.15 Optimal weave angle and transfer impedance of a two-layer spiral shield

$$Z_t \approx Z_s \approx \frac{R_1 R_2 + j\omega \frac{\mu_0}{4\pi}\left[R_2(\cos\alpha_1)^2 + R_1(\cos\alpha_2)^2\right]}{R_1(\sin\alpha_1)^2 + R_2(\sin\alpha_2)^2 + j\omega \frac{\mu_0}{4\pi}[\sin(\alpha_2 - \alpha_1)]^2} \tag{7.5}$$

Equating to zero the partial derivatives from Eq. (7.5) by α_1 and α_2, we obtain a set of two equations, which were used to plot Fig. 7.15:

$$\frac{\partial z}{\partial \alpha_1} = R_2\left[-\sin\alpha_1 \cos\alpha_1(\sin(\alpha_2 - \alpha_1))^2 + \sin(\alpha_2 - \alpha_1)\cos(\alpha_2 - \alpha_1)(\cos\alpha_1)^2\right]$$

$$+ R_1 \sin(\alpha_2 - \alpha_1)\cos(\alpha_2 - \alpha_1)(\cos\alpha_1)^2 = 0 \tag{7.6}$$

$$\frac{\partial z}{\partial \alpha_2} = R_2\left[\sin(\alpha_2 - \alpha_1)\cos(\alpha_2 - \alpha_1)(\cos\alpha_1)^2\right]$$

$$+ R_1\left[\sin\alpha_2 \cos\alpha_2(\sin(\alpha_2 - \alpha_1))^2 + \sin(\alpha_2 - \alpha_1)\cos(\alpha_2 - \alpha_1)(\cos\alpha_2)^2\right] = 0$$

♦

7.3.3 Global Braid Optimization by Cost Criterion

The essence of global optimization is that all the requirements for the shield, including system-related ones, are considered simultaneously. What is possible then is a compromise between these requirements, as determined by the adopted set of rules. As an important practical example, we will consider the optimization of a braided shield featuring two alternative criteria: technical (transfer impedance) and economic (the shield cost). Then, two outcomes are possible: the smallest transfer impedance at a reasonable cost, or a reasonable transfer impedance at the smallest cost. Just how to arrive at this "middle-ground" is the subject of the optimization theory. In the preceding section, we did touch the basics of optimization. For example, selecting the transfer impedance or shielding effectiveness and shield cost as the two alternative criteria, the Eq. (7.3) can be rewritten as follows:

- EMC optimization:

$$\begin{cases} Z_t(D_i) \to \min \\ E(D_i) < E_{\lim} \\ \Im_e(D_i) < \text{const} \end{cases} \quad (7.7)$$

- Cost optimization:

$$\begin{cases} E(D_i) \to \min \\ Z_t(D_i) < Z_{t\lim} \\ \Im_e(D_i) < \text{const} \end{cases} \quad (7.8)$$

In the EMC optimization problem, the shield is designed to have the lowest possible transfer impedance while providing a cost value not exceeding the specified value $E_{\lim}$, whereas, in the cost optimization problem, the shield is designed to have minimal cost while providing a transfer impedance value not exceeding the specified value $Z_{t\lim}$. Shield design parameters can take only certain values according to production capabilities and tolerances and specific problem requirements (constraints, $\Im$).

As an example of practical importance, consider the braided shield cost optimization problem. Expressing cost E (optimization criterion) and constraint Z_t as functions of the shield design parameters—material specific density γ, metal price p, and labor expenses on shield production T—we can write the expression for the braided shield cost as:

$$E = v\gamma p + T \quad (7.9)$$

where the metal volume in the shield is

$$v = \frac{\pi n N d^2}{4 \cos \alpha} \quad (7.10)$$

Both p and T depend on factors such as strand diameter, manufacturing equipment, wages, and so on. For approximate evaluation, the following empirical relationships can be used:

$$p = a + \frac{b}{d^2} \; ; \quad T = x + \frac{(y n \tan \alpha)}{2\pi D} \tag{7.11}$$

where coefficients a, b, x, y can be found empirically through the proper data analysis. A similar, though simpler, economic model can be obtained when the design goal is to provide a minimum volume of metal or a minimum cable weight.

Adopting Eq. (7.9) as the performance criterion and treating Z_t as a constraint, we can formulate the optimization model as follows:

$$\begin{cases} E(\alpha, n, N, d, D, a, b, x, y) \to \min \\ |Z_t| \leq \text{const} \\ K_1, K_{eq} < 0.98 \\ \alpha, n, N, d, D, K \in \text{specified domain} \\ \sigma, \gamma, a, b, x, y = \text{const} \\ 10^{low} < f < 10^{high} \end{cases} \tag{7.12}$$

From a mathematical standpoint, the model of Eq. (7.12) represents a nonlinear programming problem with both nonlinear criterion (object function) and constraints. The design parameters α, n, N, d, K are treated as independent variables whose "permitted" values are selected within discrete or continuous domains, as required by specific task. The same is true for the frequency band, f, which is limited by the lowest and the highest frequencies of interest. The surface density coefficient, K, is an example of conditions imposed by the technological limitations. The optimal values of the object function (criterion) are to be determined either at the function extremum (minimum) or at the borders of the specified area of the area delimited by constraints. This can be done using conventional optimal mathematical programming techniques.

With the use of computers, the solution of the model Eq. (7.12) does not present any difficulties in principle, as long as the values and boundaries of variables and constraints are identified and specified. In this way, the numerical values of the optimal parameters can be obtained.

522 Electromagnetic Shielding Handbook

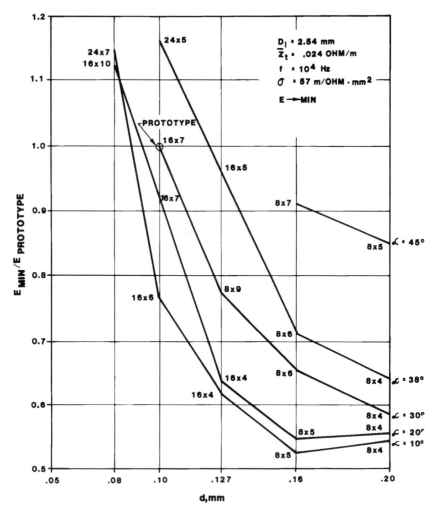

FIGURE 7.16 Low-frequency braided shield minimal cost vs. strand diameter and weave angle (required n × N values are indicated on the curves)

✦ However, for analysis purposes, it is more useful to obtain the dependence of optimal solutions on the cable design constraints as well as to follow up on the formation of optimal conditions, their stability, and sensitivity to the variations of the variables. Some of these results are shown in Figs. 7.16 through 7.20. These results were compared with a "reference" braid as presented in Fig. 3.40.

In Fig. 7.16, the dependence of the shield minimal cost on the strand diameter and weave angle is plotted. For each combination of d and α, the number of carriers, n, and strands in each carrier, N, are selected so as to achieve the lowest cable cost while satisfying the transfer impedance requirement. As to the latter, it is taken as a condition that the shield transfer impedance on frequency 10^4 Hz must be less than 0.024 Ω/m. The obtained minimal cost values are normalized to some reference cost and linked by straight segments denoting the trends. A similar set of curves is built in Fig. 7.17 under the condition of $Z_t \leq 0.280$ Ohm/m @ 10^8 Hz. On both graphs, the points are marked

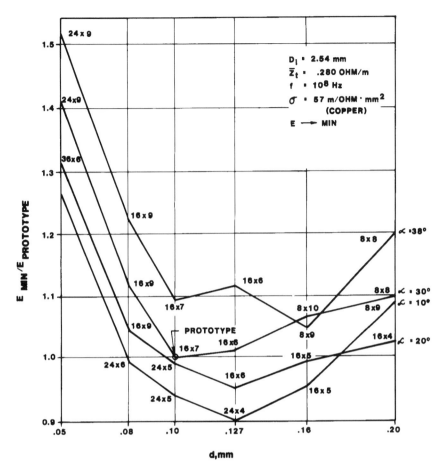

FIGURE 7.17 High-frequency braided shield minimal cost vs. strand diameter and weave angle (required n × N values are indicated on the curves)

corresponding to the reference design, i.e., manufactured prototype. (Of course, for those points, the normalized cost is 1.)

As shown, there exist sets of parameters that provide the lowest-cost shield design but still satisfy the adopted requirements within the specified boundaries of possible parameter variations. For instance, in the considered example, the low-frequency shield optimal parameters are α = 10°, d = 0.16 mm, n = 8, and N = 5, and the shield's cost is almost two times less than the reference shield. The better high-frequency economic performance of the prototype could be expected, as this particular cable was designed for a 1 GHz frequency range.

In general, for the considered example with the frequency rise, the optimal values of the number of ends (nN_{opt}) and percent coverage (K_{opt}) increase, and the strand diameter (d_{opt}) decreases, while the weave angle, α_{opt}, tends to have the lowest possible value (see Fig. 7.18).

These trends can be explained on the basis of the shield electrical performance model, Eq. (3.118); economical performance model, Eq. (7.9); and system optimization model, Eq. (7.12). In the low-frequency range, the main factor affecting the transfer impedance is shield dc resistance. If

524 Electromagnetic Shielding Handbook

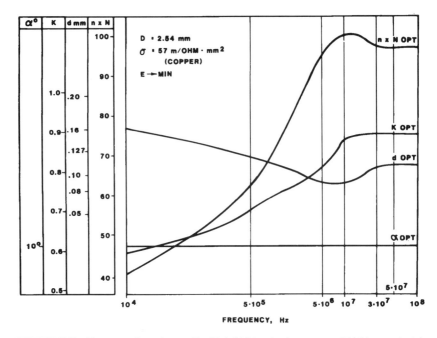

FIGURE 7.18 Frequency dependence of braided shield optimal parameters (shield cost criteria)

other parameter values are fixed, any given transfer impedance value can be kept constant by the means of simultaneous increase or decrease of d and α (see Fig. 7.12). Therefore, a multiplicity of d, α combinations will satisfy the transfer impedance requirement. On the other hand, the shield cost is less with the smaller d and α. It appears that, for the considered shield design and for required Z_t values, the shield cost drops faster with the decrease of α than d. In this case, d reaches the smallest possible (specified by the constraints) value. Then, d is selected just to achieve (but not to exceed) the required Z_t value. Therefore, the optimal shield's transfer impedance tends to be quite close to the specified limit, while the shield cost has the minimal possible value (see Fig. 7.19).

In the high-frequency range, the transfer impedance value is affected mainly by the percent coverage, which increases with the rise of both d and α, when other parameters are fixed. However, from the shield cost standpoint, it is preferable to increase the coverage by means of changing (increasing) just the number of ends n × N, while α is kept minimal and the d value is almost stabilized at some level.

For the given transfer impedance requirements on frequencies exceeding 10 MHz, the values of all shield parameters stabilize, indicating the same shield optimal design.

The optimal solution frequency dependence complicates optimal shield design for a wide frequency band. In this case, such a set of shield design parameters must be chosen to provide the lowest shield cost while satisfying the Z_t requirements at all of the frequencies simultaneously. Obviously, such a shield will not be optimal for some frequency requirements, but it will be optimal for the entire frequency range taken as a whole.

Another complicating factor is optimal solution dependence on the assumed limits of the design parameters. When the requirements and constraints change, the values of the optimal parameters, and even the trends themselves can be different from those of Figs. 7.18 and 7.19. Generally speaking, it is the shield cost criterion that defines the most expedient set of parameters in each particular case. If some values of these parameters cannot be adopted, it leads to the change of all other parameter values. For instance, according to Fig. 7.18, the optimal value of the weave angle is 10°.

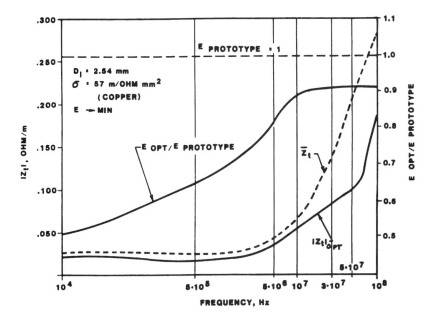

FIGURE 7.19 Frequency dependence of braided shield cost and transfer impedance (shield cost criteria)

If, due to the processing requirements or the need of flexibility improvements, the minimum possible value of α is increased, this will immediately lead to the change of the optimal values of d, n, and N (see Figs. 7.16 and 7.17). Respective trends are plotted in Fig. 7.20. It follows for the considered shield design that, with the increase of minimal admissible weave angle, the optimal values of d, K, and N rise, and n is reduced, while the optimal cost of the shield increases.✦

7.4 SHIELDING ENCLOSURE DESIGN FOR EMC PERFORMANCE

7.4.1 Shielding Enclosure Design Issues

Shielding enclosure design involves numerous factors originating from its functions:
- house and provide mechanical support to the electronic system elements
- limit the environmental pollution from the housed equipment
- protect from the ambient environmental effects: physical, chemical, weather, seismic, atomic plant, outer space or high temperature conditions, etc.
- meet EMC, ESD, ionizing and non-ionizing radiation, and electrical safety requirements and regulations
- meet aesthetical and ergonomical needs
- meet the specified standards.

526 Electromagnetic Shielding Handbook

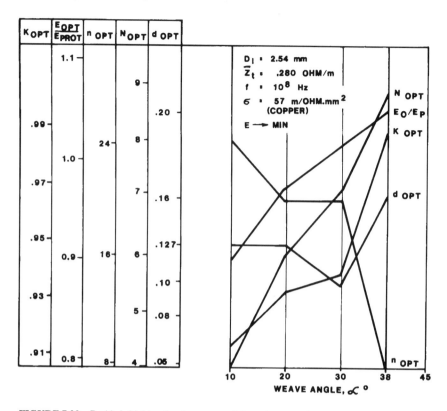

FIGURE 7.20 Braided shield optimal parameters' dependence on the minimum allowable weave angle

Previously, in Chapters 4 and 5 and at the beginning of this chapter, we have developed the methodology and approaches to address these issues. We have also covered, albeit briefly, the theoretical background and *critical parameters* of shielding enclosures. However, the practical design issues were deferred until this chapter.

Meanwhile, there is a booming industry out there, dedicated to the development and manufacturing of shielding enclosures at all levels. Many publications regularly cover the R&D-, manufacturing-, reference-, and standard-related materials on the subject. And of course numerous product catalogs are available from the manufacturers. This is a mature industry, where many practical needs can be met with off-the-shelf products meeting a set of standard sizes and materials. Along with standards, addressing specific requirements like UL, military, FDA, aerospace, and applicable also to other than enclosure products, there are "general purpose" standards developed by respective industries. However, particular selection of the applicable standards depends on the practical needs. For example, in the US, the NEMA (National Electronic Manufacturers Association) supported standards address address mainly the safety needs (e.g., UL requirements), while the EIA (Electronic Industry Association) supported standards focus more on product specifics. While it is generally not easy to satisfy all the requirements of any standards (together with the product needs — which is the main design goal, to start with), the problems start especially fast piling up when several standards must be met simultaneously!

Thus, just as it was the case with the *"theory"*, we cannot "embrace" *all* of the *practical* design problems, because of their shear number, "volume", and complexity. And like before, we will limit the following discussion to only several important shielded enclosure-related problems: enclosure and aperture resonances, EMI source placement within the enclosure, effects and treatment of doors, cable penetrations, and apertures.

To illustrate important practical problems and potential solutions, we will base this discussion on "real life" experimental studies from a number of "case histories". As usual, the particular selection of the reviewed examples is guided by three principles:
- underscore the *general* considerations
- present a practical problem / solution value on its own
- relate to the first hand experience

7.4.2 Aperture vs Cavity Resonances

As we determined in Chapter 5, the amount of electromagnetic energy "leaking" into or out of shielding enclosure through an aperture, depends on the resonances of *either* or *both* in the shielding enclosure and the aperture. As a result, two major considerations should *guide* the design of shielding enclosures:

1. *Suppress* the resonances in the cavity *and* the aperture, as much as possible.
2. *Decouple* the cavity resonances *from* the aperture resonances

The first step is to determine these resonances and evaluate their effects. In chapter 5 we have considered certain methods to perform such analysis. For example, using formulas (5.44) through (5.56) in section 5.8.2, we can evaluate the internal resonances in a rectangular cavity. If the respective enclosure is "physically" available[*], the same task can often be achieved by means of measurements.

✦ Just to "get a feel" for the phenomenon, consider results of the measurements performed in a physically large and a small enclosure.

<ins>1. Electric Field Distribution Inside A Large Tem Cell</ins>

The cell was originally rated for maximum TEM regime frequency (i.e., below first resonance and multi-moding) 85 MHz, expanded later to 120 MHz by selective placement of RF absorbers inside the cavity.

---

[*]We could have also tested a physical model of the enclosure, proportionally scaled for the given "electrical" size, as predicted by the quoted formulas

528 Electromagnetic Shielding Handbook

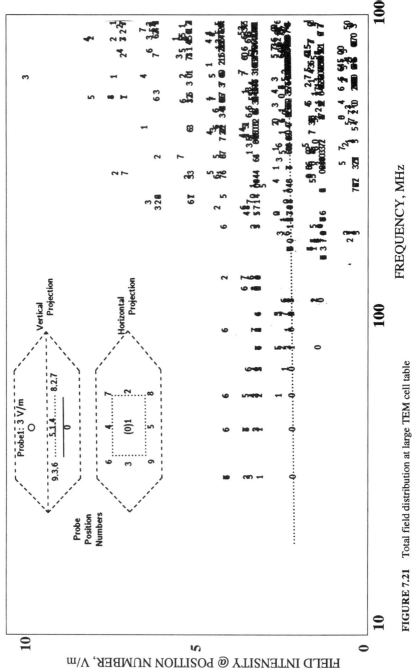

FIGURE 7.21 Total field distribution at large TEM cell table

The cell is excited with discrete CW frequency signals at 3 V/m *fixed* field (measured in the *center* of the cell *top half*) in the range 10 MHz — 1 GHz. Fig. 7.21 presents the field Amplitude (total field — a vector sum of probe's three orthogonal antennas — scaled at the vertical axis in V/m) as a function of Frequency (horizontal axis) and Position within the cell (the numbers placed at the graph coordinates according to the insert). As shown, while within "guaranteed" conditions — below 120 MHz and within the TEM cell "quiet zone" — the field distribution is generally "orderly" as expected, the data becomes *unpredictable* at higher frequencies and outside of the cell quiet zone. It is, of course, due to the internal resonances and multi-moding effects as discussed in Chapter 5. To confirm such conclusion, the data obtained by *each* of the three orthogonal antennas comprising the probe, was separated and compared.

While 120 MHz is a "low" frequency by modern standards, still lower resonant and multi-moding frequencies are observed in large enough shielded rooms (non-treated by RF absorbers), often used as EMC test facilities. Some examples are shown in Figs. 5.36 and 5.37. On the other hand, with constantly rising speed of electronic devices, even smaller enclosures become resonant at the fundamental and harmonics of the signal. Indeed, with the clock speeds approaching 1 GHz, the half-wavelength size of the enclosure is only 15 cm — well below many electronic product dimensions — see next example.

2. Small Box with a Slot

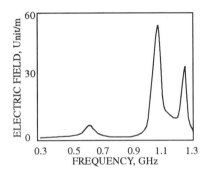

FIGURE 7.22 Calculated electric field at the center of aperture in a small box

A box of dimensions (337 × 160 × 100) mm has an aperture (228 × 2.5) mm. A rigorous analysis of the respective resonances in this box is presented in [7.14]. Fig. 7.22 presents the results of calculated electric field amplitude at the center of the aperture for an electric dipole excitation [7.14]. Three resonances are noted: about 650 MHz — at the aperture resonant frequency, about 1.1 GHz — the cavity resonance, and 1.292 GHz — the TE_{210} mode cavity resonance. Although the quolted experimental data do not completely support the theoretical analysis, it does contribute to physical understanding of the related phenomena.

✦

Section 5.8.3 also addresses the resonances in the shielding enclosure apertures. Formulas (5.61) - (5.62) and Fig. 5.39 illustrate the kind of problems encountered there. As was indicated, the aperture "performance" also depends on the nature of the illuminating field (see Fig. 5.38). Section 5 also suggests certain means to reduce the resonances and their harmful effects.

In any case, to make the aperture and cavity *inefficient* radiators, it is desirable to keep the their dimensions much smaller than the penetrating signal wavelength. To do this, a large frame or cabinet may be re-partitioned on smaller compartments, e.g., shelves, if the enclosed circuitry permits. Similarly, a large aperture can be partitioned into a set of smaller ones. A strategically placed shielding separator or a conducting "bridge" across the aperture can go a long way towards achieving this goal. By reducing the size, the resonances can be shifted to higher frequencies. If properly done, these higher resonant frequencies may be well out of the useful band and its spurious "derivatives". For example, remember the relationships (3.95) in section 3.4.5 of Chapter 3? In the author's note at the end of that section we concluded that just a twofold increase in iris diameter leads to an eightfold increase in coupling! Of course (3.95) deals with infinitesimal (with regard to the signal wavelength) irises, but we have now enough knowledge to expand to larger apertures.

Another way to dampen the resonances is to line the enclosure and aperture walls with lossy dielectrics. With reduced "Q", the resonances become weaker and less "sharp" (at the frequency scale). Just compare a non-treated shielded room with an RF absorber-lined room! Even *partial* RF absorber lining can do "lots of good". Indeed, as you see in the TEM cell example above, its frequency range was expanded from 85 MHz to 120 MHz by properly applying RF absorbers.

Obviously, the most dangerous situation arises when the resonances in the cavity, or cavities, coinside with or close enough to the resonances in the apertures. Then, the task is to decouple those resonances. One way to do that, is to re-direct the electromagnetic energy flows within the enclosure by changing the relative positions and strategically placing the radiating (or susceptible) elements and components, e.g., PCBs. This can be achieved by the proper product layout and / or packaging. Another way is to "play" with the cavity and aperture polarizations.

Although all these techniques are practically used in modern products, the field is still in its infancy. For obvious reasons, there may be serious physical and economical limitations to these techniques. On the other hand, numerous attempts to develop analytical methods to solve these problems at a system level so far were only partially successful, because of the problem inherent complexity and not always clear physical mechanisms of the involved phenomena. Some promising directions were delineated lately, using numerical techniques, e.g., see [7.15 - 17]. For certain conditions, they may provide an insight into the problems.. Then again, numerical simulations are not exactly conducive to a general analysis. In short, "business as usual" in electromagnetic shielding field!

7.4.3 Shielded Cabinets, Frames, and Shelves

"Regular" shielded telecommunications cabinets, racks, frames, shelves, and cable runs consist of metallic structures of large enough sizes. The internal resonances in such structures may be supported at relatively low frequencies. Moreover, practically it is not rare to see whole telecommunications, computer, or control offices "filled" with rows of such frames. As an example, a photo of shielded and non-shielded electronic switch frames lined up in an office, is shown in Fig. 7.23. Keep in mind, that both shielded and non-shielded frames may include separate shielded and non-shielded shelves, backplanes, cables, connectors. All of them, *together* with *"enclosed"* electronics, should be viewed as a radiating structure — active antenna array.

Shielding Engineering 531

SHIELDED ENCLOSURES   UNSHIELDED ENCLOSURES

FIGURE 7.23 Electronic telephone switch frames

✦ EMI is not the only consideration in designing or selecting a shielding enclosure. As a rule, basic requirements include static and dynamic mechanical loading, equipment mounting considerations, thermal management / air flow / cooling regimes, ease of manufacturing and assembly, economics. An efficient mechanical and physical design commands minimum weight and maximum frame stability. For example, it may incorporate aluminum die-cast base and top, interconnected by means of four extruded aluminum uprights. Because of the high initial tooling investments associated with die casting, most cabinets rely upon welded frame / upright configurations to attain structural strength. Aluminum or steel panels are used to create a closed shielding environment. All doors and and covers should permit removal without compromising structural integrity and rigidity.

To protect the structure from the ambient environment and provide for certain aesthetics, the metallic surfaces can often be painted, anodized, or lined with non-conducting panels and tiles. Unfortunately, this non-conducting "aesthetics" is often in conflict with the needs of "electrical integrity" of the enclosure.

As a matter of fact, the painted or otherwise treated joints are usually the "first suspects" of "electrical integrity" violations. ✦

What happens when an EMI source is placed inside a shielding enclosure?

Consider a general purpose shielding enclosure — "cabinet" — designed to contain emissions from diverse digital transmission systems. The cabinet has dimensions

532 Electromagnetic Shielding Handbook

(1830 × 686 × 686) mm, it contains a front and a back doors, and internal shelfs which can be placed at different height. In Fig. 7.24 this cabinet is shown installed on a turntable in an anechoic room, with the front door open. A *linearly polarized* radiated signal source (comb generator + dipole radiator of fixed length) is installed inside the cabinet (in Fig. 7.24, it is shown placed at the shelf in the lowest position within the cabinet. A receive antenna is installed at 3 m distance from the cabinet center, is connected to a spectrum analyzer to measure the radiating emissions from the cabinet.

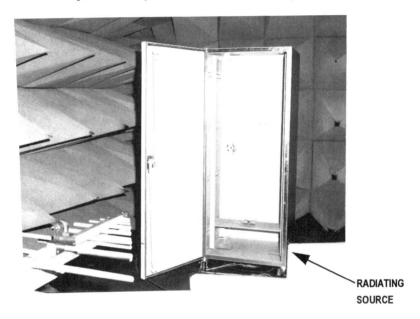

FIGURE 7.24   Testing shielding effectiveness of a telecom cabinet in anechoic room

The cabinet, *together* with the installed in it radiating source, present a radiating system — an active antenna array. To illustrate the interaction of the cabinet cavity and aperture resonances, a set of radiating patterns were measured with the cabinet door *open*. Fig. 7.25 presents several radiating patterns obtained by testing the emissions from this system at two frequencies — 70 and 528 MHz, two source polarizations — horizontal and vertical, and two receive antenna polarizations — horizontal and vertical. Just by observing the patterns in Fig. 7.25, important information comed to light:
- For the same test conditions, the difference in radiation in the direction of the *open* door (0°) and in the direction of a *solid* walls (90°, 180°, and 270°), reaches 15-30 dB — kind of "special case" shielding effectiveness.
- At the *higher* frequencies, the *multi-lobing* determines the *minimal* signal values of shielding effectiveness.
- The emissions from *horizontally* polarized source inside the cabinet are significantly *larger* than from *vertically* polarized source. For example, the field inten-

sity in the direction of the major lobe about 0° azimuth in Fig. 7.25 a (the source and receive antenna are both *horizontally* polarized) is over 20 dB larger than in Fig. 7.25 b (the source and receive antenna are both *vertically* polarized).
- Crosspolarization effects between the source inside the enclosure and the receive antenna are different at low and high frequencies: compare in Fig 7.25, patterns a and c vs patterns d and e.

As we will see later in this section, two more facts deserve to be mentioned here:
- The field intensity produced by the source placed *into* the cabinet, may be *larger* than that measured in "free space" — negative shielding effectiveness.
- A *good* contact must be provided between separate parts of the enclosure The painted surfaces do not provide the necessary contact, while special gaskets may help to improve the situation.

The conclusion is that there may be *optimal* ways to place the source within the cabinet. Of course, to do that, we are interested in explaining these facts. To start with, don't these pattern look familiar? The most interesting fact is that the patterns in Fig. 7.25 have similar shape as the respective patterns in Figs. 1.30 through 1.35! But this does not necessarily mean, that the underlying mechanisms "at work" are identical! Indeed, the patterns in Chapter 1 were "synthesized" from "physical" elementary elements: horizontally and vertically polarized dipoles and loops. The directivity of these "antenna arrays" was generated by the differences in phase of the fields radiated by the "ingredient" array elements. However, in the cabinet case two *different factors* are at work:
- In the direction of the open door, the *physical* source radiator emissions *combine* with its images' emissions
- In the opposite (to the open door) direction — the cabinet effectively eliminates "backside" radiation

With this understanding, we can explain the experimental results either by directly applying the respective field theory methods, as it was done, e.g., in [7.14-17], or using the antenna theory concepts. Consider first the low frequency results. We can easily explain the difference between the cross-polarized and matched results by "adding the images" to the actual source, e.g., see Figs. 4.20 and 5.29. The higher frequencies are more "tricky" and may need the "heavy artillery" of antenna theory. Indeed, a source radiator within a shielding enclosure can be modeled by a respective type of *reflector antenna*: plane, corner, parabolic cylinder, paraboloid, or spherical reflectors. Not that the antenna theory resolved all its problems and by referring to it we left with "nothing to do", but we can certainly benefit from this vast accumulated knowledge. Thus, modeling the cabinet with an open door by a respective reflector antenna and using the appropriate techniques (e.g., as iutlined in [1.13-17]), we can quite accurately predict the direction and width of major and minor radiating lobes, as well as other important characteristics.

(a) f=70 MHz
source inside cabinet, HRZ polarized
RCV antenna HRZ polarized

(b) f=70 MHz
source inside cabinet, VRT polarized
RCV antenna VRT polarized

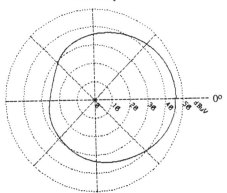
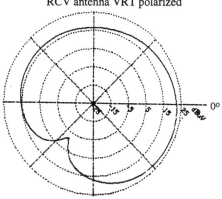

(c) f=70 MHz,
source inside cabinet, VRT polarized
RCV antenna HRZ polarized

(d) f=528 MHz
source inside cabinet, HRZ polarized
RCV antenna VRT polarized

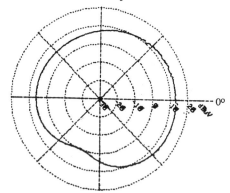
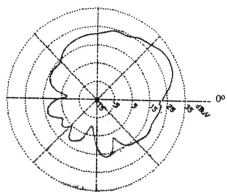

(e) f=528 MHz
source inside cabinet, HRZ polarized
RCV antenna HRZ polarized

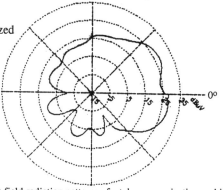

FIGURE 7.25 Electromagnetic field radiation patterns of a telecommunications cabinet dipole source in position 1, door open (at 0°)

◆ For example, take the question of *cross-polarization*. We would expect that when the source and the receive antennas are cross-polarized, a much smaller field will be measured than when these two antennas are matched (don't forget, we stated that the data in Fig. 7.25 was obtained using both source and receive antennas linearly polarized). Reviewing Fig. 7.25 we see that this is true *only* in the low frequency (70 MHz), while at 528 MHz both matched and cross-polarized results are almost identical! Antenna theory very persuasively explains this.

Indeed, consider a *y*-polarized source with a gain $G_f \cdot (\theta \cdot, \phi)$, placed at the focal point of a paraboloidal reflector. Then, at a distance r at the plane passing through the focal point, the electric field is [1.14]:

$$E_{ap} = \hat{e}_r \cdot C_1 \cdot \sqrt{G_f(\theta, \phi)} \cdot \frac{e^{-jkr \cdot (1 + \cos\theta)}}{r \cdot (1 + \cos\theta)} = \hat{a}_x E_{xa} + \hat{a}_y E_{ya} \quad (7.13)$$

where

$$\hat{e}_r = \frac{\hat{a}_x \sin\phi \cos\phi (1 - \cos\theta) - \hat{a}_y[(\sin\theta)^2 \cos\theta + (\cos\phi)^2]}{\sqrt{1 - (\sin\theta \sin\phi)^2}} \quad (7.14)$$

$\hat{a}_x$ and $\hat{a}_y$ are coordinate unit vectors, and

$$C_1 = \left(\frac{\mu}{\varepsilon}\right)^{1/4} \cdot \left(\frac{P}{2\pi}\right)^{1/2} \quad (7.15)$$

As you see, even if the source is *y*-polarized, both, x- and y-components, are present in the radiating field. ◆

Comparing the *maximum* emissions determined as the field intensity at the majop radiating pattern lobes in the frequency band of interest with that of the base line patterns of the *free standing* radiating signal source, the frequency characteristics of the enclosure *shielding effectiveness* can be obtained (note the shielding effectiveness definition used in this particular case). Fig. 7.26 shows the shielding effectiveness of the considered enclosure. Note the "negative" shielding when the cabinet door is open.

Note also the middle curve — cabinet closed and gasketed, but one gasket taped, that is, *isolated*. Of course, this violates the shielding enclosure body integrity. By no means, this is an "academic" question. Panel joints, cable penetrations, ventilation outlets, latches and locks — all create slots, seams and apertures. In Chapters 4 and 5 we have already addressed the effects of these "violations" on the coupling and radiation. But we do need ventilation, and cables entering and exiting the enclosure, and joints between the different parts of the enclosure! The answer is in sound design practices, which sometimes may explicitly or implicitly conflict with other requirements.

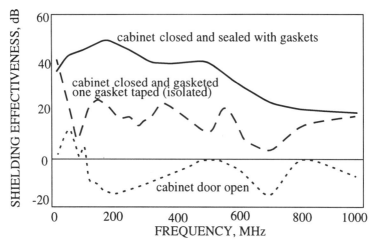

FIGURE 7.26 Telecommunications cabinet shielding effectiveness

✦ For example, to provide "aesthetical looks" the enclosures are often *painted*. However, these "looking nice" paints are as a rule good insulators, which prevent contact between adjacent conducting surfaces. A case in point with the telecommunications cabinet per Figs. 7.24 through 7.26. At the start of the cabinet testing session, the pattern measurements and "sniffing" with a magnetic probe helped to identify two leakage sources: front door lock and back door lock, each approximately 5" in diameter. After the doors were disassemblied, it was found that the locks were attached to the door walls *over paint*, so that electrical contact between the lock and door metal was realized only via the screws. After the paint was removed (do you like working with a sander on a painted metallic surface?) and hand-made conducting gaskets were applied, the leakage from these locations was reduced below the test setup sensitivity! ✦

From antenna theory we know that the radiation properties of reflector antennas are extremely critical to the *feed* placement with regard to the reflector. It could be expected then, that the radiation of PCB, or a cable, placed within shielding enclosure will depend on *where* in the enclosure it is located. And indeed, this is the case. So, the position of the source within a shielding enclosure should be accounted in the process of product and system layout and packaging.

✦ Consider an example. Fig. 7.27 shows the radiated emissions frequency characteristics of a PCB frame-carrier measured @3 m distance from the EUT. The carrier is *open* at front and back, but with *shielding walls* at the sides, top, and bottom. The emissions in Fig 7.27a correspond to the source PCB placed at 120 cm over the carrier bottom plane, and Fig. 7.27 b —with the carrier in the lowest possible position — 25 cm over the carrier bottom plane. Although the difference *in this particular case* is not that drastic, nevertheless the EUT in Fig. 7.27 b fails the class B FCC limit at around 150 MHz, while the EUT in Fig. 7.27 a — meets the limit at this frequency. The probable mechanism is that when the "feeder" (PCB) is close enough to the "reflector" (enclosure) the coupling between them increases, which results in larger radiating currents in the reflector. Since the enclosure is a

(a) source PCB at 120 cm over the carrier bottom plane

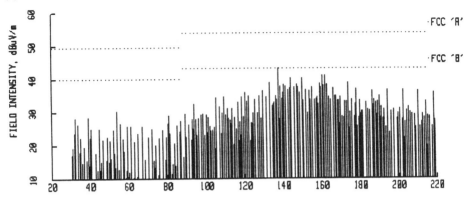

(b) source PCB at 25 cm over the carrier bottom plane

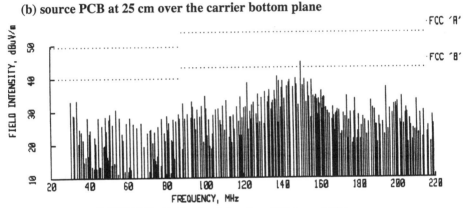

FIGURE 7.27 Radiated emissions from PCB frame carrier

much larger, and therefore much more efficient radiator than a separate PCB (at the frequencies in question), the total radiated signal increases.

Another observation made during this testing. *In this particular case*, the emissions from vertically polarized sources (PCB) were larger and more sensitive to the distance from the bottom ground plane, than emissions from the horizontally polarized sources. ✦

7.4.4 Shielding Enclosure "Hardware"

Doors, covers, windows, vents, gaskets, and shielded connectors — any of these elements can easily degrade the EMC performance of a shielding enclosure. To prevent this, there's a whole design and manufacturing industry out there! Usually, the manufacturer's specifications provide detailed information about their designs. A good source of hands-on how-to practical information, dedicated to the architechtural shielding issues, is ref. [7.18]. Of course, the *mechanical* design and specification is

538 Electromagnetic Shielding Handbook

beyond the subject of this book. The purpose of the following discussion is mainly to help EMC professionals know what and where to look for.

Doors and Covers

Take doors and covers. These are often the main "offenders" of the enclosure shielding integrity. In large shielding enclosures — like rooms, chambers, cabinets — the main problem with doors is the necessity to maintain a good contact between the door perimeter and the enclosure body. The contact is usually provided by presision fit and door balance, and the use of gaskets and strips around the door perimeter. The problems are that the doors are heavy, constantly in use, may need locks, while the gaskets are exposed to mechanical impact and environmental elements. For instance, finger stock fingers may break, while some of the elastomer- or "hollow" core-based gaskets can lose their resiliency with aging and repetitious mechanical loads. In the previous section we have discussed the effects of the gasket contact violation (see also examples in Chapter 2) and the effect of the door lock.

Similar problems may plague the shelf and enclosure covers. One of the most often used type is a clamshell (snap-on) cover. As a rule in real life applications, the joining surfaces àre painted (remember non-conductive paint effect in the previous section?) and can hardly provide a tight mechanical fit. Thus, the contact between the joined surfaces is mainly provided by screws (if any). Such practice proves an absolute failure from the EMI point of view. By the meantime, the problem may be relatively easy corrected by add-on or built-in gaskets, "bumps", dimples, hinges, special grooves and rails.. When painting the contacting surfaces must be masked. Hard pressed by the necessity to meet the requirements of EMC regulations, the industry gradually turns to sound design practices, e.g., see [7.19].

Electromagnetic Gaskets

To understand why we need a gasket, just refer to the gasket definition in section 6.5.4. And indeed, when used properly, the gaskets can provide a drastic improvement in the EMC performance. To prove the point, Fig. 7.28 shows the change in radiated emissions from a telecommunications product shelf mounted on a rack, when the rubber (nonconductive) gasket on the shelf cover is substituted by a copper tape gasket.

But doesn't it look like by using a gasket between two adjacent shields, we double the contact surface area, and thus exacerbate the problem? Unfortunately, this *is* the case when "bad" gaskets are used. Then, the question is, what is a "good" gasket?

As far as EMC is concerned, when using gaskets there are several important *electrical* objectives dealing with EMI suppresion. To these we must add electrical safety issues. From the gasket "inner workings", comes a long list of requirements to the gaskets and other contact "facilitators". It all "boils down" to establishing and maintaining the *quality* of *contact* between the four respective shield surfaces. And to be sure, we understand the contact behavior in a pretty *wide sense,* not just DC resistance — see section 5.7.3 and Fig. 5.41 in Chapter 5, and also [7.53]. To achieve "good" contact quality, several obvious conditions should be met: electrical, mechanical, physical and environmental / chemical. *mechanical* objectives.

(a) shelf with gasketed cover (shown open)

(b) emissions with cover closed - rubber gasket

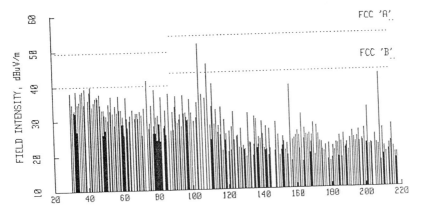

(c) emissions with cover closed - copper gasket

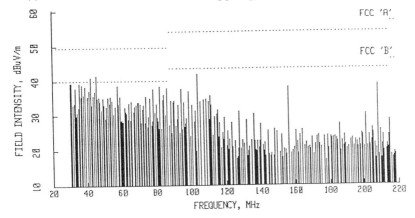

FIGURE 7.28 Effect of a copper tape gasket on radiated emissions from telecom shelf

To this we can add manufacturability, cost effectiveness, and aesthetics, when needed. And finally, to meet all these objectives, the industry presents an impressive list of gasket designs.

All these issues are summarized in the following Table 7.1 a.

Table 7.1a **Gasket Design and Selection Issues**

Electrical Objectives	Mechanical Objectives	Available Designs
• shielding effectiveness: a good gasket may be able to provide up to 100 dB or more shielding effectiveness from several tens of kHz to several tens of GHz! • highly conductive gasket surface and / or a large number of contact points • low electrical resistance (don't forget *safety* grounding)	• perfect mechanical fit • flexibility and resilient contacting surface • mechanical "endurance" — resistance to breaks and abrasion • corrosion and chemical resistance • friction (no"loose" fit) • large enough compression/deflection range, especially after long term and periodical loads • chemical safety, outgassingand flammability(especially matrix and conductive elastomer-based gaskets) • resistance to environmental factors, aging, and mechanical loads (including cyclical ones) • electro-chemical compatibility with the shield	• "good old" berillium copper and nickel copper fingerstock, • monel, tinned copper, or steel mesh or braid, • tape and wire spirals, • laminates • conductive coatings on low closure foam cores and elastomers, • conductive elastomers, • loaded plastics filled with metal or metal-plated particles or carbon, • "form-in-place" gaskets

Who could have thought that a small gasket may be that "big in work"! Really, by any standard, there is a lot to think of and to choose from! Table 7.1 b presents some experimental data of electromagnetic gasket EMC performance.

Vents

We have already discussed the effects of the size of the apertures on emissions and immunity of shielding enclosures. Here, we will briefly address the ways to prevent the electromagnetic field penetration through these these violations of the shield integrity. Actually, the vents and windows serve quite different objectives.

Proper design of vents permits to reduce the needs and simplify the design of the enclosure ventilation system — be it a computer enclosure or a shielded room.

Shielding Engineering 541

Table 7.1b **Electromagnetic Gasket Performance**

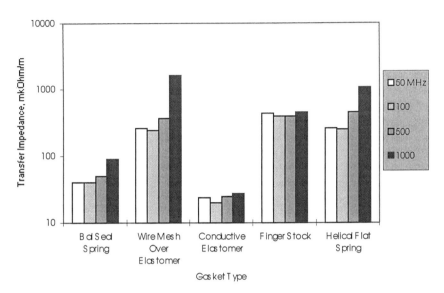

The aperture electromagnetic theory discussed above calls for a simple solution: make the vents electrically small. When this is undesirable or impossible, an other simple solution is to cover the aperture with a metallic mesh of sufficient density— and this is often done. The mesh material isn't really important, except that it must "conducive" to good continuity with the shielding enclosure body. Still another means is to provide the vents with louvres — framed openings in the shielding enclosure body fitted with slanted slats.

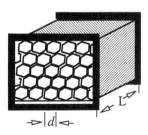

FIGURE 7.29 Honey comb vent

A specific kind of the air vent cover is honeycomb — a matrix assembly of honey comb-shaped brass or steel tubes, as shown in Fig. 7.29. The tubes are soldered to each other, forming a frame, which can be fit over the vent. Usually the honey comb assembly is either welded or soldered to the enclosure body. In some instances, it may be applied via gaskets. Considering each vent a waveguide well below the cutoff frequency ($L > d \ll \lambda$), the "rule of thumb" for comb assembly shielding effectiveness is [7.18]:

$$A_{hc} \approx 27\left(\frac{L}{d}\right) - 20\log n \qquad (7.16)$$

where n - number of cells within a square of side length λ or area of the vent — whichever is smaller.

Windows

In shielding enclosure windows, all the previously discussed for vents conditions apply, except that they must provide for light penetration. Obviously, such solutions as honey comb, are not acceptable. A window construction may have a metallic wire mesh or a thin perforated sheet bonded between transparent sheets of glass, acrylic, or polycarbonate. A vapor-deposited or sputtered very thin conductive film of vacuum deposited gold or indium oxide may be transparent, but its shielding properties are more or less acceptable only in the microwave frequency band. Still another approach is to combine non-conducting films deposited at correct thickness to convert by interference effects low light transmitting metal films into high light transmitting window glasses. There were reports [7.20, 7.57] of development of transparent laminated glasses and sealed double structures yielding up to 40-50 dB shielding effectiveness below 1 GHz, and over 60-80 dB - at the higher frequencies — up to 10 GHz (measured using MIL-STD-285C techniques).

Shielded Connectors

We finish this section with a note on connector shielding. A large diversity of connector shield designs exists: made of solid metal and metal-coated plastics, soldered or tight-fit to the cable or enclosure shield, with special berillium-copper or other springs, with different designs of front and back shells. Usually, these designs and their physical properties are meticulously described in product catalogs, so there is no need to do it here. However, the roadmap to their understanding and consequently design and selection, comes from an *electrical* perspective. First, keep in mind that "connector interconnections" are no different from "regular" lines which can be treated as "lumped" circuits when electrically small, and as transmission lines — when electrically long. By the same token, the connector shields are no different from other kinds of shielding enclosures. The same concepts of EMC performance, electromagnetic coupling, shield transfer function, including the transfer impedance and transfer admittance, and shielding effectiveness are fully applicable to connector shielding. And just as it is the case with those other enclosures, it is often* not so much the transfer function of the connector shield that affects the shielding system performance, as the ability to provide *continuity* of the shielding system between the shielded product enclosures and cables. Moreover, since as the very name implies, connectors "work" at the interface of different system elements both in an electrical and physical sense, the connector shielding continuity becomes even a higher priority. For both, "regular" lines and connectors, we will address the shield termination, grounding, and continuity issues in section 7.6.

---

*Although not always, e.g., think of some extremely thin shielding coatings which became so popular with certain connector designs. In Fig. 6.42, we took already a "glimpse" at the performance of a coating only 1000 Å thin. More about the coatings in the next chapter

7.5 TRANSMISSION EFFECTS IN SHIELDING CIRCUIT

When using shielded circuits, one often overlooked effect is that of the shield reaction on the circuit transmission performance. By no means these effects are new or little known. Abundant literature exists on the subject, including "classical" works, e.g., one of the most useful sources is [P.1]). The more surprising it is to hear sometimes even from seasoned EMC professionals, that "shielding can be applied to new and existing systems without impacting high speed circuit operation" (that's a quote!). Probably such oversight is caused by the fact that the purpose of shielding is to improve the system crosstalk, emissions, and immunity parameters, while the changes in the transmission patameters are kind of "side effects". "Unfortunately", these effects do exist and are often undesirable. Following, we will consider three manifestations of the shield effects on transmission parameters: shielded cables and lines, ground plane effect on PCB interconnections, and parallel shielding.

Shielded Cables and Connectors

♦ We will illustrate the discussion with parameters of *two-conductor* lines.

We start with *primary* transmission line parameters: resistance R, inductance L, and capacitance C. At higher frequencies and in lossy dielectrics, also of interest can be high frequency dielectric conductivity $G = \omega C \tan\delta$, where δ is dielectric loss angle. In Chapter 3, we have discussed the loss impedance of solid tubular conductors at high frequencies {see formula (3.52)}, while in Chapter 4 we "touched upon" the line inductance and capacitance {see (4.40)}. But neither of these formulas had "anything to do" with the effects of the shield! The problem is that both these formulas considered the transmission line in *free space*. However, the shield effects on these parameters may be quite dramatic and should be accounted for.

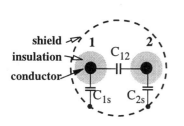

FIGURE 7.30 Partial capacitances in shielded pair

Consider capacitance of a shielded two conductor line, as shown in Fig. 7.30. In principle, this doesn't have to be a "canonical" shield, a metallic *tube* enclosing the pair, but just about *any* conductor(s) in the vicinity of a considered pair: a layer of other conductors in a multi-pair cable, a shielding enclosure of any shape — either individual or shared with other circuits, a ground plane, . Neither it should be a shielded balanced *pair*, nor even unbalanced one. For example, it could be a *single-ended* shielded line, which, when coaxially placed will result in a coaxial cable.

Fig. 7.30 illustrates *partial* capacitance C_{12} between the two insulated line conductors, and *partial* capacitances C_{1s} and C_{2s} between each conductor and the shielding enclosure. When the effects of the shield on transmission parameters are neglected, *only* C_{12} is accounted for. This is one of the main sources of errors (other sources are similar — neglecting the effect of the shield on the line losses and inductance), since C_s can be (but not necessarily is) much larger than C_{12}! Just compare the capacitance of a coaxial cable to capacitance of an *equivalantly* sized balanced pair in free space.

That's a good "incentive" to understand and account for the shield *reaction* on the transmission circuit and line parameters.

As a matter of fact, formula (4.40) yields[*] only C_{12}, which is equal to the total interconductor capacitance if the shield is absent. However, it is clear from Fig. 7.30 that with the shield present, the *total* line capacitance will be determined by all *three* partial capacitances:

$$C_{line} = C_{12} + \frac{C_{1s}C_{2s}}{C_{1s} + C_{2s}} \qquad (7.17)$$

Of course, each of the partial capacitances is a function of the dielectric constant (actually, *equivalent* dielectric constant of a composite insulation, e.g., air+dielectric).

Similar derivations are true for inductances, with the exception that the dielectric constant is substituted for magnetic permeability (remember our discussions in Chapter 4 about the orthogonality of the electric and magnetic field line of force?)

If we are not interested in great accuracy, as a first degree of approximation, the following formulas can be used, similar in structure to (4.40):

$$L \approx \mu_r \log \left| \frac{a}{r_0} \psi \right|, \mu H/m; \quad C \approx \frac{\varepsilon_r}{0.08 \cdot \log \left| \frac{a}{r_0} \psi \right|}, pF/m \qquad (7.18)$$

where coefficient ψ accounts for the effect of the shield. its value depends on the distance from the line conductors to the shield *relative* to the distance between the line conductors. The usual value varies from 0.5 to 1.0 (no shield effect).

The active losses of the line signal energy in the shield also depend on the distance from the line conductors to the shield *relative* to the distance between the line conductors. However, the accurate formulas are pretty complex. As a first approximation for high frequencies ($\gamma t > 5$, where t is the shield thickness and γ is the intrinsic propagation constant of the shield material - see formula (2.9) and Table 2.1), the following expression can be used:

$$R_s \approx 8 \frac{Z_a}{2\pi r_s} \cdot \frac{a^2 r_s^2}{r_s^4 - a^4} \qquad (7.19)$$

where the the real part of the wave impedance Z_a is determined from the Table 2.1, r_s and a — are the shield radius and half the distance between the line conductors.

---

[*]With only marginal accuracy, because we don't account for the finite conductor diameters, as well as "several other effects". However this problem can often be solved analytically, and almost always, with a necessary accuracy, using numerical techniques. It's a "favorite" example with many texts on electromagnetics.

At higher frequencies, the line insertion loss related to the energy losses in the shield and characteristic impedance, can be determined from a simple formuls:

$$\alpha \approx \frac{R_s}{2}\sqrt{\frac{C}{L}} + \frac{G}{2}\sqrt{\frac{L}{C}}; \quad Z \approx \sqrt{\frac{L}{C}} \qquad (7.20)$$

How large could be the losses in the shield? Suppose we want to compare the line insertion losses @100 MHz, related to two shields: copper and steel. From (7.20), substituting (7.19) and neglecting the second item in α, follows that the ratio of the shield-related losses will be proportional to the ratio of the respective wave impedances Z_a. Then, using the values in Table 2.1, obtain:

$$20\log\left(\frac{\alpha_{steel}}{\alpha_{copper}}\right) \approx 20\log\left(\frac{Z_{steel}}{Z_{copper}}\right) \approx 30\,\mathrm{dB} \qquad (7.21)$$

Of course, to be of importance the calculated losses must constitute a significant part of total losses, however that's often could be the case. How to protect from excessive losses in the steel shield? One way is to "shield" the line conductors from a steel shield by a low loss shield, like aluminum or copper. It follows also that, for example, in a cable with a double-layer copper-steel shield, the copper layer better be "looking inside" the cable towards the signal-carrying conductors, with the steel layer — "looking outside". Any cable catalog will confirm that this is indeed the design of choice!✦

Shielded PCB Interconnects

✦
Two main circuit configurations are used at the PCB level, which are related to the ground plane — a shield per our definition: microstrip and strip line. In Chapters 4 and 5 we have devoted time to the discussion of EMI-related effects in such circuits (see Sections 4.4.7, 4.5.3, and 5.3.3). Here, we will briefly address the transmission issues. Due to the relatively small length of these interconnections, the insertion losses are usually of a secondary value, while the most important parameters affected by the ground plane are the characteristic impedance and line delay. These parameters depend on the width of the line traces and their distance to the ground plane(s), as well as on the substrate dielectric constant and the presence of other conductors in the line proximity. In general, these dependencies are pretty complicated [5.7, 7.21], however computer simulations usually provide the necessary accuracy.

For estimates and to evaluate the trends, approximate formulas can be used. Unfortunately, different sources quote different approximations, depending on the adopted set of assumptions and accounted for physical mechanisms. What's even worse, often these assumptions are not stated in enough detail. So, before using the formulas, it would be wise to check their applicability to the problem at hand. Of course in the context here, this is "not our problem": we want only to underscore the effects of shielding on the transmission characteristics. Just to provide the "flavor" of the issue, the following expressions are often used to calculate the characteristic impedance and time delay of the PCB traces, "narrow enough" comparative to the trace distance from the ground plane:

microstrip:

$$Z_o \approx \frac{87}{\sqrt{\varepsilon_r + 1.41}} \cdot \ln\left(\frac{6h}{0.8w + t}\right), \text{Ohm}$$

$$t_d \approx 3.3\sqrt{0.5\varepsilon_r + 0.7} \quad , \text{nsec/m}$$

(7.22)

stripline:

$$Z_o \approx \frac{60}{\sqrt{\varepsilon_r}} \cdot \ln\left(\frac{4h}{0.67\pi w(0.8 + t)}\right)$$

$$t_d \approx 3.35\sqrt{\varepsilon_r} \quad , \text{nsec/m}$$

(7.23)

Similar formulas and their analysis are widely available for many layout configurations (e.g., see [1.22, 5.7] and other sources).

Is this shield reaction on the transmission parameters really important? You bet! As a "trule fo thumb", given similar dimensions and dielectric properties, the characteristic impedance of a stripline can be 30-50% lower, while time delay — 30-50% larger, than those of a microstrip. ✦

Transmission Effects of Parallel Shielding

✦ While parallel shielding at a cable-, transmission line-, connector, or PCB- level, provides essential EMI mitigation benefits (e.g., see sections 4.4.7 and 5.3.3), it also affects the respective transmission and propagation performance. As an example, Fig. 7.31 presents the line characteristic impedance dependence on the typical parallel shielding-related circuit configuration. Similar effects can be produced by "shielded traces" on a PCB. As you see, the difference can be quite significant. ✦

7.6 SHIELDING SYSTEM GROUNDING, TERMINATION, AND PARTITION

7.6.1 What Is Grounding Really About: Myth and Reality

The *ground* (noun) is usually understood as a conducting surface used as a common reference point for circuit returns and electric signal potentials. To ground (verb) means to connect the circuit element to the ground. Any conductor contacting the ground acquires the ground potential and thus is "grounded".

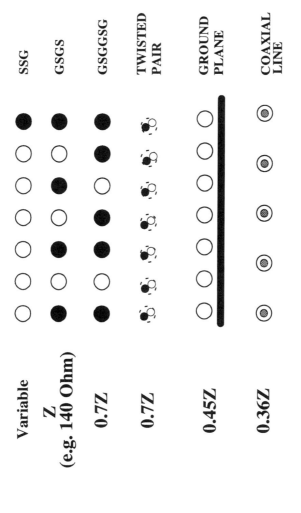

FIGURE 7.31 Comparative characteristic impedance of transmission cables

548 Electromagnetic Shielding Handbook

As many textbooks will state, there are several grounding objectives:
- Create a common potential reference
- Isolate the system from ambient environment
- Isolate sub-systens from each other
- Separate circuit components from mutual interference
- Provide for the system stability

Looks almost like a magic cure from all EMI and signal integrity troubles! But what's the magic that enables the ground to do all these good things? The same textbooks explain, how are these functions accomplished. An ideal ground, e.g., Earth, is neutral, they say, but it stores an "infinite" number of positive and negative charges, which are always readily available "on request" to whoever or whatever need them. For instance, you remember how these charges were "called upon" in our electrostatic shielding model (see Fig. 5.4 in Chapter 5). Also, from the circuit theory point of view, the "ideal" ground does not present any impedance to the currents propagating in it. The ground also serves as a "mirror" for electric charges and currents.

As convenient and widely accepted such explanations are, "electromagnetically" speaking, they may be highly questionable!

✦ To start with, let's disspell one misconception which treats Earth (or any conductive "ground" for that matter — we do assume Earth to be conductive) as an infinitely large reservoir of charges. While conductors do "contain free electrons", in the absence of *external* fields the conductive bodies are electrically neutral. To "access" the charges in a conductor, an electric field E must be applied and a potential difference V created, which will move the free electrons against the electric field forces. Thus, the charges are available not "just for asking — for free", but for a "price": a certain amount of work must be done by *external* (with regard to conductor) forces. Since this work is determined by V, the charge Q supplied to (or "taken from") a conducting body is proportional to the potential difference V. From the "opposite" perspective and as a matter of metrics, a conductive body or a system of bodies can be characterized by the amount of charge which must be supplied to (or taken from) this body(ies) to change its potential by a given amount, say 1 volt. If this brings to your mind an association with the definition of capacitance, you are absolutely right: C = Q / V, coulombs / volt, or farads. In electrostatic environment, a metallic physical body capacitance depends only on its geometry and the dielectric constant of the adjacent to the body media.

Now, let us return to the Mother Earth and tackle the problem at hand. If we embark on the concept of "infinitely large number of charges in the body", we could expect an "astronomically" large capacitance of the Earth. So, armed with the ages' old knowledge described in the previous paragraph, let's perform a simple calculation to determine the capacitance of the Earth in free space ($\varepsilon_0 = 1/(36\pi 10^9)$, F/m). Assuming Earth to be a conductive spherical body with radius $R \approx 6.37 \times 10^6$ m and applying the familiar from electrostatics formula for the capacitance of a sphere immersed in a dielectric, obtain:

$$C_{Earth} = 4\pi\varepsilon_0 R \approx 7.1 \times 10^{-3} \text{ F} = 710 \text{ μF}$$

Not impressively large, to say the truth. Probably, you've seen "much smaller than the Earth" electrolytic caps which proudly carry labels with larger capacitance values!

Keep in mind, that the capacitance of a "solo" body corresponds to bringing the charge in from or out to "infinity". Now, if you consider a system consisting of two conducting bodies, say, our shielding enclosure and Earth at some finite distance from each other, the capacitance of such system may be even much smaller. Returning back to "astronomy" and borrowing an example (and a curious one!) from Bob Pease column (*Electronic Design* magazine), many readers will find much to their surprise, that the capacitance of the Earth - Moon system is ... what would you expect? — make a "wild" guess and, probably you will be wrong!* ✦

But then, if "infinite charges" explanation does not hold, why did we use it with regard to Fig. 5.4? Does that mean that our model of the electrostatic shielding was wrong? And in this case, why do we need grounding, anyway?

Well, remember that we actually never declared, nor used, the "infinite charge reservoir" assumption. We have only stated, that the charges from the "ground" would come and neutralize an uncompensated charge at the external shield surface. Since this uncompensated charge at the shield surface is in principle of a finite volume (it is exactly equal to the original "offending" charge), *given enough time* it could always be compensated with the opposite sign charge from our ground, even if its capacity to "contain" charges is *limited*. The only thing that matters, is the existence of a potential difference between our shield and the ground, which will be the "moving engine" for the compensating charges. Of course, with the smaller "charge capacity" of the ground, it will take longer for the charges to reach the shield and "neutralize" it, but in this case of statics the charges have "all the time in the world" to arrive at their destination. Because of this infinite time available, we could *disregard* the "levelling-off" transient currents from the ground to the shield and consider only the "final effect": an equipotential surface of the shield.

Now, we have come to the heart of the problem:

it is not the "infinite charge" of the Earth, but the "infinite time" associated with electrostatics, that "did the trick" of electrostatic grounding!

There are also certain other, additional, considerations confirming our conclusions. They can be found in available literature, for example, see Refs. [7.22-23].

Of course, in fast enough time-varying fields, we cannot disregard and should account for these "levelling-off" transient currents from the ground to the shield and the concept of equipotential surface does not hold any more. In such applications, the ground must be considered as just a conductor that conducts electrical currents, exhibits certain impedance, generates voltage drops, and can radiate. But then, do we still need grounding at all? And the answer is "may be", but for *different* reasons, in different circumstances, and in different context than quoted at the beginning of this section.

In fact, after we are done with this section, you will see that the "real topic" of *grounding in electronics* (as opposed to electrical safety) is how to use the *unbalanced* devices connected by single-ended lines, in a *least EMI-harmful* way. Except for the safety considerations and certain issues of electrostatic shielding,

---

*It is only around 159 µF!

electronic circuit conductor connection to the ground plays no other role, as to provide the signal and EMI current returns (whether desired ones or not).

If only balanced interconnections were used, there would be no need at all for grounding, from the signal point of view. However, the balanced lines occupy almost twice the "real estate" compared to single-ended lines, and besides, it is almost impossible to preserve their balanced nature in a typical electronic environment. Also, under certain conditions it may be necessary to provide the *common mode* currents return even in balanced circuits.

7.6.2 To Ground or Not To Ground ?

"It depends. . . ." While everyone would usually agree with such answer, that's about where the consensus ends. How will a grounded or nongrounded shield behave in any particular system configuration? Should the shield be terminated and/or grounded at one enclosure point or cable end, two points / ends, at intervals, continuously, or not at all? How are the grounding and termination recommendations affected by the shield parameters? What if the ground connections have an impedance of their own? Is it worthwhile to have "decoupled" grounds for different parts of the system, e.g., with regard to cables? The confusion is often exacerbated by different effects of grounding on the functions of a shield: as a line conductor, an energy barrier, and a radiator. Another important cause of disagreements (and problems) is definition-related.

Indeed, when points of a circuit with potentials different from that of the ground are connected to the ground, corresponding "leveling" currents are generated in the circuit (including the ground itself), which can create additional interference. Similar "levelling" currents are generated in the circuit which is connected to the points of ground which have differenct potentials caused by other circuits, than the considered one. This is the "infamous" ground loop, which often renders the approach of grounding at several different points to be less than a great idea.

Along with the ground loops, there are two more aspects of shield grounding and termination that should be considered, especially at higher frequencies. First, the shield grounding and termination define the EMI current propagation conditions and distribution on the shield by providing certain electrical loads. These loads are mainly at the shield's ends but, in specific cases, they exist continuously along the shield (e.g., cables in soil or water). Second, grounding and termination interconnects are realized as electrical circuits which themselves have propagation and radiation properties, produce a voltage drop that changes the shield potential with regard to the ground and other system elements, and may significantly affect system performance. We will review the first aspect in this section, while the second one will be considered in the next section.

✦ Once more about the potential "conflict" between the grounding objectives of electronics and safety — this is really important! Our discussion of grounding problems is strictly limited to EMI.

On the other hand, for safety reasons, grounding and earthing are often not only useful but *necessary* and *mandated* by the authorities and regulations. Often, the necessity to ground the shield for safety results in ground loops and is in a direct conflict with EMI considerations. Since safety considerations *always* take priority over EMI, technical ways and means must be found to "fight" the negative effects of grounding. One of the ways to satisfy conflicting safety and EMI grounding requirements is by using so-called *hybrid grounding,* whereby contact with the ground is realized in respective circuit points via capacitors or inductors, *along* with a direct connection in other points. There is a whole industry out there, dedicated to breaking the ground loops: use of transformers, filters, balancing, optical couplers, etc. We will return to the subject of ground loops in shielding in sections 7.6.3 - 7.6.5.✦

In real life, the terminations are different for the transmission line and the shield. For example, an unshielded single-ended line formed by a conductor will often be connected to the ground via a resistive load equal to the characteristic impedance of the line (e.g., 50 Ohm). If this line is shielded, the shield may be shorted (or so we think—because the "short" has its inductance) to the ground, or left open (or so we think—because the "open" has its capacitance). However, a single-ended line may also be connected to a very high or a very low impedance load (e.g., CMOS or ECL gate, respectively), which further complicates the picture. Even more uncertain are the loads of the common-mode currents flowing in the conductors of balanced and unbalanced circuits that do not contact the ground plane directly, but are capacitively or inductively coupled to it.

Here's the problem. Allegedly the "facts" prove that shield grounding is absolutely necessary. Consider, for example, the reflection of an electric field from the shield in the electrostatic regime. Figure 5.4 tells us that the shield performance is almost ideal. But as we saw (Fig. 5.8), this ideal performance takes place only when the shield is properly connected to an ideal ground, which results in the neutralization of the charges at the shield's surface. As a matter of fact, we showed that when the shield is not grounded at least at one point, it is absolutely transparent to the static electric field! In the magnetostatic regime, both ends of the shield must be grounded. The same is true, at least to a certain degree, in the quasi-static regime. In the quasi-stationary regime, when the shield is not connected to the ground, the residual current difference, $I_i - I_{si}$, is large and degrades the shielding effectiveness. Everything seems simple and logical: yes, we need to ground the shield!

But let us take a second look at the same problems. Except for "pure" statics, is it really the grounding that "did the trick"? For example, see Fig. 5.8 and 5.9. Suppose that, instead of the ground, there is just a conductor that provides the necessary connections and is coupled by the respective inductances and capacitances to the interfering circuits and the shield. (O.K., make it a low-impedance conductor, if that's how you perceive the ground). It is easy to see that, in principle, nothing really changed, as our "fake ground" conductor will perform just as well as the "real" ground! Or in the quasi-stationary regime, when the line load is connected not to the ground but terminates only the line (e.g., via a transformer as in a balanced line), no "ground" is present per se. In this case, the shield can be connected directly to the transformer's center tap. You can verify experimentally that the shield will work just as fine (and sometimes even better) than if connected to the ground.

Thus, it is not some "magical" ground properties that determine the shielded line performance, but a "regular" circuit function of the ground conductor that provides "mundane" current paths. Indeed, how could it be otherwise, for example, in the aircraft or satellite grounding systems? Another way to look at the same problem is by considering the images to reduce the analysis of ground effects to the analysis of circuits, including galvanic connection ("a short," or via load impedance) of the circuit ends to the ground. Also, note that nowhere during the crosstalk derivations or analyses did we use the concept of ground—just different regimes that originated from different shield terminations.

But then, should or should we not ground the shield? As you see now, the question can be reformulated in terms of the shield termination: how should we terminate the shield? The previous discussion permits us to formulate *two* important general *rules*. More rules addressing the shield terminations, may or may not be related to the shield grounding and, for this reason, they are considered in the next sections.

Rule No. 1: *To suppress the radiation from or the susceptibility of the line enclosed in the shield, the shield must provide an efficient internal current return.*

This is achieved by properly terminating the shield to the line, which may or may not be related to the shield grounding. For example, when the line is itself terminated to ground (via the line loads, as in a single-ended circuit), the shield termination to the line coincides with the shield grounding at the ends. In this case, you may say that you need to ground the shield at the ends, but what is really required is that the shield be connected to the line at these ends, securing the return of the line current through it. The grounding-related "side effects" of such a connection may be positive or negative, depending on the resulting antenna current distribution at the shield surface. However, even the negative effects seldom can outweigh the benefits of terminating the shield to the line. Also, certain techniques are available to terminate the shield to the line without terminating it to ground, even if the line's loads are connected to the ground. One such electromagnetic shielding-related technique is to use a triaxial line with isolated shields.

Rule No. 2: *To suppress the radiation from or susceptibility of a shield that encloses a line, the current in the shield external circuit (antenna current) should be reduced as much as possible.*

Of course, the "straightforward" way to reduce this current is to reduce the shield's transfer impedance [see Eqs. (5.40) through (5.42)]. But having stated this, what should we look at while considering grounding schemes for the same shield? Connections to the ground change the loads and configuration of the external shield circuits and thus change the current distribution at the shield external surface, which is a decisive factor in the shielded line radiation. Indeed, in the "regular" antenna regime, the radiator's ends are open (for antenna current), the standing wave pattern is formed, and the ends of antenna always would have carried a zero current node.

Shielding Engineering 553

By contrast, grounding may provide a mechanism for re-distributing this current or storing electrical charges at certain points, e.g., the antenna ends, and facilitating the uniform current distribution. For instance this way, the grounding may even result in a change of antenna radiation regime, e.g., from a resonant dipole to a traveling wave current distribution pattern. As if the situation weren't complex enough, there is an additional important variable associated with the presence of the ground plane: we have already seen that the ground plane exerts a fundamental influence on the radiating fields and radiation resistance of current-carrying conductors. All these factors seriously affect the emission and susceptibility characteristics of the shielded cables and must be accounted for when the shielding-related grounding philosophy and architecture are considered.

Bringing the two rules together, we see that the ideal result would have been achieved if we could close the internal circuit with the least possible obstacles facilitating the internal differential-mode current, and interrupt the external circuit reducing the antenna current. While the first rule is kind of "absolute" in nature originating from the necessity to provide an unimpeded signal propagation path, the second rule is associated with numerous complex effects. Since the change in grounding conditions of the shield results in different current distributions and different functions for the radiation resistance, the tool to evaluate the related effects can be derived on the basis of the item A_r in Eq. (5.42) — which denotes numerous complex and system- and line-specific properties of a radiating system.

Consider now a line that carries signal current I_i and antenna current I_o. [As we saw, these currents are not the same—see Eq. (5.38) and related text]. The line is placed at height h over a ground plane. It is considered matched as far as the internal circuit (differential transmission mode) is concerned, but this does not automatically mean that the line's antenna circuit (common-mode) is matched too! We will look into two cases of the line loading: first, where the line's loads are connected to the ground and, second, where the line's loads are not connected to the ground. The line load connections are extremely important because they determine the necessity of grounding the shield (see rule no. 1) and affect the shield modeling by respective antennas.

First consider a line over a ground plane with ground connections of the load (Fig. 4.20). When nonshielded, such a line with its image can be modeled by a loop (e.g., a rectangular loop) antenna configuration. The shield, if applied, must be connected to ground by the virtue of its connection to the line (according to rule no. 1). Thus, in this case, the shield is always grounded, unless special techniques are used (e.g., separation of the high- and low-frequency grounding). Then, assuming that the shield diameter is comparable to the line diameter, antenna current conditions at the shield surface and on the unshielded line surface are identical. Per Eqs. (5.41) and (5.42), this amounts to $\xi_i = \xi_s$, and so the shielding effectiveness is almost completely determined by the shield transfer impedance (the "almost" relates to the effect of changing the characteristic impedance of the shielded line, but this effect is limited).

When the line loads are not connected to ground (Fig. 7.32), we have different options of terminating the shield: no ground at all, ground at both ends (using "zero" impedance grounding, "matched" impedance grounding that ensures a traveling antenna wave at the shield external surface, or via any impedance including the impedance of the ground connections), ground at one end (at the beginning or end of the line), or continuously ground the shield. Each case results in different ratios of items ζ and R_r in Eqs. (5.41) and (5.42). A typical example of such a line is a balanced pair.

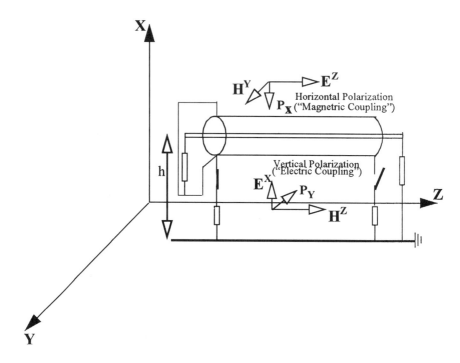

FIGURE 7.32 Shielded cable in radiating field in free space (compare with Fig. 4.20)

In terms of calculation, the simplest situation occurs when a nongrounded line is protected by a nongrounded shield. A nongrounded line over a ground plane can be modeled by a two-antenna array (the line and its image), and so it is an ungrounded shield. In this case, when the shield is connected to the line (e.g., via the center taps of the transformer) we can again assume that $\xi_i = \xi_s$, and the shielding effectiveness is determined only by the shield transfer impedance.

The most interesting case (although not necessarily the best in an EMC sense) is when a line runs over the ground plane with the line loads not connected to the ground, but with the shield grounded. A nongrounded line is modeled by a two-antenna array (the line and its image), while the shield is modeled either by a bent wire antenna over the ground plane (e.g., see Ref. [7.24]) if grounded only at one end, or by a loop if grounded at both ends.

To calculate the shielding coefficient and shielding effectiveness, Eqs. (5.41) and (5.42) can be used. The radiation resistance, R_{ri}, and the radiating field intensity, E_j,

Shielding Engineering 555

generated by a nonshielded line at some point in space is calculated using the "regular" antenna equations. Thus, for example, Eqs. (4.67) and (4.66) can be used for horizontal (that is, parallel to the ground plane), and Eqs. (4.69) and (4.68) can be used for vertical (that is, perpendicular to the ground plane) electrically short line of length, l. For a traveling wave signal along the line, Eqs. (4.64) and (4.62) can be used, and so on. These are the reference models to calculate the shielding effectiveness per Eqs. (5.41) or (5.42). Formulas for other practical configurations can be found elsewhere in the literature, or derived using respective analytical or numerical techniques.

When the line is shielded, the shield may be regarded as a radiating antenna, excited by voltage dV_{ex} and respectively modeled as required. Then, comparing these models with the reference models, the shielding effectiveness can be determined, and respective recommendations on the shield grounding can be made. The practical situations to consider here are so numerous and diverse with regard to the line electrical length, shield transfer function, and the value of the terminations, that only some general considerations can be spelled out here.

How important are the grounding-related changes of the antenna current distribution at the shield? Before we review some experimental data, consider the effect that a high magnetic permeability sleeve (say, a ferrite core) placed over a radiating conductor will have on the antenna current distribution at its surface. Since such a core presents a large impedance to this current, we can assume that at its placement location the current nodes (zeroes) appear. Thus, by moving the ferrite core along the line, we can continuously change the current distribution. If we install a current probe at some point along the line, we can directly measure the antenna current at this point. (If this reminds you of the absorber clamp, you are right—see Chapter 6). Of course, when the ferrite is extremely lossy, all the energy at its location is absorbed, and the proper point of the line can be viewed as matched. In this case, by placing two such ferrites at the ends of the wire, we can excite a traveling radiating wave along it. Obviously, in practical applications we won't often use such ferrites over a nonshielded line (because they attenuate the useful signal, as well), but their use over the line shield to reduce the radiating current (according to *Rule No. 2*) is a common practice.

Now to the "promised" experimental data — in fact, we already saw it. Revisit Fig. 6.14 in Chapter 6, which presents a set of curves obtained for the same line: shielded and nonshielded. There, we saw that the magnitude of the measured current is 50 to 60 dB smaller in shielded line than in the nonshielded one. However, at that time (that is, in Chapter 6) we did not mention that almost identical curves were obtained for both the non-shielded line and the shielded line when the latter was measured with the shield *not connected* to the line. (Isn't that what *Rule No. 1* aims to prevent?) The curves in Fig. 6.14 a correspond to the same line as Fig. 6.14 b, but with the shield connected to it. This proves once more that the shield connected to the line really "works"!

In Chapter 6, we have also noted the relative independence of the shape of the curves with regard to the position of the absorbing clamp along radiating line and and made a parallel with the maximum emissions from lines of different length. The conclusion was that varying the cable or assembly length will *not result* in the changes of maximum emissions (susceptibility), viewed in a wide frequency band. Now we can develop this concept further and simultaneously obtain a deeper insight into the grounding effects.

Indeed, one way to change the effective radiating current distribution at the radiator surface is by means of grounding. So, sometimes "ideas" are "generated" and circulated that by selecting the proper grounding points, the EMC performance of assemblies and cables can be improved. We see now, that in a general case of testing in a *wide band of frequencies*, this method will *not work*! Only in certain specific situations, when some *particular* frequencies or line lengths and positions (or quite *narrow ranges* of their variations) are critical, may we attempt to use different grounding schemes to vary the current distribution and corresponding emissions (susceptibility)—hopefully, with a positive result.

Similar conclusions were arrived at in Ref. [7.25], based on the so-called topological approach to shielding (see the next sections). There, grounding is considered in the context of the whole system shielding integrity, including the line shield.

Returning now to Fig. 2.10 in Chapter 2, we see that the effect of using a shield (connected to the line according to *Rule No. 1*) and / or changing its transfer function reaches about 50 dB, which is much more significant than the 2 to 10 dB resulting from varying the grounding conditions at the radiator ends.

7.6.3 Challenges of Shielded Cable Assembly

In this section, we will address the shield interconnection and termination in shielded enclosure / cable / connector assemblies. We will also formulate the shield termination Rule No. 3.

> ✦ In this regard it is appropriate to quote a familiar cliche: a shielded line is only as good as its connectors. And that may be true, because even the best shield EMC performance can be degraded by improperly designed, installed, or maintained shield terminations. Alternative statements are also true: a shielded line is only as good as its cable and terminal enclosure shields, as well as the whole shielding system integrity. However, the shield termination is often the "weakest" link in the shielding system.
> Also note, when the separate element contributions are evaluated into the whole assembly system EMC performance, there is kind of a paradox here: while the enclosure, cable, and connector shields are joined "*in series*", their interactions can be modeled and described by a *parallel* shielding model, as presented in Chapter 2. ✦

Now, to the "business". Have you ever twisted a pigtail? — the majority of electronics and cable professionals did!* The most often, a pigtail is an end section of cable braided shield, separated from the cable core and jacket, twisted to shape as a flexible conductor, and used for shield termination†. A pigtail can also be created by any sec-

---

*Please pardon the expression — it has nothing to do with the venerable animal, but rather with a shield termination technique, and besides, as you'll see we don't really recommend to do this. Although the way the pigtail is introduced here is rather a joke, in fact, we are discussing a very serious issue.

tion of usually stranded flexible conductor: cable drain wire in foil shields (see Chapter 3), or even by screws attaching the connector shielding enclosure to the product housing shield. By now, it is a known fact that pigtails mean "bad influence", that is, strong EMI coupling. However, to understand the problem, it is *not sufficient* to consider a separate element (cable or connector) shield which can be modeled at extremely *general* levels. We need provide a certain level of system specifics as determined by the model *relevancy*. A minimal relevancy model *level,* necessary to consider the pigtail effects, is that of a whole *assembly*: shielded enclosure / cable / connector.

There are several kinds of shield termination / shielded connector-related problems:
- connection effect on the line shield performance.
- connector attachment to the shield
- "connector-proper" performance as a radiating object
- coupling and crosstalk within the connector itself.

The majority of shield termination faults can usually be traced down to providing an *improper internal* current path, creating an *efficient external* (antenna) current path, and emitting/receiving signals *on their own*. However, there also may be more subtle, less obvious effects.

◆ As an example of such "subtle" effects, consider how the shield termination conditions affect the performance of multilayer shields. In Chapter 3, we determined the transfer function of multilayer shields by considering an infinitely small line element in an infinitely long and uniform (no variation with the line length) line. In principle, we could have solved this problem similarly to that of crosstalk, by viewing each of the shield layers as an element of the respective third circuit (tertiary). However, since our assumptions implied that any third circuit formed by the shield layers was matched (that is, was loaded on its characteristic impedance), we were able to arrive at the solution in a simpler way by equating the voltages at the adjacent surfaces of the shield layers. Thus, we were able to write, almost without derivations, many equations in Chapter 3.

With the integration of the transfer function into a more system-specific model, which accounts for the line finite length and different shield termination and grounding, we can no longer assume the matching conditions. In this case, the arbitrary loads at the ends of the shield interlayer tertiaries result in resonant phenomena (e.g., see Figs. 5.21 and 5.22), which may significantly affect the shielding effectiveness. Indeed, a rigorous solution for double- and triple-shielded cables [7.25] reveals the existence of such resonances.

As a rule, the presence of the dielectric between the shield layers leads to an improvement in shielding performance, compared to a single-layer shield of the same thickness as the total thickness of all multiple layers, but at the resonances this improvement is eliminated. The resonant frequencies are inversely proportional to the product of line length by the square root of the dielectric constant of the insulation between the layers, and the amplitude of the resonances increases with the insulation thickness. The resonances can be damped using a lossy dielectric. In practice, even with regular dielectrics, the interlayer circuits appear quite lossy anyway, which limits the effect of the resonances to only several decibels.◆

---

†There are also more general interpretations of a pigtail as *any* short length of metallic, or even fiber optic line connecting the "main" line to other line or terminal — but we will limit our interpretation to shielding applications only

558 Electromagnetic Shielding Handbook

Even a superficial look at the Fig. 7. 32, reveals potential basic shield termination / connector-related problems:

- The termination is not a perfect contact but presents an impedance (Z_g). Especially harmful can be the connector connection inductance, which creates an impedance increase with frequency. In the most severe cases, the connection of the shield to the shielded line is completely violated, which may lead to extremely high emissions and system malfunction (see *Rule No. 1*).
- Connector design and installation techniques can facilitate the external (antenna) current at the line shield.
- Connector geometry can violate the coaxial structure of the shield. For example, considering Fig. 4.8, Kaden [P.1] showed that such violations (excentricity) in a signal-carrying line will result in the external EMI field.
- If the connector does not provide an "electrically hermetic" enclosure, the connector itself may radiate.

Some of these problems are illustrated with the experimental data. A 3 m long flat line assembly was placed at an open-area test site (OATS) horizontally over the ground plane, and a GSG line in the line was powered by a sweep generator in the frequency range of 30 to 1000 MHz. The generator and the load were connected to the line using connectors with metallic front- and backshells. The contact between the connector shielding shells and the line shield was secured by special metallic springs. The radiating field intensity was measured by a horizontally polarized receive antenna at 3 m distance from the line and fed into the spectrum analyzer.

The envelopes of the maximum values of the field intensity obtained in different connector configurations are presented in Figs. 7.33 and 7.34.

Also shown are the nonshielded assembly emissions. As shown, when the shielding system continuity from the line to connector is violated (by removing the contact either between the connector backshells and the line shield, or between the connector frontshells and the terminal enclosure shield), the emissions are almost as large as in nonshielded assembly (see *Rule No. 1*). That means, that when improperly implemented, the shielding is *useless* (and actually may be harmful — as manifested by *negative* shielding effectiveness †we saw examples of this in Chapter 2 and in this Chapter). By improving the shield continuity, the emissions are consistently reduced, resulting in the shielding effectiveness about 50 to 70 dB, especially in the lower part of the frequency band.

With the frequency rise, even the smallest inconsistencies in the shield continuity may drastically *erode* the assembly shielding system performance. Just look how the performance of a "regular" shielded assembly in Fig. 7.33 is effectively degraded at frequencies over 100-200 MHz (the radiating emissions increase). The data in Fig. 7.34 identify one of important reasons for such degradation: the front shell connection between the connector and the terminal housing.

As we saw in the previous section, the shielded line-connector assembly problems are usually complicated by the "overlapping" functions of the shield termination to the

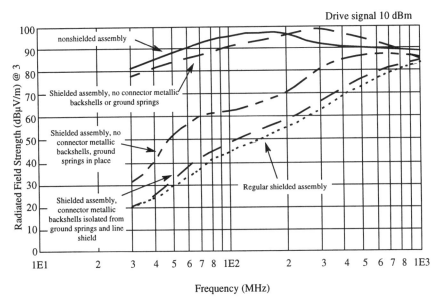

FIGURE 7.33 Radiating emissions from shielded flat line/D-connector assemblies, GSG powered

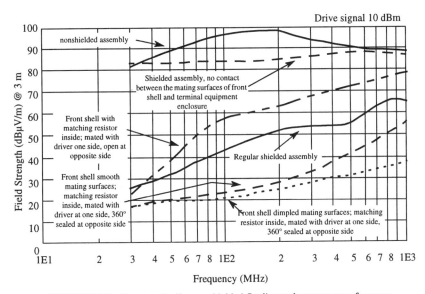

FIGURE 7.34 Front shell effect on shielded flat line and connector performance

line and grounding. Indeed, for safety and/or EMI reasons, the equipment metallic housings may be grounded. By the virtue of terminating the line shield to this housings the ground potential is supplied to the line shield, independent of whether it is "useful" or not for the line EMI. For this reason, when the shield is connected to the housing, it is not immediately obvious whether it is its termination or grounding (mainly, the created ground loop), or both, that determine the line worse case EMC performance. If the shield must be connected to a grounded housing but such grounding negatively affects the line EMI, often a hybrid grounding scheme or a double-shield with isolated layers can improve the EMC performance of the line.

By now, "even homemakers and babysitters" know that a TV line or a personal computer interconnection shield must be "360° terminated." Another question is, why do we need to do this, and what happens when it is not done? To illustrate potential problems with the shield termination, first consider a device that not only does not have the 360° shield termination but whose action is based on the *absence* of such termination. This is a magnetic field probe (Fig. 7.35), widely used for EMC troubleshooting. While the body of the probe is shielded, its tip provides a non-compensated line current. While such a probe is an extremely useful instrument, system engineers are hardly interested in reproducing it in the electronic system!

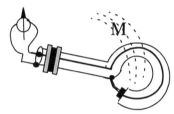

FIGURE 7.35   Magnetic field probe

A notorious unintended equivalent of such a probe is a "pigtail" line shield termination. The term pigtail originally designated a length of the shield (usually braid) separated from the line at one end and twisted to form a wire conductor. However, this designation later was expanded to mean any wire (including, e.g., the drain wire in foil shields) that terminates the line or connector shield.

Although a pigtail termination is the simplest to provide, its effects are far from simple. Ample research conducted on the subject, including the latest works (e.g., see Refs. [4.12-13, 7.26-31, 5.18] and many others), serves as a solid proof of the complexity of the problem. We have already had an experimental "taste" of the pigtail effects in Chapter 2 (see the descriptions and Figs. 2.11 and 2.12). Because of the extensive literature available on the effects of pigtail connections, we will concentrate only on the most general considerations and several important applications.

At first glance, the problem is not all that "bad." Return to Fig. 7.35. Neglecting the radiation from the shielded part, the radiating (sensing) part of the probe can be approximated by the loop antenna, as a rule an electrically infinitesimal one. Equations (4.46) through (4.49) provide the necessary "ammunition" to calculate the field of such an antenna, which is proportional to frequency square, the loop area, and the current in it (again, don't forget that this is the antenna current). However, more thorough research shows that we cannot neglect the effect of the pigtail on the radiation from the shielded part of the radiator. Indeed, it follows from Fig. 7.35 that all the voltage V_p generated in the pigtail contour is applied to the shield external surface. But we know that such voltage is a source of antenna currents on the shield. In a shielded line, this voltage produces antenna currents simultaneously, and atop of the voltage dV_{se} penetrating from the shielded line throughout the shield {see Eq. (5.39)}. However, unlike dV_{se}, which is reduced by the shield transfer function, voltage V_p does not depend on the shield design; rather, it is proportional only to the area of the pigtail loop (i.e., for an infinitesimal loop; for electrically large loops the processes are much more complicated, but the formulas are also available—see the previous sections).

It follows that the pigtail-related emissions from a line assembly can be modeled with an antenna array consisting of two parts: the pigtail itself ("loop" antenna) and the emissions from the line shield to which the pigtail voltage is applied. Since the emissions from both array ingredients are proportional to the voltage generated by the pigtail, we can expect the line assembly emissions to be proportional to the area of the loop. With regard to the frequency dependence of the pigtail effects, we should remember that while the voltage developed by a small pigtail is proportional to the frequency, the radiation resistance of linear and loop radiators is also proportional to the frequency, but in respective power {e.g., see Eqs. (4.58) and (4.59)}. This dependence is further modified by the dependence of the radiating properties, at a given antenna current, on the size of the radiator, as well as the presence of other conductors and ground planes.

In Chapter 1 we presented several examples of radiation patterns and maximum emission boundaries of some typical single radiators and combined linear/loop arrays.(For example, see Figs. 1.30 through 1.35). Here, with experimental data, we will illustrate the effect of the pigtails on the emissions from a flat line shielded assembly. A shielded flat line line (GSG) was terminated with the ends to BNC connectors inside a shielded box (see Fig. 7.36). The BNC ends of the assembly were connected with good quality coaxial cables to the signal generator and to the matched load. The line shield at the generator end was 360° connected to the housing, while at the opposite end the connection was either 360°, or different length pigtails, or completely severed.

The test results are presented in Fig. 7.37 and are obvious. Other experiments (not shown here) indicated that, with the increase of the number of pigtails, the emissions are reduced. In this case, the uniform distribution of the pigtails around the shield perimeter was essential.

Now we are ready to formulate:

Rule No. 3: *A necessary condition for suppressing the radiation from or susceptibility of a shielded electronic system is as electrically tight as possible whole shielding system, including the line shield terminations to the connector's shield and connector shield termination to the shields of other electronic system elements. Ideally, the 360° termination should be approached as closely as possible.*

Using the formulated three *Rules*, we can apply the developed so far principles to real life shielding systems. But this is easier to say than to do and practically such analysis should be done for each particular application. We will follow up this statement with several important illustrations.

562 Electromagnetic Shielding Handbook

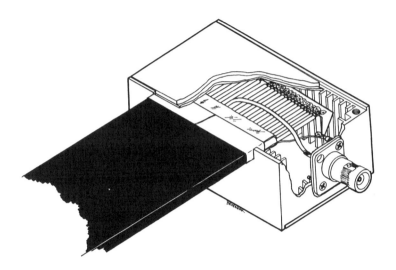

FIGURE 7.36 Shielded flat cable line terminated to BNC connectors

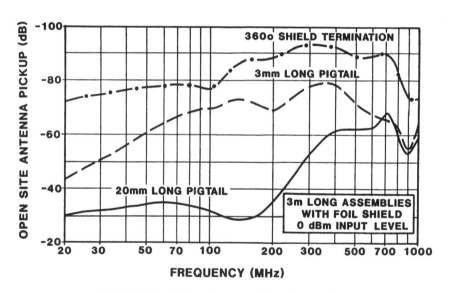

FIGURE 7.37 EMC performance of flat cable assemblies

Fig. 7.38 illustrates some of the related concepts. A shielded single-ended (or a coaxial) line carries high frequency signal to a PCB placed inside a shielding enclosure.

Shielding Engineering 563

In Fig. 7.38 a, almost no signal current returns through the ground (only what "leaks" through the shield as determined by the shield transfer function, in particular, by transfer impedance). This is the closest to ideal situation. Provided that all the element shields' transfer functions are good, it yields excellent protection, *independent of whether the shield is grounded or not*! Indeed, if the shield is grounded, say, at the ends of the line, *and* even if there exist a potential difference between the grounding points (that is, a ground loop present), the generated in the shield current will have an *external* return path, and thus its effect on the shielded conductors will be reduced as determined by the shield's transfer function. That's a nice protection from the ground loop! At the diagram, two grounding mechnisms are illustrated: galvanic and capacitive coupling. The galvanic coupling corresponds to the switches at the diagram — no coupling when the switches are open. In the low frequencies, this is often the main, if not the only, grounding mechanism. At high frequencies the shield may be effectively grounded via the respective capacitive coupling, which *cannot* be "turned of"!

In Fig. 7.38 b, the shield is terminated to the housing (at the point of the cable entrance *inside* the housing) by a pigtail. This spells trouble, because now not only the assembly EMC performance is affected by "leakage" from the housing aperture, but also a part of the signal current returns via "ground" even with the "best" shields, i.e., the lowest transfer impedance of the shield "*per se*" will not result in the higher system shielding effectiveness. Practically, you can use the transfer impedance parameter to characterize the leakage from the pig tail-terminated shield by measuring (or calculating) an "integral" transfer impedance parameter, with the pigtail termination effectincluded!

By all accounts, the worse shield termination situation is shown in Fig. 7.38 c, where the shield is terminated by a pigtail *inside* the housing. Such scheme can be a source of extremely bad EMC performance. First, as shown in the diagram, the EMI current at the shield's external surface (which originally had an external return path) is brought inside the shielding enclosure where it couples into the shielding enclosure walls creating in them current i with *internal* return path. Current i is associated with large radiating loops inside enclosure which efficiently "crosstalk" with the shielded lines! Thus, the shield does not provide any protection from current i for the shielded circuits, because both current i and the shielded sircuits are located *at the same side* of the shielding enclosure. The second "bad news" is that under certain conditions, the shield also may act as a secondary EMI *radiator*. The common mode signal from the *exposed* section of the line is coupled to this radiator in two ways: the common mode *potential* at the end of the line is *galvanically* transfered to the enclosure, and common mode potentials and currents are capacitively and magnetically coupled to the enclosure walls. As a rule, the enclosure will be a *large*r and often a *more efficient* radiator than the exposed section of the line. Just imagine what happens when the enclosure dimensions correspond to the resonant wavelength!

No wonder, such shield termination may result even in *negative shielding effectiveness* leading to extremely large product emissions and susceptibility. We have seen the examples of this effects in the preceding chapters (e.g., see Figs. 2.15-16).

564 Electromagnetic Shielding Handbook

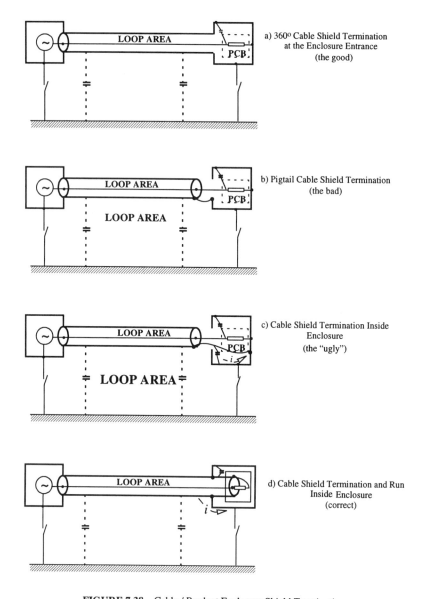

FIGURE 7.38 Cable / Product Enclosure Shield Termination

But what do we do, when we *need* to bring a shielded cable inside the product shielding enclosure? Obviously, if we bring a cable shield inside the enclosure without terminating it, the situation will be no different from that in Fig. 7.38 c — which is absolutely unacceptable! Of course, the solution is to terminate the cable shield to the enclosure at the *point of entrance*, as shown in Fig. 7.38 d. As you can see, now the EMI current i is returning via the external shielding enclosure surface (that is, with an external return path) and the shielded circuits are protected by the shield. Also, by now we know that such termination should be 360° - "tight". Only then we can continue to lead the cable lines *inside* the enclosure, *shielded or unshielded* — depending on the amount of protection for the lines that we need *within* the enclosure.

So, here comes

Rule No. 4: *At the penetration into the product shielding housing, the cable shield must be 360° bonded to the housing at the <u>entrance</u> to the housing.*

✦ While our analysis of pigtail contribution to the assembly emissions is based on the application of antenna theory, other ways of tackling the same problems are also possible. Lately, we see abundant applications of numerical techniques to resolve *specific* (as opposed to general analysis) problems. For example, in Ref. [7.27], the external problem of electrically small pigtail geometries was modeled by lumped voltage thin wire circuits and computed using the moment method. The major, essentially "antenna," characteristics of different pigtail connections were calculated: antenna current distributions at the shield, radiated pattern directivity, radiated power, and shielding effectiveness (as a power ratio of emissions from nonshielded line to emissions from shielded line with different length of pigtail connection). The main conclusions are:

- The type of pigtail connection is crucial with respect to the shielding effectiveness.
- Multiple pigtails are more effective than single one.
- A long pigtail may enhance radiation. (Negative shielding effectiveness!)
- Different load impedances may cause changes in shielding effectiveness.
- The directivity is insensitive to variations in pigtail dimensions.

These conclusions are favorably comparable to respective experimental data obtained in the literature [7.26].

7.6.4 Shielding and Ground Loops

This is one of the most "obscure" and controversial subjects in shielding. But it doesn't have to be such, because there is nothing "mysterious" in ground loops. More so, that the subject really is not that complex, much simpler than some "other" shielding problems! In this section we will review several typical ground loop-related situations. We will illustrate at least some of them with experimental data, as well as potential mitigation techniques.

Fig. 7.39 is based on certain simplification and generalization. It presents circuit diagrams of three lines corresponding to practical applications: non-shielded single-

566 Electromagnetic Shielding Handbook

ended line, shielded single-ended line (a coaxial shielded line is shown), and balanced line.

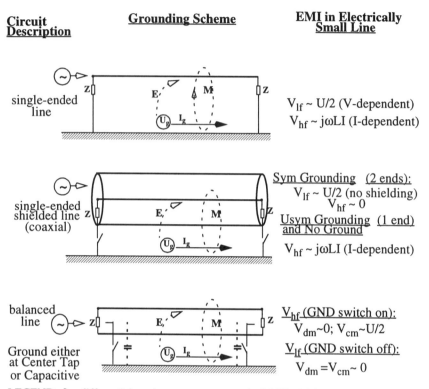

FIGURE 7.39 Ground loops in interconnections

Common feature for all the diagrams is the "ground" which carries *externally* generated EMI current I and voltage drop U. As we saw, this "ground" can be either a ground plane (e.g., as in PCB), or a second line conductor which is common to both lines (remember GSGS circuit configurations?), or just about any metallic conductor in "close proximity" (to be able to provide for galvanic or reactive electric and magnetic coupling), or even Earth. What is important in this case, that the source of voltage u and current i is "extraneous" with regard to the considered lines, that is, the EMI current i and voltage u are generated by the "other" circuit returns or just by crosstalk between these other circuit and this "ground".

The ground between the grounding connections of the line in Fig. 7.39 is part of the ground loops formed by the respective lines. Brief clarifications in Fig. 7.39 provide the necessary explanations.

A single-ended nonshielded line in Fig. 7.39 *must* incorporate the load ground connection. Otherwise, the circuit will not be closed and there will be no return at all: so the line will not be functional. Thus, the line ends are shown *galvanically connected* to the ground.

In this case, the voltage U is the source of galvanic coupling between the "other" line (which generated this voltage) and "our" single-ended line, as described in section 4.4.3. The circuit diagram of such coupling is similar to that of Fig. 4.4 c.

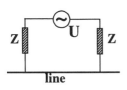

FIGURE 7.40 Galvanic coupling via ground loop

The coupling model corresponds to Fig. 7.40. It corresponds to the case when the ground and the line resistances are *relatively small* (comparing to that of the line load — thus the internal resistance of the source u is neglected and not shown). Practically, this is always the case, if the line length is small enough. It is clear from Fig. 7.40, why the voltage V_{lf} at each of the line end loads Z (they are assumed identical) is equal to U/2. This is true for both, high and low frequencies. However at low frequencies this voltage, and not the current, is the *main* EMI result in our single-ended line. This is, because the respective current-induced EMI is proportional to the frequency of the current variations (di/dt, or for sinusoidal variations — $j\omega LI$), and at low frequencies this item is negligibly small. However at high frequencies, the current I generated EMI will certainly exceed the item U/2, and this EMI mode becomes dominant.

✦ Fig. 7.41 illustrates the effect of the ground loop on electromagnetic emissions from a printed circuit board. Remember Figs 5.32 and 5.33 in Chapter 5? There we witnessed how the addition of a ground plane to a single-layer PCB resulted in drastic reduction of the radiated emissions over the whole investigated frequency band, and in meeting the FCC limits for the previously failing card. There we also "promised" to return to the same test case in chapter 7 and clarify the issue with the "10 ground points". Now the time has come! Indeed, why did we need the "10 ground points", not more nor less?

First of all, some clarifications are appropriate. In order to perform, the single-ended lines at the single-layer PCB were provided with adequate return traces in the form of a *grid*. However, by now we know that even better return path is provided by a ground plane, as long as the ground plane is placed close enough to the direct

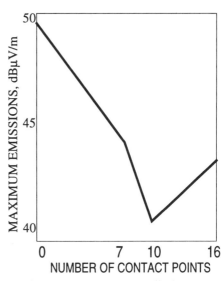

FIGURE 7.41 Ground plane effectiveness

paths and galvanic connections are made between the line loads and the ground plane. By connecting the points of the PCB ground grid to the retrofit ground plane, such galvanic connections were secured. It seems, that the more grounding points are connected to the ground plane, the better returns will be provided and the less will be radiation from the board.

As Fig. 7.41 shows, in the tested PCB this was really the case, however only to a certain number of points! A possible explanation of this fact can be arrived upon, if we remember that different points of the ground grid may have different potentials. Then, when increasing the number of grounding points (which indeed leads to the *improved* return path), there is a possibility that some grounding points with large enough potential difference will be connected. The generated by this levelling-off current becomes *a source* of additional radiation, which would not be present without such connection. Of course, practically such points with large potential can be either calculated (in simple enough circuits) or tested experimentally, by measuring the voltage differential between different grounding points. ✦

Consider now a shielded single-ended line in Fig. 7.39. Such a line *doesn't have*, but *can* incorporate the ground connections at line ends, because the line load can be *connected* either to the ground or to the shield (or both). Now, two options must be considered: when the shield is grounded and when it floats. Obviously, for the voltage U to affect the line, both its ends must be grounded. On the other hand, the magnetic field generated by current I couples to the line whether it is grounded or not.

I. Consider first the shield grounded at both ends - the so called *symmetric grounding*. In our equivalent diagram this corresponds to the switches in the ground connections being closed.

- In the low frequencies, the transfer impedance of the shield can be equated to its DC resistance (see Chapter 3). Assuming that the ground resistance is *relatively small* (comparing to that of the line *and* the shield), an identical to the previous case situation arises with regard to the low frequency voltage: $V_{lf} = U/2$. This means, that the shield *does not* protect from galvanic coupling via the ground loop, *independent* of whether the line is "grounded" or not! (This *is* the "scary" situation which "instills terror" in those users and even designers, which may not understand the origins of this trouble).

- At higher frequencies, both U- and I-generated EMI should be accounted for. When the shield is grounded at two points (symmetric grounding), the circuit diagram related to the effect of voltage U in the circuit ground-shield, is identical to the definition of transfer impedance in Fig. 3.2 of Chapter 3. This directly relates the corresponding effects of the voltage u to the transfer impedance and the transfer admittance of the shield. On the other hand, the EMI effects of the current I at high frequencies the ground loop effect is determined by the magnetic coupling. . When the shield is grounded at two points (symmetric grounding) the magnetic coupling to the shield/ground and line/ground circuits is identical. This happens because this coupling is determined by the same magnetic flux — from the circuit theory point of view, this is a reflection of the fact that the *line external inductance* is equal to the *mutual inductance* between the line and the shield. Therefore, the shield is *ideal*: $V_{hf} = 0$. If only one end of the line is grounded, there is no closed circuit for the current in the shield/ground system — and the shield again is useless: $V_{hf} = j\omega Li$

II. When the shield is grounded at one end (asymmetric grounding), the circuit diagram related to the effect of voltage U in the circuit ground-shield, is identical to the definition of transfer admittance in Fig. 3.2 of Chapter 3. Only when a metallic shield will have apertures, will the electric field associated with voltage U penetrate inside the shielding enclosure. However, the EMI currents generated by magnetic field of the current I will not have a return in the shield, and thus will return via the shielded line — a very undesirable situation. Now, in the case of asymmetric grounding, we have a "tag of war" between two "evils": the EMI associated with the voltage U will be smaller, but EMI associated with the current I will be larger! Since the first factor is relatively independent on the frequency and the second — grows with frequency, we can expect that as the frequency rises, the current-associated interference becomes more significant. However, don't forget about the existance of capacitive coupling of the shield to the ground: at high enough frequencies, even a non-grounded shield will behave as grounded! At higher frequencies, the shield "grounding" can, and almost always will be realized by capacitive coupling between the shield and ground (whatever it is: plane, conductor, etc). Since the "capacitive grounding" takes place whether we "want it or not", the shielding effect from magnetic loop will always take place, of course, according to obvious frequency patterns.

III. Probably by now, there is no need for explanations of the ungrounded shielding: the same considerations can be used as in the previous two cases — symmetrical and asymmetrical grounding.

✦ But suppose that by properly grounding the shielded cable we have provided EMI protection to our shielded circuits from ground loop effects. Is our EMI protection job done? Not yet! Because EMC performance is determined not only by the EMI reduction in particular system elements, but by immunity and emissions of the system as a whole.

A case in point is presented in Fig. 7.42. In a multi-media computer system, the interconnection between the host computer and the VCR is made using two coaxial cables: audio and video. In this particular design, while at the plug-in card in the host computer shields of both cables are connected to the respective logic grounds, in the VCR they are both connected the chassis. Because potential difference existed between the two logic grounds in the host computer, a ground loop was generated consisteng from the shields of both coaxial cables. That this was the case, was confirmed by disconnecting the cables (see insert in Fig. 7.42).

As soon as the cause is identified, the solution to the problem is relatively easy: two 1000 pF capacitors connecting the cable shields to the computer chassis was all that's needed! ✦

Consider now a non-shielded balanced line. This line is either not grounded all together, or is grounded at the center taps. Indeed, a balanced line, by definition, does not have a galvanic connection to the ground plane (or rather it *may* have such a connection, but made in a "balanced" way, through the line "center taps", as shown in Fig. 7.39).

570 Electromagnetic Shielding Handbook

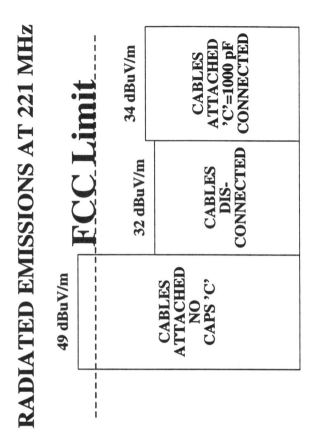

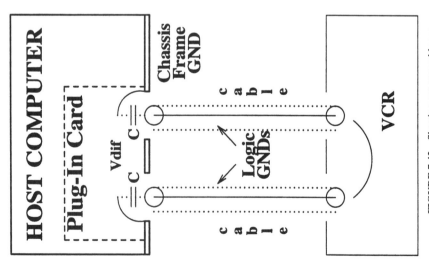

FIGURE 7.42 Shorting out ground loops

Since a non-grounded balanced line has no ground connections, the question is in order: where "in the world" the ground loop comes from? At higher frequencies, the balanced line "grounding" can, and almost always will be realized by *capacitive coupling* to the "ground" (see Fig. 4.4). "Almost" relates to the case, when there is no ground or grounded conductors in the vicinity of the balanced line — practically "almost" an impossible case. Of course, we must assume that both line conductors have *identical* coupling to ground — otherwise the line will not be balanced. the shielding effect from magnetic loop will always take place, of course, to a certain degree. This means, that the differential mode coupling will *always be zero:* $V_{dmlf} = V_{dmhf} = 0$).

The same is true (i.e., $V_{dmlf} = V_{dmhf} = 0$), when the ground connection in balanced line is provided at center taps. However, for common mode currents, the center tap connection results in $V_{cm} = u/2$, that is the shield is useless to protect from such ground loop. That should be clear! Indeed, isn't it how we have introduced common and differential modes in Fig. 1.17 of Chapter 1?

What do we do to reduce the common mode EMI and break respective ground loops associated with shielding? In principle, the same means are used, as in non-shielded interconnections. These include transformers (balanced-to balanced or baluns), inductive and lossy longitudinal chokes, line balancing and use of amplifiers with high common mode rejection ratio (CMRR), use of optocouples — the "usual" means of an EMC engineer.

✦ As an illustration, consider the use of a shielded balanced cable in electrocardiogram (EKG) monitor [7.32] — see Fig. 7.43. Practically, the EKG signal is extremely small: about 1 mv in adults with down to 0.05 mV in an unborn baby. The problem is that the frequency of EKG signal is around 50-70 Hz, while the electric power supplied to the cardiac monitor is 110-240 V of 50-60 Hz! Thus to provide a modest enogh 20 dB of signal-to-noise ratio, at least 120 dB noise protection margin is necessary. For this reason, not only shielded balanced cables must be used, but some means should be applied of preventing the common mode currents induced in the cable shield shield to close via ground loops in the grounding system and power network. In particular, in [7.32] the application of an isolation amplifier is suggested based on the use of a linear optocoupler to reduce the common mode and break the ground loop.

7.6.5 Shielding and Grounding System Topology and Partition

So far in this section, we have considered *separate* issues of shield design and grounding. But our goal, as defined in Chapter 2, is the *shielding system as a whole.* Therefore, it is appropriate to return to the section 2.5 of Chapter 2 and revisit our shielding models: "Series" model of system electromagnetic shielding and "Parallel bypass" model of energy transfer through the shield. As you see, *in principle,* our task of shielding system integration, synthesis, and analysis (e.g., see section 1.1.3 in Chapter 1for basic concepts) isn't really that complex, because these models do address the system issues. In fact, we have used these models as a blueprint to device roadmaps for selecting the topics and approaches which we have covered in the book so far.

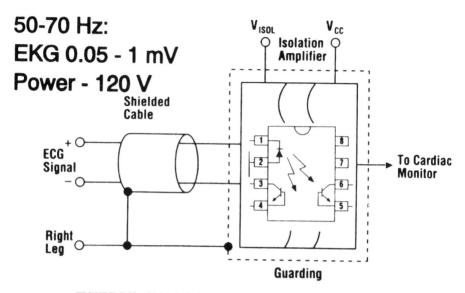

FIGURE 7.43 Using isolation amplifier and grounding in EKG monitor

However, this does not mean that our work is done: there may be a large distance between the "principles" and their realization — often, "the devil is in the detail!"

So, return to Fig. 2.21. The diagram indicates at several important shielding system imperfections (by far, not all!), as well as related physical processes which may have negative effects on the shield performance, including the ground loop "bypass" path and cable-"carried" interference. Along with these negative effects, several appropriate mitigation techniques are shown: gaskets, filtering, and "EMI hardened" system elements, e.g., shielded cables. Now, after five more chapters of the book, we can widely expand both the list of shielding system imperfections and the respective mitigation techniques, as well as explain and evaluate them.

In the course of our discussions, we analyzed and illustrated them with numerous examples. Often we had to simplify the task to emphasize the problems at hand, to make our points clear. However in many real life complex systems, such simplification can be a challenge, and special techniques are necessary to bring-in some order and make the problems more manageable. One such technique which permits to decompose a system into a set of simpler systems is based on the application of topological approach. The basic principles of topological approach were formulated originally for EMP and lightning protection applications, but then developed in a more or less independent discipline [7. 33-38].

Shielding Engineering 573

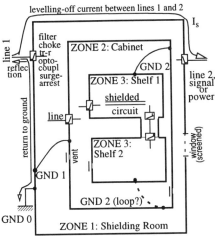

FIGURE 7.44 Shielding zone layout for EMC

Briefly, here's what this is about (see Fig. 7.44). Often, a single enclosure cannot adequately resolve all the EMI issues in a complex system (and indeed, you witnessed the often far from desired shielding performance as well as possible "disasters" with "real life" shielding in the previous chapters). Then, to achieve a greater degree of interference control, the space can be partitioned with additional *nested* shields into several levels of electromagnetic environmental zones. Each zone possesses a certain degree of shielding system integrity with regard to the adjacent zones. And so are combinations of zones — e.g., Shelves 1 and 2 joined by a shielded interconnection. The zones may contain seams, apertures (cable penetrations, vents), windows — all must be properly treated to secure the shielding system integrity. If the zone levels are identified starting from Zone 0 for "external" non-shielded space and up to the Nth Zone, which is enclosed into N-1 lower level zones, then, the higher is the zone order, the more electromagnetic protection is expected to provide to the enclosed in it circuits.

The basic premise is that at each level the shielding enclosure must provide a reliable electromagnetic barrier that *mutually* isolates the enclosed and external space. However, the inherent and structural imperfections, combined with bringing in (or out) power and signal, as well as electromechanical services, ventilation, fire protection, etc., compromises the shielding system integrity. This is being prevented by respective "remedies". Some of them are indicated in Fig. 7.44, while in general, the picture corresponds to that in Fig. 2.21.

A separate important issue is with *grounding* (GND 1-2) or *earthing* (GND 0, per notation in Fig. 7.44). As we know, we may need grounding for electrical safety, to support electrostatic shielding, and to provide the signal current returns in certain designs, as we have discussed above. However, by now we know enough about the subject to all but, by and large, agree with the following statement [7.36]:

"... Grounding has little to do with interference control, although improper grounding may, indeed, introduce additional sources of interference."

In this respect, Fig. 7.43 illustrates the potential for a ground loop between the ground connections "GND 2" and "GND 2 (loop)?".

Another potential problem indicated in Fig. 7.44, originates when there is a potential difference between the shields of the lines entering the shielding enclosure. As shown in Fig. 7.44, a levelling-off current I_s is generated between the shields of lines

574 Electromagnetic Shielding Handbook

1 and 2, terminated at different parts of the enclosure. Of course, the current I_s "flowing at the external surface" of the enclosure (that is, it has an external return path) is a source of radiation. Practically, such situation often have place, when the signal lines are placed at the front panel of the device, while the power supply line (together with a filter and a switch) is mounted at the devices back (just look at some oscilloscopes and spectrum analyzers!). From the topological point of view (as contrasted to the pure electrical means, say the use of chokes over the shields carrying EMI currents), a way to reduce this radiation is to place *all* the entering shield terminations as close as possible. For instance, in the just quoted example with an ascilloscope or spectrum analyzer, both the signal lines and power supply lines should be placed in close proximity. However, in this case, we increase the potential for crosstalk between these lines — as always, EMC is fraught with technical conflicts!

As a result, we can now formulate

> **Rule No. 5**: *The shielding system topology should provide electromagnetic isolation between environmental zones adequately defined to provide for the whole shielding system integrity, to minimize the EMI current amplitude, return path length and associated loop area.*

7.6.6 Designing Shielding System

Based on the preceding discussions, we can spell out three fundamental premises of a practical shielding system design and grounding:

1. The general system design and optimization philosopy which were considered in the beginning of this chapter in sections 7.1 - 7.2
2. The process interaction per "blueprint" of Fig. 2.21, as suported by sections 7.3 - 7.6
3. The *topological* approach to shielding system analysis, as outlined in the previous subsection.

Using these premises in the context of different applications and objectives, the appropriate shielding system design methodologies can be developed — just as it is the case with other shielding-related issues discussed previously. With this in mind, we can all but repeat the five shielding design and grounding rules:

Summary: Five Rules of Shielding System Grounding and Termination

> **Rule No. 1**: *To suppress the radiation from or the susceptibility of the line enclosed in the shield, the shield must provide an efficient internal current return.*

> **Rule No. 2**: *To suppress the radiation from or susceptibility of a shield that encloses a line, the current in the shield external circuit (antenna current) should be reduced as much as possible.*

Rule No. 3: *A necessary ingredient for suppressing the radiation from or susceptibility of a shielded electronic system is as electrically tight as possible whole shielding system, including the line shield terminations to the connector's shield and connector shield termination to the shields of other electronic system elements. Ideally, the 360° termination should be approached as closely as possible.*

Rule No. 4: *At the penetration into the product shielding housing, the cable shield must be 360° bonded to the housing at the <u>entrance</u> to the housing.*

Rule No. 5: *The shielding system topology should provide electromagnetic isolation between environmental zones adequately defined to provide for the whole shielding system integrity, to minimize the EMI current amplitude, return path length and associated loop area.*

Example

Fig. 7.45 illustrates the recommended approach with a case study example. An instructive detail in *this particular case*, is that when *only* either the product shielding enclosure or the filter are used, the system performance is almost no different from the top curve — original design with no "fixes". *Only* when *both*, the shielding enclosure *and* the filter are used *together*, the positive effect is achieved.

Of course, this might be different for other designs, depending on the relative significance of the involved EMI mechanisms.

7.7 SHIELDING PERFORMANCE STABILITY AND RELIABILITY

7.7.1 Electromagnetic Shield in Physical Environment

During manufacturing, construction, and installation, as well as their service life, product shields are subjected to all kinds of mechanical, chemical, and physical impacts that affect their performance. The electrical reliability of a shield can be defined as its ability to retain the required shielding properties under the effect of specified environmental conditions. It implicitly follows that, depending on the goals of the shielding application, there also may be suggested alternative definitions of shield reliability: mechanical reliability, if the shield is mainly designed for mechanical protection (e.g., in a telephone pedestal or submarine cable); chemical reliability; nuclear radiation reliability; and so on. By and large, these aspects of shield performance will not be considered here.

At the beginning of this chapter, along with the shielding and transmission performance, Figs. 7.1, 7.3-4 list several important environmental parameters that determine the shield reliability and stability in the process of manufacturing, installation, and use (by no means a complete list). Although the tables addressed the cable braid in partic-

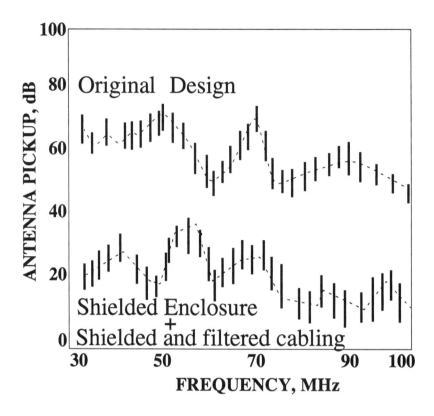

FIGURE 7.45 Combined effect of enclosure and cable shielding and cable penetration filtering on radiated emissions from a telecommunications system

ular, similar characteristics can be associated with any shield. Just which of these and/or other parameters should be used, how they affect the shield performance, and their limits are determined by the application specifics. For instance, the following shield classifications mimic the respective general electronic product classification with regard to their service environment characterized by the environment specifics and also may be overlaid by construction and maintenance conditions:

- Outside plant. In the outside plant, special requirements must be accommodated for extreme temperature zones, chemically aggressive or radioactive media, military applications, subway train vibrations, aerial wind and snow loads, endurance to corrosion and rodent attacks, underwater pressure, UV and nuclear irradiation, and others — you name it.
- Mobile environment with associated need for vibration tolerance, high and low temperature endurance (satellites), random mechanical loads

- Office plant: free stand, wall, rack-installed products. Some applications may require crush resistance (e.g., for elevator or undercarpet cables) and / or extreme flexibility.
- Special applications: outer space, industrial ovens, nuclear plants, and so on.

Obviously, such classification can be expanded, extended, and modified, depending on particular circumstances, goals and objectives.

To ensure the required performance in these conditions of the products in general, and their shielding in particular, a set of environmental performance parameters is used:

- tensile strength
- "cold flow"
- average and low temperature flexibility (cold bend)
- vibration, mechanical impact, crush, and peel-off resistance
- weather and moisture resistance
- abrasion resistance
- chemical resistance (corrosion in liquid and gas aggressive media)
- irradiation resistance (light, UV, EMP, nuclear)
- metal compatibility at joints.

We don't need to go into the details, because these are all too well known parameters of general physical environment, and they have been covered extensively in available sources. Just to name several of such sources:

- manufacturer and end-user catalogs and technical notes
- industrial, national, and international standards (e.g., ASTM, MIL-STD)
- books (e.g., Refs. [7.40-41])
- original articles, and particularly those about shielding reliability (e.g., see Refs. [7.42-45])

These sources provide a wealth of reference data and treat specific topics of shielding reliability with different degrees of detail. And, of course, the general principles of product reliability also apply (e.g., see Refs. [7.46–48]). For this reason, we will limit the discussion here to the illustration of only several important aspects of shielding reliability and performance stability.

7.7.2 Environmental Stability and Aging of Electromagnetic Shielding

A major source of shield performance variability is change in the physical environment. To appreciate the effects of the ambient environment on shield performance, just think of the Curie temperature when ferromagnetics change to paramagnetics—there goes our magnetostatic shielding, along with very essential changes in shield performance in quasi-stationary and electrodynamic regimes. A similar effect is produced by excessive and/or fast bending or a strong mechanical impact on a ferromagnetic shield.

Variations of shield parameters are observed under the impact of various environmental factors during product quality testing in laboratory conditions and in actual service. Some of these variations are temporary, vanishing when their cause is eliminated, whereas other variations result from steady degradation of the shield performance. For example, it is a well known fact that the shielding effectiveness of ferromagnetic shields varies with a change of the ambient temperature and/or applied field intensity (see Figs. 7.46 and 7.47), which affect the shield material magnetic permeability. However, when the environment changes are contained within certain limits, the shield performance variations are reversible—unlike the ferromagnetic's properties after the exposure to the Curie point. Of course, the processes in ferromagnetics are extremely complex, and the reader will be well advised to refer to the abundant original literature.

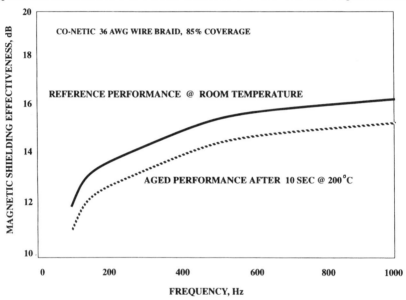

FIGURE 7.46 Temperature aging stability of magnetic shielding effectiveness

By and large, aside from the shield physical disintegration (e.g., by corrosion or gophers) one of the most harmful effects of the shielding enclosure exposure to environmental factors is caused by the increase of contact resistance at the joints and seams. As we have seen previously, the increase in the contact resistance at the enclosure joints can result in significant degradation of shielding performance. Just revisit Figs. 2.3 and 7.26. There, large contact resistance between the telecommunications cabinet body and its door had all but abated any shielding effect. Of course, in that particular case the abrupt violation of the contact was caused by inserting an isolating strip between the mating surfaces. However, a similar effect is often "achieved" by improperly (from the EMC standpoint) painting the product enclosure.

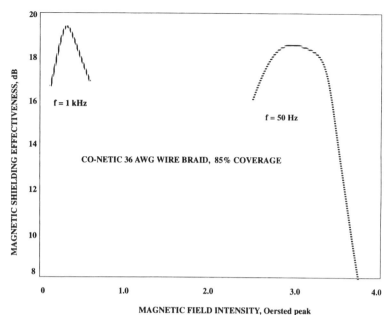

FIGURE 7.47 Field strength and frequency dependence of magnetic shielding effectiveness

In contrast, environmental effects usually result in gradual degradation of the contact resistance and related shielding properties. Such degradation is modelled and experimentally tested in laboratory conditions, by exposing the product to certain agents in a controlled environment. As an example, Figs. 7.48 and 7.49 illustrate the changes in point-to-point conductivity of several types of conducting coatings, exposed to the temperature / humidity and salt spray tests [7.49]. In this respect, one of the most "favorite" topics related to the shielding performance aging is that of the electromagnetic gasket performance stability (e.g., see [7.50-53]).

Table 7.2 **Copper Foil Test Results after 20 Years of Service**

Test Parameter	Units	Experimental Tape	Product Specification
Shear (constant force)	mm/hr@10.3 N/cm	<1.75	<1.00
Peel 180° (constant force)	mm/min@6.2 N/cm	<12	<250
Peel 90° (constant velocity)	N/cm@30 cm/min (avg)	<8.2	<4.8
Peel 180° (constant velocity)	N/cm@30 cm/min (avg)	<10.9	<4.4
Surface resistivity	Ohms/sq	20–150	10–100
Test temperature	°C	40°	23°

It is of great practical value to draw a parallel between the test data obtained in the laboratory and actual performance of the shield in the real life conditions. Table 7.2 presents results (see Ref. [7.44]) of tests conducted on a shielding tape after 20 years

580 Electromagnetic Shielding Handbook

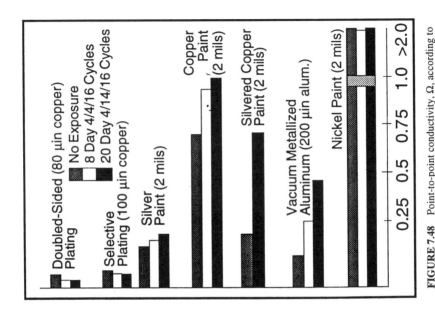

FIGURE 7.48 Point-to-point conductivity, Ω, according to temperature humidity test

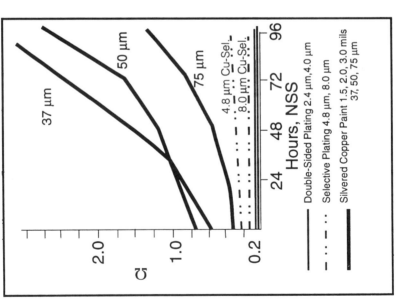

FIGURE 7.49 Point-to-point conductivity, Ω, according to salt spray test

of service in comparison with typical values obtained by measuring newly manufactured tapes.

The shielding tapes were 1" or 1/2" wide copper foils with a conductive adhesive layer, applied over the 1/8" seams between galvanized steel panels. In areas that had been dry throughout the building history, construction materials appeared corrosion-free, while the areas exposed to water over a prolonged time period were corroded.

According to these results, the tape and its adhesive performed well enough, in spite of the potentially aggressive zinc/copper combination. These results are indicative of the shield performance in an indoor environment in the presence of humidity and temperature cycling (in the range –4°C to +65°C between day and night, summer and winter) and in the absence of varying mechanical loads at the shield.

Much more damage to shield performance is inflicted by varying mechanical loads. Figure 7.50 presents the results of flexing the shielded portable cordage (foil shield) over a 3.8 cm mandrel. As shown, degradation of the transfer impedance is in direct progression with the number of flexes. Similar degradation is observed with multilayer shields (see Fig. 7.51).

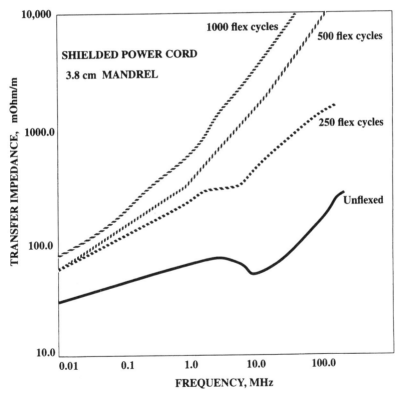

FIGURE 7.50 Shielding performance flex stability

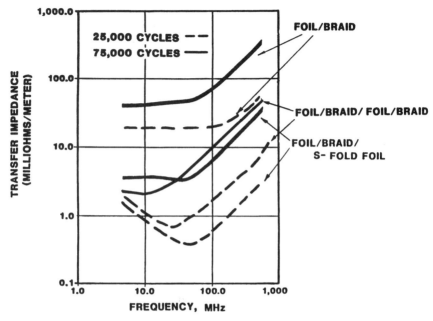

FIGURE 7.51 Transfer impedance frequency characteristics, flex testing effects

As in the case of shields with static mechanical loads (or no loads), the presence of adhesive also appears to be beneficial for the shield exposed to multiple flexes (see Fig. 7.52).

✦ Statistical data of CATV drop cable shield performance aging over a period of time up to 23 years was reported in Ref. [7.43]. Fifty drop cables were removed from actual cable systems and tested in the laboratory. By comparing the performance of the "naturally" aged shields with the performance of those flexed in the laboratory, it was concluded that each 15,000 flex cycles correspond to one year of cable service. Based on this correlation and obtained flex test results, predictions were made relating to transfer impedance variations versus years of service (see Fig. 7.53).✦

7.7.3 Effect of Manufacturing Tolerances on the Shield Performance Variability

While the reference sources usually give a single curve of shield performance, in reality, a significant spread of shielding effectiveness values can be observed among different manufacturers and even among sample cables made by the same manufacturer—even at the same plant and in the same lot. As an illustration, Fig. 7.54 presents some "typical" transfer impedance variability as measured on samples of different CATV drop cables.

Of course, there are different reasons for such variability, including construction tolerances, production inconsistencies, test sample preparation techniques, and instrumentation errors. The study of all these factors presents a large and challenging problem. As an example, we will investigate the effect of tolerances on braided shield performance variability. Carrier width, contact resistance between the individual strands, and weave angle variance are among the main affecting factors.

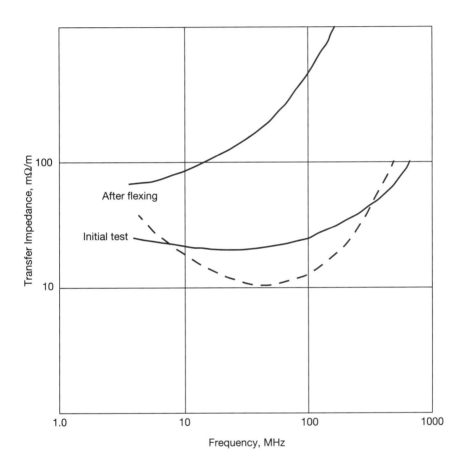

FIGURE 7.52 Shielding performance vs. flexing for dry (solid line) and sealed (dashed line) tape

Carrier width directly affects equivalent coverage parameter, which is one of the major contributors to the high-frequency transfer impedance value. The carrier width depends on the number of strands in the carrier, their gauge, and spacing. Yet, in practice, only the last of these items is an important instability source that directly affects

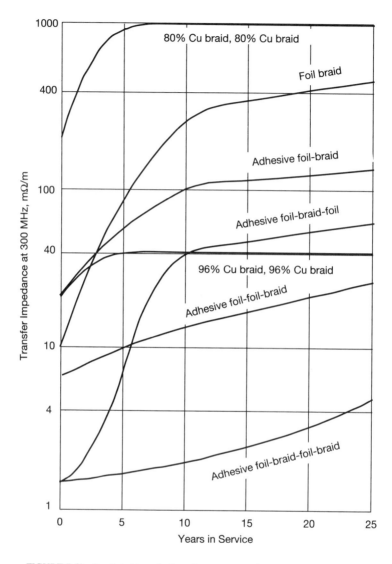

FIGURE 7.53 Predicted transfer impedance vs. years in service for drop cable

the volume of parameter τ [see Eqs. (3.113) and (3.114)]. Comparison of curves ωM_{12} and ωM_{12} ($\tau=1.1$) in Fig. 3.40 gives an idea of this factor's impact on the transfer and transmission impedance. Some general considerations about τ dependence on the shield density have been given previously.

The instability of the contact resistance between different strands and shield layers could have been a problem both in the low- and the high-frequency bands. Fortunately, for "regular" copper and aluminum shields, the impact of this factor is quite limited and often can be disregarded. This is because the actual contact resistance between the strands is so much larger than the strand resistance that it can be neglected altogether, let alone its variation. Some experimental data indicates that for commonly used copper braids, the contact resistance between the strands is about an order or magnitude larger than the strand resistance. This justifies the assumption that the currents from different strands do not "mix." Some indirect confirmation of this fact is that the dc resistance formula of Eq. (3.106), derived under this assumption, invariably corresponds to experimental data. If there were current paths straight along the cable axis in "detour" of the strand direction, the braid resistance value would have been much smaller, approaching the resistance of a homogeneous tube of respective diameter and thickness.

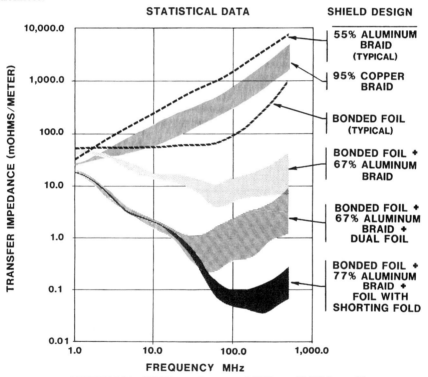

FIGURE 7.54 Shielding performance of 59/U type CATV drop cable

The change of the weave angle has a strong impact on braid performance, as proved by Figs. 7.9-7.11. Yet, even more important is the opposite direction carrier weave angle unbalance

$$\Delta Z_t = \frac{\mu_o}{4\pi}(\tan|\Delta\alpha|)^2 \qquad (7.24)$$

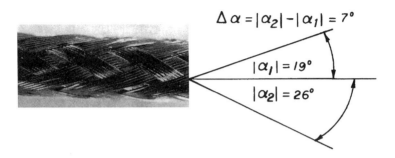

FIGURE 7.55 Braided shield with unbalanced carriers

The larger the angles' "unevenness" $\Delta\alpha$, the larger ΔZ_t will be.

Figure 7.55 presents a magnified photo of a sample braided shield that exhibited an anomalously large discrepancy between the theoretically calculated and experimentally measured transfer impedance data. As shown, the difference between the opposite direction carrier weave angles is about 7°. The shield characteristics and its measured transfer impedance (solid line) are presented in Fig. 7.56.

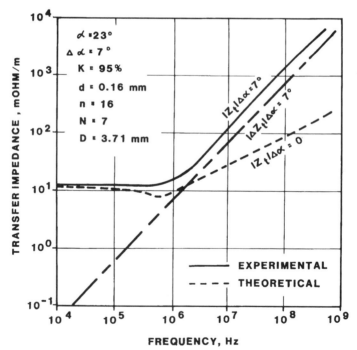

FIGURE 7.56 The impact of weave angle unbalance on transfer impedance

Also shown are theoretically calculated characteristics of the same shield transfer impedance for $\Delta\alpha = 0$ according to Eq. (3.118) and the additional inductive item according to Eq. (7.24). As shown, the sum of the two theoretical curves is quite close to the experimental one. Above 10 MHz, the item addressed by Eq. (7.24) becomes dominant, increasing the transfer impedance value 10 times, and even more in the high-frequency range.

7.7.4 Sneaky Problem of a "Rusty Bolt": Intermodulation

We have already mentioned the "nasty" consequences of contact violation between the shielding enclosure panels, product housing and connector shields, gasket and the "main" shield body. Along with "human-inflicted" problems — non-conducting paints, not accurate fit, "sloppy" design — there are also environmental causes of such problems: temperature expansion mismatch, lose of form and shape with aging, oxidation under the influence of weather and industrial agents. In particular, the oxidation presents one especially difficult problem because many metals exhibit non-linear impedance properties, when oxidized. In the EMC discipline, this phenomenon is known as "rusty bolt" [7.54-56].

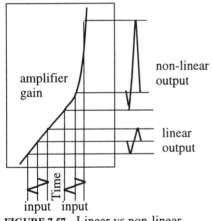

FIGURE 7.57 Linear vs non-linear response

Really, it is often perplexing to find in the EMI signature of a product some frequencies which are *not* harmonics of *any* "legitimate" signal in the system. But this is exactly the effect of a non-linearity in the current path. Fig. 7.57 illustrates the generation of a linear and non-line00ar response, when a sinusoidal signal is passing through a linear and non-linear areas in the amplifier characteristics. In the linear response, only the amplitude, and possibly phase, of the original signal will change. However, a reference to Chapter 1 will tell you that the non-linear signal contains not only the signal frequency as a fundamental, but also the infinite number of its harmonics.

As bad as the distortion of the non-linear signal itself is, this is even not the main problem from EMI point of view. Indeed, the "fun" starts when two or more narrow band signals, or even one broadband signal (that is, containing a numberof harmonics) are simultaneously applied to the non-linear input. Then non-linearity, combined with intermodulation (IM) phenomena, contribute to unwanted emission harmonics and audio-rectification susceptibility.

Even "good old" coaxial connectors may present a source of non-linearity, and must meet respective specifications. Thus, at +43 dBm (20 W) output levels of GSM, DCS1800, and PCS1900 base stations, the typical IM specification for connectors located in a post-filtering section of the system, is better than -160 dBc with two +43 dBm tones.

The power of IM products depends on the shield design. Table 7.3 presents experimental data [7.54] for several connector and cable shields, measured @+43 dBm carriers:

Table 7.3 **Measured third order IM levels (dBm)**

Product	IM Level, dBm
Nickel plating	-83
Low pressure contact, inner	-85
Low pressure contact, outer	-96
Low IM design	-120
Foil/Braid cable assembly @1775 MHz	-77
Seamweld corrugated cable assembly @1775 MHz	-113
Foil/Braid cable assembly @900 MHz	-91
Seamweld corrugated cable assembly @900 MHz	-117

8

This Brave New Old World Of Electromagnetic Shielding

8.1 SHIELDING UNLIMITED

Where is electromagnetic shielding technology heading? This book is dedicated mainly to what we may call traditional shielding: the use of metallic enclosures and gaskets, plane sheets, tubes, tapes, and wires for EMI protection. While this traditional field is progressing on its own in terms of addressing the higher and higher, as well as the lower and lower, frequencies, high magnetic permeability alloys, clad-metals, non-corrosive materials, and so on, the field of shielding at large encompasses a much broader perspective. In particular, during the last several decades, the discipline has advanced toward non-metallic shielding materials, innovative manufacturing processes, new shield designs, and novel applications of "old" and "new" shields.

We will briefly review some of these "non-traditional", or even "old and traditional", applications and innovations:
- "leaky" shielding
- "plastic" shielding, including composites and inherently conductive plastics
- shielding coats
- ferromagnetic absorptive shielding
- superconductive shields
- chiral shields
- low- and high-frequency shielding trends
- shielding specifics of high voltage discharge
- "smart" shielding "on-demand", including electromagnetic and optical shielding

Even if this list looks impressive, we still could hardly expect to cover all the relevant subjects in the selection. In this respect, in the chapter on "unlimited shielding" we must note at least three limiting factors:
1. Space and time constraints: many mentioned topics themselves deserve a book, or two, or three ..., and for some of them extensive libraries do exist, while others still need original research and study methods, mathematical apparatus, and even philosophical comprehension.
2. The ever accelerating progress: some of the topics are at the forefront of the ever accelerating modern technological progress. Even as the available material is discussed, many innovations are leaving the "breadboard". So that by the time the book is published it will undoubtedly miss many new ideas. And of course after publication the next updates are possible only with new editions.
3. Bibliographical and proprietary restrictions: not everything is published "at large" and not everything published is available to this author.

The ironical conclusion is that the readers may "rest assured" that the provided here information is not complete! But that's actually is a "good news", because it will stimulate their "natural curiosity" and desire to expand on these and related topics. It's a never-ending process and there is always room for improvement!

8.2 LEAKY SHIELDING

> ✦ Not being involved with the mining industry, when we first encountered the term "leaky cable" back in the early 70s (of the XX century), we made it a favorite topic for jokes within the engineering community! Indeed, after investing so much time and labor in *improving* the shielding performance of the electronic product enclosures and cables shields, why would anybody, in their "sound mind", dream about compromising the shielding? Besides, the term "leaky" sounded so "ugly" (isn't it, indeed? —the only reason we use it here is that it was standardized)
> Of course, at that time nobody would think that one day a leaky cable will support communications services in the Channel. Neither anybody could envision its use as a backbone for indoor PCS! Then again, who could have thought at that time that there may be needs, and means, to use superconductive shielding or regulate "in real time" the shielding effectiveness "on-demand"? ✦

By now, leaky communications feeders are a pretty "old" subject. Both theory and applications are covered in numerous literature sources, e.g., see [8.2-9]. The technology was also addressed in the latest standards, e.g., [8.1]. Keep in mind that the referenced sources are the "tip of the iceberg". For example, the Ref. [8.2} provides about eight pages of references up to the date of publication — and it was published back in 1982!

Historically, the studies of *leaky feeders* (as the technology was identified in [8.1]), were first approached as a generalization and expansion of the transfer impedance concepts for electromagnetic shields with multiple apertures [8.2]. In Chapter 3, we have dedicated enough attention to the transfer impedance of the shields with apertures — meshed, braided, foil with slots. There, our goal was mainly to characterize the *existing*, "naturally occurring" arrays of apertures. However, with the *specially designed* leaky cables, the goal is different: select the *aperture shape and spacing* to support a desired field radiation pattern outside of a given cable and in a given environment, e.g.,

This Brave Old New World of Shielding 591

in a tunnel or within a building. In a way, the latter goal is an *inverse* to the first one. Although both problems do have similarities, they also have enough differences to often warrant *different* simplifications and approximations, and even completely different approaches to obtain the solutions.

Applications
Although the majority of applications of electromagnetic shielding deal with EMI protection, there are also applications of shielding with an "opposite" goal, i.e., to provide controlled radiation in and/or out of the system. But why would anybody want a "leaky" line, albeit with "controlled" leakage? Actually, applications abound, from wireless access by mobile users to telecommunications networks, to communications in tunnels and mines, to security and detection functions, to sophisticated sensors responding to the changes in the environment and automatic control signals.

Leaky feeders guide and radiate/receive electromagnetic waves along the designated route. There are three main areas for their potential application:

1. Wireless access to industrial communications networks. These can be in tunnels, subways and mines, along the highways and railroads, or used in buildings for paging and telemetry. One of the latest significant use of leaky cable line is in the channel tunnel between France and Great Britain.
2. Wireless access to personal communications networks (PCN) and personal communications services (PCS). These can be for voice and data transmission in buildings and vicinities, for access to telecommunications networks in a mobile environment with blocked free-space propagation, and for paging and telemetry.
3. Guided radar systems. These can be utilized for obstacle detection in transportation, surveillance, and security systems.

In all these cases, leaky lines are used as a means of electromagnetic signal transport into confined spaces that are blocked from outside transmission and distribution, with simultaneous wireless access to this signal in telecommunications networks. Also, unlike a radio broadcast, the leaky lines can have coverage that is limited to the desired areas. This is a major advantage with regard to spectrum conservation.

Leaky Feeder Design
The most specific element of a leaky feeder system is the leaky cable which provides for both signal transmission along the line and signal exchange in the space around the line. In a way, all cable shielding is "leaky" (except superconductive shielding, and even that is non-leaky only to some extent). Indeed, we saw that even a solid-wall tube "leaks" energy via diffusion. However, the term leaky cable implies useful emissions/susceptibility, *intentionally* created for telecommunications and other purposes.

Several different cable designs are presently used. For example, the ITU-R (CCIR) gives the following classification of "leaky feeders" for land mobile system applications [8.1]: bifilar (as a rule, balanced) lines, continuously leaky coaxial cables, coaxial cables with periodic apertures, and cables with mode converters. Since the CCIR Report publication, another system can be suggested that potentially belongs to the mode conversion type, or it may constitute a class in itself. It consists of a fiber optic distribution cable with periodic taps providing the signal to electro-optical transducers, and then to electromagnetic field radiating/receiving elements, which is to say, antennas. Here we will concentrate on coaxial leaky cables.

In the leaky shield, the continuous or discretely distributed apertures of a certain form and orientation shape the radiation pattern and ration the radiated power. Several of the most common leaky coaxial cable shield designs are shown in Fig. 8.1. They range from "loose" braids (about 40 to 50 percent coverage), to continuous longitudinal or transversal slots in the shield, to pretty "fancy" slot orientations specially designed to obtain the desired radiation/reception leaky cable performance. These cables are used in the frequency range from 30 MHz to 1.5 GHz, with the latest trend toward higher frequencies. The typical line lengths vary from several meters to tens of kilometers, with the radio transmission distances from tens of centimeters to tens of meters.

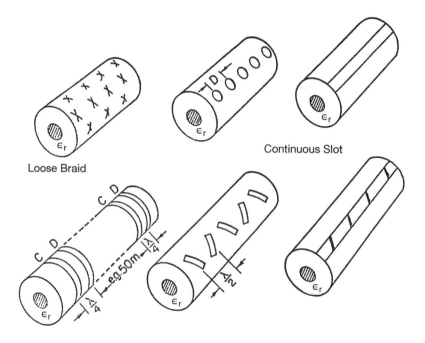

FIGURE 8.1 Common leaky coaxial cable shield designs

How It Works?

Essentially, the leaky cable acts as a transducer: it transforms a part of the travelling *TEM wave* energy propagating along the cable circuit *inside* the cable into the *radiating modes* propagating *outside* the cable. The basic equation of a leaky line can be written as follows:

$$P_r = P_t - K - \alpha L \tag{8.1}$$

where P_t and P_r are, respectively, the input power to the cable and power received by the mobile antenna (or vice versa), K is the coupling loss between the traveling wave power at a point on the line and the corresponding power received at this point, $\alpha = \alpha_i + \alpha_e$ is the cable attenuation represented by the sum of respective circuit losses and radiation losses, and L is the line length.

From this equation, the two fundamental parameters of a leaky cable are its *attenuation* and *coupling loss*. While the circuit losses are determined by the cable overall geometry and materials [e.g., see Eq. (A.54)], the coupling losses of leaking coaxial cables depend upon the "inside-to-outside" (or vice versa) coupling which is a function of the size and distribution of the radiating slots in the shield.

The theory of leaky coaxial cables can be developed according to the solutions to Maxwell's equations with proper boundary conditions. It can be also treated as a specific case of leaky-wave antenna theory. In particular, IEEE Standard 145 defines a leaky-wave antenna as an "antenna that couples power in small increments per unit length, either continuously or discretely, from a traveling wave structure to free space." In a leaky coaxial cable, the traveling wave is generated in the coaxial circuit, while the coupling is realized via a system of slots and/or apertures in the cable outer conductor which act as antennas in an antenna array. There are two major modes (mechanisms) of coupling the energy from the traveling wave propagating inside the cable to the receiving antenna outside the cable: radiated mode and coupling mode.

The coupling mode is generated by a continuous longitudinal slot, or a set of small slots or apertures spaced at distances much smaller than the signal wavelength. In this design, the radiated energy forms a surface wave that tends to concentrate in the proximity of the shield and quickly diminishes with the distance from the shield. This quick reduction can be explained by the character of summation of the contributions from different apertures: at a large enough distance from the cable, the path lengths of these contributions are almost equal, while the phases are systematically shifted so that, in the case of an infinitely long line, for each aperture contribution there is a similar contribution, but with an opposite phase. Of course, theoretically, the end result is a total cancellation. Obviously, such cables are not efficient at large enough distances. But this is not necessarily "bad".

To generate the radiated mode, the slots or apertures in the shield must be spaced periodically at distances comparable to the signal wavelength. Then, with proper spacing of the apertures and at a certain, large enough, optimal angle, ϕ, relative to the cable axis (the angle is measured with regard to the perpendicular to the axis), there may exist a group of apertures or slots that have contributions with identical phase and therefore add, forming a major lobe. Beyond this group, the distances from each aperture to the observation point (in which the communicating antenna is placed) cannot be assumed to be identical, and therefore the phases of the arriving contributions are different. However, for these other apertures, the observation angle is different from the optimal one, and their contributions can be neglected.

Since the optimal angle depends on the slot geometry and the signal frequency, it is the major radiated lobe. It was shown [8.7] that the electrical field intensity radiated in the direction ϕ, by a slot of length s, with the tilt angle τ with regard to the cable axis is proportional to:

$$E^\phi \sim s^2 \sin 2\tau \tag{8.2}$$

Thus, by varying the s and τ, the leaky cable coupling loss can be controlled. The analysis shows that having only one type of slot (that is, identical s and τ) results in the formation of several major radiation modes. As could be expected, their interaction leads to a Rayleigh-type fading that deteriorates with frequency. However, by properly choosing the slot length, width, tilt angle, and spacing pattern, it is possible to reduce the fading to reasonable limits while creating near optimal conditions for coupling in a wide frequency band [e.g., see Fig. 8.2 [8.5]].

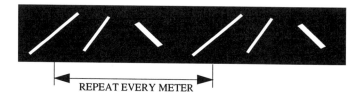

FIGURE 8.2 Example of a three-slot pattern copper strip

For example, Figs. 8.3 and 8.4 present characteristics of a wideband cable developed in Japan [8.7].

Another specific for leaky cable problem arises due to the dependence of the radiating power on the cable length. Indeed, since the signal propagating in the cable incurs losses, the radiating power at the beginning of the line is larger than at the end of the line. This "injustice" can be corrected by correspondingly reducing the coupling loss with the cable length. A leaky shield design was suggested [8.6] in which the number of slots periodically increases when the cable circuit losses reach a certain value.

8.3 "PLASTIC" SHIELDING

Is There a Shield Beyond Metal?
Yes, there is, and we have proof of it! The "post-metal" shielding era has come and includes such technologies like conductive plastics and fabrics, absorptive ferrite-type materials, superconductive ceramics, chiral shields, and many others. The burgeoning innovations gradually but surely penetrate the industry, both in" niche" and in "mainstream" applications.

Just review the following (incomplete) list of several "non-traditional" alternative technologies making its way into the electronic product shielding:

- plastic composites for EMI shielding applications, including reinforced thermoset conductive composites
- development and use of non-metallic shielding materials combining electrical *and* thermal conductivity

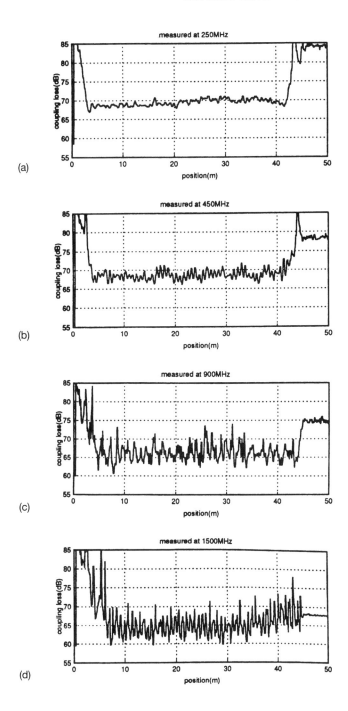

FIGURE 8.3 Wideband leaky cable characteristics

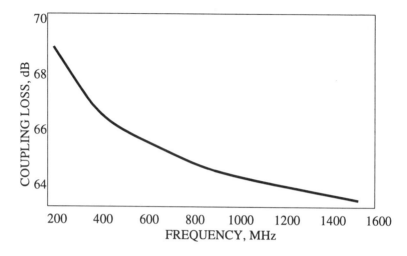

FIGURE 8.4 Wideband cable coupling loss vs frequency

- development and use of intrinsically electrically conductive plastics
- EMI/RFI shielding
- using paint for EMI shielding on plastics
- development of plastic plating techniques for EMI shielding: ion plating, electroless plating, vacuum deposition, arcspray
- formable metalized plastic laminates for EMI/RFI shielding

And even this list is quite incomlete, and could be complemented with new items with every issue of trade magazines and conference proceedings. Generally speaking, these and other new "plastic-related" shielding techniques can be classified in three categories:

1. *Composite* materials consisting of a plastic matrix and *metallic filler*
2. *Coated* material consisting of a plastic body with a *metallic layer*
3. *Intrinsically conducting non-metallic* materials

In recent years, the field has expanded significantly. An array of conductive plastics, metallized fabrics, sprays and paints, along with numerous methods of their application, have flooded the shielding market. Correspondingly, abundant information on the subject is available, e.g., see Refs. [8.10-18]. An almost comprehensive compendium of electromagnetic material descriptions, including conductors and dielectrics, as well as many new and innovative materials, is given in [A.25]. Thus, the main purpose of the brief review in the following sections is develop an introductory awareness of the related subjects and disciplines.

Shielding Composites

Originally in "ancient times" (to be sure, the XXth century 70s and 80s), the main means of conductive plastic engineering were to chemically blend a dielectric matrix with conductive powders or fibers, or to plate the plastic with a metallic layer, or combine plastic and metallic fibers in textile fabrics. Since then, these materials became even better. They can provide as much as 30 to 100 dB plane wave shielding effectiveness, depending on the properties, quality, and the amount of filler (e.g., see Fig. 8.5, Fig. 8.6, and Table 8.1).

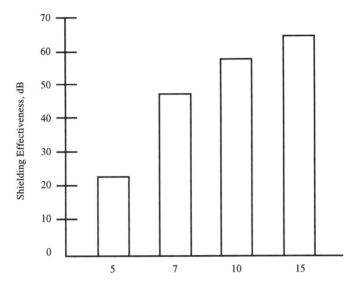

FIGURE 8.5 Conducted plastic composite shielding effectiveness

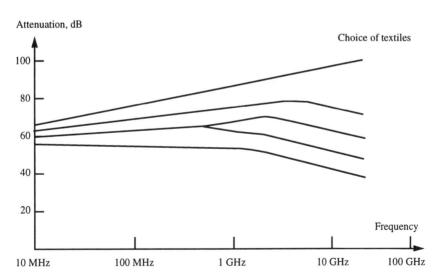

FIGURE 8.6 Attenuation vs. frequency in metallized textiles

Table 8.1 **Comparisons of Several Shielding Systems [8.16]**

EMI/RFI Shielding System	Conductive Medium	Surface Resistivity (¾/sq)	Volume Resistivity (¾/cm)	Shielding Effectiveness @		
				30 MHz	100 MHz	1 GHz
Zinc arc spray[a]	zinc	0.020		106	92	98
Sprayed acrylic coating[b]	nickel	3.0	0.01	35	47	57
	silver	0.004		67	93	97
Panels of:						
• Thermoplastic[c]	5% NCG	20.0	0.5	32	30	34
• Polycarbonate[c] and	10% NCG	2.0	0.1	36	36	44
• polyetheramide[c]	20% NCG	0.8	0.5	50	52	59
Nickel (0.4 µm on 7.0 µm graphite fiber)[d,f]	5%	10^{13}	1.0	40	35	40
Stainless steel fibers (7.3 µm) in polycarbonate	10%	10^2	0.5			
Aluminum flake (1 mm × 1.2 mm, 30 µm thick, from rapid solidification)[e]	18 to 22% (volume)		1.0			35

(a) M. Thorpe, Plastics Engineering 38(4), April 1982.
(b) Al-Technology Inc., Technical Bulletin, 1986.
(c) A. Luxon and M. Murthy, SPE antec., 1986, p. 233. R. Evans et al., SAMPE Quarterly 17(18), July 1986.
(d) S. Gerteisen, Northcon/85, October 1985. R Tolokan and J. Noblo, Plastics Engineering 41(31), August 1985.
(e) A. Holbrook, Intl. Journal Powder Met. 22(1): 40, 1986.
(f) S. Kidd, SPI 42nd Annual Conference, paper 25F. Feb. 2–6, 1987, Composites Institute of S.P.I.

However, when reviewing the shielding effectiveness data in Figs. 8.5-6 and Table 8.1, the reader will "hopefully" ask *the question*: how this performance was tested. As a rule, such data is derived from the *plane wave* (or equivalent, e.g., TEM cell, etc. — see Chapter 6) shielding effectiveness measurements. But by now we know that although the *plane wave shielding effectiveness* numbers may run high, in itself this does not yet guarantee satisfactory EMC performance. As we have seen, in many shielding regimes the plane wave penetration is *not* the *main*, or not the *only* EMI mechanism. What is often required for EMC shielding performance, is the *high conductivity* of the shield.

Nevertheless, there are reports of successful applications of anisotropically conductive polymers in product shielding, high density connectors, sockets and VLSI packaging.

Shielding Coats

High conductivity of plastic shielding enclosures — electronic product housings, cable jackets, gaskets — can often be achieved by applying a relatively thin coat of metal at the enclosure surface. Several techniques to accomplish this are included in the bullet listing in section 8.3.1. Usually, a very thin layer of metal, 10 to 100 µm thick, is deposited on the plastic enclosure walls. Also, much thinner and thicker layers can be applied, when needed, but their inherent mechanical problems are obvious. Of course, the thick The electronic product EMC performance enhancements achieved by applying the coating shielding, depend on the specific techniques and can be dramatic.

As an example, Fig. 8.7 presents comparative results of immunity testing of a typical hearing aid product [5.48]. The immunity was evaluated as improvement in signal-to-noise ratio at the device output.

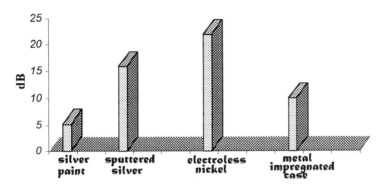

FIGURE 8.7 Immunity improvement of hearing aid devices with coated shielding

Of course, the achieved shielding effect depends not only upon the used materials and techniques, but also on their implementation: product design features (shape, size, proximity to other elements within the shielding enclosure, etc), as well as shield physical and electromagnetic integrity, grounding and termination, treatment of apertures and cable penetrations. Also, along with the shielding effectiveness, the physical and environmental factors should be accounted for.

For example, as shown in Table 8.1, the metal sprayed shielding systems that provide high conductivity, as a rule, have extremely poor flexibility and other mechanical properties. On the other hand, low-content metal-impregnated thermoplastics that have acceptable mechanical properties also have low conductivity and perform poorly in quasi-stationary and near radiating fields. We have addressed many of these problems in the preceding chapters.

Using special coating materials, a *combination* of useful properties can be achieved, *combined* with EMI shielding performance. For example, a tin-oxide coating exhibits a useful combination of solar absoptance, thermal emittance, and electrical conductivity, and does not lose electrical conductivity during long exposure to vacuum. The material of this type is made highly electrically conductive by incorporating *antimony and indium via chlorides or oxalates in concentrations* of 1 to 4 weight percent relative to the amount of tin oxide (don't blame the author of this book for all this chemical "slang" — I didn't invent it!).

The material can be sprayed to coat the desired surface. Such coatings were developed for use as thermal control + shielding properties for spacecrafts [8.19], but they can also be utilized in that consist

Inherently Conductive Plastics

Great strides have been made in developing new polymer conductors. By doping the originally dielectric polymer with p-type or n-type materials (e.g., iodine), its conductivity can be increased by several orders. Although even several years ago we were "not yet there" with regard to electromagnetic shielding, the latest reports indicate that conductivities of the newest polymers run from 10^{-4} to 10^7 S/m, as compared with 1.1×10^7 S/m for copper.

The main advantages of inherently conductive plastics stem from the absense of lossy fillers in their matrix: such materials may be easier and cheaper to produce and apply, they can have smaller weight, better mechanical properties, and very attractive performance parameters. In certain applications the use of inherently conductive plastics may pay out right away. For example, during comparative testing, balanced and coaxial cables enclosed in conductive vynil tubing exhibited significantly better electrostatic coupling protection than similar cables with copper braids (braid does have apertures!). To terminate and ground the tubing a drain wire was used running parallel the tube along the whole cable length, just as in "regular" shielded cables. In radiating field regime, up to 30-40 dB plane wave shielding effectiveness (TEM horn-based testing) was achieved on flat plaque test samples made from the materials under test.

In section 8.7, we will also briefly comment on the millimeter-wave performance of an inherently conductive plastic sample.

So get ready to add doped *polyacetylene*, *polyaniline*, and even *tetracyanoquinodimethane* shields to your professional vocabulary. These are the basic materials used to manufacture intrinsically conductive plastics.

8.4 FERROMAGNETIC ABSORPTIVE SHIELDING

When analyzing Fig. 7.2, we have compared the use of cable shielding vs system filtering (we have also considered the use of fiber optics). We noted that when each of these techniques could solve the EMI problem in its own application domain, the particular choice was determined by the system technical and economical objectives, parameters, and constraints. This is good. However, wouldn't it be better, if we could find a technique which combines the positive features of both, the shielding *and* the filtering? In fact, such technique does exist and we even mentioned it, albeit at a flash, when we alluded to a "more elegant and practical solution" — the possibility to continuously apply a flexible absorptive ferromagnetic enclosure (jacket) along and around the entire length of the cable core.

Just this has been achieved in cables with ferrite elastomer extruded jacket and / or insulation (or coat over insulation, if it is desirable to retain the original, non-absorptive insulation adjacent to the conductor). Made from a properly designed dielectric matrix of ferromagnetic powder composites, such extruded tubes have sufficient flexibility in conjunction with high magnetic permeability and absorption losses. Unfortunately, because of low conductivity, a single ferroelastic jacket does not perform well as a shield in the quasi-stationary regime. (To see why, just figure out the magnitude of the transfer impedance of a ferrite tube!) But this problem can be solved by using a

This Brave Old New World of Shielding 601

combination of a ferroelastic jacket with "regular" metallic layers. A set of such cables was developed for applications in CATV plants, antenna feeders, and power cords [8.20].

The applications of such "stealth" technology are abound. Just to name a few (see Fig. 8.8):

- telecommunications
- test & measurements, e.g., receive antenna feeders
- electric utilities: power mains and high voltage applications
- medical electronics
- industrial control
- secure communications

(a) Power, data & signal cables

(b) International filtered cord sets

(c) RF cord sets & cables

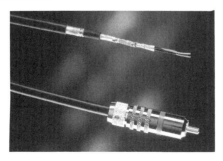

FIGURE 8.8 Ferroabsorptive cables for EMI protection (courtesy EMC / Eupen)

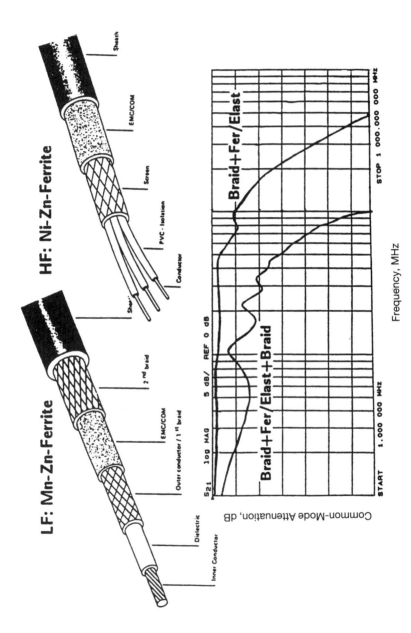

FIGURE 8.9 Ferrite-elastomer cable shielding

This Brave Old New World of Shielding 603

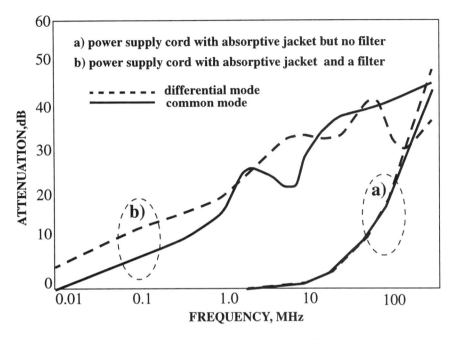

FIGURE 8.10 Losses in power supply cords with absorptive shielding

The absorptive cables obliterate EMI by transforming it to heat. So no reflected energy will appear at unexpected radiators / sensors. When the individual conductors of a cable are insulated with a ferro-absorptive compound, the cable acts as a *distributed* low-pass filter. The longer the cable, the more it filters. The high frequencies are absorbed through the magnetic and dielectric losses of the ferrite granules, transforming this RF energy into heat. This prevents the EMI from being conducted back down the line only to cause problems elsewhere in the equipment.

The cables can be also constructed with a common coating (jacket) of the ferro-absorptive compound extruded over an *entire* bundle of conductors, rather then on each individual wire. To improve the common mode suppression, the absorptive layers can be combined with metallic layers, e.g., braid. This type of construction offers a marked attenuation of the common mode interference.

Both types of cable are illustrated in Fig. 8.9. Fig. 8.9 also shows the effect of addition of an extra braid layer to the shield.

In power supply cords, combining the absorptive shield with metallic layers and filters, the desirable common and differential mode EMI suppression frequency characteristic of power supply cords can be achieved (Fig. 8.10).

✦ Two practical examples illustrate the use of ferro-absorptive cables.

1. Frequency inverters for asynchronous motor speed control are well known and troublesome sources of conducted and radiated interference. This is due to the short pulse rise time of the inverter currents. The fast voltage variations generated by the inverter, are conducted by the cable to the motor. There, the coupling capacitances of motor windings to the ground cause high current

pulses which return back to the source via general grounding system, generating common mode interference. And since the motor-inverter distance easily reaches 100 m and longer, a large area may be exposed to this interference. As the invertor cannot be modified (generator), the interference suppression of the system is possible only by external means, that is, at a radiator level.

Using the ferro-absorptive cables, the interference can be mitigated in the interconnection itself. This is achieved by using a combination of a double-shielded low-pass absorptive cable, or for more demanding applications, a combination of such a cable with a filter.

2. Typical sensors (pressure, temperature, strain gauges, etc.) used in manufacturing locations are especially sensitive to both high frequency EMI disturbances and low frequency magnetic fields, because the analog signals generated by the sensors are small, in order of millivolts / milliamps. Once the receiving circuit rectifies this noise and adds it to the measured (intelligent) signal, a false reading can occure causing the system error.

By using the ferro-absorptive low-pass sensor cable, the common mode EMI noise is attenuated in the cable, while the intelligent differential mode signal is left undisturbed.

✦

8.5 SUPERCONDUCTIVE SHIELDS

Although discovered relatively recently (to be exact, in 1986), high-temperature superconductors have already started making their way into practical applications, including electromagnetic shielding [8.21-28, A.25]. Most of these materials are copper oxide-based ceramics: Y-Ba-Cu-O (yttrium-barium-copper-oxide), Bi-Sr-Ca-Cu-O (bismuth-strontium-calcium-copper-oxide), Tl-Ba-Ca-Cu-O (thallium-barium-calcium-copper-oxide), Hg-Ba-Ca-Cu-O (mercury-barium-calcium-copper-oxide). To improve performance, they may also be treated with small quantities of other elements, e.g., Ag (silver). The "magic" property of superconductors is that they lose their resistance to current flow when cooled below some critical temperature T_c, which so far (as of 1994) is from 77K to 160K (–196° C to –113° C). These critical temperatures hold until the current and/or magnetic fields in the superconductor reach certain critical values.

At these relatively high temperatures, the industrial use of superconductors makes economic and logistic sense. Such applications as low-loss power transmission lines, transformers, generators and motors, and stable magnets needed for MRI (magnetic resonance imaging) are already in the R&D stage in the laboratories of several companies around the world. On the horizon are such applications as magnetic levitation (for transportation), electromagnetic pumping (e.g., of liquid metals), energy storage (the currents circulating in a superconductor do not encounter any losses), magnetic suspension (bearings), high-speed digital interconnections and electronic devices.

Electromagnetic shielding is one of many potential applications of superconductors. In particular, the use of these materials for power cable shielding looks especially promising because this can be done simultaneously with the reduction of the current-carrying wire resistance, which determines the losses of transmitted energy. Power transformer shields based on the application of both low-T_c and high-T_c superconducting materials are used in such applications as overcurrent protection, current limiting, arc welding transformers [8.28].

How do superconductors work in shielding? Because of nearly infinite conductivity, both the dc and ac resistances of solid superconductive tubes approach zero. Therefore, the transfer impedance of such tubes is also zero — perfect shielding! It has been shown [8.24] that monocrystalline thin films (~1μm thick) can be used to shield electromagnetic fields ranging from dc to ultraviolet frequencies, with a critical power density ~10^4 W/cm^2. Shielding factors in polycrystalline materials are somewhat worse, but still pretty good.

To date, a large enough body of experimental data has been accumulated on the manufacturing and performance of superconductors. This data illustrates both the capabilities and the related problems and limitations of these materials. Figure 8.11 is a plot of the magnetic field insertion loss ("shielding effectiveness") of a Y-Ba-Cu-O superconductor cylinder serving as a separator between the source and pickup windings in a transformer-type measurement setup [8.22]. Also, for comparison, the performance of a copper shield is shown. The designations in the figure are as follows: RT and LN2 = room temperature and liquid nitrogen (77K) respectively; S and Cu = superconductor and copper conductor, respectively; G = applied magnetic field strength in Gauss; and BKGND = system "noise."

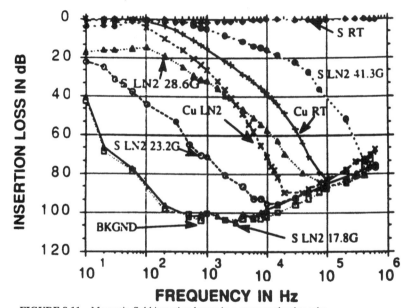

FIGURE 8.11 Magnetic field insertion losses in a superconductive tube

As shown, at the room temperature, the superconductive tube produces no shielding effect whatsoever. However, when the temperature falls below T_c, the shielding effectiveness is below well below the system sensitivity (theoretically, the shielding coefficient is "zero"). Also from the plots, it is obvious that the superconductor shielding effectiveness depends on the applied field intensity: the larger the field intensity, the worse the shield performance will be. However, this is only partially true. Research shows that there exists a critical field intensity, H_c, or a critical current density, J_c, through the superconductor below which it retains its superconductive properties independent of the applied field (as long as it is below H_c) and the frequency. In Fig. 8.32, this is about 17.8 G. On the other hand, at a high enough field intensity and/or current, and depending on the signal frequency, the material may completely lose its superconductivity. For example, in Fig. 8.11 there was no shielding detected in the 41.3 G field at frequencies below 1000 Hz, but at higher frequencies some shielding occurs. There also exists a transitional zone (called a vortex by some researchers) at field intensities just above H_c where the shield behaves nonlinearly.

Another specific property of superconductor shielding performance is hysteresis: different performance is obtained at the same applied field intensity when it is achieved by increasing or decreasing the originally set field (see Fig. 8.12 [8.23]).

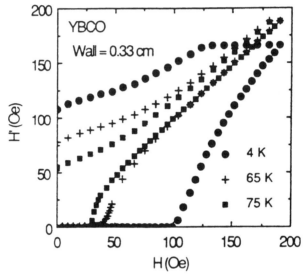

FIGURE 8.12 Hysteresis in a superconducting shield

The manufacturing problems and physical properties of the investigated superconductor shields are "far from satisfactory." In this respect, it is informative to look into the sample preparation for testing as described in [8.23]. Commercially available $YBa_2Cu_3O_7$ powder, cold-pressed to 50,000 psi, was fired in oxygen and annealed. The annealing cycle consisted of a ramp to 965° C at 200° C/hour, holding at 965° C for 10 hours, cooling to 400° C at 25° C/hour, holding at 400° C for 20 hours, and then cooling to room temperature at 25° C/hour. Final tube density was about 4.5 gcm^{-3}. The room temperature resistivity of this material is about 2600 $\mu\Omega-cm$.

Table 8.2 [8.25] lists potentially available techniques for fabricating superconducting shields. A spray-deposited superconductor/polymer coating was reported in Ref. [8.26].

✦ While we are at the *low temperature shielding*, it is worth to mention the cryoelectronical applications of *non-superconducting* shielding. The importance of such shields for aerospace applications is obvious. For example, in [8.29] the development of a compound is reported, based on aluminum and amorphous soft magnetic alloys. Such compounds are used in aerospace applications to manufacture metallic or fibrous laminates for construction of magnetic shielding working from DC to 100 kHz. It appears that magnetic properties of such material are not affected by mechanical working processes. ✦

Table 8.2 Selection of Possible Techniques for Fabricating Superconducting Shields

Technique	Requirements and limitations
Bulk ceramic parts	Self-supporting. Cylinders: very limited size. Plates: very thick, very costly.
Electrophoretic coating	Low density of shielding material. High-temperature sintering required. Small to medium modules. Silver, thin sheet substrate.
Laser ablation	Thin layers. No subsequent heat treatment required. Very small targets.
Sputtering	Thin layers. No subsequent heat treatment required. Small modules only. Very costly.
Laser deposition	Extremely critical parameters. Critical high-temperature heat treatment required. Small modules.
Plasma spray	Very high coating temperature. Large-area coating possible. High-temperature heat treatment required.
Ultrasonic spray (CDS)	Coating of large areas possible with little grain melting. High layer density. HIgh-temperature sintering required. Structural materials can be used as substrate.
Ink coating	Low density of shielding material. High-temperature contamination possible from dispersing media. High-temperature sintering required, limiting module size.
Filled resin, paint	Very low density of the shielding material. No subsequent heat treatment required. No constraints on module size.

8.6   CHIRAL SHIELDS

What is Chirality?

Although the term *chirality* is relatively new to RF and microwave electromagnetics, the phenomena it describes date back to the beginning of the nineteenth century (this again brings up the question: is there anything new in shielding, anyhow?). At that time, Arago (contemporary of Napoleon!), Biot, Fresnel, and Pasteur set the foundations to study the optical activity, also known as optical rotational power, or rotational polarization. A brief listing of these contributions and respective references are given in Ref. [8.30]. These studies related to the polarization plane rotation of a linearly polarized light which passed along a quartz crystal axis. Lately, other materials with these qualities have been discovered, or naturally or artificially synthesized, including amorphous solids and liquids.

Briefly, here's what it is all about. Optically active molecules can exist in two asymmetrical mirror image forms which have either a left-handed or right-handed D-type structure. Hence, the term *chirality*, which means "*handedness*" (from the same root as chiropractor, chiromancy, chirography, and so on) of asymmetric objects which can exist in two modifications with regard to a certain quality. All such materials may be called *chiral*, while the materials that do not exhibit chiral properties are called *achiral*. Geometrically, the two chiral objects present incongruent mirror images of each other, i.e., they cannot be superimposed, while otherwise they are identical with regard to considered properties. They are also called *enantiomers*, or *enantiomorphs* (check with Webster).

A passing electromagnetic wave interacts differently with each type of a chiral object or structure. Per Fresnel, a ray of light traveling along the quartz crystal axis is resolved into two circularly polarized rays of opposite polarization direction and different phase velocities. On a microscopic level, the difference in velocities can be explained by the different 3D structures of the molecules of the chiral matter: they can be pictured as *helices* with opposite direction of rotation — "right-handed" and "left-handed" (the quotes are of course because the "right" and "left" are "in the eyes of the beholder". Now, the difference between the two wave velocities is the cause of optical activity. Then the optical activity may be modelled as a spatial distribution of coupled oscillators, each of which moves in a dyssymmetric field.

The renewed interest in chiral phenomena in modern times is associated with the expansion of their principles to explain behaviors of electromagnetic waves in the whole electromagnetic spectrum, including RF and microwave regions. On a "macro" level, we can illustrate chirality using the subject at hand, that is electromagnetic shielding, as it is treated in this book. Consider a spiral shield (remember its "hideous" performance, as discussed in Chapter 3?) Then, the left-handed and right-handed phenomena can be represented by a serve or a spiral tape cable shields which we have analyzed previously, e.g., as shown in Figs. 3.17 and 3.18. Now, imagine a spiral wound in the opposite direction of the one in Fig. 3.18, and you have an enantiomer spiral. When the current flows in a spiral, its circular (tangential) components produce a longitudinal (solenoidal) magnetic field of respective direction, i.e., polarization. As you remember, this solenoidal field is the "cause" of unfavorable EMC performance of a spiral shield.

Several important chirality-related phenomena follow from a simple inspection of our illustration.

- To start with, if we look just at the shielding effectiveness parameter of a single spiral, it really *doesn't matter*, which way the spiral is wound — left or right! That is, judging by the shielding effectiveness, one left- or right-handed wound shield does not *manifest* their handedness!
- However, if two spirals are combined, than the relationship between their "handedness" is of vital importance!
- Another important property is related to the "connection" between the electric and magnetic fields in a spiral shield, which is very much different to the process in a longitudinal shield. In the case of the longitudinal conductor (placed along "longitudinal" axis Z), the *longitudinal only* current component I_z can flow in the conductor, accompanied with respective (tangential) magnetic field. But, as we saw in section 3.4.2 of Chapter 3, the total current, I, returning in the spiral (not just "circular"!) shield can be decomposed on two components: axial I_z (along the cable axis) and circular I_ϕ (around the cable)*. Now, suppose, a longitudinal electric field is illuminating both conductors, linear and spiral. In the case of the linear

---

*The basic relationships between these two components follow the shield geometry and can be expressed by Eqs. (3.83).

conductor, the *longitudinal only* current component will be induced in the conductor accompanied with respective (tangential) magnetic field. Similarly, in the case of a *spiral conductor*, its *"longitudinal part"* will generate the *longitudinal only* current, while the spiral conductor's "circular part" will be "insensitive" to the longitudinal electric field. However the current in the spiral *must* be continuous! Then in the spiral, the induced *longitudinal only* current will flow in the *whole conductor, including* its circular part and producing the respective *tangential electric* field and *longitudinal magnetic* field! Now you see, that the spiralshield produces the field components which *never existed* in the original field! Thus, an electric field may generate magnetic field of *the same direction*, e.g., E_z generates H_z, etc.
- A similar "scenario" could have been presented with regard to tangential electric field. There, the "linear longitudinal" conductor will be "insensitive" to the illuminating field, while the spiral conductor will produce both longitudinal and circular electric fields.

Modeling Chiral Electromagnetic Shield

As you will see, chiral propagation media can "play similar tricks" on the electric and magnetic fields, just like the spiral shield.

First, one left- or right-handed wound shield does not *manifest* a chirality phenomenon yet! Indeed, optical activity is caused not by particular "magnitude" of the wave polarization at a given point but by the polarization change at this point; i.e., the optical activity is related to the spatial dispersion of the polarization. To generate this polarization dispersion in an electromagnetic shielding environment, the chirality model should consist not of one spiral but of a multiplicity of spirals distributed in space. Then, the change of polarization of the wave propagating from one spiral to another may result in chirality phenomena, assuming the necessary conditions are met.

How can such material be modeled and realized? Two main approaches deserve consideration. The first approach is to identify in nature, or to artificially synthesize, materials which have chiral molecular structure and correspondingly perform in the RF and microwave spectra. Although this approach has many advantages, such work has yet to be done. The second approach is to artificially introduce chiral elements in some neutral (achiral) matrix (see Fig. 8.13). This approach has been realized on an experimental scale, and the first test results are published in Ref. [8.33]. In principle, both right- and left-handed elements can be introduced. Obviously, chiral properties will be exhibited if there is a prevalent number of one kind of the elements. If both kinds of elements are introduced in identical numbers, and if they are uniformly distributed in the matrix, their action will be mutually compensated, and such material will behave as achiral. To distinguish between "naturally achiral" materials that do not contain any chiral elements and materials that contain and equal number of oppositely polarized chiral elements, the latter is called equichiral material.

The available studies (e.g., see Refs. [8.30-35]) develop the theory of electromagnetic wave propagation in chiral media and confirm the potential usefulness of chiral

610 Electromagnetic Shielding Handbook

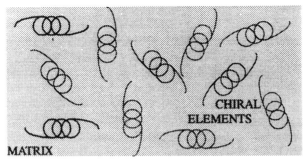

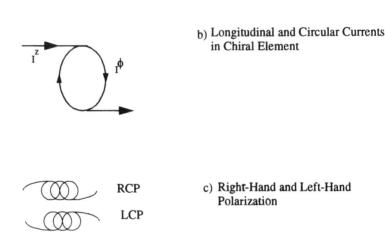

FIGURE 8.13 Chiral shield model

materials for RF and microwave applications: electromagnetic shielding, RF absorbers, and controlled wave transmission.

✦ Chirality was first discovered with regard to the optical phenomena. How does the chiral material "work" in RF and microwave domains?

As could be expected, the theory of electromagnetic wave propagation in *any* media, including chiralic, is based on Maxwell's equations. Then the chirality effects are determined by the media characteristics. The fundamental premise of the theory is based on the special relationship between the electric and magnetic field components, which is caused by the chiral material behavior in electromagnetic field. Indeed, when discussing the spiral shield illuminated by electromagnetic field, we have emphasized the generation of the circular current component (along with the longitudinal current) even if the source field had only longitudinal electric field component, or the generation of the longitudinal current component (along with the circular current), even if the source electric field had only circular component. Similarly of course, respective considerations could have been put forward for magnetic fields: the longitudinal current component was generated (along with the circular current), even if the source magnetic field had only longitudinal magnetic component.

This mechanism can be interpreted and modeled as generation of both electric and magnetic dipole moments in a spiral under the impact of electromagnetic field. Now, the averaged through the chiral material polarization $\vec{P}$ and magnetization $\vec{M}$ vectors can be expressed as functions of the electric and magnetic self-susceptibilities ς_{ee} and ς_{mm} and the electric and magnetic cross-susceptibilities χ_{em} and χ_{me}:

$$\vec{P} = \varsigma_{ee}\varepsilon_o\vec{E} + \chi_{em}\vec{B} \tag{8.3}$$

$$\vec{M} = -\chi_{me}\vec{E} + \varsigma_{mm}\frac{1}{\mu_o}\vec{B} \tag{8.4}$$

Now you see, that in chiral media the electric polarization is a function *not only* of electric field, but *also* of a magnetic field. Similarly, in chiral media the "magnetic polarization" is a function *not only* of magnetic field, but *also* of a electric field. Of course, in achiral media the electric and magnetic polarizations are independent (this is not to say that *the electric and magnetic fields* themselves are independent, at least in the time-varying fields — that's what Maxwell's equations are all about!)

Now consider a *lossless, reciprocal, isotropic* medium composed of *chiral objects* of arbitrary shape. It follows from the above, that in such medium the relations between the electric and magnetic time-harmonic varying field vectors are:

$$\vec{D} = \varepsilon\vec{E} + j\chi\vec{B} \tag{8.5}$$

$$\vec{H} = j\chi\vec{E} + \frac{1}{\mu}\vec{B} \tag{8.6}$$

Here $\varepsilon = \varepsilon_o(1 + \varsigma_{ee})$, $\mu = \mu_o/(1 - \varsigma_{mm})$, and χ are real quantities, and for a reciprocal chiral medium $j\chi = \chi_{em} = \chi_{me}$

When Eqs. (8.5) and (8.6) are substituted into Maxwell's Eqs. (3.13) and (3.14), then instead of Eq. (3.18) (this *type* of equation is common for both electrical and magnetic fields) for isotropic achiral media, we obtain for *chiral* lossless isotropic and reciprocal medium:

$$\nabla^2\vec{E} + k^2\vec{E} + 2\omega\mu\chi\nabla\times\vec{E} = 0 \tag{8.7}$$

Solving Eq. (8.20), we obtain expressions for two propagation constants K_+ and K_- (for two propagation modes in a chiral medium), and for the wave impedance Z_{cc} of the chiral medium:

$$K_\pm = \pm\omega\mu\chi + \sqrt{\omega\mu\chi + k^2} \tag{8.8}$$

$$Z_{cc} = \sqrt{\frac{\mu}{\varepsilon}}\frac{1}{\sqrt{1 + (\mu/\varepsilon)\chi^2}} \tag{8.9}$$

where $k^2 = \omega^2\mu\varepsilon$, and the phase velocities of the modes propagating in the chiral medium are ω/K_+.

Now, it is informative and instructional to compare the results (8.8) and (8.9) with those for achiral medium (e.g., see Table A1 in Appendix). Easy to see, that the results (8.8) and (8.9) are more general, reducing to those of the achiral medium when substituting $\chi = 0$. Also, easy to see that in the statics ($\omega = 0$) no chirality is manifested!

By changing the chiral media parameters ε, μ, χ, the desirable values of the wave impedance and propagation constants can be achieved. In particular

> to reduce the reflections ("RF absorber" application), the medium impedance can be matched to that of the contiguous medium, and to increase the reflections ("shielding" application) the chiral medium impedance should be mismatched with that of the contiguous medium.

Again, the chiral medium possesses broader capabilities to affect the signal propagation, than the achiral medium. In this respect, the chiral medium can be controlled by variations of *three* parameters ε, μ, γ, whereas an achiral medium has *only two* variable parameters, ε, μ.

To obtain a deeper insight into the properties of chirality, it is interesting to compare the behavior of the meadia parameters when the direction of the coordinate system changes. In particular, when a right-handed coordinate system changes to the left-handed system, the ε and μ stay unchanged, while χ changes it sign to reverse.

✦

A Chiral RF and Microwave Shield

Unfortunately, only a limited amount of experimental data related to RF and microwave applications of chiral materials is available so far. We will illustrate these with the influence of chirality on the reflection of electromagnetic waves by planar dielectric stabs [8.33]. The composite samples were manufactured by embedding a large number of randomly oriented copper coated stainless steel helices into a "neutral" plastic matrix with the dielectric constant $\varepsilon/\varepsilon_o = 2.5 + j0.1$ (see Fig. 8.13).

Each helix contained 3 turns of wire of diameter 0.15 mm wound with the pitch 0.53 mm. Helix diameter was 1.3 mm, and height 1.6 mm. Both left- and right-wound springs were made. The prepared chiral material samples had (10×10) cm area and were "sandwiched" between two pure matrix layers, with overall sample thickness 2.7 cm. The samples were irradiated by an antenna in the frequency range of 14.5 to 17.5 GHz, and reflected signal were compared with the source signal to obtain the reflection coefficients.

Fig. 8.14 presents the results of five materials:

1. Pure metal plate.
   This was taken as a reference, 0dB.
2. Metal plated pure plastic matrix, 2.7 cm thick.
   Close to metal reflections were observed.
3. Metal plated with 2.7 cm thick matrix, filled to 3.2 percent of the volume with stainless steel spheres of the same diameter as the helices.
   This was tested to separate the effect of the helices' chirality from their presence as metal objects.
4. Metal plated with equichiral material containing a mix of equal quantity of the right- and left-handed helices, totalling 3.2 percent of the sample volume.

5. Metal plated with chiral material containing the right-handed helices, totalling 3.2 percent of the sample volume.

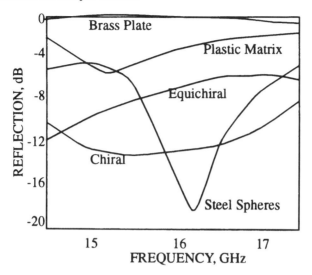

FIGURE 8.14 Reflectivity of chiral and achiral materials

As shown, the chirality effects produced a meaningful change of reflections from the metal plate. However, many questions are yet to be answered regarding the material optimal properties for particular design and application problems: shape, size, and concentration of the chiral elements, the absorption properties of the chiral medium, interaction of the matrix losses with the chiral effects, and many other.

"Chiro-Static" Shielding?

♦ One of the main questions is: can the chirial shields be used in the low frequencies? As we have already stated, the chirality effects do not manifest in statics. However, a *quasistatic* solution can still be considered describing chirality at low enough frequencies, as long as the chiral structures are much *smaller* compared to the signal wavelength. Just such problem is considered in [8.35] for spherical chiral chield. As a result, explicit expressions were obtained for "electric" and "magnetic" shielding coefficients:

(8.10)

$$\frac{E_2}{E_0} = \left\{ \frac{9\delta}{\Delta_0} [2K^2(\mu_r - 2) - 2\varepsilon_r(\mu_r - 1)^2 + \delta[(2\mu_r + 5)(\varepsilon_r\mu_r - K^2) + 2\varepsilon_r]]u_x \right.$$
$$\left. + \quad 2jK(\delta - 1)(K^2 - \varepsilon_r\mu_r + 1)u_y \right\}$$

$$\frac{H_2}{E_0/Z_0} = \frac{9\delta}{\Delta_0}\Big\{2jK(\delta-1)(K^2 - \varepsilon_r\mu_r + 1)u_x + [2K^2(\varepsilon_r - 2) - 2\mu_r(\varepsilon_r - 1)^2$$

$$+ \delta[(2\varepsilon_r + 5)(\varepsilon_r\mu_r - K^2) + 2\mu_r]]u_y\Big\} \qquad (8.11)$$

$$\Delta_0 = ([(\mu_r + 2)(\varepsilon_r + 2) - K^2]\delta + 2[(\mu_r - 1)(\varepsilon_r - 1) - K^2]) \qquad (8.12)$$
$$\times ([(2\mu_r + 1)(2\varepsilon_r + 1) - 4K^2]\delta + 2[(\mu_r - 1)(\varepsilon_r - 1) - K^2])$$
$$- 18\delta(\varepsilon_r\mu_r - K^2 - 1)^2$$

All the field components depend on all three material parameters, includeing the chirality parameter K. However, for achiral medium this parameter drops out and formulas (8.11) and (8.12) reduce to familiar formulas for electrostatic and magnetostatic shielding, similar to (5.11). ✦

8.7 "MOVING UP" THE FREQUENCY SPECTRUM

✦ The millimeter waves largely retain the positive properties of optical and infrared transmission systems, while being not as sensitive to atmospheric weather elements and pollution. In particular, the existence of atmospheric windows with minimum attenuation at 35, 94, 140, and 220 GHz boosted the interest to these bands as transmission media of choice.

With practical applications the need for EMI protection becomes an important issue, including electromagnetic shielding. And while the same shielding principles and theory are appropriate, as in the microwaves, their practical implementation does have certain specifics. This is true for all facets of the shield analysis, design, evaluation, and measurements. In this section, we will illustrate this maxim with two examples based on literature sources.

Shielding Effectiveness Measurements In The Millimeter Wave Band

While many of the measurement techniques described in Chapter 6, can be used to test the mm-wavelength shielding, it is informative, indeed, to review several conceptual measurement setups [8.36], as presented in Fig. 8.15. In principle, any of these setups can be used for measurements. Using a parallel-plate transmission line with center dielectric rods placed in line and separated by the gasket under test (Fig. 8.15 c), several types of electromagnetic gaskets were evaluated.

In this setup, one rod works as a transmitter and the second rod — as a receiver. To provide for measurements in such a wide band, the separation distance between the parallel plates is selected small enough to excite the TE propagation mode. For a given dielectric constant , the wider is the rod, the lower are the working frequencies. By selecting $\varepsilon_r = 2$ and rod width 1.5 mm, the working frequency band from 13 GHz to 110 GHz was achieved.

Fig. 8.16 shows some experimental results presented in [8.36].

(a) measurement for plate samples

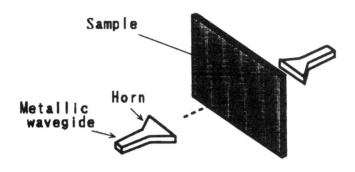

(b) measurement for gasket samples

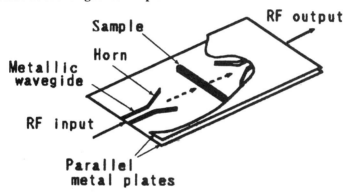

(c) measurement for gasket samples

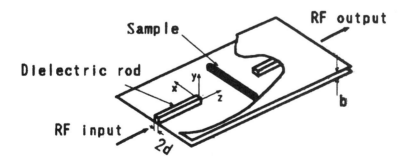

FIGURE 8.15 Conceptual configurations of shielding measurement setup

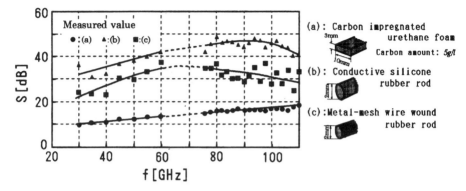

FIGURE 8.16   Shielding effect measurement results for different gaskets

Conductive Lossy Polymer Performance At Millimeter Waves

We have already briefly mentioned the shielding advantages and rationale behind the inherently conductive plastics. Here, we will complement the previous discussion with conclusions reached in [8.37] based on experimental data. In particular, for thin planar samples of a tested conductive polymer the transmission loss in a given thickness of a sample was found about 5.3 dB/mm in the 26.5-60 GHz band.

In the referenced source, this particular example deals with transmission and reflection performance of conductive lossy polymers in the frequency band of 26.5 to 60 GHz [8.37]. In the context, the first function implies shielding transfer function applications, while the second function is pertinent to the electromagnetic wave absorbers ("stealth" technology). ✦

8.8   SHIELDING FROM HIGH VOLTAGE DISCHARGE

✦ High voltage discharge (HVD), including such natural and man-made phenomena as lightning, electromagnetic pulse (EMP), electrostatic discharge (ESD), corona, and several other pohenomena has been for many years and still is a source of EMI and a threat to the electronics and biological safety. As a result, the descriptions and the means to cope with these problems received an extensive coverage in the industry and in the literature: books, papers, standards, conferences, etc. They are so widespread that there isn't even need to provide any references here.

The only reason that we even mention this subject is the necessity to create a *link* between the electromagnetic shielding discipline at large, as treated in this handbook, and the electromagnetic shielding as one of the means to protect from HVD consequences. Therefore, while referring the reader to all this vast body of available information, we will provide here only a brief comment.

As far as electromagnetic shielding is concerned, the HVD alludes to three main phenomena:

1. Electric charge accumulation *before* the discharge. This process starts with the surrounding medium (e.g., air) ionization leading to a *quiet discharge*. For DC and low enough frequencies the generated high intensity electric field can be effectively shielded by *any* metallic grounded enclosure. Of course, the enclosure integrity should be in place and the design of the grounding system should be extremely careful. In the previous Chapters 2 throuh 7 of this book you can find ample illustrations of the related problems and solutions.
2. After the discharge occurs*, the current in the discharge channel, *including* the may vary with pretty high speed. For instance, an IEC standard specifies the ESD test pulse rise time down to 100 picoseconds. At such speeds, the magnetic near field and electromagnetic far field EMI becomes a serious issue. On the other hand, there is nothing "new" here from the point of view of the shielding working regime. For example, revisit section 2.2.8 and Fig. 2.12 in Chapter 2.
3. Another aspect of the HVD discharge current is its amplitude. Indeed, the same IEC standard specifies the ESD test voltage reaching 8 kV. Or take the lightning protection regulations: the subjected conductors are tested up to 200 kA! Now we are talking serious threat of *physical damage* to the conductors in the discharge current path and to the dielectrics — from electrical overstress. Of course, this includes the shield and its insulation. . On one hand, the conductors' *ampacity* must be sufficient. On the other hand, in order not to damage the shielded circuits, a much larger *"amount of shielding"* is required, than that dictated by EMI generated by the low powered circuits.

In summary, phenomena listed in items 1 and 2 above hold nothing "new" from the point of view of the shielding working regime. However, item 3 presents a very particular challenge, dealing with keeping the shield physically safe and intact. This last problem is more "insidious", than may seem. Suppose, we decide to use a *metallic shield* in the discharge current path to protect the shielded circuits. Then, the first impetus is to use a *large thickness low resistance* shield. Given proper shield integrity and termination, it provides for better shield transfer function and can withstand larger discharge currents, without being destroyed. However, just because such a shield possesses low resistance, it may support larger *amplitudes* and *faster rise times* of the discharge current. But we know what that means: more EMI!

This is an obvious conflict since we want the shield to be *simultaneously* low and high resistance. The conflict is resolved by the industry by typically "engineering means". First, the amount of accumulated charge and the amplitude of a discharge current is reduced by using *static-dissipating* conductors. Such conductors possess *moderate* conductivity of 10^5 - 10^{11} Ohms/square, which secures a relatively slow discharge path to the ground. For example, a whole class of composite materials was created for such application: polyethylene, polypropylene, polyurethane, or other plastics loaded with conductive particles or fibers, e.g., carbon. This is also a very promising field for inherently conductive plastics. A more resistive material permits the charges to build up to dangerous levels, a more conductive material facilitates fast discharge pulses of larger amplitude. The static-dissipating materials are used for floors, mats, cots, garments, gloves, shoe soles etc. Another application example is a special jacket coating of high voltage power cables, which increases their corona discharge voltage.

When *along* with the static dissipation, higher shielding effectiveness is needed, multi-layer protective enclosures are used. One important example is a shielding bag, used to store and/or ship sensitive electronic components. Such bags may consist of up to five layers, including the static-dissipator, metallic shield, and mechanical protection. They may be also required maintain the shield integrity, be reusable (e.g., with ziptop locks), be transparent, abrasion resistant, have buble cushioning, and even aesthetically pleasing — quite a challeng for an allegedly "low tech" product! Not only they provide protection from *external* discharge, but their *internal* layers are often treated

---

*By the way, do you know "the rule of thumb": *in air,* every mm of distance between the discharging surfaces accounts for 1 kV of breakdown voltage?

with non-corrosive, amin-free antistat to limit triboelectric charging as the parts move inside the bag.

8.9 SHIELDING "ON-DEMAND"

A class of active materials can be identified which change their electromagnetic field reflection and absorption properties under the impact of different applied stimuli: electric potential, magnetic field, light, and heat. In particular, experiments were described in Refs. [8.38, A.25] that prove the potential performance of an active element composed of a lossy polyacrylamide medium dispersed with a ferroelectric material, e.g., barium titanate (see Fig. 8.17).

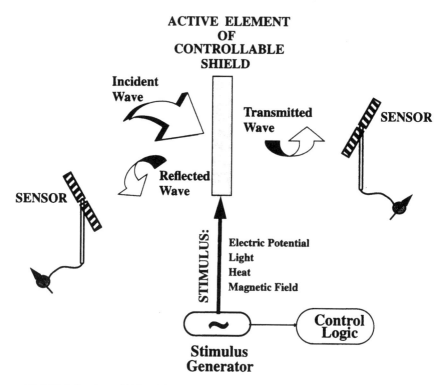

FIGURE 8.17 Controllable shieldings

When a several hundred volt potential was applied to such an element, its reflection coefficient would change from approximately 0.2 to 0.3. In a similar experiment, the photo-sensitive element, consisting of AgBr, would change its reflection coefficient from 0.4 to 0.6 under the stimulus from an optical fiber. Also, certain types of superionic compounds were found to be sensitive to the thermal energy.

The controllable "smart" shielding is "at large" and still awaits applications in shielding. For example, one may think of a leaky cable or a sensor cable based on this principle. An antenna-related practical application of a light-sensitive shielding in the so-called synaptic antenna array was described in Refs. [8.39-40]. The array contains a matrix of conducting line segments joined at the nodes with photo-conductive cells. When the cells are stimulated with optical signal, they reconfigure the antenna to generate the desirable pattern.

One of the most promising applications of "shielding on-demand" is in "optical shielding". By using a system of two or more sheet polarizers which mutual positions can change, the effect of variable light transmission is achieved. The polarizers are formed by coating a usually transparent base with light polarizing tranparent film. For example, Polaroid corporation advertizes two types of polarizing materials: HNVL-36 acrylic laminated neutral, ultra-violet absorbing linear polarizer, and HNIR-27 acrylic laminated neutral, ultra-violet and near-infra-red absorbing polarizer. Such polarizer already find application in aircraft, office, or even house windows, where they permit to adjust not only the amount of penetrating through the window visible light, but also infra-red and ultra-violet radiation. Looks like the shielding discipline is "never tired" of bringing surprises. But light propagation is also an electromagnetic process, isn't it!

As innovative, promising, and even "exotic" are the described controllable shields, in principle, the concept is pretty old. Just think of the shielding performance of ferromagnetic materials in a nonlinear regime, stimulated by the variations of the transmitted current or incident magnetic field, or even the temperature (e.g., see Figs. 7.46 and 7.47).

EPILOGUE

"Dear Reader:"

This is how this book started 600-something pages ago. Now, *if* you came to this page *after* thoroughly working through the book, that's an accomplishment.

CONGRATULATIONS!

Let's recognize: this was not an easy reading, but such was the "nature of the beast" — the discipline of Electromagnetic Shielding. You saw, how fast shielding problems may pile up, when addressed without proper background, misleading you *down* to hundreds dB of errors. No wonder, tough problems required strong medicine. In the course of reading you've probably got more of that than you bargained for. Even so, to ease the learning burdens this author devoted major efforts to create a discussion-like environment, to put the complex and sometimes controversial subjects in the real engineering life perspective, to be as logical and illustrative as possible, and to inject some humor. You may have even enjoyed the reading and the very learning process!

But mostly, the book was written for you to use it, with your practical applications in mind. Presumably for your labors, you have been rewarded by the results of successful applications of the acquired knowledge to your technical problems. If this is the case, we have achieved our goal.

Then again, more such problems, as well as successes in solving them could well wait ahead. Really, one can hardly fail to be amazed by this exciting old new world of electromagnetic shielding. Inspite of a century-old history, the innovative shielding products, processes, and theories keep emerging with literally every technical conference, scientific report and proceedings, and trade magazine, dealing with the subjects of electronics, telecommunications, medicine, military technology — you name it. But now you are well equipped to handle all this information. What else, if not all of these — the developed definitions, classifications, and roadmaps, the theoretical and experimental analyses, the mathematical and physical background — will guide you through sometimes deep, sometimes rocky, but always troubled waters of electromagnetic shielding?

This *is* the purpose of this book: it *was* conceived as a hands on guide into the practice and theory of shielding. Hopefully, you've learned to trust the suggested ideas, approaches, and techniques. Then, after the first reading, you will continuously return to it as to a reference source: to look for solutions to EMC predicaments, to analyze related phenomena, to come up with new ideas, and just to have "technical fun". These reasons being what they are, it seems quite appropriate to end this book by referring back to its start, to the very first paragraph of the preface.

Appendix

Selected Topics in Electromagnetics and Circuit Theory and Practice

A.1 "HE WHO WOULD SEARCH FOR PEARLS, MUST DIVE BELOW"*

Of course, Dryden was not likely referring to the theory of electromagnetism, and cable shielding in particular (on the other hand, who knows?). But this quotation certainly fits the subject: from a serious research study to just a working knowledge of cable shielding, it is impossible to obtain consistent, practical results without resorting to the theoretical basics. As much as we try to limit the use of mathematics in order to make this book easy to read for a practitioner in the field, it is absolutely necessary to understand at least the basics behind the shielding phenomena. As an encouragement, just think of the need to explain and avoid the abysmal gap between the experimental and theoretical data we discussed in Chapter 2.

This leads us to two important theoretical topics: Maxwell's equations and circuit theory, plus the "related issues."

✦ All macroscopic electromagnetic processes can be described by Maxwell's equations, which are based on and have been proved by a vast body of experimental data. These equations are fundamental to the theory of electromagnetism and its applications. In reality, this is quite a complex subject. Maxwell's equations specify electric and magnetic fields in terms of time variations and space distributions of their amplitudes and phases, and material properties. The analysis leads to differential equations in partial derivatives, which must be solved with respect to certain boundary conditions. The notation and comprehension are made much more convenient by utilizing the concepts of vector algebra. Although, in principle, this is not necessary, the use of vector operators significantly

---

*John Dryden (1631–1700).

formalizes the derivations. An equivalent set of equations describing the same laws of electromagnetism can be formulated with regard to specified closed surfaces and volumes in space, which are characterized by certain media properties. Such formulations substitute the operations on partial derivatives by integration over the surfaces and volumes, which leads to a set of equations in integral form. As a rule, these equations, both in differential and integral forms, can be solved analytically only for simple geometries. Because of the complexity, many important practical problems could not be tackled until the creation of modern numerical methods based on digital computers. Nevertheless, the understanding of theoretical foundations is necessary to correctly formulate the problems and interpret the obtained results, even if numerical methods are used.

Theoretical derivations often can be simplified, especially when the problems can be reduced to currents and voltages in known circuits and certain items in the Maxwell's equations are disregarded. In this case, applying the laws of circuit theory can produce practical results in a much simpler way, albeit at the price of simplifications and approximations. Therefore, not all problems can be solved using this method. As a result of the advantages and difficulties associated with field and circuit theories, it is often beneficial to combine the methods based on both of them.

Along with the original "classic" works, there is abundant literature treating electromagnetic theory and circuit theory at different levels of complexity (e.g., see Refs. [A.1–6]). Therefore we will provide here, mainly without derivations, only a "survival kit"; that is, a very brief review of the most vital principles we will use to address the problems at hand. The word "we" is emphasized because there are alternative ways to tackle the problems to be discussed here. However, we often will not even mention these alternative techniques (e.g., methods involving Hertz's vectors, retarded potentials, Green functions, and so on). Readers will be well advised not to limit themselves to the information presented here.✦

A.2 MAXWELL'S EQUATIONS SURVIVAL KIT

A.2.1 Differential Form of Maxwell's Equations

Maxwell's equations can be presented as a set of differential equations in partial derivatives. These equations link the vectors of electric and magnetic field intensities and the field propagation media parameters, using two operators:

1. The dot-product of two vectors is a scalar $\vec{A} \cdot \vec{B} = |\vec{A}||\vec{B}|\cos\phi_{(\vec{A}\vec{B})}$.

2. The cross-product of two vectors is a vector perpendicular to the plane containing these vectors: $\vec{A} \times \vec{B} = \hat{x}|\vec{A}||\vec{B}|\sin\phi_{(\vec{A}\vec{B})}$.

In the above relationships,

$$\phi_{(\vec{A}\vec{B})}$$

is the smaller angle between the vectors $\vec{A}$ and $\vec{B}$, and the direction of the unit vector $\vec{x}$ is that of advance of a right-hand screw when rotated the shortest way from $\vec{A}$ toward $\vec{B}$.

When these relationships are applied to physical units, the SI unit system is adopted, as elsewhere in this book.

We will adopt here the following formulations:

$$\text{Maxwell's Equation Set ``D''} \tag{A.1}$$

Equation I: the circulation (in mathematical terms, rotor or curl) of electric field intensity at any point is equal to the time rate of decrease of the magnetic flux density (also called magnetic induction) at the point of space

$$\nabla \times \vec{E} = -\frac{\partial \vec{B}}{\partial t} \tag{I}$$

Equation II: the circulation (rotor, curl) of magnetic field intensity at any point is equal to the total current density at this point.

$$\nabla \times \vec{H} = \frac{\partial \vec{D}}{\partial t} + \vec{J} \tag{II}$$

Equation III: the total electric displacement passing through a closed surface (in mathematical terms, displacement divergence) is equal to the total charge within the volume enclosed. In free space, volume charge density $\rho = 0$.

$$\nabla \cdot \vec{D} = \rho \tag{III}$$

Equation IV: the total magnetic flux passing through any closed surface (magnetic induction divergence) is zero. In other words, there exist no magnetic charges.

$$\nabla \cdot \vec{B} = 0 \tag{IV}$$

In the above equations,

$\vec{E}$ = electric field strength (intensity) vector

$\vec{D}$ = electric flux density vector

$\vec{J}$ = current density vector

$\vec{B}$ = magnetic flux density

The vectors of electric and magnetic field intensities are related to electric and magnetic flux densities and current density, respectively, via the parameters of the media. In a homogeneous linear media, the relations are pretty simple:

$$\vec{D} = \varepsilon \vec{E} \qquad (A.2a)$$

Here, the permittivity, or dielectric constant of the field propagation medium $\varepsilon = \varepsilon_0 \varepsilon_r$, where free space permittivity $\varepsilon_0 = (1/36\pi)10^{-9}$ F/m, and ε_r is the relative permittivity of the medium.

$$\vec{B} = \mu \vec{H} \qquad (A.3a)$$

Here, the permeability of the field propagation medium $\mu = \mu_0 \mu_r$, where free space permeability $\mu_0 = 4\pi \times 10^{-7}$ H/m, and μ_r is the relative permeability of the medium.

Equation (A.4a) relates the current density in a conductor to its conductivity, σ.

$$\vec{J} = \sigma \vec{E} \qquad (A.4a)$$

It represents Ohm's law in differential form (for a conductor).

✦ In the general case of anisotropic (but linear) media, the permittivity, permeability, and conductivity functions are described by *tensors* (e.g., see Ref. [A.1]), or *matrices* (e.g., see Ref. [A.2]), reflecting variations of the signal propagation properties in different directions. Thus in compact matrix form, the expressions (A.2a) through (A.4a) can be written:

$$[D] = [\varepsilon][E] \qquad (A.2b)$$

$$[B] = [\mu][H] \qquad (A.3b)$$

$$[J] = [\sigma][E] \qquad (A.4b)$$

Of course, to be of any use, expressions (A.2b) through (A.4b) must be expanded. For example, assuming a Cartesian coordinate system, (A.3b) becomes:

$$\begin{bmatrix} B_x \\ B_y \\ B_z \end{bmatrix} = \begin{bmatrix} \mu_{11} & \mu_{12} & \mu_{13} \\ \mu_{21} & \mu_{22} & \mu_{23} \\ \mu_{31} & \mu_{32} & \mu_{33} \end{bmatrix} \begin{bmatrix} H_x \\ H_y \\ H_z \end{bmatrix} \quad (A.3c)$$

which results in three algebraic equations:

$$B_x = \mu_{11} H_x + \mu_{12} H_y + \mu_{13} H_z$$

$$B_y = \mu_{21} H_x + \mu_{22} H_y + \mu_{23} H_z \quad (A.3d)$$

$$B_z = \mu_{31} H_x + \mu_{32} H_y + \mu_{33} H_z$$

◆

◆ The operator ∇ (nabla, or del) is treated as a special kind of vector. Its application to the field intensity vectors results in certain mathematical operations which can be expressed in different coordinate systems. The simplest expressions are usually obtained in the Cartesian coordinate system, while the most popular in cable shielding is the *cylindrical* coordinate system, which conveniently reflects the cable geometry. The following are some important vector operations given in Cartesian coordinates:

Del:

$$\nabla \equiv \vec{x}\frac{\partial}{\partial x} + \vec{y}\frac{\partial}{\partial y} + \vec{z}\frac{\partial}{\partial z} \quad (A.5)$$

Gradient:
When directly applied to a scalar function F(x,y,z), operator ∇ produces a vector:

$$\nabla F(x, y, z) \equiv \vec{x}\frac{\partial}{\partial x}F + \vec{y}\frac{\partial}{\partial y}F + \vec{z}\frac{\partial}{\partial z}F \quad (A.6)$$

Laplacian:

$$\nabla^2 F(x, y, z) \equiv \vec{x}\frac{\partial^2}{\partial x^2}F + \vec{y}\frac{\partial^2}{\partial y^2}F + \vec{z}\frac{\partial^2}{\partial z^2}F \quad (A.7)$$

Divergence:
Scalar product ("dot" product) of "nabla" (which is a vector) by a vector function is a scalar:

$$\nabla \cdot \vec{F}(x, y, z) \equiv \frac{\partial}{\partial x}F_x + \frac{\partial}{\partial y}F_y + \frac{\partial}{\partial z}F_z \quad (A.8)$$

Rotor:
Vector product ("cross" product) of "nabla" by a vector function is a vector:

$$\nabla \times \vec{F}(x, y, z) \equiv \vec{x}\left(\frac{\partial F_z}{\partial y} - \frac{\partial F_y}{\partial z}\right) + \vec{y}\left(\frac{\partial F_x}{\partial z} - \frac{\partial F_z}{\partial x}\right) + \vec{z}\left(\frac{\partial F_y}{\partial x} - \frac{\partial F_x}{\partial y}\right) \qquad (A.9a)$$

Equation (A.9a) can be written more conveniently as a determinant:

$$\nabla \times \vec{F} = \begin{bmatrix} \vec{x} & \vec{y} & \vec{z} \\ \dfrac{\partial}{\partial x} & \dfrac{\partial}{\partial y} & \dfrac{\partial}{\partial z} \\ F_x & F_y & F_z \end{bmatrix} \qquad (A.9b)$$

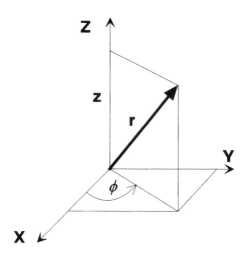

FIGURE A.1 Cartesian and cylindrical coordinate systems

The Cartesian coordinate system can be *transformed* to a cylindrical coordinate system (Fig. A.1), using the following simple formulas:

$$x = r\cos\phi$$

$$y = r\sin\phi$$

$$r = \sqrt{x^2 + y^2}$$

$$\phi = \tan^{-1}(y/x) \tag{A.10}$$

Coordinate z is identical in Cartesian and cylindrical systems.

Thus, the same vector operations {Eqs. (A.5) through (A.9)} can be expressed in cylindrical coordinates:

Del:

$$\nabla \equiv \vec{r}\frac{\partial}{\partial r} + \vec{\phi}\frac{1}{r}\frac{\partial}{\partial \phi} + \vec{z}\frac{\partial}{\partial z} \tag{A.11}$$

Gradient:

$$\nabla F(r, \phi, z) \equiv \vec{r}\frac{\partial}{\partial r}F + \vec{\phi}\frac{1}{r}\frac{\partial}{\partial \phi}F + \vec{z}\frac{\partial}{\partial z}F \tag{A.12}$$

Laplacian:

$$\nabla^2 F(r, \phi, z) \equiv \frac{1}{r}\frac{\partial}{\partial r}\left(r\frac{\partial F}{\partial r}\right) + \frac{1}{r^2}\frac{\partial^2 F}{\partial \phi^2} + \frac{\partial^2 F}{\partial z^2} \tag{A.13}$$

Divergence:

$$\nabla \cdot \vec{F}(r, \phi, z) \equiv \frac{1}{r}\frac{\partial}{\partial r}(rF_r) + \frac{1}{r}\frac{\partial F_\phi}{\partial \phi} + \frac{\partial F_z}{\partial z} \tag{A.14}$$

Rotor:

$$\nabla \times \vec{F}(r, \phi, z) \equiv \vec{r}\left(\frac{1}{r}\frac{\partial F_z}{\partial \phi} - \frac{\partial F_\phi}{\partial z}\right) + \vec{\phi}\left(\frac{\partial F_r}{\partial z} - \frac{\partial F_z}{\partial r}\right) + \vec{z}\left(\frac{1}{r}\frac{\partial}{\partial r}(rF_\phi) - \frac{1}{r}\frac{\partial F_r}{\partial \phi}\right) \tag{A.15a}$$

Similar to Eq. (A.19b), the previous equation can also be written in a determinant form:

628 Electromagnetic Shielding Handbook

$$\nabla \times \vec{F} = \frac{1}{r} \begin{vmatrix} \vec{r} & r\vec{\phi} & \vec{z} \\ \frac{\partial}{\partial r} & \frac{\partial}{\partial \phi} & \frac{\partial}{\partial z} \\ F_r & rF_\phi & F_z \end{vmatrix} \quad (A.15b)$$

It should be noted that in the telecommunication applications, vector $\vec{F}$ is also a function of time. Thus, the differential form of Maxwell's equations specify the electric and magnetic field intensities as functions of points in space and time. They can be applied to any macroscopic electromagnetic problem, and in this sense are universal.◆

When sinusoidal signals are considered, Eqs. (A.1) can be significantly simplified by introduction of phasor fields, currents, and voltages; e.g.,

$$\dot{E} = Re(\vec{\dot{E}}e^{j\omega t}), \quad \dot{H} = Re(\vec{\dot{H}}e^{j\omega t})$$

and so on. Here ω is the clock frequency of signal variation. The item $e^{j\omega t}$ (alternatively, $e^{-j\omega t}$ is also used) are usually omitted from the expressions so that the phasors do not explicitly contain the time dependence. It is still implied mathematically that this can be done because these items are present at the both sides of the equations. Hence, the phasors represent complex-value functions of spatial coordinates only. In phasor notation, the operation of differentiation is substituted by operation of multiplication, and operation of integration is substituted by operation of division. Thus, using phasor notation, Maxwell's equations as shown in Eq. (A.1) can be written in a much simpler form, without involving the time derivatives:

Maxwell's Equations, Differential Form, in Phasor Notation (A.16a)

$$\nabla \times \dot{E} = -j\omega \dot{B} \qquad \nabla \times \dot{H} = j\omega \dot{D} + \dot{J}$$

$$\nabla \cdot \dot{D} = \rho \qquad \nabla \cdot \dot{B} = 0$$

In the top two equations of (A.16a), both the left and right parts are vectors. Since vector equality requires that each component of both vectors be equal, each of the top vector equations reduces to three scalar equations dealing with variable components. Substituting the component expressions per Eq. (A.9) and expressing

$$\vec{B}, \vec{D}, \text{ and } \vec{J}$$

via the electric and magnetic field components per Eqs. (A.2) through (A.4), we obtain for Cartesian coordinates:

$$\frac{\partial E_z}{\partial y} - \frac{\partial E_y}{\partial z} = -j\omega\mu H_x \qquad \frac{\partial H_z}{\partial y} - \frac{\partial H_y}{\partial z} = (j\omega\varepsilon + \sigma)E_x$$

$$\frac{\partial E_x}{\partial z} - \frac{\partial E_z}{\partial x} = -j\omega\mu H_y \qquad \frac{\partial H_x}{\partial z} - \frac{\partial H_z}{\partial x} = (j\omega\varepsilon + \sigma)E_y$$

$$\frac{\partial E_y}{\partial x} - \frac{\partial E_x}{\partial y} = -j\omega\mu H_z \qquad \frac{\partial H_y}{\partial x} - \frac{\partial H_x}{\partial y} = (j\omega\varepsilon + \sigma)E_z \qquad (A.16b)$$

In a similar way using (A.15), we obtain for cylindrical coordinates:

$$\frac{1}{r}\frac{\partial}{\partial \phi}(E_z) - \frac{\partial E_\phi}{\partial z} = -j\omega\mu H_r \qquad \frac{1}{r}\frac{\partial H_z}{\partial \phi} - \frac{\partial H_\phi}{\partial z} = (j\omega\varepsilon + \sigma)E_r$$

$$\frac{\partial E_r}{\partial z} - \frac{\partial E_z}{\partial r} = -j\omega\mu H_\phi \qquad \frac{\partial H_r}{\partial z} - \frac{\partial H_z}{\partial x} = (j\omega\varepsilon + \sigma)E_\phi$$

$$\frac{1}{r}\frac{\partial r E_\phi}{\partial r} - \frac{\partial E_r}{\partial \phi} = -j\omega\mu H_z \qquad \frac{1}{r}\frac{\partial r H_\phi}{\partial r} - \frac{\partial H_r}{\partial \phi} = (j\omega\varepsilon + \sigma)E_z \qquad (A.16c)$$

In a complex environment, containing different kinds of dielectrics and conductors of various geometrical shapes, the electromagnetic shielding problem solution involves defining the regions with different field propagation properties, formulating the field behavior within these regions using the Maxwell's equations, and solving the obtained differential equations to meet the boundary conditions. In general, the integration of differential equations leads to a set of functions described in terms of integration constants which can assume infinite number of values. Therefore, the same solution can correspond to (i.e., describe) an infinite number of processes. To relate the mathematical solutions to specific physical problems, the integration constants are determined by applying the boundary conditions—that is, by comparing the function behavior at the interface between regions with different media parameters. This is true for both numerical and analytical methods of solving the problems in any disciplines using differential equations: electromagnetics, circuit theory, mechanical engineering, economics, social sciences, and so on.

In electromagnetics, the boundary conditions are defined by the electric and magnetic field vector behaviors at the interface between adjacent physical media. We will consider here two groups of such conditions:

1. At the boundary between two different media, 1 and 2, the components E_t of electric field intensity and H_t of magnetic field intensity parallel to the surface of the boundary (they are called tangential components) are equal.

$$E_{t1} = E_{t2} \tag{A.17}$$

$$H_{t1} = H_{t2} \tag{A.18}$$

In other words, tangential components of electric and magnetic field intensity vectors are continuous across the boundary. This is true only in the absence of source current J_s at the boundary.

2. Corresponding conditions at the boundary between two different media for the components D_n of electric flux density vector and B_n of magnetic flux density normal to the surface of the boundary (they are called normal components) are as follows:

The normal components of electric flux density vectors are in general discontinuous by the amount of the free charge present at the interface.

$$D_{n1} - D_{n2} = \rho, \ C/m^2 \tag{A.19a}$$

However, between two good dielectrics, the normal components of the electric flux density are also continuous (because there are no free charges in a perfect dielectric).

$$D_{n1} = D_{n2} \tag{1.19b}$$

Between a perfect conductor and a good dielectric

$$D_{n1} = \rho \tag{A.19c}$$

because the time-varying electric field cannot penetrate into the conductor, and thus the electric field is zero behind the surface of the conductor.

Across the boundary, the normal components of magnetic flux density vectors are continuous.

$$B_{n1} = B_{n2} \tag{A.20}$$

Some concluding remarks are appropriate to this section: Is there really a need for this "heavy" mathematical stuff? Yes, you say, we may include Maxwell's equations "because all authors do, so we'll do it too." But why on earth do we need all these vector operators, matrices, tensors?

The answer is simple: they all are useful, and sometimes necessary, to formulate, describe, and solve our shielding problems because they incorporate the experience, knowledge, and wisdom of many generations of scientists and engineers. Thus, Max-

Selected Topics in Electromagnetics and Circuit Theory and Practice 631

well's equations formulate, in a very compact form, all known macroscopic electromagnetic phenomena, so that every time we need to solve an electromagnetic problem, we don't have to reinvent all the laws of electromagnetism. As a result, using "advanced" math helps to achieve this goal in the most effective way. For example, the tensor analysis was applied quite resourcefully and with down-to-earth practical output to determine the performance of coaxial cable multilayer spiral shielding. The research is presented in Ref. [3.15] and briefly described at the end of the chapter 3. In this work, tensor form expressions were used to derive basic formulas for shield engineering and minimizing crosstalk between the coaxial cables. As a rule, the manufactured "hardware" (products, systems, cables) does confirm the validity of this "high power" math by matching the theoretical and experimental data with relatively adequate accuracy. Doesn't this make more sense than the results we achieved (see Section 2.3.3) using a much simpler approach?

A.2.2 Integral Form of Maxwell's Equations

The laws of electromagnetism can be also described by experimentally obtained integral relationships between electric and magnetic fields, which historically preceded the differential equations formulated by Maxwell. Four laws describe the electromagnetic phenomena in terms of the current and charge distribution in space and their time variations.

$$\text{Maxwell's Equation Set "I"} \tag{A.21}$$

Faraday's Law shows that a time-varying magnetic field generates an electric field.

$$\int_L \vec{E} \cdot \vec{dl} = -\frac{d}{dt} \int_S \vec{B} \cdot \vec{ds} \tag{I}$$

Ampere's Circuital Law states that the line integral of the magnetic flux density along a closed contour is equal to the total current (means, conductivity + displacement currents) crossing the area enclosed by this contour. Thus both, the time-varying electric field and electric currents are sources of magnetic fields.

$$\int_L \vec{H} \cdot \vec{dl} = \int_S \vec{J} \cdot \vec{ds} + \frac{d}{dt} \int_S \varepsilon \vec{E} \cdot \vec{ds} = \sum I \tag{II}$$

where Σ is the algebraic sum of all the currents embraced by the contour.

Gauss's Law for electric field defines the electric field as a function of electric charge distribution.

$$\int_S \varepsilon \vec{E} \cdot \vec{ds} = \int_V \rho_v dv \qquad (III)$$

Gauss's Law for magnetic field states the fact that there are no magnetic charges in nature, therefore the total magnetic flux from a closed surface should be equal to zero.

$$\int_S \vec{B} \cdot \vec{ds} = 0 \qquad (IV)$$

◆Faraday's and Ampere's laws can be obtained from Maxwell's equations in differential form [see Eq. (A.1, I and II)], by integrating them over a surface and applying Stoke's theorem. According to this theorem, the line integral of a vector taken along a closed contour is identically equal to a surface integral extended over a surface bounded by the contour:

$$\int_L \vec{F} \cdot \vec{dl} \equiv \int_S (\nabla \times \vec{F}) \cdot \vec{ds} \qquad (A.22)$$

Gauss's laws for electric and magnetic fields can be obtained from the respective differential form equations [see (A.11, III and IV)], by integrating them throughout a volume and using the divergence theorem, which equates the volume integral of the divergence of any vector $\vec{F}$ throughout a volume, V, to the surface integral of that vector flowing out of the surrounding surface.

$$\int_V \nabla \cdot \vec{F} dV = \int_S \vec{F} \cdot \vec{ds} \qquad (A.23)$$

Thus, the Stoke's and divergence theorems establish the link between the linear, surface, and volume quantities, pertaining to the fields. The very fact that such transformations are possible proves that both differential and integral forms of Maxwell's equations are equivalent: they lead to the same results. Therefore, both forms can be used interchangeably, or in combinations, depending on what's more expedient for particular problems.

The ease of application of the integral form equations is determined by the geometry of the evaluated systems. When geometries are simple and symmetrical, the integral equations may relatively quickly yield useful results. However, in case of complex geometries, the integration difficulties may be enormous, and it may be preferable to use the Maxwell's equations in the differential form in conjunction with numerical calculation techniques. Since electronic cables as a rule have a relatively simple and symmetrical geometry, they are "prime candidates" to use the integral form equations. However, when these cables are evaluated within a complex electronic system, a correct selection of the theoretical tools may present a serious challenge.◆

A.3 POYNTING VECTOR AND POYNTING THEOREM

Propagation of radio waves is associated with the energy transfer. Therefore, it is useful to define the quantitative relationship between the transmitted energy and electric and magnetic field intensities. Indeed, noting that the dimensions of the electric and magnetic field vectors are V/m and A/m, it is easy to see that the vector product of these two vectors is also a vector with dimensions W/m², which is power density.

$$\vec{E} \times \vec{H} = \vec{P}_r \quad (A.24)$$

The parameter $\vec{P}$ was introduced in the theory of electromagnetics by Poynting (it is a specific case of a more general approach to the energy flux determinations, developed by Poynting predecessors), and therefore is called the Poynting vector. Mathematically, the "cross"-product of two vectors is a vector orthogonal to these vectors. The Poynting vector expresses the energy flux associated with electromagnetic field propagation. The flux direction is according to the "cross"-product laws.

For example, in Cartesian coordinate system

$$(\vec{x}, \vec{y}, \vec{z})$$

a plane wave with the field components

$$\vec{E}_x, \vec{H}_y$$

will generate the Poynting vector $\vec{P}_z$ in the direction of the Z-axis and "aimed" such as a right-thread screw would move when rotated from E to H.

♦ ♦ The Poynting vector deals only with a fraction of the total generated energy. In general, the *Poynting theorem* expresses the balance between the total generated, P_{tot}; transmitted, P_r; stored (electrical, P_{Ec} and magnetic, P_{Hs}); and dissipated, P_d, electromagnetic power [A.3].

$$P_{tot} = P_{Es} + P_{Hs} + P_d + P_r \quad (A.25)$$

As we see, the Poynting vector describes the transmitted (radiated) power, which is of the most interest in shielding study problems. As such, the Poynting vector represents the time rate of energy flow in certain direction per unit area.

Being a complex number, the Poynting vector has both *real* and *imaginary* components, and both affect the instantaneous value of the transmitted energy. If only the *time-averaged* energy is of interest (which is often the case in studying shielding phenomena), the Eq. (A.24) simplifies to

$$P_{av} = \frac{1}{2}\text{Re}\left(\vec{E} \times \vec{H}^*\right) \qquad (A.26)$$

where $\vec{H}^*$ is a complex conjugate of the vector $\vec{H}$.

A.4 CIRCUIT THEORY SURVIVAL KIT

An electrical circuit is a set of electric sources and loads interconnected by impedance and conductance networks which provide for energy transfer between its parts. The energy transfer processes in an electrical circuit can be described using the concepts of electrical currents and voltages applied to electrical impedances and conductances. The basic elements of impedances and conductances are resistance, inductance, and capacitance. We will discriminate between two classes of electrical circuits: lumped circuits with geometrical dimensions much smaller than the transmitted signal wavelength, and distributed circuits, with elements distributed within dimensions comparable or larger than the signal wavelength.

A.4.1 Lumped-Element Networks

Circuit theory establishes relation between the voltages and currents, impedances, and sources in the circuits. At the basis of the circuit theory are two Kirchhoff's laws and electrical network theorems. Kirchhoff's first law (voltage law) states that for any closed loop of a circuit, the algebraic sum of the voltages for the individual branches of the loop is zero:

$$\sum_i V_i = 0 \qquad (A.27)$$

Kirchhoff's second law (current law) states that the algebraic sum of the currents flowing out of a junction is equal to zero:

$$\sum_n^N I_i(t) = 0 \qquad (A.28)$$

The relationship between the voltage $\vec{V}$ and the current $\vec{I}$ at any point of an electrical circuit is defined by the circuit impedance,

$$\vec{Z} = \frac{\vec{V}}{\vec{I}} \qquad (A.29)$$

which, of course, is the Ohm's law for the circuit. The angle of the $\vec{Z}$ defines the phase shift between the voltage and current in the circuit. The relationship between the voltage $\dot{V}$ and the current $\dot{I}$ at different points of an electrical circuit is defined by the network theorems.

Consider a two-port circuit network (Fig. A.2a). Only linear networks will be considered, for which the superposition principle is true. In such networks, the voltages and currents are related by Eqs. (A.30) or (A.31).

$$\dot{V}_1 = Z_{11}\dot{I}_1 + Z_{12}\dot{I}_2$$

$$\dot{V}_2 = Z_{21}\dot{I}_1 + Z_{22}\dot{I}_2 \tag{A.30}$$

$$\dot{I}_1 = Y_{11}\dot{V}_1 + Y_{12}\dot{V}_2$$

$$\dot{I}_2 = Y_{21}\dot{V}_1 + Y_{22}\dot{V}_2 \tag{A.31}$$

Here, the input impedances and input admittances at the network ports are determined by ratios of voltages and currents at the same ports:

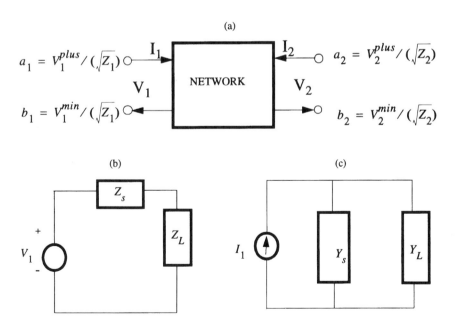

FIGURE A.2 Equivalent circuit networks

$$Z_{11} = \frac{\dot{V}_1}{\dot{I}_1}, \quad Z_{22} = \frac{\dot{V}_2}{\dot{I}_2}, \quad Y_{11} = \frac{1}{Z_{11}}, \quad Y_{22} = \frac{1}{Z_{22}} \tag{A.32}$$

and the transfer (or mutual) impedances and transfer admittances are determined by ratios of voltages and currents at the opposite ports:

$$Z_{12} = \frac{\dot{V}_1}{\dot{I}_2}, \quad Z_{21} = \frac{\dot{V}_2}{\dot{I}_1}, \quad Y_{12} = \frac{1}{Z_{21}}, \quad Y_{21} = \frac{1}{Z_{12}} \tag{A.33}$$

Equations (A.30) and (A.31), are equivalent. It follows from Eq. (A.30) through (A.33) that each of the input and transfer impedances and admittances are a measure of the open-circuit conditions at the respective ports.

Consider important network theorems. As always in this book, we will present without proof those theorems that we will use to describe cable shielding phenomena.

Compensation Theorem

Any impedance in a network may be replaced by a generator of zero internal impedance, whose generated voltage is equal to the instantaneous potential difference that existed across the impedance.

Thevenin Theorem (see Fig. A.2b)

If between any two terminals of a linear network the impedance equals Z_s and there exists a voltage, V, then the current that will flow through an impedance, Z_L, connected to these two terminals will equal

$$\dot{I} = \frac{\dot{V}}{Z_s + Z_L} \tag{A.34}$$

Norton Theorem (see Fig. A.2c)

This is similar to the Thevenin theorem but, instead of impedances, admittances are considered:

$$\dot{I}_s = \dot{V}(Y_s + Y_L) \tag{A.35}$$

A.4.2 Reciprocity Theorem

See Fig. A.2a

$$Z_{12} = Z_{21}, \quad Y_{12} = Y_{21} \tag{A.36}$$

Selected Topics in Electromagnetics and Circuit Theory and Practice 637

This is a basic principle of electromagnetics. It is important to remind at this point, that it is applicable only to linear circuits. We will return to the reciprocity principle many times, while discussing cable shielding.

✦ Expressions (A.30) and (A.31) are not the only way to present the relation between the voltage and current in electrical networks. Before we mention several alternatives, we will rewrite these equations respectively using matrix notation:

$$\begin{bmatrix} V_1 \\ V_2 \end{bmatrix} = \begin{bmatrix} Z_{11} & Z_{12} \\ Z_{21} & Z_{22} \end{bmatrix} = \begin{bmatrix} I_1 \\ I_2 \end{bmatrix} \qquad (A.37)$$

$$\begin{bmatrix} I_1 \\ I_2 \end{bmatrix} = \begin{bmatrix} Y_{11} & Y_{12} \\ Y_{21} & Y_{22} \end{bmatrix} = \begin{bmatrix} V_1 \\ V_2 \end{bmatrix} \qquad (A.38)$$

In certain applications, it is convenient to express the input voltage and current in terms of the output parameters:

$$\begin{bmatrix} V_1 \\ I_1 \end{bmatrix} = \begin{bmatrix} A & B \\ C & D \end{bmatrix} = \begin{bmatrix} V_2 \\ -I_2 \end{bmatrix} \qquad (A.39)$$

where coefficients A, B, C, D can be expressed via Z_{11}, Z_{22}, Z_{12} or Y_{11}, Y_{22}, Y_{12} parameters (e.g., see Ref. [A.2]).

For telecommunication signals, it is often more convenient to formulate the network transformation properties in terms of incident and reflected waves. In this case, the *total* voltage and/or current at the network terminals are substituted by the voltage and/or current of *incident* and *reflected* waves. The incident and reflected waves can be normalized with respect to the input impedances at the ports. As it was the case with voltages and currents, there exist different ways to express the dependence between input and output incident and reflected waves. As shown in Fig. A.2a, in a linear medium the two reflected waves are related to the two incident waves via a *scattering coefficients* S_{mn} arranged into a *scattering matrix*:

$$\begin{bmatrix} b_1 \\ b_2 \end{bmatrix} = \begin{bmatrix} S_{11} & S_{12} \\ S_{21} & S_{22} \end{bmatrix} = \begin{bmatrix} a_1 \\ a_2 \end{bmatrix} \qquad (A.40)$$

The two output waves are related to the two input waves via *transmission coefficients* T_{mn} arranged into a *transmission matrix*:

$$\begin{bmatrix} b_2 \\ a_2 \end{bmatrix} = \begin{bmatrix} T_{11} & T_{12} \\ T_{21} & T_{22} \end{bmatrix} = \begin{bmatrix} a_1 \\ b_1 \end{bmatrix} \quad (A.41)$$

Again, as in the case of currents and voltages, the scattering and transmission coefficients are interrelated.✦

A.4.3 Differential Equations of a Transmission Line

A transmission line can be viewed as consisting of a distributed network of small two-port elements with impedances Z and admittances Y. The impedances are modeled by circuit resistance R and inductance L connected in series, and the admittances are modeled by connected in parallel circuit conductance G and capacitance C. Parameters R, L, G, C are called sometimes primary transmission parameters of a transmission line. They are measured per unit of the line length.

Consider the transmission line element in Fig. A.3.

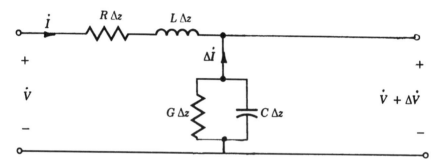

FIGURE A.3 An incremental length of a uniform transmission line

Assuming sinusoidal signals (that is, $\dot{V}$ and $\dot{I}$ are phasors) and the length of the element Δz "shrinking" to 0, the following expressions can be written:

$$\frac{d\dot{V}_s}{dz} = -(R + j\omega L)\dot{I}_s \quad (A.42)$$

$$\frac{d\dot{I}_s}{dz} = -(G + j\omega C)\dot{V}_s \quad (A.43)$$

These are basic differential equations for a transmission line. In the next section, we will use them to compare the wave propagation in transmission lines and free space.

A.5 ANALOGY BETWEEN WAVE PROPAGATION IN TRANSMISSION LINES AND FREE SPACE

There is an intimate relationship between field theory and circuit theory. Indeed, the basis for the first of Kirchhoff's laws is Faraday's law for a closed path [e.g., see Eq. (A.21, I)], and the basis for the second of Kirchhoff's laws is Ampere's law. To this effect, the primary transmission parameters of a cable circuit can be determined by applying the Maxwell's equations to the circuit transmission. If only a part of the circuit is considered, the link between Kirchhoff's first law and the electromagnetic field equations is established by the definition of the voltage between two reference points, a and b, of the electrical circuit:

$$V_{ba} = -\int_a^b \vec{E} \cdot \vec{dl} \tag{A.44}$$

By moving one of the reference points to "infinity," the absolute potential is obtained. On the other hand, if a closed loop is considered, then the sum of ingredient voltage drops in the loop should be equal to the sum of all the electromotive forces (emf) generated by all sources affecting the loop (both internal and external, e.g., including mutual coupling between circuit elements), which is Kirchhoff's first law.

◆ Now, consider a plane wave in a *non-lossy* medium (e.g., free space). By definition, a plane wave exhibits variations only in one direction. For simplicity, assume a Cartesian coordinate system in which the wave components are $\dot{E}_x$ and $\dot{H}_y$. The respective Poynting vector shows that the energy propagates in the direction of z-axis (remember the motion of a right-hand screw when it is rotated from E toward H by the shortest route). This is a typical TEM (transverse electromagnetic) wave.

Assuming sinusoidal signals, we use a set of Maxwell's equations (A.16) in phasor notation where the vector operators are defined by Eqs. (A.5) through (A.9). Under the specified conditions, the components of E_y, E_z, H_x, and H_z are absent (i.e., equal to 0), and the remaining component partial derivatives in the directions x and y (that is, $\partial/\partial x$ and $\partial/\partial y$) are also equal to 0. Also, assuming linear isotropic medium, Eqs. (A.2a) through (A.4a) are applied. Thus, the first two equations in (A.16) are reduced to

$$\frac{\partial \dot{E}_x}{\partial z} = -j\omega\mu\dot{H}_r \tag{A.45a}$$

$$\frac{\partial \dot{H}_y}{\partial z} = -\dot{E}_z(\sigma + j\omega\varepsilon) \tag{A.46a}$$

Also, since there exists only one component of each phasor, $E_x = E$ and $H_y = H$ and the partial derivatives can be substituted by full derivatives:

$$\frac{d\dot{E}}{dz} = -j\omega\mu\dot{H} \tag{A.45b}$$

$$\frac{d\dot{H}}{dz} = -\dot{E}(\sigma + j\omega\varepsilon) \tag{A.45b}$$

Comparing now Eqs. (A.45) and (A.46) with Eqs. (A.42) and (A.43), we see that they have identical structure:

$$\frac{d\dot{X}}{dz} = A\dot{Y} \tag{A.46}$$

$$\frac{d\dot{Y}}{dz} = B\dot{X} \tag{A.47}$$

Therefore, the solutions of these equations will have the same form, with the respective variables substituted for the $\dot{X}$ and $\dot{Y}$, and respective coefficients substituted for A and B. The results of such "parallel processing" are presented in Table A.1, with the explanations in the following text.

Finding $\dot{Y}$ from Eq. (A.47) and substituting it into Eq. (A.48), and finding $\dot{X}$ from Eq. (A.48) and substituting it into Eq. (A.47), two differential equations of the second order can be written, each one dealing with only one variable, $\dot{X}$ or $\dot{Y}$:

$$\frac{d^2}{dz^2}\dot{X} = (AB)\dot{X} = \gamma^2\dot{X} \tag{A.48}$$

$$\frac{d^2}{dz^2}\dot{Y} = (AB)\dot{Y} = \gamma^2\dot{Y} \tag{A.49}$$

where $\gamma^2 = AB$

Each of these equations can be used to find the variable functions $\dot{X}$ and $\dot{Y}$. The solution to this type of equations is well known. For example, solving Eq. (A.49) with regard to $\dot{X}$, we obtain

$$\dot{X} = \zeta_1 e^{-\gamma z} + \zeta_2 e^{\gamma z} \tag{A.50}$$

[You can check the correctness of this solution by substituting it into Eq. (A.49)]. Then, substituting the expression for $\dot{X}$ into Eq. (A.47), we obtain:

Table A.1 Results of Parallel Processing

"GENERIC" EQUATIONS Eq. (A.47) and (A.48)	TRANSMISSION LINE Eq. (A.42) and (A.43)	PLANE WAVE Eq. (A.45) and (A.46)
$\dot{X}$	$\dot{V}$	$\dot{E}$
$\dot{Y}$	$\dot{I}$	$\dot{H}$
A	$-(R+j\omega L)$	$-j\omega\mu$
B	$-(G+j\omega C)$	$-(\sigma+j\omega\varepsilon)$
$\gamma = \sqrt{AB}$	$\sqrt{(R+j\omega L)(G+j\omega C)}$	$\sqrt{j\omega\mu\sigma - \omega^2\mu\varepsilon}$
$Z_c = \sqrt{\dfrac{A}{B}}$	$\sqrt{\dfrac{(R+j\omega L)}{(G+j\omega C)}}$	$\sqrt{\dfrac{j\omega\mu}{j\omega\varepsilon + \sigma}}$

$$\dot{Y} = \sqrt{\frac{B}{A}}(-\zeta_1 e^{-\gamma z} + \zeta_2 e^{\gamma z}) \qquad (A.51)$$

♦

The relationship between the voltages $\dot{V}$, currents $\dot{I}$, electric $\dot{E}$ and magnetic $\dot{H}$ fields at different points of space or an electrical circuit is defined by the propagation constant

$$\vec{\gamma} = \alpha + j\beta \qquad (A.52)$$

where α is the loss constant in decibels per unit length, and β is the phase constant in radians per unit length. The expressions for the propagation constant of a transmission line and a plane wave are obtained, by substituting A and B with their values in Table 2.1. Thus, for the transmission line we obtain

$$\gamma_{tl} = \alpha + j\beta = \sqrt{(R+j\omega L)(G+j\omega C)} \qquad (A.53)$$

At frequencies above several tens of kilohertz (where $R \ll \omega L$ and $G \ll \omega C$), the Eq. (A.54) simplifies to

$$\gamma_{tl} = \alpha + j\beta = \frac{R}{2}\sqrt{\frac{C}{L}} + \frac{G}{2}\sqrt{\frac{L}{C}} + j\omega\sqrt{LC} \qquad (A.54)$$

For plane wave propagation,

$$\gamma = \sqrt{j\omega\mu\sigma - \omega^2\mu\varepsilon} \qquad (A.55)$$

If the propagation medium is predominantly conductive or dielectric, separate expressions can be found from Eq. (A.56). Thus, in conductors

$$\gamma_c = \sqrt{j\omega\mu\sigma} \qquad (A.56)$$

For dielectrics,

$$\gamma_d = j\omega\sqrt{\mu\varepsilon} \qquad (A.57)$$

Also, since γ_d is imaginary and does not contain a real part, from Eq. (A.58) it follows that for the plane wave propagating in the dielectric $\alpha = 0$ (that is, a plane wave in a non-lossy dielectric) does not incur any losses. Parameters γ_c and γ_d are intrinsic propagation constants of the respective materials.

For conducting materials, parameter γ_c is related to the eddy currents in the conductor and thus sometimes is called eddy current constant. A phenomenon associated with alternate current propagation in conductors is skin effect. Skin effect manifests itself in the reduction of current density, as the electromagnetic wave penetrates into the conductor. It is characterized by equivalent penetration depth:

$$\delta = \sqrt{\frac{2}{\omega\mu\sigma}} \qquad (A.58)$$

If a wave propagates in a homogeneous conductor, at a distance Δ from any reference point, the field intensity is reduced by 2.72 times (which is the basis for natural logarithms). Consider a conductive cylinder with thickness $t \ll r$, where r is the cylinder radius. In this case, the equivalent penetration depth Δ equals the wall thickness of another cylinder of identical material and diameter, and which dc resistance is equal to the ac resistance of the first cylinder at a given frequency ω. It is easy to see that the equivalent penetration depth and the intrinsic propagation constant are related:

$$\gamma_c = \frac{1+j}{\delta} \tag{A.59}$$

The presence in Eqs. (A.51) and (A.52) of items with $e^{\pm \gamma z}$, corresponds to the waves moving in opposite directions, and this indicates at the possibility of propagation of direct (sign "–") and reflected (sign "+") waves. And, indeed, the formation of direct and reflected waves occurs when the wave encounters a mismatch in the impedance of the propagation medium. If the wave propagation occurs in the direction of the axis z, the amplitude of the direct wave will diminish with the increase of z, and the amplitude of the reflected wave should be larger with the increase of z.

The characteristic impedance that the propagation wave encounters in a uniform and homogeneous medium is identical for direct (Z_{dir}) or reflected (Z_{refl}) waves. Thus, dividing Eq. (A.51) by (A.52) yields the medium impedance:

$$\frac{\dot X}{\dot Y} = \sqrt{\frac{A}{B}} \tag{A.60}$$

Applying the substitutions per Table 2.1, we obtain:
- For the transmission line,

$$Z_{tl} = \sqrt{\frac{(R+j\omega L)}{(G+j\omega C)}} \tag{A.61}$$

- For a plane wave in the general case,

$$Z_{cw} = \sqrt{\frac{j\omega\mu}{j\omega\varepsilon + \sigma}} \tag{A.62}$$

- For a plane wave propagating in conductors,

$$Z_{cm} = \sqrt{\frac{j\omega\mu}{\sigma}} \tag{A.63}$$

- For a plane wave propagating in a dielectric,

$$Z_{cd} = \sqrt{\frac{\mu}{\varepsilon}} \tag{A.64}$$

Note that Eqs. (A.63) through (A.65) are identical to Eqs. (2.4) through (2.5)!
Obviously, for free space, $\mu = \mu_0$ and $\varepsilon = \varepsilon_0$, which results in

$$Z_{c0} = 120\pi \approx 377 \ \Omega \tag{A.65}$$

While discussing wave propagation conditions in the medium (propagation constant and characteristic impedance) we did not need to address the meaning and value of the equation integration constants ζ_1 and ζ_2. It is clear by now that these constants describe the initial (i.e., at z = 0) amplitudes of the propagating wave, direct and reflected, respectively. Since the selection of the reference points in the transmission line or wave propagation medium is arbitrary, the value of ζ should not have affected the propagation parameters. However, this is true only in a linear and homogeneous medium, which is often (but not always) the case.

Another factor determining the wave propagation conditions is impedance mismatch. In general, when a wave W_+ propagates in the medium with impedance Z_1 and encounters a medium with impedance Z_2, the reflected wave, W_-, is generated, and the reflection coefficient is

$$\vec{\rho} = |\vec{\rho}|e^{j\varphi} = \frac{W_+}{W_-} = \frac{Z_2 - Z_1}{Z_2 + Z_1} \tag{A.66}$$

The standing wave ratio is

$$SWR = \frac{1 + |\vec{\rho}|}{1 - |\vec{\rho}|} \tag{A.67}$$

In this way, if the loads Z_L at the line ends are not matched (i.e., are not equal to the line characteristic impedance Z_c), reflections occur at the points of mismatch, which leads to standing waves in the line.

✦ Consider a transmission line. In general, the input impedance, Z_n, of a line with characteristic impedance Z_c and loaded at an arbitrary load of Z_L is equal to

$$Z_{in} = \frac{Z_c \sinh \gamma l + Z_L \cosh \gamma l}{Z_c \cosh \gamma l + Z_L \sinh \gamma l} \tag{A.68}$$

Equation (A.68) can be written in a more compact form by designating $\frac{Z_L}{Z_c} = \tanh \chi$:

$$Z_{in} = Z_c \tanh(\gamma L + \chi) \tag{A.69}$$

The resonant phenomena drastically change the impedance the line presents to the propagating wave. In general case, if the line with characteristic impedance Z_c is terminated into an impedance Z_1 at the start and impedance Z_2 at the end, then the oscillation processes in the line at a distance l from the start can be described by the following equation [2.5, A.4-6]:

$$\frac{Z_1}{Z_c} = \frac{Z_c \sinh \gamma l + Z_2 \cosh \gamma l}{Z_c \cosh \gamma l + Z_2 \sinh \gamma l} = 0 \quad (A.70)$$

In a *quasi-stationary* regime, the hyperbolic functions of complex arguments in Eq. (A.22) can be reduced to trigonometric functions. Thus, for example, the input impedance of the line loaded at its end on a mismatched impedance Z_L becomes:

$$Z_{in} = Z_c \frac{Z_L + jZ_c \tan \beta l}{Z_c + jZ_L \tan \beta l} \quad (A.71)$$

◆

Another important parameter of the wave propagation media is the speed of wave propagation:

$$v = \frac{\omega}{\beta} \quad (A.72)$$

Thus, the wave propagation speed in a dielectric is

$$v = \frac{c}{\sqrt{\varepsilon \mu}} \quad (A.73)$$

where $c = 3 \times 10^8$ m/s = the speed of light in vacuum

The wave propagation speed in a circuit is

$$v = \frac{1}{\sqrt{LC}} \quad (A.74)$$

Parameters $\vec{Z}$, α, β, and V are sometimes called the secondary transmission parameters. The circuit secondary transmission parameters are determined by its primary transmission parameters: resistances, R; inductances, L; capacitances, C; and conductivities, G.

A.6 NUMERICAL TECHNIQUES: ARE WE THERE YET?

On a very general note: wouldn't it be good to perform a computerized EMC analysis of an electronic system before it is actually built? Or to optimize a system for EMC performance at the early design phase? Because of the modern system complexity, this is not always possible using "traditional" analytical methods and, naturally, engineers look at the computer "to the rescue". That's why now-a-days no discussion of any engineering discipline is complete without at least a brief mention of the use of numerical techniques. Of course, by virtue of its complexity, electromagnetic shielding is one of the most felicitous such candidates for "computerization". The following brief note is an attempt to observe this modern "tradition" and to make this book no exception from the "general rule" herefore (just a joke).

Indeed, a large variety of "standard" off-the-shelf and custom design software packages presently exist and are used to solve the Maxwell's equations and circuit theory problems. And modern computers facilitate their efficient utilization. Coincidentally, all this is perfectly applicable to the electromagnetic shielding problems. To be sure, in the course of the book we have quite extensively used computer simulations, data processing, and graphics to support our derivations. Also, we have often referred to specific problems and examples which were analyzed and illustrated using numerical techniques — usually in "fine print" of the author's notes. The respective references to these examples frequently contained quite detailed descriptions of the corresponding techniques and their implementation, as well as related bibliography.

While it would have been quite desirable to complement our discussion with a detailed, "ready to use" description of the major numerical techniques, this is much, much beyond the scope, as well as outside the goals of *this* book. First, just because these numerical techniques are *general purpose,* they deserve *dedicated independent* coverage. Second, this is a vast and complex field of knowledge which we are in no position to give justice in several pages. And last, but not least, these descriptions are presently available from numerous literature sources, as well as workshops, seminars, tutorials, lectures and courses, dedicated to the subject, e.g., see [A.7 - 24]. The quoted bibliography is just the "tip of the iceberg". Open any of these references, and chances are you will "stumble upon" a very detailed additional bibliography on the subject.

Fortunately enough (for the topic of this book), the electromagnetic shielding is often a "favorite" subject in numerical technique applications. Consequently, existing sources provide excellent coverage of many important shielding problems, examples, and illustrations. Thus, there is no place, no time, and no need to dedicate the "precious shielding space" in this book to the subject described in many details elsewhere (of course, this doesn't mean to discourage endeavours in this potentially promising area elsewhere). Instead, we will concentrate (and even that, very briefly) on general considerations about the "cons and pros" of using the numerical techniques in solving electromagnetic shielding problems.

Based on the practical needs, at least four major objectives can be formulated for applications of numerical techniques in EMC and in electromagnetic shielding in particular:

 a. System EMC fundamental theoretical analysis and generalization
 b. System EMC modeling and problem formulation in terms facilitating numerical analysis

c. Development and/or selection of adequate numerical techniques and respective algorithms for numerical analysis of the problem
d. Calculation and visualization of separate equations.

Let's start with the last objective: this is the "easy" part. In principle, the engineering community at large, including EMC, has already arrived "there" several decades ago. The whole process became even simpler now, with the advent of powerful personal computers, work stations, and software packages. In fact, in this book, we have used extensively several "standard" packages for type setting, illustrations, data processing and graphing.

The situation with other objectives is not that clear.

Take the objective listed under a). Why do we need to analyze the system, anyway? Couldn't we just enter the basic data in the computer, call out a software program, and ... voila? Unfortunately (or fortunately?) numerous experiences confirm that the "brute force" approach rarely works! Although some simplified EMC problems can be resolved in such a way, for majority of practical, "real life" applications the experience is often not as rewarding. Why? Just look at the footprint pattern in the ground plane in Fig. 4.18: this is only one of *many layers* in a *multi-layer* PCB in a *multi-board* shelf in a *multi-shelf* frame!

Long gone are the times, when the "software people" claimed that it is enough to perform a simple computer simulation, and "all practical problems would be magically resolved". Only recently, with the full realization of the complexities of actual EMC problems, serious development had started, and more and more complex practical problems yield themselves to simulation techniques.

Somewhat greater progress was made with system modeling and problem formulation (objective b)), although even more is to be done. As soon as basic EMI mechanisms and interactions are discovered and system EMC analysis is performed, *in principle,* the system models *can* be developed in formalized terms. However, these models are often necessarily approximate, with many uncertainties and dependencies which are difficult to account for in numerical analysis.

With regard to objective c), the development of adequate numerical techniques and respective algorithms is presently the most advanced area of "computerized" EMC. Even a brief, and probably incomplete, listing of the available techniques shows how active the field is:

- Boundary Element Method (BEM)
- Conjugate Gradient Method (CGM)
- Finite Difference Time Domain (FDTD)
- Finite Difference Frequency Domain (FDFD)
- Finite Element Method (FEM)
- Generalized Multipole Technique (GMT)
- Geometrical Optics (GO)
- Geometrical Theory of Diffraction (GTD)
- Hybrid MOM/GTD (HMG)
- Method of Equivalent Currents (MEC)
- Method of Lines (MOL)

- Method of Moment (MOM)
- Physical Optics (PO)
- Physical Theory of Diffraction (PTD)
- Transmission Line Matrix Method (TLM)
- Uniform Theory of Diffraction (UTD)

Of course as we have already mentioned at the beginning of this section, there is no sense, nor need to discuss these techniques *here*: first, such a discussion would be way out of this book's "charter", and second, many excellent books treat the respective subjects in all the necessary details. Anyway, the purpose of the preceding discussion was different: to note the existence and availability of eligible numerical techniques and emphasize the extent of their present and future practicality to resolve shielding problems.

We will therefore proceed directly to the "bottom line" of our discussion:

- With *any* shielding problems, the numerical analysis must be *always preceded by and followed with* a serious analysis of the system EMC.
- When properly "administered", numerical analysis can be an important part of the shield and shielding system evaluation and design
- With regard to the first two listed objectives, a) and b), we are positively not "there yet". However, with the fast progress of computers (both, hardware and software), more and more practical problems yield themselves to numerical evaluation
- Numerical analysis has its limitations. In the end, it is still a long way to the point where *each and every complex* problems will be quickly and accurately enough resolved by computer simulations.

Thus we see, that presently there exist significant gaps between the numerical simulations, practical evaluations, design, and testing. In a way, this book is a contribution towards filling at least some of such "gaps".

Bibliography

Preface / Foreword

[P.1] Schelkunoff S. A., "The electromagnetic theory of coaxial transmission lines and cylindrical shields," BSTG, v. 13, 1934, pp.537-579.

[P.2] Kaden H., Wirbelströme und Schirmung in der Nachrichtentechnik. Springer-Verlag, Berlin, 1959, 345 pp. (The First edition of this book was published in 1949).

[P.3] Grodnev I.I., Elektromagnitnoye Ekranirovaniye v Shirokom Diapazonye Chastot (Electromagnetic Shielding in Wide Frequency Band). "Svyaz'" Publishers, Moscow, 1972, 112 pp.

[P.4] White D. R. J., A Handbook on Electromagnetic Compatibility, vol.3, EMI Control Methods and Techniques. Interference Control Technology, 1981. (The First edition of this book was published in 1973)

[P.5] Vance E. F., Coupling to Shielded Cables. John Wiley & Sons, 1978, 183 pp.

[P.6] Morrison R., Grounding and Shielding Techniques in Instrumentation. John Wiley & Sons, 1986, 172 pp.

[P.7] Tsaliovich, A., Cable Shielding for Electromagnetic Compatibility. Van Nostrand Reinhold, 1995, 469 pp

Chapter 1

[1.1] CFR 47, Parts 15, 18, 21, 94

[1.2] CISPR Publication 22 (also EN 55 022), Limits and Methods of Measurement of Radio Interference Characteristics of Information Technology Equipment, IEC, Geneva, 1985.

[1.3] VDE 0871, "Radio Interference Suppression of Radio Frequency Equipment for Industrial, Scientific, and Medical (ISM) and Similar Purposes. Part 1: RFI Measurement Set with Quasi-Peak Detector and Accessory Equipment."

[1.4] Voluntary Control Council for Interference by Data Processing Equipment and Electronic Office Machines (VCCI), Enacted on March 27, 1986 (Translation from Japanese).

[1.5] MIL-STD-461, Requirements for the Control of Electromagnetic Interference Emissions and Susceptibility
MIL-STD-462, Measurements of Electromagnetic Interference Characteristics

[1.6] IEC 801, Parts 1 through 6

[1.7] CISPR 24,ITE Immunity, (Not yet developed) Other related standards: EN 55 082-2, EN 55 101-2

[1.8] Tsaliovich A., "Modeling Electronic System Radiated Field Patterns", Proceedings of the 8th International Symposium and Technical Exhibition on EMC, .Zurich, Switzerland, 1989, pp.323-328.

[1.9] Papoulis A., Signal Analysis, McGraw-Hill Book Co., 1977

[1.10] Jong M. T., Methods of Discrete Signal and System Analysis, McGraw-Hill Book Co., 1961

[1.11] Oppenheim A., Shafer R., Discrete Time Signal Processing, Prentice-Hall, 1989

[1.12] Reference Data for Engineers: Radio, Electronics, Computer, and Communications, Seventh Edition, Howard W. Sams & Company, 1988, Chapter 7.

[1.13] Kraus J.D., Antennas, McGraw-Hill, 1988 (Second Edition).

[1.14] Balanis C.A., Antenna Theory, Harper & Row, 1982.

[1.15] Stuzman W.L. and Thiele G.A., Antenna Theory and Design, John Wiley & Sons, 1981.

[1.16] Jasik H., Antenna Engineering Handbook, McGraw-Hill Book Co., 1961.

[1.17] Fujimoto K. et al. Small Antennas, Research Studies Press, 1987.

[1.18] Tsaliovich A., "Statistical EMC: A New Dimension in Electromagnetic Compatibility of Digital Electronic Systems, " Proceedings of the 1987 IEEE International Symposium on EMC, Atlanta, GA, 1987, pp.469-474.

[1.19] Sanchez-Hernandez D., Robertson I.D., " A Survey of Broadband Microstrip Patch Antennas," Microwave Journal, September 1996, pp.60-84.

[1.20] Ott H. W., Noise Reduction Techniques in Electronic Systems, John Wiley & Sons, 1988

[1.21] Paul C., R., Introduction to Electromagnetic Compatibility, John Wiley Interscience, 1988

[1.22] Johnson H. W., Graham M., High Speed Digital Design. A handbook of Black Magic. PTR Prentice Hall, 1993

[1.23] Williams T., EMC for Product Designers, Newnes, 1996 (second edition)

[1.24] Bennett, W., S., Control and Measurement of Unintentional Electromagnetic Radiation, John Wiley & Sons, 1997, 260 p.

Chapter 2

[2.1] Handbook on Electromagnetic Compatibility, R.Perez (Editor), Academic Press, 1995, 1098 p.

[2.2] Hemming, L.,H., Architectural Electromagnetic Shielding Handbook.IEEE Press,1992, 222 p.

[2.3] Tsaliovich A., "Anechoic Room vs Open Area Test Site - A Case for EMC Study", Proceedings of the 7th International Symposium and Technical Exhibition on EMC, Zurich, Switzerland, 1987, pp.359-364

[2.4] Slattery, K. P., "Interior Resonances in Passenger Vehicles," EMC Test & Design, September, 1994, pp. 24-25
See also, Porter R. C., "Measurement of the Sceening Effectiveness of a Large Motor Vehicle,", IERE Seventh International Conference on Electromagnetic Compatibility Proceedings, York, England, 1990, pp. 261-263

[2.5] Schelkunoff S.A., Electromagnetic Waves, Van Nostrand, 1943, 530 p.

[2.6] Schultz R. B., Plantz V. C., and Brush D. R., "Shielding Theory and Practice," IEEE Transactions on Electromagnetic Compatibility, EMC-30, 1988, pp. 187-201

[2.7] Moser J., R., "Low-Frequency Shielding of a Circular Loop Electromagnetic Field Source," IEEE Transactions on Electromagnetic Compatibility, EMC-9, 1967, pp. 6 - 18.

[2.8] Moser J., R., "Low-Frequency Low-Impedance Electromagnetic Shielding," IEEE Transactions on Electromagnetic Compatibility, EMC-30, 1988, pp. 202-211.

[2.9] Levy S., "Electromagnetic Shielding Effect of an Infinite Plane Conducting Sheet Placed Between Circular Coaxial Cables," Proceedings IRE, vol. 21, 1936, pp. 923-994.

[2.10] Bannister P. R., "New Theoretical Expressions for Predicting the Shielding Effectiveness for the Plane Shield Case," IEEE Transactions on Electromagnetic Compatibility, EMC-10, 1968, pp. 1-7.

Chapters 3

[3.1] Scott K., J., "An Equivalent Circuit Model for Arbitrary Shaped Metallisation Areas on Printed Circuit Boards," IEE Seventh International Conference on Electromagnetic Compatibility, York, England, 1990, pp.42-49

[3.2] Fowler E. P., "On the Performance of Joints in small screening enclosures," IEE Seventh International Conference on Electromagnetic Compatibility, York, England, 1990.

[3.3] Vance E. F., "Shielding Effectiveness of Braided Shields," IEEE Transactions on Electromagnetic Compatibility, EMC-17, No. 2, 1975, pp. 71-77.

[3.4] Latham R. W., "Small Holes in Cable Shields," Interaction Notes, Note 118, AFWL, Kirtland AFB, NM, Sept, 1972.

[3.5] Cathey W. T., "Approximate Expressions for Field Penetration Through Circular Apertures," IEEE Transactions on Electromagnetic Compatibility, EMC-25, No. 3, 1983, pp. 339-345.

[3.6] Fowler E. P., "Superscreened Cables," The Radio and Electronic Engineer, vol. 49, N1, 1979, pp.213-228.

[3.7] Madle P., "Contact Resistance and Porpoising Effects in Braid Shielded Cables" Proceedings of the 1980 IEEE International Symposium on Electromagnetic Compatibility, Baltimore, MD, pp.206-210.

[3.8] Tyni M., "Transfer Impedance of Coaxial Cables with Braided Outer Conductors," Research Proceedings of Wroclaw Institute of Telecommunications and Akusticcs, 1975, pp.410-419

[3.9] Sali S., "An Improved Model for the Transfer Impedance Calculations of Braided Coaxial Cables, " IEEE Transactions on Electromagnetic Compatibility, vol. 33, No. 2, 1993, pp. 139-143

[3.10] Zhou G. and Gong L., "An Improved Analytical Model for Braided Cable Shields," IEEE Transactions on Electromagnetic Compatibility, vol. 32, No. 2, 1990, pp. 161-1993, pp.

[3.11] Tsaliovich A., "Braided Shield Engineering Model: Shielding and Transmission Performance, Optimal Design," Proceedings of the 30th International Wire and Cable Symposium, Cherry Hill, NJ, 1981, pp. 294-309.

[3.12] Tsaliovich A., "Shielding and Crosstalk in Coaxial Telecommunications Cables and Lines," PhD Dissertaion, 1966.

[3.13] Tsaliovich A., "Calculation of the Coupling Impedance of a Trunk Communication Coaxial Cable," Elektrosvyaz, No. 3, 1967, pp.68-73 (English translation in "Telecommunications and Radio Engineering (USSR)", No. 3, 1967, pp. 49-53.)

[3.14] Sellers J., et al, "Flexible Braids for Improved Magnetic Shielding of Cables," 1978 EMC Symposium, Atlanta, 1978

[3.15] Kumamary Hiroyuki, "The Crosstalk Characteristics of Coaxial Cables Shielded with Electromagnetically Anisotropic Metallic Layers," Sumitomo Electric Technical Review, No.3, 1964 (For those who prefer reading in Japanese, the original is: The Journal of the Institute of Electrical Communication Engineers of Japan, 46, No.9, 1963)

[3.16] Bethe, H., A., "Theory of Diffraction by Small Holes," Physical Review, vol. 66, 1944, pp. 163-182

[3.17] van Helvoort, M.,J.,A.,M., van Deursen, A.,P.,J., van der Laan, P.,C.,T., "The Transfer Impedance of Cables with a Nearby Return Conductor and a Noncentral Inner Conductor," IEEE Transactions on Electromagnetic Compatibility, vol. 37, No. 2, May, 1995, pp. 301-306.

[3.18] Broyde, F, Clavelier, E., Comparison of Coupling Mechanisms on Multiconductor Cables," IEEE Transactions on Electromagnetic Compatibility, vol. 35, No. 4, November, 1993, pp. 409-416

[3.19] Broyde, et al., Discussion of the Relevance of Transfer Admittance and Some Through Elastance Measurement Results," IEEE Transactions on Electromagnetic Compatibility, vol. 35, No. 4, November, 1993, pp. 417- 422

[3.20] Broyde, F., Clavelier, E., ," Definition, Relevance and Measurement of the Parallel and Axial Transfer Impedances," Proceedings of the 1995 IEEE International Symposium on Electromagnetic Compatibility, Atlanta, GA, 1995, pp.490-495
Also, same authors, "Parallel and Axial Transfer Impedances: Theoretical Summary and Local Measurement Methods," Proceedings of the 11th International Symposium and Technical Exhibition on EMC, Zurich, Switzerland, 1995, pp.501-506

Chapter 4

[4.1] Capacitive, Inductive and Conductive coupling: Physical Theory and Calculation Methods, CCITT Directives, vol. III, Geneva, 1989, 270 p.

[4.2] Smith A., Coupling of Electromagnetic Fields to Transmission Lines, John Wiley & Sons, 1977, 132 p.

[4.3] King R, W., P., Mimno H., R., Wing A., H., Transmission Lines, Antennas, and Wave Guides. Dover Publications, NY, 1965, 347 p.

[4.4] Schelkunoff S., A. and Friis H., T., Antennas, Theory and Practice, John Wiley & Sons, 1952, 639 p.

[4.5] Silver S., Microwave Antenna Theory and Design, McGraw-Hill, NY, 1949, 1984, 623 p.

[4.6] Rikitake T., Magnetic and Electromagnetic Shielding. Terra Scientific Publishing Co., Tokyo, and D. Reidel Publishing Co, Dordrecht, Holland, 1987, 226 p.

[4.7] Tsaliovich A., "Crosstalk in Computer and Communications Cables." Belden Innovators", Part I, November, 1981, pp.5-23, and Part II, July, 1982, pp. 25 - 39.

[4.8] Shwartzman V., Mutual Interference in Telecommunications Cables, "Svyaz" Publishers, Moscow, 1966, 432 p.

[4.9] Mohr R., J., "Coupling Between Open and Shielded Wire Lines Over a Ground Plane," IEEE Transactions on Electromagnetic Compatibility, vol. EMC-9, No. 2, September 1967, pp.34-44.

[4.10] Paul C., R., "Prediction of Crosstalk in Ribbon Cables: Comparison of Model Predictions and Experimental Results," IEEE Transactions on Electromagnetic Compatibility, vol. EMC-20, No. 3, August 1978, pp.394-406.

[4.11] Paul C., R., "Prediction of Crosstalk Involving Twisted Pairs of Wires - Part I: A Transmission Line Model for Twisted-Wire Pairs," IEEE Transactions on Electromagnetic Compatibility, vol. EMC-21, No. 2, May 1979, pp.92-114.

[4.12] Demoulin B., et al., "Shielding Performance of Triply Shielded Coaxial Cables," IEEE Transactions on Electromagnetic Compatibility, vol. EMC-22, August 1980, pp.173-180.

[4.13] Paul C., R., " Transmission-Line Modeling of Shielded Wires for Crosstalk Prediction," IEEE Transactions on Electromagnetic Compatibility, vol. EMC-23, No. 4, November 1981, pp.345-351.

[4.14] Klein W., Die Theorie des Nebensprechens auf Leitungen, Springer, Berlin, 1955.

[4.15] Barrow J., " Avoiding Ground Problems in High-Speed Circuits," RF Design, July 1989, pp.32-34.

[4.16] Leferink F., B., J., "Inductance Calculations; Methods and Equations," Proceedings of the 1995 IEEE International Symposium on Electromagnetic Compatibility, Atlanta, GA, 1995, pp.435-439

[4.17] "Impedance of Ground Conductors," by ICT Staff, " EMC Technology", May-June, 1991, pp.29-30.

BIBLIOGRAPHY 655

[4.18] Paladian, F. and Le Fevre, D., " Measuring the Impedance of Grounding Systems at High Frequencies," Proceedings of the 7th International Symposium and Technical Exhibition on EMC, .Zurich, Switzerland, 1987, pp.579-582.

[4.19] Adams, A., T., Leviatan, Y. and Nordby K. S., " Electromagnetic Near Fields as a Function of Electrical Size," IEEE Transactions on Electromagnetic Compatibility, Volume EMC-25, No. 4, November 1983, pp. 428-432

[4.20] Adams A., T. et al, "Electromagnetic Field-to-Wire Coupling in the SHF Frequency Range and Beyond," IEEE Transactions on Electromagnetic Compatibility, vol. EMC-29, No. 2, May 1987, pp.126-131.

[4.21] Ari N. and Blumer W., "Analytic Formulation of the Response of a Two-Wire Transmission Line Excited by a Plane Wave," IEEE Transactions on Electromagnetic Compatibility, vol. EMC-30, No. 4, November 1988, pp.437-448.

[4.22] Cellozzi, S. and Feliziani, M., " FEM Analysis of the Plane-Wave Electromagnetic Field Coupling to a Multiconductor Line," Proceedings of the 9th International Symposium and Technical Exhibition on EMC, .Zurich, Switzerland, 1991, pp.127-132.

[4.23] Ehrich M. and Mrozynski G., "Transient Shielding of Conducting Screens," Proceedings of the 1985 IEEE International Symposium on Electromagnetic Compatibility, Wakefield, MA, 1985, pp.38-48

[4.24] Aguet, M., Ianovici, M., Chung-Chi Lin, "Transient Electromagnetic Field Coupling to Long Shielded Cables," IEEE Transactions on Electromagnetic Compatibility, vol. EMC-22, No. 4, November 1980, pp.276-282.

[4.25] Stern, R, B., "Time Domain Calculation of Electric Field Penetration Through Metallic Shields," IEEE Transactions on Electromagnetic Compatibility, vol. 30, No. 3, August 1988, pp.307-311.

[4.26] Voronkov, A.A., Sidnev, S.A.,Timofeyev, S.A.,., "Protection Properties of Cable Sheaths under the Impact of the Impulse Currents," Electrosvyaz'", No. 2, 1988, pp.32-36.

[4.27] Wolff E. A., Antenna Analysis, Artech House, Inc., 1988, 534 p

[4.28] Clemmow,, P.,C., The Plane Wave Spectrum Representation in Electromagnetic Fields. London, Pergamon Press, 1966

[4.29] Yang, R., Mittra, R., "Coupling Between Two Arbitrarily Oriented Dipoles Through Multilayered Shields," IEEE Transactions on Electromagnetic Compatibility, vol. EMC-27, No. 3, August 1985, pp.131-136.

[4.30] Nishikata, A., Sugiura, A., "Analysis for Electromagnetic Leakage Through a Plane Shield with an Arbitrarily-Oriented Dipole Source," IEEE Transactions on Electromagnetic Compatibility, vol. 34, No. 1, February, 1992, pp.284-291.

[4.31] Bucci, O., M., et. al., "Far-Field Pattern Determination from the Near-Field Amplitude Measurements on Two Surfaces," IEEE Transactions on Antennas & Propagation, vol. 38, No. 11, November, 1990

[4.32] Laroussi, R., Costache, G., I., "Far-Field Predictions from Near-Field Measurements Using an Exact Integral Equation Solution," IEEE Transactions on Electromagnetic Compatibility, vol. 36, No. 3, August, 1994, pp.189-195

Chapter 5

[5.1] Kaden H., Das Nebensprechen Zwischen parallelen koaxialen Leitungen," "ENT", November, 1936

[5.2] Schelkunoff S. A., Odarenko T.,M., "Crosstalk Between Coaxial Transmission lines," BSTG, XVI, No. 2, 1937

[5.3] "Tsaliovich A., Mutual Interference Between Coaxial Lines with Contacting Outer Conductors," Elektrosvyaz, No. 12, 1971, pp.74-75

[5.4] Gould K., E., "Crosstalk in Coaxial Cables - Analysis Based on Short-circuited and Open Tertiaries," BSTG, XIX, No. 3, 1940

[5.5] Roch R., Badellon R. et Orsini J., Parturbations de la propagation dans les paires coaxiales. "Cables et Transmission", 1967, October

[5.6] Tsaliovich A., "Crosstalk in Telecommunication Coaxial Cable Lines," Elektrosvyaz, No. 5, 1971, pp.49-53

[5.7] Goel, A. K., *High-Speed VLSI Interconnections: Modeling, Analysis, and Symulation*," John Wiley & Sons, Inc., 1994, 622 p

[5.8] Farhat N.H., Yung-Ping Loh, Showers R.M., "Effects of Partial Shields on Transmission Lines at Low Frequencies," IEEE Transactions on Electromagnetic Compatibility, vol. EMC-10, No. 1, March 1968, pp.44-51.

[5.9] Myung-Kul Kim, "Crosstalk Control for Microstrip Circuits on PCBs at Microwave Frequencies," Proceedings of the 1995 IEEE International Symposium on Electromagnetic Compatibility, Atlanta, GA, pp.459-464

[5.10] Tsaliovich A., "Electromagnetic Compatibility Performance of Electronic Cables." Belden Innovators", November, 1983,pp.7-23.

[5.11] Tsaliovich A., "Defining and Measuring EMC Performance of Electronic Cables." Proceedings of the 1983 IEEE International Symposium on Electromagnetic Compatibility, Crystal City, VA, pp.214-219.

[5.12] Wilson, P. ,F., " A Comparison Between Near-Field Shielding-Effectiveness Measurements Based on Coaxial Dipoles and on Electrically Small Apertures," IEEE Transactions on Electromagnetic Compatibility, Volume 30, No. 1, February 1988, pp. 23-28

[5.13] A.J. Maddocks, Ground Plane on Printed Circuit Boards Improve the Immunity of Circuits to Electromagnetic Fields, ERA Technology Ltd.

[5.14] Montrose, M., I., Printed Circuit Board Design Techniques for EMC Compliance," IEEE Press, 1995

[5.15] Grignon, N. "Shielding of an Electric Line Source by a Finite Ground Plane and a Narrow Metal Strip," Proceedings of the 1995 IEEE International Symposium on Electromagnetic Compatibility, Atlanta, GA, 1995, pp.313-317

[5.16] Rusek, A., "Ground Currents Tests in Metal Planes,"Proceedings of the 1995 IEEE International Symposium on Electromagnetic Compatibility, Atlanta, GA, 1995, pp.318-319

[5.17] Smith, T.,S. and Paul, C., R., "Effect of Grid Spacing on the Inductance of Ground Grids,","Proceedings of the 1991 IEEE International Symposium on Electromagnetic Compatibility, Cherry Hill, NJ, 1991, pp.72-77

[5.18] Paul C., R., "Prediction of Crosstalk in Ribbon Cables: Comparison of Model Predictions and Experimental Results," IEEE Transactions on Electromagnetic Compatibility, vol. EMC-20, No. 3, August 1978, pp.394-406.

[5.19] German R., F., Ott H.,W., and Paul C.,R., "Effect of an Image Plane on Printed Circuit Board Radiation," Proceedings of the 1990 IEEE International Symposium on Electromagnetic Compatibility, Washington, DC, 1990, pp. 284-291

[5.20] Swainson A.,J.,G., "Radiated Emission, Susceptibility and Crosstalk Control on Ground Plane Printed Circuit Boards," IERE Seventh International Conference on Electromagnetic Compatibility Proceedings, York, England, 1990, pp. 37-41

[5.21] Dockey, R.,W., "Asymmetrical Mode Radiation from Multi-Layer Printed Circuit Boards," EMC/ESD International Symposium, Denver, 1992, pp.247-251

[5.22] Dockey, R.,W. and German R., F., "New Techniques for Reducing Printed Circuit Board Common-Mode Radiation," Proceedings of the 1993 IEEE International Symposium on Electromagnetic Compatibility, Dallas, TX, 1993, pp. 334-339

[5.23] Quine, J.,P., et al., "Distortion of Radiation Patterns for Leakage Power Transmission Measurements Through Attenuating Cover Panels and Shielding Gaskets - Need for Reverberation Chamber Measurements of Total Leakage," Proceedings of the 1994 IEEE International Symposium on Electromagnetic Compatibility, Chicago, IL, 1994, pp. 285-290

[5.24] Johnson, D., M., Hatfield, M., G., "Shielding Effectiveness Measurements of a Shielded Window: Comparative Results Obtained Using Mode-Stirred and Anechoic Chambers," Proceedings of the 1995 IEEE International Symposium on Electromagnetic Compatibility, Atlanta, GA, 1995, pp. 378-382

[5.25] Alexander, S., E., "Characterising Building for Propagation at 900 MHz," Electronic Letters, 29th September 1983, Vol. 19, No.20, p. 860

[5.26] Walker, E.,H., "Penetration of Radio Signals Into Buildings in the Cellular Radio Environment," The Bell System Technical Journal, Vo. 62, No.9, November 1983, pp.2719-2734

[5.27] Cox, B. D., Murray, R.,R.,, Norris, A.,W., "800-MHz Attenuation Measured in and Around Suburban Houses," The Bell System Technical Journal, Vo. 63, No.6, July-August 1984, pp.921-954

[5.28] Kozono, S., Watanabe, K., "Influence of Environmental Buildings on UHF Land Mobile Radio Propagation," IEEE Transactions on Communications, Vol. COM-25, No. 10, October 1977, pp.1133-1143

[5.29] Turkmani, A., M., D., de Toledo, A., F., "Radio Transmission at 1800 MHz into, and Within, Multistory Buildings," IEE Proceedings-I, Vol. 138, No. 6, December, 1991, pp. 577-584

[5.30] Molkdar, D., "Review On Radio Propagation into and Within Buildings," IEE Proceedings-H, Vol. 138, No. 1, February, 1991, pp. 61-73

[5.31] Seidel, S.,Y., Rappaport, T.,S., "914 MHz Path Loss Prediction Models for Indoor Wireless Communications In Multifloored Buildings," IEEE Transactions on Antennas and Propagation, vol. 40, No. 2, 1992, pp.207-217

[5.32] Kraft, C., H., "Modeling Leakage Through Finite Apertures with TLM," Proceedings of the 1994 IEEE International Symposium on Electromagnetic Compatibility, Chicago, IL, 1994, pp. 73-76

[5.33] Booker, H. G., "Slot Aerials and Their Relation to Complementary Wire aerials (Babinet's Principle)," Journal of the IEE, vol. 93, III-A, 1946, pp.620-626

[5.34] Harrington,R., F., Auckland, D., T., "Electromagnetic Transmission Through Narrow Slots in Thick Conducting Screens," IEEE Transactions on Antennas and Propagation, vol. 28, No. 5, 1980, pp.616-622

[5.35] Warne, L.,K., Chen, K.,C., "Slot Apertures Having Depth and Losses Described By Local Transmission Line Theory," IEEE Transactions on Electromagnetic Compatibility, Vol.32, No. 3, August 1990, pp. 185-196

[5.36] Warne, L.,K., Chen, K.,C., "A Simple Transmission Line Model for Narrow Slot Apertures Having Depth and Losses," IEEE Transactions on Electromagnetic Compatibility, Vol.34, No. 3, August 1992, pp. 173-182

[5.37] Reed, E., K., "Time-Domain Electromagnetic Penetration Through Arbitrarily Shaped Narrow Slots in Conducting Screens," IEEE Transactions on Electromagnetic Compatibility, Vol.34, No. 3, August 1992, pp. 161-172

[5.38] Hill, D., A., et al., "Aperture Excitation of Electrically Large, Lossy Cavities," IEEE Transactions on Electromagnetic Compatibility, Vol.36, No. 3, August 1994, pp. 169-178

[5.39] Chen, C., C., "Transmission Through A Conducting Screen Perforated Periodically With Apertures," IEEE Transactions on Microwave Theory Techniques, vol. MTT-18, No. 5, September 1970, pp. 627-623

[5.40] Ko, W., I., Mittra, "Scattering by a trunkated periodic array," IEEE Transactions on Antennas and Propagation, vol.36, No. 4, April 1988, pp. 496-503

[5.41] Criel, S., Martens, L., De Zutter, D., "Theoretical and Experimental Near-Field Characterization of Perforated Shields," IEEE Transactions on Electromagnetic Compatibility, Vol.36, No. 3, August 1994, pp. 161-168

[5.42] Criel, S., et al., "Near Field Penetration Through a Perforated Flat Screen," Proceedings of the 1993 IEEE International Symposium on Electromagnetic Compatibility, Dallas, TX, 1993, pp. 200-201

[5.43] Miyake Shinto, et. al., "Investigation Related to Construction Method and Performance of an Electromagnetic Shielded Enclosure," Proceedings of the 1991 IEEE International Symposium on Electromagnetic Compatibility, Cherry Hill, NJ, 1991, pp. 120-125

[5.44] Hill, C., Kneisel, T., "Portable Radio Antenna Performance in the 150, 450, 800, and 900 MHz Bands "Outside" and In-Vehicle," IEEE Transactions on Vehicular Technology, Vol.40, No. 4, November 1991, pp. 750-755

[5.45] Gandhi, O.,P., Editor, Biological Effects and Medical Applications of Electromagnetic Emergy, Prentice Hall, 1990, 573 p

[5.46] Kuster, N., Balzano, Q., Lin, J.C., Editors, Mobile Communications Safety, Chapman & Hall, 1996 288 p

[5.47] Nakauchi, E., Downs, J., "Potential Interference Problems Between Wireless Products and Personal Computers," Proceedings of the 1995 IEEE International Symposium on Electromagnetic Compatibility, Washington, DC, 1995, pp. 507-509

[5.48] Interference to Hearing Aids by the Digital Mobil Telephone System, Global System for Mobil Communications (GSM), National Acoustic Laboratories (NAL) Report No. 131, May, 1995, Australia

[5.49] Moreno, P., Olsen, R., G., "A Simple Theory for Optimizing Finite Width ELF Magnetic Field Shields for Minimum Dependence on Source Orientation," IEEE Transactions on Electromagnetic Compatibility, Vol.39, No. 4, November 1997, pp. 340-348

[5.50] Yildrim, B.,S., "Analysis of a Magnetically-Shielded Cellular Phone Antenna Using Finite-Difference Time-Domain Method," 1996 IEEE MTT-S International Microwave Symposium Digest, vol. 2, 1996, pp. 979-982.

Chapter 6

[6.1] Testing Methods and Measuring Apparatus, CCITT Directives, vol. IX, Geneva, 1989, 273 p.

[6.2] MIL-STD-285, "Attenuation Measurements of Enclosures, EM Shielding for Electronic Test Purposes, Method of," 1956.

[6.3] IEEE Std 299-1991 "Standard Method of Measuring the Effectiveness of the Electromagnetic Shielding Enclosures," IEEE, Inc., July 26, 1991.

[6.4] MIL-STD-1377 "Effectiveness of Cable, Connector, and Weapon Enclosure Shielding and Filters in Precluding Hazards of Electromagnetic Radiation to Ordnance, Measurement of," 1971.
MIL-C-85485 "Military Specification. Cable, Electric, Filter Line, Radio Frequency Absorptive," Septembert 1981

[6.5] IEC Publication 96.1 3-d Edition "Radio Frequency Cables: General Requirements and Measuring Methods," 1971.

[6.6] MIL-STD-1344A, "Test Methods for Electrical Connectors", Method 3008, Shielding Effectiveness of Multi-Contact Connectors. 1981.
See also, Jesh, R.,L. " Measurement of Shielding Effectiveness of Different Cable and Shielding Configurations by Mode-Stirred Techniques," NBSIR 87-3076, NBS (NIST), October, 1987, 24 p.

[6.7] IEC Publication 1196.1:1995-05 ,"Radio Frequency Cables, Part 1: Generic Specification — General, Definitions, Requirements and Test Methods."

[6.8] EIA -364-66, EMI Shielding Effectiveness Test Procedure for Electrical Connectors," 1991.

[6.9] DIN 47250 (Teil 6), "Shielding Attenuation of RF Cables in the Frequency Range from 30 to 1000 MHz," 1983. (Absorbing clamp)

[6.10] IEEE-PES "Guide on Shielding Practice for Low Voltage Cables," Project P1143/D4 (Draft), 1991.

[6.11] ANSI C63.4-1992 "Methods of Measurement of Radio-Noise Emissions from Low-Voltage Electrical and Electronic Equipment in the Range of 9 kHz to 40 GHz," IEEE, Inc., 1992.

[6.12] Tsaliovich A., "Absorber Lined Open Area Test Site - A New Type of EMC Test Facility, " Proceedings of the 1988 IEEE International Symposium on EMC, Seattle, WA, 1988, pp.106-111.

[6.13] Berger, S., Tsaliovich, A., "Unlicensed PCS Product EMC Compliance Measurement Rationale and Alternatives," Proceedings of the 1996 IEEE International Symposium on EMC, Santa Clara, CA, 1996, pp.396-401.

[6.14] Bersier, R., "Mesure de l'efficacite du blindage de cables coaxiaux en ondes metriques an moyen de la pince absorbante MDS," Publie par l'Enterprise des postes, telephones et telegraphes Suisses Tirage a part du "Bulletin technique PTT," No.5, 1971

[6.15] Carter, N., J., Bull, D., A. Evolution of Aircraft Clearance Techniques," IERE International Conference on Electromagnetic Compatibility, Guilford, UK, 1984

[6.16] Sultan, M., F. ., "Modeling of a Bulk Current injection setup for Susceptibility Threshold Measurements, " Proceedings of the 1986 IEEE International Symposium on EMC, San Diego, CA, 1986 .

[6.17] Bronaugh, E.,L., Helmholtz for Calibration of Probes and Sensors: Limits of Magnetic Field Accuracy and Uniformity," Proceedings of the 1995 IEEE International Symposium on EMC, Atlanta, GA, 1995, pp. -

[6.18] Bergovoet, J., R., Van Veen, H. "A Large Loop Antenna for Magnetic Field Measurements," Record of the 8th International Symposium and Technical Exhibition on EMC, Zurich, Switzerland, 1989, pp.29-34

[6.19] IEEE Std. C95.3-1991, IEEE Recommended Practice for the Measurement of Potentially Hazardous Electromagnetic Fields — RF and Microwave, 1991, 107 p

[6.20] CENELEC, Considerations for Human Exposure to EMFs from Mobile Telecommunication Equipment (MTE) in the Frequency Range 30 MHz — 6 GHz, Secretariat SC 211/B, WGMTE, February, 1997, 84 p

[6.21] ANSI C63.19-199X "American National Standard for Methods of Measurement of Measurement of Compatibility between Wireless Communications Devices and Hearing Aids," Draft (as of this writing)

[6.22] Simons, K., A., "Calibrating the Belden S.E.E.D.," 1973 (Private Correspondence with author, 1983)

[6.23] Knowles, E., D., Olson, L. W., "Cable Shielding Effectiveness Testing," IEEE Transactions on Electromagnetic Compatibility, Volume EMC-16, February 1974, pp. 16-23

[6.24] Simons, K., A., "A Review of Measuring Techniques for Determining the Shielding Efficiency of Coaxial Cables," IEC Document SC46A/WG1(Simons) 1, May, 1973 (Private Correspondence with author, 1983)

[6.25] Fowler E. P., "Test Methods for Cable Screening Effectiveness - A Review," 6th IERE Conference on EMC, York, 1988. IERE Publication No. 81.

[6.26] Halme L. and Szentkuti B, "The Background for Electromagnetic Screening Measurements of Cylindrical Screens," Technishe Mitteilungen PTT, 3/1988, pp. 105-115.

[6.27] Fowler, E. P. and Halme, L. K. " State of Art in Cable Screening Measurements ," Record of the 9th International Symposium and Technical Exhibition on EMC, Zurich, Switzerland, 1991, pp.151-158.

[6.28] Szhentkuti, B., T. "Shielding Quality of Cables and Connectors: Some Basics for Better Understanding of Test Methods," Proceedings of the 1992 IEEE International Symposium on Electromagnetic Compatibility, Anaheim, CA, pp.294-231

[6.29] Demoulin, B., Kone, L., Rochdi, M., Degauque P., " Comparative Study of Some Methods to Measure the Transfer Impedance of Coaxial Cables in the Few kHz - Few GHz Frequency Range," Proceedings of the 9th International Symposium and Technical Exhibition on EMC, Zurich, Switzerland, 1991, pp.

[6.30] Tsaliovich, A. "Cable and Connector Shielding Test: A Blueprint for a Standard," Proceedings of the 1992 IEEE International Symposium on Electromagnetic Compatibility, Anaheim, CA, pp.315-320

[6.31] Halme, L. "Development of IEC Cable Shielding Effectiveness Standards," Proceedings of the 1992 IEEE International Symposium on Electromagnetic Compatibility, Anaheim, CA, pp.321-328

[6.32] Frankel, S. "Terminal Response of Braided-Shield Cables to External Monochromatic Electromagnetic Fields," IEEE Transactions on EMC, vol. EMC-16, pp.4-16, Feb. 1974

[6.33] Oakley, R., J., "Surface Transfer Impedance Measurements - a Practical Aid to Communication Cable Shielding Design," 18 IWCS, Dec. 1969

[6.34] Zorzy, J., Muehlberger, R., F., R. F. Leakage Characteristics of Popular Cables and Connectors, 500 MC to 7.5 GC," Microwave Journal, Nov. 1961, pp.80-86

[6.35]   Madle, P., J., "Cable and Connector Shielding Attenuation and Transfer Impedance Measurements Using Quadraxial and Quintaxial Test Methods, "1975 IEEE Electromagnetic Compatibility Symposium Record", 1975, pp.4B1b1-4B1b5.

[6.36]   Zimmerman, W.,R. and Wellems, L.,D. "Measuring the Transfer Impedance and Admittance of a Cylindrical Shield Using a Single Triaxial or Quadraxial Fixture," IEEE Transactions on Electromagnetic Compatibility, Volume 35, No. 4, November 1993, pp. 445-450

[6.37]   Simons, K., A., "The Terminated Triaxial Test Fixture," IEC Document SC46A/WG1(Simons) 21, 1973 (Private Correspondence with author, 1983)

[6.38]   Smith K., L., " Cable Transfer Impedance Testing and Performance," EMC Technology, vol.2, No. 4, October-December, 1983, pp.31-37.

[6.39]   Tsaliovich A. , "The transfer Impedance Test Clamp - Improvements in Electronic Cable Shield Effectiveness Measurement Techniques," Proceedings of the 1982 IEEE International Symposium on Electromagnetic Compatibility, Santa Clara, CA, pp.342-347.

[6.40]   Hoeft, L., O. and Hofstra, S. "Measurement of Surface Transfer Impedance of Multi-Wire Cables, Connectors and Cable Assemblies," Proceedings of the 1992 IEEE International Symposium on Electromagnetic Compatibility, Anaheim, CA, pp.308-314

[6.41]   Chalk, D., Hoeft, L., O. and Hofstra, S. "Comparison of Surface Transfer Impedance Measured in Quadraxial and Triaxial (IEC 96-1 and MIL-C-85485) Test Fixtures" Proceedings of the 1983 IEEE International Symposium on Electromagnetic Compatibility, Washington D.C, pp.521-525

[6.42]   Tsaliovich et al. Method and Apparatus for Measuring The Surface Transfer Impedance Of A Piece Of Shielded Cable. Unites States Patent No. 4,425,542. Jan 10, 1984.

[6.43]   Madle P., "Transfer Standards for Cable and Connector Shielding Test Fixtures," Prepared for the TC-4 Committee of the IEEE EMC Society, May, 1989.

[6.44]   Eicher B., et al, "Simple and Accurate Screening Measurements on RF Cables up to 3 GHz," Technishe Mitteilungen PTT, 4/1988, pp. 166-173.

[6.45]   Eicher, B. " Cable Screening Measurements in the Frequency Range 1-20 MHz: Line Injection Method versus Mode-Stirred Chamber," Record of the 9th International Symposium and Technical Exhibition on EMC, Zurich, Switzerland, 1991, pp.159-162.

[6.46] Eicher, B. and Boillot, L. "Very Low Frequency to 40 GHz Screening Measurements on Cables and Connectors; Line Injection Method and Mode Stirred Chambert," Proceedings of the 1992 IEEE International Symposium on Electromagnetic Compatibility, Anaheim, CA, pp.302-307

[6.47] Merewether, D., E. and Ezell, T., F. "The Effect of Mutual Inductance and Mutual Capacitance on the Transient Response of Braided Shield Coaxial Cables," IEEE Transactions on EMC, vol. EMC-18, No. 1, pp.15-20, Feb. 1976

[6.48] Fowler, E., P. "Screening Measurements in the Time Domain and Their Conversion into the Frequency Domain," Jpirnal of the Institution of Electronic and Radio Engineers, Vol. 55, No. 4, pp.127--132, April 1985

[6.49] Bruns, H. D. and Gonschorek, K. D."Efficient Determination of the Complex Cable Transfer Impedance Using Arbitrary Outer Circuits," Record of the 9th International Symposium and Technical Exhibition on EMC, Zurich, Switzerland, 1991, pp.

[6.50] Demoulin B., Duvinage P., Degauque P., " Measurements of Transfer Parameters of Shielded Cables at Frequencies Above 100 MHz," Proceedings of the 6th International Symposium and Technical Exhibition on EMC, Zurich, Switzerland, 1985, pp.521-524.

[6.51] Garbe, H. and Hansen, D. "EMI Response of Cable Systems to Transients Considering Complex Transfer Impedance,"

[6.52] ASTM Standard D 4935-89, "Standard Test Method for Measuring the Electromagnetic Shielding Effectiveness of Planar Materials,"ANSI, September 1989 (watch for possible later issues)

[6.53] Kinnigham, B., A. and Yenni, D.,M. "Test Methods for Electromagnetic Shielding Materials," Proceedings of the 1988 IEEE International Symposium on EMC, Seattle, WA, 1988, pp.223-230.

[6.54] Simon, R. M. and Stutz D., " Test Methods for Shielding Materials," EMC Technology, vol.2, No. 4, October-December, 1983, pp.39-48.

[6.55] Scheps R., D., " Shielding Effectiveness Measurements Using a Dual-TEM Cell Fixture, " EMC Technology, vol.2, No. 3, July-September, 1983, pp.61-65.

[6.56] Wilson, P.,F., Ma, M.T., Adams, J.W., "Techniques for Measuring the Electromagnetic Shielding Effectiveness of Materials: Part I - Far-Field Source Simulation," IEEE Transactions on Electromagnetic Compatibility, Volume 30, No. 3, August 1988, pp. 239-250

[6.57] Wilson P.,F., Ma M.T., "Techniques for Measuring the Electromagnetic Shielding Effectiveness of Materials: Part II - Near-Field Source Simulation," IEEE Transactions on Electromagnetic Compatibility, Volume 30, No. 3, August 1988, pp. 251-260

[6.58] Wilson P.,F., Ma M.T., "A Study of Techniques for Measuring the Electromagnetic Shielding Effectiveness of Materials," NBS Technical Note 1095, NBS, May 1986, 72 pp.

[6.59] Wilson P.,F., Ma M.T., "Shielding Effectiveness Measurements with a Dual TEM Cell," IEEE Transactions on Electromagnetic Compatibility, Volume EMC-27, No. 3, August 1985, pp. 137-142

[6.60] Allen, J.,L. "Electromagnetic Shielding Effectiveness for Isotropic and Anisotropic Materials", Phase Report RADC-TR-81-162, Rome Air Development Center, Griffiss AFB, NY 13441, June 1981

[6.61] Wyatt, K. "An Important Technique for Measuring Shielding Effectiveness,", EMC Test & Design, September/October 1992, pp.15-21

[6.62] ´Catrysse, J.,A. et al. "Correlation Between Shielding Effectiveness Measurements and Alternatice Methods for the Characterization of Shielding Materials," IEEE Transactions on Electromagnetic Compatibility, Volume 35, No. 4, November 1993, pp. 440-444

[6.63] Palasciano, J.,D.,, Dike, S.,B., "Measurement Method Evaluates Cable Shielding Effectiveness," Microwaves and RF, January 1997, pp. 105-108

[6.64] Schwab, A., Herold, J., "Electromagnetic Interference in Impulse Measuring Systems," IEEE Transactions on Power Apparatus and Systems, Vol. PAS-93, No. 1, January/February 1974, pp.333-339

[6.65] Bernauer, J., Schwab, A., "Surface Transfer Impedance of Shielding Enclosures," Proceedings of the 1995 IEEE International Symposium on EMC, Atlanta, GA, 1995, pp.387-391.

[6.66] Bernauer, J., Weis, R., Schwab, A., "Shielding Effectiveness of A 19"-Case Through Direct Current Injection (DCI)," Proceedings of the 1997 IEEE International Symposium on EMC, Austin, TX, 1997, pp.567-572.

[6.67] Tsaliovich, A., Method to Determine Magnetic Permeability of Electricaql Ferromagnetic Conductors, USSR Patent No. 304526, Nov. 13, 1967.

[6.68] Madle, P.,J., "Transfer Impedance and Transfer Admittance Measurements on Gasketed Panel Assemblies, and Honeycomb Air-Vent Assemblies," Proceedings of the 1976 IEEE International Symposium on EMC, Washington, DC, 1976

[6.69] Freyer, G.,J., Hatfield, M.,O., "Comparison of Gasket Transfer Impedance and Shielding Effectiveness Measurements, Parts I and II," Proceedings of the 1992 IEEE International Symposium on Electromagnetic Compatibility, Annaheim, CA, pp.139-148

[6.70] Hatfield, M.,O., "Shielding Effectiveness Measurements Using Mode-Stirred Chambers: a Comparison of Two Approaches," IEEE Transactions on Electromagnetic Compatibility, Volume 30, No. 3, August 1988, pp. 229-238

[6.71] Kunkel, G.,M., "Design of Transfer Impedance Test Fixture Accurate Through 10 GHz," Proceedings of the 1990 IEEE International Symposium on Electromagnetic Compatibility, Washington, DC, pp.628-633

[6.72] Kunkel, G.,M., "Introduction to the Testing for the Shielding Quality of EMI Gaskets and Gasketed Joints," Proceedings of the 1992 IEEE International Symposium on Electromagnetic Compatibility, Annaheim, CA, pp.134-138

[6.73] Adams, J.,W., "Electromagnetic Shielding of RF Gaskets Measured by Two Methods," Proceedings of the 1992 IEEE International Symposium on Electromagnetic Compatibility, Annaheim, CA, pp.154-157

[6.74] Catrysse, J.,A. "A Measuring Method for the Characterization of Shielding Gaskets," Conference Publication No. 362 of IEE 8th International Conference on Electromagnetic Compatibility, Edinburgh, UK, 1992, pp. 251-255

[6.75] Quine, J.,P., "Shielding Effectiveness of an Enclosure Employing Gasketed Seams — Relation Between SE and Gasket Transfer Impedance," Proceedings of the 1995 IEEE International Symposium on EMC, Atlanta, GA, 1995, pp.392-395.

Chapter 7

[7.1] Satler, F. and Gonschorek, K.-H. "Measurement and Computation of Cable Coupling in Arbitrary Environment," Proceedings of the 1994 IEEE International Symposium on Electromagnetic Compatibility, Chicago, IL, 1994, pp.5-10

[7.2] Fowler, E., P. "Cables and Connectors - Their Contribution to Electromagnetic Compatibility," Proceedings of the 1992 IEEE International Symposium on Electromagnetic Compatibility, Anaheim, CA, pp.329-333

[7.3] A. Tsaliovich, The methods for Communications Cable Line Parameter Optimization," "Sviaz" Publishers, Moscow, 1973, 96 pp.

[7.4] Harrington G.,J., Shultz R., B,. "Design of Minimum Weight and Maximum Effectiveness of Very-Low-Frequency Shielding," IEEE Transactions on Electromagnetic Compatibility, vol. EMC-10, No.1, March 1968, pp. 152-157.

[7.5] Goto E. and Soma T., "Optimization of Anisotropic Magnetic Shielding," IEEE Transactions on Electromagnetic Compatibility, vol. EMV-29, No. 3, Aug. 1987, pp. 237-241.

[7.6] Homann, E. "Geschirmte Kabel mit optimalen geflechtschirmen," Nachrichtentechnische Zeitschrift, vol 21, No.3, 1968, pp.155-161

[7.7] Sali, S., "Screening Efficiency of Triaxial Cables with Optimum Braided Shields, " IEEE Transactions on Electromagnetic Compatibility, vol. 32, No. 2, 1990, pp. 125-136

[7.8] Kley, T., "Optimized Single- Braided Cable Shields, " IEEE Transactions on Electromagnetic Compatibility, vol. 35, No. 1, 1993, pp. 1-9

[7.9] Mason, L. G. and Combot J.-P., "Optimal Modernization Policies for Telecommunications Facilities," IEEE Transactions on Communications, vol. COM-28, No.3, March 1980, pp. 317-324.

[7.10] Holland, L.D., et al., "System Analysis for Millimeter-Wave Communication Satellites," Microwave Journal, June 1980, pp. 35-43.

[7.11] Kraft, F. B. "Three Fine Wire Drawing Systems - an Economic Comparison," Wire Journal, July 1980, pp. 35-43.

[7.12] Sioshansi, F. P. and Whinston, A. B. "A Two-Period Model of Choice Under Life and Death Uncertainty," IEEE Transactions on Systems, Man, and Cybernetics, vol. SMC-10, No.10, October 1980, pp. 616-623

[7.13] Lightner, M. R.. and Director, S. W. "Multiple Criterion Optimization for the Design of Electronic Circuits," IEEE Transactions on Circuits and Systems, vol. CAS-28, No.3, March 1981, pp. 169-179.Kley, T., " Optimized Single- Braided Cable Shields, " IEEE Transactions on Electromagnetic Compatibility, vol. 35, No. 1, 1993, pp. 1-9

[7.14] Cerri, G., De Leo, R., Primiani, V., M., "Theoiretical and Experimental Evaluation of the Electromagnetic Radiation from Apertures in Shielded Enclosures," IEEE Transactions on Electromagnetic Compatibility, vol. 34, No. 4, 1992, pp. 423-432

[7.15] Radu, S., et al., "Investigation of Internal Partitioning in Metallic Enclosures for EMI Control," Proceedings of the 1997 IEEE International Symposium on Electromagnetic Compatibility, Austin, TX, 1997, pp. 171-176

[7.16] Li, M., et al., "EMI from Apertures at Enclosure Cavity Mode Resonances," Proceedings of the 1997 IEEE International Symposium on Electromagnetic Compatibility, Austin, TX, 1997, pp. 183-187

[7.17] Min Li, et al, "Numerical and Experimental Corroboration of an FDTD Thin-Slot Model for Slots Near Corners of Shielding Enclosures," IEEE Transactions on Electromagnetic Compatibility, vol. 39, No. 3, August 1997, pp. 225-232

[7.18] Hemming, L., H., Architectural Electromagnetic Shielding Handbook, A Design and Specification Guide, IEEE Press, 1992, 222 p

[7.19] Davis, C., A., "Improving the Clamshell: New Cover Design Makes EMI/RFI Shielding Standard," Electronic Manufacturing, February 1988, pp.27-29.

[7.20] Goodall, D.,R., "Shielding Architectural Windows,", Electromagnetic News Report, July-August 1991, pp. 18-20

[7.21] Gupta, K. G. et al, "Microstrip Lines and Slotlines," Second Edition, Artech House, Inc., 1996, 470 p.

[7.22] Van der Laan, P., C., T., Van Houten, M. A. and Van Deursen, A. P., J., " Grounding Philosophy," Proceedings of the 7th International Symposium and Technical Exhibition on EMC, .Zurich, Switzerland, 1987, pp.567-568.

[7.23] Montandon, R. and Szentkuti, B., " The Rationale of Earthing and EMC Requirements for the Swiss PTT's Digital Exchanges," Proceedings of the 7th International Symposium and Technical Exhibition on EMC, .Zurich, Switzerland, 1987, pp.573-578.

[7.24] Walter, C., H., Traveling Wave Antennas. McGraw-Hill Book Company, 1965, 429 pp.

[7.25] Vance E.,F., "Cable Grounding for the Control of EMI," EMC Technology", January-March, 1983, pp.54-58.

[7.26] Moser R., J., "Peripheral Cable Shield Termination: The System EMC Kernel," IEEE Transactions on Electromagnetic Compatibility, vol. EMC-28, No. 1, February 1986, pp.40-45.

[7.27] Hejase H. et al, "Shielding Effectiveness of 'Pigtail' Connections," Compliance Engineering, Summer 1993, pp.55-67.

[7.28] Demoulin, B., Degauque, P. Gabillard, R. " Transient Response of Braided-Wire Shield," Proceedings of the International Symposium and Technical Exhibition on EMC, Zurich, Switzerland, 19 , pp.19-26.

[7.29] Demoulin B. and Degauque., "Effect of Cable Grounding on Shielding Performance," EMC Technology", October-December, 1984, pp.65-73.

[7.30] Jones,J., W. E., " Grounding, Bonding and Inter-Unit Wiring," Proceedings of the 7th International Symposium and Technical Exhibition on EMC, .Zurich, Switzerland, 1987, pp.559-564.

[7.31] Shang Fang , "Electromagnetic Leakage From Shielded Cables by Pigtail Effect," Proceedings of the 1992 IEEE International Symposium on Electromagnetic Compatibility, Anaheim, CA, pp.278-282

[7.32] Tassone, A., R., Jr., Isolation Amplification in EKG Monitor, EDN Product Edition, May 15, 1996, p. 14

[7.33] Teshe, F. M. et al., "Internal Interaction Analysis: Topological Concepts and needed model improvements," Interaction Notes, Note 248 , AFWL, Kirtland AFB, NM, July, 1975.

[7.34] Vance E. F., "Shielding and Grounding Topology for Interference Control," Interaction Notes, Note 306 , AFWL, Kirtland AFB, NM, April, 1977.

[7.35] Teshe, F. M., "Topological Concepts for Internal EMP Protection," IEEE Transactions on Antennas and Propagation, vol. AP-26, No. 1, January 1978.

[7.36] Vance E. F., "Electromagnetic-Interference Control, " IEEE Transactions on Electromagnetic Compatibility, vol. EMC-22, No. 4, November 1980, pp. 319-328.

[7.37] Lee, K.,S.,H., "EMP Interaction: Principles, Techniques, and Reference Data," Hemisphere

[7.38] Baker, G., Castillo, J.,P., Vance E. F., "Electromagnetic - Interference Control," IEEE Transactions on Electromagnetic Compatibility, vol. 34, No. 3, August 1992, pp. 267-274.

[7.39] Hansen R.,C., "Fundamental Limitations in Antennas,", Proceedings IEEE, vol. 69, No.2, February, 1981

[7.40] Electronic Cable Handbook, by the Engineering Staff of Belden Manufacturing Company, Howard W. Sams & Co, 1966, 224 pp.

[7.41] A. Tsaliovich et. al., Watertight Telephone Cables," "Sviaz" Publishers, Moscow, 1977, 87 pp.

[7.42] Plantz, V.C., Schultz , R.B., Goldman, R. "Temperature Stress and Nuclear Radiation Effects on Electromagnetic Shielding," IEEE Electromagnetic Compatibility Symposium Record, Washington, D.C., July 18-20, 1967, pp.413-432

[7.43] Smith, K., L. "Drop Cable RF Leakage Throughout 20 Years of Service,", EMC Technology,July, 1982, pp.80 - 88

[7.44] Jackson, R. H.,. " 20-YearEvaluation of Shielding Tape", RF Design, November, 1987, pp. 85 - 86

[7.45]   Amato T., Jr., Mis D., J., Willard B., B., "Shielding Effectiveness Before and After the Effects of Environmental Stress on Metallized Plastics," IEEE Transactions on Electromagnetic Compatibility, vol. 30, No.3, August 1988, pp. 312-325.

[7.46]   Tobias, P. and Trindade, D., "Applied Reliability," Van Nostrand Reinhold, NY, 1986

[7.47]   Comerford, R.,. " The Basics of Environmental Testing", Electronics Test, November, 1989, pp.30 - 35

[7.48]   Eyring, H., Glasstones, S., and Laidler, K.,J., "The Theory of Rate Processes," McGraw Hill, NY, 1941

[7.49]   Bastenbeck, E., et. al., "A Comparison of Conductive Coatings for EMI Shielding Applications," ITEM, 1995, pp. 100-106,278-280, 287.

[7.50]   Das, S.K., Nuebel, J., Zand, B., "An investigation on the Sources of Shielding Degradation for Gaskets with Zinc Coated Steel Enclosures," Proceedings of the 1997 IEEE International Symposium on Electromagnetic Compatibility, Austin, TX, pp.66-71

[7.51]   Lessner, P., Inman, D., "Quantitative Measurement of the Degradation of EMI Shielding and Mating Flange Materials After Environmental Exposure," Proceedings of the 1993 IEEE International Symposium on Electromagnetic Compatibility, Dallas, TX, pp.207-213

[7.52]   Denny, H.,W., Shouse, K.,R., "EMI Shielding of Conductive Gaskets in Corrosive Environments," Proceedings of the 1990 IEEE International Symposium on Electromagnetic Compatibility, Washington, DC, pp.20-24

[7.53]   Bates, R., et al, "EMI Gaskets — Disspelling the Technology Folklore," Conference Publication No. 362 of IEE 8th International Conference on Electromagnetic Compatibility, Edinburgh, UK, 1992, pp. 246-250

[7.54]   Nudd, H.,R., "Optimized Performance from Coaxial Feeder Cables and Connectors for Wireless Systems," Microwave Journal, July 1997, pp.88-98; see also Nudd, H.,R., in Wireless Design and Development, 11/96

[7.55]   Carlson, B., "RF/Microwave Connector Design for Low Intermodulation Generation," Interconnection Technology, July 1993, pp.23-27

[7.56]   Tromp, L.,D., Rudko, M., "Rusty Bolt EMC Specification Based on Nonlinear System Identification," Proceedings of the 1985 IEEE International Symposium on Electromagnetic Compatibility, Boston, MA, 1985, pp.419-425

[7.57]   Bright, C., I., "Broadband EMI Shielding for Electro-Optical Systems," Proceedings of the 1994 IEEE International Symposium on Electromagnetic Compatibility, Chicago, IL, pp.340-342

Chapter 8

[8.1] CCIR, "Leaky Feeder Systems in the Land Mobile Service," Report 902-1, Annex 1 to Vol. VIII, Geneva, 1990, pp.256-265

[8.2] Delogne, P., Leaky Feeders and Subsurface Radio Communication, Peter Peregrinus Ltd., 1982, 283 p.

[8.3] Gale, D.,J. and Beal, J.,C., "Comparative Testing of Leaky Coaxial Cables for Communications and Guided Radar," IEEE Transactions on Microwave Theory and Techniques, vol. MTT-28, No.9, September 1980

[8.4] Hill, D.,A. and Wait, J.,R., "Electromagnetic Characteristics of a Coaxial Cable with Periodic Slots," IEEE Transactions on Electromagnetic Compatibility, November 1980, pp. 303-307. See also,
Wait, J.,R., Electromagnetic Radiation from Cylindrical Structures, Peter Peregrinus Ltd, 1988, 202 p

[8.5] Levisse, A., "Leaky or Radiating? Radiation Mechanisms of Radiating Cables and Leaky Feeders - Channel Tunnel Applications," Proceedings of the 1992 International Wire and Cable Symposium, Reno, NV, pp.739-747.

[8.6] Coraiola, A. et al, "Leaky Coaxial Cable with Length Independent Antenna Receiving Level," Proceedings of the 1992 International Wire and Cable Symposium, Reno, NV, pp. 748-756.

[8.7] Watanabe, K. et al, "New Widebandwidth Microwave Leaky Coaxial Cables," Proceedings of the 1993 International Wire and Cable Symposium, St. Louis, MO, pp. 596-600.

[8.8] T. Tamir, "Leaky-Wave Antennas" in Antenna Theory, part 2, editors Collin, R.,E. and Zucker, F.,J., McGraw-Hill Co, 1969, pp.259-295

[8.9] Burt, D., J., "Radiating Cable Enhances In-Building And Tunnel Coverage," Wireless System Design, September, 1996, pp.40-45

[8.10] Proceedings of the SPE Regional Technical Conference on EMI/RFI Shielding Plastics, June, 1982, Chicago, IL., 297 p.

[8.11] Blanchard J.,P., et al., "Electromagnetic Shielding by Metallized Fabric Enclosure: Theory and Experiment," IEEE Transactions on Electromagnetic Compatibility, vol. 30, No.3, August 1988, pp. 282-288.

[8.12] Schoch, K.,F. and Saunders, H.,E., "Conducting Polymers," IEEE Spectrum, June 1992, pp. 52-55

[8.13] Chamberlain, G., "Conducting Polymers Open New Worlds," Design News, 1-21-91, pp. 60-66

[8.14] Naishadham, K., "Shielding Effectiveness of Conductive Polymers," IEEE Transactions on Electromagnetic Compatibility, vol. 34, No.1, February 1992, pp. 47-50.

[8.15] Colaneri, N.,F. and Shacklette, L.,W., "EMI Shielding Measurements of Conductive Polymer Blends," IEEE Transactions on Instrumentation and Measurement, vol. 41, No.2, April 1992, pp. 291-297.

[8.16] Delmonte, J., Metal/Polymer Composites, Van Nostrand Reinhold, NY, 1990, 250 pp.

[8.17] Huan-Ke Chiu, Ming-Shing Lin, Chun Hsiung Chen, Near-Field Shielding and Reflection Characteristics of Anisotropic Laminated Planar Composites," IEEE Transactions on Electromagnetic Compatibility, vol. 39, No. 4, November 1997, pp.332-339.

[8.18] Rowberry, P.,J., "Intrinsically Conductive Polymers for Electromagnetic Screening," Conference Publication No. 396 of IEE 9th International Conference on Electromagnetic Compatibility, Manchester, UK, 1994, pp. 132-137

[8.19] Marshall Space Flight Center, Alabama, "Electrically Conductive Thermal-Control Coating Materials,", NASA Tech Briefs, November 1997, p. 65

[8.20] Halbach, L. and Kirshvink, M. "RF-Absorptive Cables," Euro EMC '90 Conference, London, UK, October 1990

[8.21] Malozemoff, A. P. "Superconducting Wire Gets Hotter," IEEE Spectrum, December 1993, pp. 26 -30

[8.22] Feinberg, A.,A., Johnson, D., W., and Rhodes, W.,W., "Low-Frequency Magnetic Shielding Studies on High Tc Superconductor Y-Ba-Cu-O," IEEE Transactions on Electromagnetic Compatibility, vol. 32, No.4, November 1990, pp. 277-283.

[8.23] .Willis, J.,O., et al, "Magnetic Shielding by Superconducting Superconductor Y-Ba-Cu-O Hollow Cylinders,"IEEE Transactions on Magnetics, vol. 25, No.2, March 1989, pp. 2502-2505

[8.24] Bourdillon, A. and Bourdillon, N.,X., High Temperature Superconductors, Academic Press, 1994, 277 pp.

[8.25] Advances in High Temperature Superconductivity, Edited by I.S.I. - Andreone, D., Connelli, R., S., and Mezzetti, E., Worlds Scientific Publishing Co., 1991, 348 pp.

[8.26] Langley Research Center, Hampton, VA, "Spray-Deposited Superconductor/Polymer Coatings," NASA Tech Briefs, November 1993, pp.80-82

[8.27] Vendik, O. and Ter-Martirosyan, L., "Superconductors Spur Applicationss of Ferroelectric Films," Microwaves & RF, July 1994, pp. 67-70

[8.28] Bashkirov, Yu., A., et al, "Application of Superconducting Shields in Current-Limiting and Special-Purpose Transformers," IEEE Transactions on Applied Superconductivity, vol. 5, Issue 2, Part I, June 1995, pp. 1075-1078

[8.29] Weyand, K., Yi Zhang, Bousack, H., "Compound Material for Electromagnetic Shielding in Electrical and Cryoelectronical Applications," Digest of 1994 Conference on Precision Electromagnetic Measurements, June-July 1994, pp. 140-141

[8.30] Recent Advances in Electromagnetic Theory, edited by Kriticos, H., N. and Jaggard, D.,L., Springer Verlag, 1990, 392 pp

[8.31] Engheta, N. and Jaggard, D., L. "Frequency Spectrum and Time Domain Response of Chiral Coated Shields," Proceedings of the 9th International Symposium and Technical Exhibition on EMC, Zurich, Switzerland, 1991, pp.93-96.

[8.32] Jaggard, D., L. and Engheta, N. "Novel Use of Layered Chiral Materials for Control of Absorption, Reflection and Transmission," Proceedings of the 9th International Symposium and Technical Exhibition on EMC, Zurich, Switzerland, 1991, pp.97-100

[8.33] Guire, T., Varadan, V., V., and Varadan, V., K., "Influence of Chirality on the Reflection of EM Waves by Planar Dielectric Slabs," IEEE Transactions on Electromagnetic Compatibility, vol. 32, No.4, November 1990, pp. 300-303.

[8.34] Lakhtakia, A., "Stratified Planar Chiral Shields," Proceedings of the 1993 IEEE International Symposium on Electromagnetic Compatibility, Dallas, TX, pp.179-182

[8.35] Sihvola, A.,H., Ermutlu, M.,E., "Shielding Effect of Hollow Chiral Sphere," IEEE Transactions on EMC, Vol. 39, No.3, August 1997, pp. 219-224

[8.36] Hatakeyama, K., Togawa, H., "Evaluation Method for Shielded Gasket at Microwave and Millimeter Waves," IEEE Transactions on EMC, Vol. 39, No.4, November 1997, pp. 349-355

[8.37] Williams, N., et al, "Measurement of Transmission and Reflection of Conductive Lossy Polymers at Millimeter-Wave Frequencies," IEEE Transactions on EMC, Vol. 32, No.3, August 1990, pp. 236-240

[8.38] Neelakanta, P.,S., "'Smart' Shielding May Modify Performance To Fit," EMC Technology, May/June 1990, pp.25-28

[8.39] Bevensee, R.,M. and Dempsey, R.,C., The Synaptica Antenna for Reconfigurable Array Applications - Description," 1989 IEEE AP-S, International Symposium Digest, Vol. 2, pp.760-761

[8.40] Bevensee, R.,M. and Dempsey, R.,C., The Synaptica Antenna for Reconfigurable Array Applications - Behavior," 1989 IEEE AP-S, International Symposium Digest, Vol. 2, pp.764-765

Appendix

[A.1] Stratton J.A., Electromagnetic Theory, McGraw-Hill, 1941, 615 p.

[A.2] Ramo S., Whinnery J. R., Van Duzer T., Fields and Waves in Communication Electronics, John Wiley & Sons, 1984, 817 p.

[A.3] Kraus J.D., Electromagnetics, McGraw-Hill, 1992 (Fourth Edition), 847 p.

[A.4] King R., W., P., Fundamental Electromagnetic Theory, Dover Publications, 1963.

[A.5] Jordan E.C., Balmain K.G, Electromagnetic Waves and Radiating Systems, Prentice Hall, 1968, 753 p.

[A.6] Hayt W.,H., Jr., Engineering Electromagnetics, McGraw-Hill, 1989, 472 p.

[A.7] Booton, R., C., Jr., Computational Methods for Electromagnetics and Microwaves, John Wiley & Sons, 1992, 182 p.

[A.8] Chari, M.,V.,K., Silvester, P., P., Finite Elements in Electrical and Magnetic Field Problems, John Wiley & Sons, 1980, 219 p

[A.9] Christopoulos, C., The Transmission-Line Modeling Method TLM, IEEE Press, 1995, 220 p

[A.10] Harrington, R.,F., Field Computations by Moment Methods, IEEE Press, 1993, 229 p (an IEEE edition of the book previously published by the Macmillan and the Krieger Publishers)

[A.11] Itoh, T., (editor), Numerical Techniques for Microwave and Millimeter Wave Passive Structures, John Wiley & Sons, 1989, 707 p.

[A.12] James, G., L., Geometrical Theory of Diffraction for Electromagnetic Waves, Peter Peregrinus Ltd, 3d Edition, 1986, 293 p

[A.13] Jin, J., The Finite Element Method in Electromagnetics, John Wiley & Sons, 1993

[A.14] Kunz, K.,S., Luebbers, R., L., The Finite Difference Time Domain Methods for Electromagnetics, CRC Press, 1993

[A.15] Miller, E.,K., Medgyesi-Mitschang, L., Newman, E.,H., (editors), Computational Electromagnetics, IEEE Press, 1992, 508 p

[A.16] Scott, C., The Spectral Domain Method in Electromagnetics, Artech House, 1989, 141 p.

[A.17]   Silvester, P.,P., Ferrari, R.,L., Finite Elements for ElectricalEngineers, Cambridge University Press, 1983, 209 p

[A.18]   Sorrentino, R., (editor), Numerical Methods for Passive Microwave and Millimeter Wave Structures, IEEE Press, 1989, 485 p

[A.19]   Taflove, A., Computational Electrodynamics: The Finite Difference Time Domain Method, Artech House, 1995, 596 p

[A.20]   Ufimtsev, P., Y., Volnovoy Metod v Fizicheskoy Teorii Diffraktsii (Method of Waves in the Physical Theory of Diffraction). Sovetskoye Radio Publishers, Moscow, 1962, 243 pp

[A.21]   Volkov, E.,A., Numerical Methods, Hemisphere Publishing Corporation, 1990, 238 p (original published in Russian by Nauka Publishers, Moscow, as "Chislennye Metody", 1986

[A.22]   Archambeault, B., The World of EMI Modeling, Interference Technology Engineer's Master (ITEM), 1995, pp.47-54, 246-248

[A.23]   Perini, J., "Numerical Methods in EMC - an Update," Tutorial Lectures and Workshop Record of the 11th International Symposium and Technical Exhibition on EMC, Zurich, Switzerland, 1995, pp.3-19

[A.24]   Computer Analysis Techniques for EMC Problems, Workshop, 1991 IEEE International Symposium on Electromagnetic Compatibility. Cherry Hill, NJ

[A.25]   Neelakanta, P.,S., "Handbook of Electromagnetic Materials," CRC Press, 1995, 591 p.

Subject Index

A

absorbing clamp 438, 440
absorptive shielding 600
achiral shields 607
aging
 cable shielding, of 577
ambients
 analog 2
anisotropy
 shield parameters, of 242
antenna
 applying analogy 301
 currents 305
antenna
 active antenna array 15, 55
 current measurements 438
 shielding 343
aperture
 coupling 311
 large 312
 small 311
 resonances 397

B

Babinet
 principle 314
biological safety 412
boundary design. See design by constraints
braid
 cost optimization 519
 design 511
 design options 498
 engineering model 214
 geometry and design 209
 performance 219
 physical processes in 210

C

cable
 coaxial 192
 designs 23
 assembly performance 558
capacitive coupling
 impedance 142
chiral shields 607

cigarette wrap 201
circuit theory
  differential equations 638
  lumped-element networks 634
circuit type
  balanced 23
clock
  pulse shape 29
coaxial 23
computer
  emissions 2
  monitor 2, 335
conetic braid 335
connectors
  effects of shielding 87
coupling
  crosstalk 282
  energy flux 300
  external and internal 175, 177
  galvanic 278
  in multi-dimensional shields 178
  in shielding 245, 249
  path 18
  reduction 284
  regimes 256
  vs electromagnetic environment 246
  common impedance 278
  definitions 249
  effect on transmission mode 28
  electric 272
  electromagnetic 245
  in radiating fields 295
  induction fields, in 264
  magnetic 272
  mechanisms 252
  to shield 245
  to the shield 288
  via third circuit 279
crosstalk 6
  interactions 269
  reducing by shielding 80
  between microstrip lines 358
  between shielded lines 344
  coupling mechanisms 269

  line length dependence 363
  measurement 64
  pulse 64
  reducing 80
  vs cable design 67
  vs transfer impedance 360

D

decoupling
  using shielding, to 245
  vs economics 284
design
  static shields, of 338
  by constraints 506
  criteria 502
  parameters 502
  requirements and objectives 499

diffusion 166
distance/power gradient 394

E

Einstein 306
elastance 142
electromagnetic gasket 191
electromagnetic interference 4
EMC
  definition 6
  limits 10
  modeling 44
  parameters 8
  performance 14
  phenomena 6
  standards 8
EMC performance
  electronic cables of 374
  modeling 45
EMI
  channel 17
  elementary source 53
  elementary sources 15

Index 679

emitters / sensors 18
environment 2, 4
environment, deterministic 367
environment, statistical 367
generators and radiators 15
hearing aid 413
signature 2, 14
study approaches 18
suppression 59
synthesis and analysis 14
system 1
units 10

emission spectrum 2

emissions

computer spectra 6
radiated and conducted 8
system spectra 3

enclosure

audio transformer 335

environment

performance parameters 577
performance stability 577
physical 575

F

fading 298
FEXT 363
field

quasi-stationary 266, 295
static and stationary 264, 324
radiating 300
reactive 270

time variations 253, 254

foil shield 195
Fourier series 35

G

gap

in shield 198
effect of 199

generality

definition 131
evaluation levels 419
model of 133

ground plane 291

effects 304
in PCB 91
test site 378

ground plane

external, at PCB 383
return effects 385

grounding

shielding traces 355
effect on transmission mode 28
general rules 552

GTEM cell 430

H

Helmholtz coils 433

I

impedance

characteristic, plane wave 643
characteristic, transmission line 643
loss 141

integrity

shielding system, of 89

interconnections 18

L

layout

to reduce coupling 286

leaky feeder 590
line 641

injection 461
separation 284

loop antenna system 434
loss constant

M

maximum bounds
 signal spectra 38
 transfer parameter of 163

Maxwell's equations
 differential 623
 integral 631

measurement
 correlation 482
 in time domain 479
 multi-dimensional problem 420
 procedure requirements 418
 statistics 22
 vs evaluation 417

media
 signal propagation, of 253

mesh 205

multi-path 298

multipath propagation
 Rayleigh fading 367
 Rician-distributed 368

N

narrowband 2

near zone 299

NEXT 363

O

optimal
 design 508
 weave angle 518

overlap 200

P

pair
 balanced and twisted 286
 balanced 356

penetration
 radio waves, of 394

phase constant
 line 641

pollution
 environmental 4

propagation constant
 free space 642
 intrinsic 99
 line 641

Q

quadraxial 454

quintaxial 454

R

radiation resistance 305, 307–310

radiators
 dominant 22
 intentional 2, 19

reciprocity
 in shielding 127

reference standards 488

reflection loss 98

relevancy
 definition 131
 evaluation levels 419
 model of 133

return path
 of high frequency currents 180
 external 138
 internal 140

reversibility
 in shielding 128

S

S.E.E.D. 445

seam
 effect of overlap 195
 in enclosure 194
 longitudinal 191

seam
 longitudinal 193
 longitudinal, with overlap 204
secondary 22
shield
 electrostatic 336
 foil 195
 multilayer 221
 multilayer nonhomogeneous 232
 parallel bypass model 119
 periodically perforated 403
 spherical 337
 spiral 188
 derivative 240
 homogeneous solid 149–172
 multilayer 80
 multilayer homogeneous 221
 multilayer nonhomogeneous 228, 516
 open 355
 solid homogeneous 149
shielding
 architectural 81
 electrostatic 326
 enclosures 74
 from currents 173
 from ESD 89
 non-parallel lines 353
 parallel 91
 performance of 319
 separation 292
 series model 121
 system 84
 system definition 129
 taxicab 84
 transmission theory, of 96
 "negative" 101
 as mitigation technique 71
 compensation 340
 definitions 92
 electrostatic 326
 from crosstalk 340
 from radiating fields 295
 magnetostatic 331
 power cord 87
 system definitions 129

system ingredients 125
system view 116
telecommunications cabinet 77
transmission theory 96, 254
wireless product 410
shielding effectiveness
 system specific 134
shielding engineering
 "Black Box" model of 497
shielding measurements
 "absolute" 424
 magnetostatic 433
 roadmap 426
 spiral 185
 non-uniform 182
shorting fold 199, 201
signal
 aircraft 2
 bandwidth 42
 frequency domain 35
 reflection in shielding enclosure 388
 time domain 32
single-ended 23
skin depth 99
slot
 in shield 198
 radiation 398
spectrum
 bandwidth 41
spiral shields 186
 design 511
superconductive shields 604
susceptibility. See **immunity**
system
 configuration 305

T

tape
 longitudinally applied 192

TEM cell 430
tensors
  application to shield study 242
terminated triaxial 455
theorem 1 147
theorem 2 149
theorem 3 186, 189
  solenoidal 186
time / space domain 298
time domain 32
tolerances
  effect on shield performance 582
transfer admittance 141
transfer effectiveness 142
transfer function 137
  shield of 133
transfer impedance 134, 141
  real and imaginary part 168
  test clamp 458
transfer parameters
  vs coupling 174
  universal set 137

transient
  response 316
transient effects 316
transmission
  differential and common mode 26
  mode 26
triaxial line 446

U

unbalanced 23
unintentional 2, 19

V

validation
  test procedure, of 487

W

wave regime
  in conductors 259
  in dielectrics 258
wideband 2